A Specialist Periodical Report

Nuclear Magnetic Resonance
Volume 5

A Review of the Literature published between
June 1974 and May 1975

Senior Reporter
R. K. Harris, *School of Chemical Sciences, University of East Anglia*

Reporters
P. S. Allen, *University of Nottingham*
W. Derbyshire, *University of Nottingham*
P. Diehl, *Physikalisches Institut der Universität Basel, Switzerland*
R. Ditchfield, *Dartmouth College, New Hampshire, U.S.A.*
T. Drakenberg, *Lund Institute of Technology, Sweden*
M. I. Foreman, *Organon Laboratories Ltd., Newhouse, Lanarkshire*
D. G. Gillies, *Royal Holloway College, Egham*
R. Grinter, *University of East Anglia*
M. Holz, *University of Karlsruhe, West Germany*
D. I. Hoult, *University of Oxford*
R. G. Jones, *University of Essex*
I. D. Robb, *Unilever Research Ltd., Port Sunlight, Cheshire*
D. Shaw, *Varian Associates Ltd., Walton-on-Thames*
H. Wennerström, *Lund Institute of Technology, Sweden*
M. D. Zeidler, *University of Karlsruhe, West Germany*

© Copyright 1976

The Chemical Society,
Burlington House, London, W1V 0BN

ISBN: 0 85186 292 6
ISSN: 0305-9804

Library of Congress Catalog No. 72-78527

Printed in Great Britain
at the Alden Press, Oxford

Foreword

Volume 5 of Specialist Periodical Reports on Nuclear Magnetic Resonance ostensibly deals with literature up to 31 May 1975 (in practice, as is usual, this means up to the journals received by libraries at the time the Reports were being written). The volume is the usual mixture of annual and biennial chapters. In the former category are chapters on Shielding, Spin Coupling, Relaxation, Bandshapes in Liquids, Macromolecules, The Solid State, and Medium Effects; these all cover the 1974–5 literature. Chapters which report on two years (1973–5) of publications are those on Spectral Analysis, Fourier Transform N.M.R., and Oriented Molecules (though for the last case it should be noted that there was a chapter in Volume 4 on Liquid Crystals which had a rather different emphasis). The chapter on Experimental Techniques also covers the two-year period 1973–5, but this topic will normally be dealt with annually. The only innovation in the present volume is the inclusion of a separate chapter on Heterogeneous Systems. Previously, work on adsorption was included in the chapter on the Solid State, but the present report also draws on material previously classified under Macromolecules or Liquid Crystals.

As before, coverage of the literature is, for many chapters, based on the U.K. Chemical Information Service Macroprofile 'N.M.R. – Chemical Aspects', which operates from *Chemical Abstracts* magnetic tapes. The quantity of n.m.r. literature is still increasing. In order to keep this volume of Specialist Periodical Reports to a reasonable size the Reporters have been forced to become increasingly selective. In particular, publications of conference proceedings or articles in house journals of companies or institutes are often neglected, since their quality, availability, and topicality often leave a lot to be desired. It is hoped that these Specialist Periodical Reports remain authoritative and as critical as is feasible.

Fourier Transform techniques continue to gain in popularity, and the literature is consequently changing in at least three ways, *viz.* (i) There is much less emphasis on ^{1}H n.m.r., but far more on ^{13}C. This trend is now spreading to 'other nuclei', especially ^{2}H, ^{15}N, ^{17}O, ^{29}Si and certain metals, together with the old stalwarts ^{11}B, ^{19}F, and ^{31}P; (ii) There is more work on relaxation properties, compared with chemical shift and coupling constant evaluations; (iii) Although the liquid (solution) state is still much the most popular there is an increasing amount of solid-state work, especially using the recently developed high-resolution methods. Information about chemical shift anisotropies is becoming more common. All three of these trends are likely to continue, and this may necessitate some variation in the organization of these Reports in future. It is interesting to note, however, that the upsurge in gas-phase work, expected by this Senior Reporter, has not yet materialized.

As usual this volume includes a table of definitions of the symbols and abbreviations used in the individual chapters, plus a list of recently published reviews on n.m.r. topics. The prompt appearance of Volume 5, together with its style and presentation, owes much to the Editorial Staff of the Chemical Society.

October 1975   R. K. HARRIS

Contents

Chapter 1 Nuclear Shielding   1
By R. Ditchfield

  1 **Introduction**   1

  2 **Basic Aspects of Nuclear Shielding**   2
    A General Theory   2
    B Basic Physical Aspects   5

  3 **Calculations of Nuclear Shielding**   8

  4 **Transmission of Shielding Effects within Molecules**   16
    A Introduction   16
    B Inductive, Resonance, and General Substituent Effects   17
    C Electric Field Effects   32
    D Magnetic Anisotropy Effects   33
    E Ring-current Effects   36
    F van der Waals and Steric Effects   41
    G Intramolecular Hydrogen Bonding   44
    H Isotope Effects   45
    I Chemical Shift Anisotropy   49

  5 **Shieldings of Particular Nuclear Species**   51
    A Introduction   51
    B Carbon Chemical Shifts   51
    C Fluorine Chemical Shifts   53
    D Phosphorus Chemical Shifts   54
    E Chemical Shifts of other Nuclei   55

Chapter 2 Nuclear Spin-Spin Coupling   58
By R. Grinter

  1 **Introduction**   58

  2 **Theoretical Work**   58
    A *Ab Initio* Calculations   58
    B Semi-Empirical Calculations (Excluding Those on π-Electron Systems)   59
    C Semi-Empirical Calculations on π-Electron Systems   62

D	Calculations of Through-space Coupling	64
E	Other Topics of Theoretical Interest	64

3 Coupling of Directly-bonded Nuclei   64
- A   Coupling to Hydrogen   64
- B   Coupling to Elements of Groups IIIB and IVB   67
- C   Coupling to Elements of Group VB   70
- D   Coupling to Elements of Group VIB   73
- E   Coupling to Elements of Group VIIB   73
- F   Miscellaneous One-bond Couplings   73

4 Coupling between Nuclei Separated by Two Chemical Bonds, ^2J 74
- A   Geminal Proton–Proton Coupling   74
- B   Two-bond Coupling of Other Nuclei to Protons   75
- C   Two-bond Coupling to Carbon [Excluding $^2J(^{13}CX^1H)$]   77
- D   Two-bond Coupling to Group VB Atoms   78
- E $^2J(^{31}PC^{19}F)$ 79
- F   Two-bond ($^{19}F,^{19}F$) Coupling   79
- G   Miscellaneous Two-bond Couplings   79

5 Coupling between Nuclei Separated by Three Chemical Bonds, ^3J 80
- A   Vicinal Proton–Proton Coupling *via* Carbon, $^2J(^1HCC^1H)$ 80
- B   Three-bond Coupling of Protons *via* Other Atoms   81
- C   Three-bond Coupling of Other Nuclei to Protons   82
- D   Three-bond Coupling to Carbon   85
- E   Three-bond Coupling to Phosphorus   86
- F   Three-bond Coupling to Fluorine   87

6 Coupling between Nuclei Separated by Four Chemical Bonds, ^4J 88
- A   Proton–Proton Coupling *via* Carbon, $^4J(^1HCCC^1H)$ 88
- B   Four-bond Coupling of Protons *via* Other Atoms   89
- C   Four-bond Coupling of Other Nuclei to Protons   89
- D   Four-bond Coupling to Carbon   90
- E   Four-bond Coupling to Fluorine   90

7 Coupling between Nuclei Separated by Five Chemical Bonds, ^5J 90
- A   Proton–Proton Coupling *via* Carbon, $^5J(^1HCCCC^1H)$ 90
- B   Five-bond Proton-Proton Coupling *via* Other Atoms   90
- C   Five-bond Coupling of Other Nuclei to Protons   91
- D   Five-bond Coupling to Carbon   91
- E   Five-bond Coupling to Fluorine   92

8 Coupling between Nuclei Separated by Six Chemical Bonds, ^6J 92

9 Coupling between Nuclei Separated by Seven Chemical Bonds, ^7J 92

10	Coupling in π-Electron Systems	93
	A Homoaromatic Molecules	93
	B Heterocyclic Molecules	95
	C Annulenes	96
	D Linear Conjugated Systems	96
11	Through-space Coupling	97
12	Recent Experimental Advances	97

Chapter 3 Nuclear Spin Relaxation in Fluids 99
By M. Holz and M. D. Zeidler

1 Introduction 99

2 Theoretical Work 99

3 Relaxation Studies in Pure Liquids 102
   A Proton and Deuteron Relaxation 102
   B ^{13}C Relaxation 105
   C Relaxation of Other Nuclei: ^{14}N, ^{15}N, ^{17}O, ^{19}F, ^{29}Si, ^{77}Se 107
   D Studies of Several Resonant Nuclei: ^{2}H, ^{13}C, ^{14}N, ^{15}N, ^{19}F 109

4 Relaxation Studies in Liquid Mixtures of Non-electrolytes 110
   A Proton and Deuteron Relaxation 110
   B ^{13}C Relaxation 113
   C Studies of Several Resonant Nuclei: ^{1}H, ^{2}H, ^{13}C, ^{14}N 116

5 Relaxation Studies in Electrolyte Solutions 117
   A Proton and Deuteron Relaxation 117
   B Relaxation of the Nuclei: ^{13}C, ^{17}O, ^{69}Ga, ^{71}Ga, ^{113}Cd 118
   C Relaxation of Alkali-metal and Halogen Nuclei: ^{7}Li, ^{23}Na, ^{39}K, ^{41}K, ^{87}Rb, ^{133}Cs, ^{35}Cl, ^{79}Br, ^{81}Br 120

6 Relaxation Studies in Gas–Liquid Solutions 122

7 Relaxation Studies of Gases 123
   A Theoretical Work 123
   B Mono- and Di-atomic Gases 125
   C Polyatomic Gases 127

Chapter 4 Experimental Techniques 129
By D. I. Hoult

1 Introduction 129

2 Diverse Applications of N.M.R. 129
    A General 129
    B Neutron Resonance and Diffraction 131
    C Image Formation by Zeugmatography 132

 3 Temperature and Pressure Studies 134

 4 Measurement of Relaxation Times 135
    A Systematic Errors 135
    B Quantization and Other Errors 137
    C Technique 138

 5 Broad-line and Solid-state Spectroscopy 139

 6 Instrumentation 141
    A Signal Processing 141
    B The Probe Area 142
    C The Transmitter 145
    D Modulation Control 145
    E Complete Spectrometers 146
    F Modification of Commercial Instruments 148

 7 Miscellaneous 150

Chapter 5 Spectral Analysis 152
By R. G. Jones

 1 Introduction 152

 2 New Approaches to Known Spin Systems 152
    A $[A]_2X$ and ABX Systems 152
    B $[AB]_2$ Systems. 153
    C The X-{A} INDOR Spectrum of the $[AX_n]_2$ Spin System 153
    D $[[A]_2X_n]_2$ and $[ABX_n]_2$ Spin Systems, with $n = 1, 2$ or 3 154

 3 More Examples of Known Spin Systems 154
    A Three-spin systems 154
    B Four-spin systems 155
    C Five-spin systems 156
    D Six-spin systems 158
    E Seven-spin systems 159
    F Eight-spin systems 160
    G Spin systems of nine nuclei or more 161

 4 Conclusions 162

Chapter 6 Bandshape Phenomena in Liquids   163
By T. Drakenberg and H. Wennerström

  1 Introduction   163

  2 General Theory   163

  3 Exchange of Magnetic Sites   165
    A   General Theory   165
    B   Intramolecular Exchange Processes in Organic Molecules   166
    C   Complex Rearrangements   179
    D   Intermolecular Exchange Processes   182
    E   Biochemical Applications   185

  4 Relaxation Effects: Linewidths   185
    A   Chemical Exchange Rates Determined by the Swift–Connick Approach   185
    B   Linewidth Determinations in Paramagnetic Systems   187

Chapter 7 Fourier Transform N.M.R.   188
By D. Shaw

  1 Introduction   188

  2 New Applications   189

  3 Instrumentation   192
    A   Multinuclear Spectrometers   192
    B   Quadrature Phase Detection (QPD)   193
    C   Solvent Elimination   194

  4 Computer Simulation of Pulsed N.M.R.   195
    A   Multipulse Experiments   195
    B   Diffusion and Field Gradient Effects   196

  5 The Equivalence of Fourier and Continuous Wave Spectra   197

  6 Correlation Spectroscopy   198

  7 Hadamard Spectroscopy   200

  8 Double Resonance in FT N.M.R.   200

  9 The Measurement of Relaxation Times   202

  10 Miscellaneous   204

Chapter 8 Macromolecules 205
By I. D. Robb

1 Introduction 205

2 Reviews 205

3 Synthetic Macromolecules 206
A High-resolution ^{1}H N.M.R. 206
B Broadline N.M.R. 209
C Spin-echo N.M.R. 211
D ^{13}C N.M.R. 211
E Other Nuclei 213

4 Biological and Related Macromolecules 214
A ^{1}H N.M.R. of Synthetic Polypeptides 214
B ^{1}H N.M.R. of Proteins 215
C ^{1}H N.M.R. of Enzymes 217
D ^{1}H N.M.R. of Natural and Related Polypeptides and Polysaccharides 219
E ^{1}H N.M.R. of Polynucleotides 220
F ^{13}C N.M.R. 221
G Other Nuclei 223

5 Small Molecules 224
A ^{1}H N.M.R.: Hydration 224
B ^{1}H N.M.R.: General Studies 228
C ^{13}C N.M.R. Studies 230
D Other Nuclei 232

Chapter 9 The Solid State 235
By P. S. Allen

1 General Introduction 235

2 Half-integral Spins in Diamagnetic Solids 235
A Introduction 235
B Theoretical Aspects 236
C Structural Studies (Broadline Techniques) 237
D High Resolution in Solids 238
E Motional Studies (Classical Regime) 241
F Quantum Effects in Molecular Solids 246
G N.M.R. Studies of Helium 248

3 Nuclei of Spin $> \frac{1}{2}$ in Diamagnetic Solids 248
A Introduction 248

B	^{2}H N.M.R.	249
C	The Halogen Nuclei	251
D	The Alkali-metal Nuclei	252
E	Transition-element Nuclei	252
F	Other Nuclei	252

4 Systems with Unpaired Electrons in the Absence of Magnetic Ordering — 253
- A Non-metallic Paramagnetic Systems — 253
- B Metals — 254
- C Alloys and Intermetallic Systems — 256
- D Hydrides — 256
- E Semiconductors — 256
- F One-dimensional Conductors — 258

5 Ordered Magnetic Materials — 258
- A Ferromagnetic Materials — 258
- B Antiferromagnetic Materials — 259
- C Ferrimagnetic Materials — 260

6 New Techniques and Apparatus — 261
- A Techniques — 261
- B Data Analysis — 262
- C Apparatus — 262

Chapter 10 Heterogeneous Systems 264
By W. Derbyshire

1 Introduction — 264

2 Adsorption Studies — 265
- A General — 265
- B Cryogenic Studies — 267
- C Zeolites — 269

3 Polymers, Exchange Resins, and Gels — 271

4 Protein Hydration — 274

5 Cellular Systems — 277

6 Membranes — 283

7 Micellar Systems — 289

Chapter 11 Medium Effects on Chemical Shifts and Coupling Constants 292
By M. I. Foreman

    1 Introduction 292

    2 Non-specific Solvent Effects 292

    3 Specific interactions with the Medium 294
        A General 294
        B Solvent Effects on Conformational Equilibria 294
        C The ASIS Effect 295
        D Hydrogen Bonding 297
        E Ionic Solutions 302
        F Micellar Systems 304

    4 Shift Reagents 305
        A General 305
        B The Stoicheiometry of the Shift-reagent–Substrate Complex 306
        C Dipolar, Contact, and Complexation Shifts 307
        D Calculation of the Substrate–Shift Reagent Complex Geometry 309
        E Structural Assignments 310
        F Preferred Binding Sites 311

Chapter 12 Oriented Molecules 314
By P. Diehl

    1 Introduction 314

    2 Structure Determination 314
        A Structures of Rigid Molecules 314
        B Corrections for Vibrational Motion 318
        C Molecules with Internal Motion 318

    3 Determination of Indirect Couplings 320

    4 Double-resonance Techniques 320

    5 Order Parameters 321

    6 Anisotropies in Indirect Couplings 322

    7 Anisotropies in Chemical Shifts 323

    8 Nuclear Quadrupole Coupling Constants 324

Author Index 325

Symbols and Abbreviations

These lists contain the symbols and abbreviations most frequently used in this volume, but they are not expected to be exhaustive. Some specialized notation is only defined in the relevant chapter. An attempt has been made to standardize usage throughout the volume as far as is feasible, but it must be borne in mind that the original research literature certainly is not standardized in this way, and some difficulties may arise from this fact. Trivial use of subscripts *etc.* is not always mentioned in the symbols list below. Some of the other symbols used in the text, *e.g.* for physical constants such as h or π, or for thermodynamic quantities such as H or S, are not included in the list since they are reckoned to follow completely accepted usage (see M. L. McGlashan, 'Physico-Chemical Quantities and Units', Royal Institute of Chemistry Monographs for Teachers, No. 15, second edition, 1971). The spin-system notation used in the present volume of Specialist Periodical Reports on n.m.r. is that suggested by Haigh [*J. Chem. Soc.* (*A*), 1970, 1682].

Symbols

a	hyperfine (electron–nucleus) interaction constant
A	(i) hyperfine (electron–nucleus) interaction constant
	(ii) parameter relating to electric field effects on nuclear shielding
B	(i) magnetic induction field (magnetic flux density)
	(ii) parameter relating to electric field effects on nuclear shielding
B_0	static magnetic field of an n.m.r. or e.s.r. spectrometer
B_1, B_2	r.f. magnetic fields associated with ν_1, ν_2
C_X	spin–rotation coupling constant of nucleus X (used sometimes in tensor form): $C^2 = \frac{1}{3}(C_\parallel^2 + 2C_\perp^2)$.
$C_\parallel, C_\perp$	components of **C** parallel and perpendicular to a molecular symmetry axis
$\bar{C}$	one third of the trace of **C**: $\bar{C} = \frac{1}{3}(C_\parallel + 2C_\perp)$
D	(i) self-diffusion coefficient
	(ii) doublet state
D	rotational diffusion tensor
$D_\parallel, D_\perp$	components of **D** parallel and perpendicular to a molecular symmetry axis
e	magnitude of the charge on the electron

xiv   *Symbols and Abbreviations*

E	electric field
E_n	eigenvalue of $\hat{\mathcal{H}}$ (or of a contribution to $\hat{\mathcal{H}}$)
f	modulation frequency
$\hat{F}_G$	nuclear spin operator for a group, G, of nuclei
$\hat{F}_{Gx}, \hat{F}_{Gy}, \hat{F}_{Gz}$	components of $\hat{F}_G$
F_G	magnetic quantum number associated with $\hat{F}_G$
g	nuclear or electronic g-factor
G	magnetic field gradient
H_{ij}	element of matrix representation of $\hat{\mathcal{H}}$
$\hat{\mathcal{H}}$	Hamiltonian operator (in energy units) – subscripts indicate the nature of the operator
$\hat{I}_i$	nuclear spin operator for nucleus i
$\hat{I}_{ix}, \hat{I}_{iy}, \hat{I}_{iz}$	components of $\hat{I}_i$
$\hat{I}_{i+}, \hat{I}_{i-}$	'raising' and 'lowering' spin operators for nucleus i
I_i	magnetic quantum number associated with $\hat{I}_i$
I	(i) ionization potential
	(ii) moment of inertia
$\mathscr{I}_{mn}$	intensity of a transition between nuclear states m and n
^nJ	nuclear spin–spin coupling constant through n bonds (in Hz). Further information may be given by subscripts or in brackets. Normally subscripts are only used for algebraic symbols for nuclei in spectral analysis cases, *e.g.* J_{AX}. Brackets are used for indicating the species of nuclei coupled, *e.g.* $J(^{13}C,^{1}H)$, or, additionally, the coupling path, *e.g.* $J(POCF)$
J	rotational quantum number
$\mathscr{J}$	electron exchange interaction
^nK	reduced nuclear spin–spin coupling constant (see the notes concerning ^nJ)
K	Knight shift
m_i	eigenvalue of $\hat{I}_{iz}$ (magnetic component quantum number)
m_T	total magnetic quantum number for a spin system (eigenvalue of $\sum_i \hat{I}_{iz}$)
M_0	equilibrium macroscopic magnetization of a spin system in the presence of B_0
M_x, M_y, M_z	components of macroscopic magnetization
M_n	moment of a spectrum (M_2 = second moment, *etc.*)
p_A	valence p orbital of atom A
P_i	fractional population (of rotamers *etc.*)
$P_{s_A s_B}$	molecular orbital bond order between s_A and s_B
$\mathscr{P}$	probability of a transition between two nuclear levels
q	electric field gradient
Q	(i) McConnell's constant relating to a
	(ii) nuclear quadrupole moment
	(iii) quality factor for an r.f. coil
R	(i) Redfield relaxation matrix
	(ii) magnitude of reaction electric field
s_A	valence s-orbital of atom A

Symbols and Abbreviations

$S_A^2(0)$	electron density in s_A at nucleus A
S	(i) singlet state
	(ii) electron (or, occasionally, nuclear) spin – cf. I
	(iii) ordering parameter for oriented systems
	(iv) overlap integral between molecular orbitals
t	elapsed time
t_{cp}	180° pulse separation in a Carr–Purcell sequence
T	(i) temperature
	(ii) triplet state
T_c	(i) coalescence temperature for an n.m.r. spectrum
	(ii) critical solution temperature
	(iii) phase transition temperature
T_1^X	spin–lattice relaxation time of the X nuclei (further subscripts refer to the relaxation mechanism)
T_2^X	spin–spin relaxation time of the X nucleus (further subscripts refer to the relaxation mechanism)
T_2'	inhomogeneity contribution to dephasing time for M_x or M_y
T_2^*	total dephasing time for M_x or M_y; $(T_2^*)^{-1} = T_2^{-1} + (T_2')^{-1}$
T_3	decay time following 90_0-τ-90_{90} pulse sequences
$T_{1\rho}^X, T_{2\rho}^X$	spin–lattice and spin–spin relaxation time of the X nuclei in the frame of reference rotating with B_1
T_{1D}	dipolar spin–lattice relaxation time
T_d	pulse delay time (in FT n.m.r.)
T_{ac}	acquisition time (in FT n.m.r.)
T_p	period for repetitive pulses (= interpulse time = $T_{ac} + T_d$ if τ_p is negligible)
T_N	Néel temperature
X_i	mole fraction of compound i
Z_A	atomic number of atom A
α	(i) nuclear spin wavefunction (eigenfunction of $\hat{I}_z$) for a spin-½ nucleus
	(ii) polarizability
α_A^2	s-character of hybrid orbital at atom A
β	nuclear spin wavefunction (eigenfunction of $\hat{I}_z$) for a spin-½ nucleus
γ_X	(i) magnetogyric ratio of nucleus X
	(ii) activity coefficient of compound X
Γ_N, Γ_M	polarization (net or multiplet, respectively) in Kaptein's rules for CIDNP
δ_X	chemical shift (for the resonance) of nucleus of element X (positive when the sample resonates to high frequency of the reference). Usually in ·p.p.m. Further information regarding solvent, references, or nucleus of interest may be given by superscripts or subscripts or in brackets.
δ_{ij}	Kronecker delta (= 1 if $i = j$, and = 0 otherwise)
Δ	(i) time between field gradient pulses
	(ii) spectral width
ΔC	anisotropy in **C** ($\Delta C = C_\parallel - C_\perp$)

Symbols and Abbreviations

ΔJ	anisotropy in J ($\Delta J = J_\parallel - J_\perp$)
Δn	population difference between nuclear states
$\Delta \delta$	change or difference in δ
$\Delta \nu_{\frac{1}{2}}$	full width (in Hz) of a resonance line at half-height
$\Delta \sigma$	(i) anisotropy in σ ($\Delta \sigma = \sigma_\parallel - \sigma_\perp$)
	(ii) differences in σ for two different situations
$\Delta \chi$	(i) susceptibility anisotropy ($\Delta \chi = \chi_\parallel - \chi_\perp$)
	(ii) difference in electronegativities
ε_r	relative permittivity
ε_0	permittivity of a vacuum
η	(i) nuclear Overhauser effect
	(ii) asymmetry factor (*e.g.* in e^2qQ/h)
	(iii) refractive index
	(iv) viscosity
μ	(i) magnetic dipole moment
	(ii) electric dipole moment
μ_0	permeability of a vacuum
μ_B	Bohr magneton
μ_N	nuclear magneton
ν_i	Larmor precession frequency of nucleus i (in Hz)
ν_0	(i) spectrometer operating frequency
	(ii) Larmor precession frequency (general, or of bare nucleus)
ν_1	frequency of 'observing' r.f. magnetic field
ν_2	frequency of 'irradiating' r.f. magnetic field
Ξ_X	resonance frequency for the nucleus of element X in a magnetic field such that the protons in TMS resonate at exactly 100 MHz
$\Pi_{s_A s_B}$	mutual polarizability of s_A and s_B
$\rho^2_{s_A}$	spin density in s_A
σ_i	(i) shielding constant of nucleus i (used sometimes in tensor form). Usually in p.p.m. Subscripts may alternatively indicate contributions to σ.
	(ii) collision cross-section
	(iii) spin density matrix
$\sigma_\parallel, \sigma_\perp$	components of σ parallel and perpendicular to a molecular symmetry axis
τ	(i) pre-exchange lifetime of molecular species
	(ii) cage lifetime (in CIDNP work)
	(iii) time between r.f. pulses (general symbol)
τ_c	correlation time
τ_{coll}	mean time between molecular collisions in the liquid state
τ_1, τ_2	correlation time for the spherical harmonics Y_{1m}, Y_{2m}
$\tau_\parallel, \tau_\perp$	correlation time for molecular rotation parallel and perpendicular to a molecular symmetry axis
τ_J	angular momentum correlation time
τ_p	pulse duration
τ_t	translational magnetic relaxation correlation time
χ	(i) magnetic susceptibility

Symbols and Abbreviations

	(ii) electronegativity
	(iii) nuclear quadrupole coupling constant ($= e^2qQ/h$)
ω_c	carrier frequency in rad s^{-1}
$\omega_1, \omega_0, \omega_1, \omega_2$	as for v_1, v_0, v_1, v_2 but in rad s^{-1}
ω_m	modulation angular frequency (in rad s^{-1})

Abbreviations

(a) *Physical properties*

a.c.	alternating current
a.f.	audiofrequency
a.u.	atomic unit
a.m.	amplitude modulation
b.c.c.	body-centred cubic
c.d.	circular dichroism
c.m.c.	critical micelle concentration
d.c.	direct current
e.d.	electron diffraction
e.f.g.	electric field gradient
e.s.r.	electron spin resonance
f.c.c.	face-centred cubic
f.m.	frequency modulation
h.c.p.	hexagonal close-packed
h.f.	hyperfine
i.d.	inside diameter
i.f.	intermediate frequency
i.r.	infrared
l.c.	liquid crystalline
m.w.	microwave
mol. wt.	molecular weight
n.m.r.	nuclear magnetic resonance
n.q.r.	nuclear quadrupole resonance
o.d.	outside diameter
p.p.m.	parts per million
r.f.	radiofrequency
r.m.s.	root mean square
s.h.f.	super-high frequency
u.h.f.	ultra-high frequency
u.v.	ultraviolet
ADC	analog-to-digital converter
ARP	adiabatic rapid passage
ASIS	aromatic solvent-induced shift
BCD	binary coded decimal
CAT	computer of average transients

CHF	coupled Hartree–Fock
CIDEP	chemically induced dynamic electron polarization
CIDNP	chemically induced dynamic nuclear polarization
CNDO	complete neglect of differential overlap
CPMG	Carr–Purcell pulse sequence, Meibom–Gill modification
CSA	chemical shift (shielding) anisotropy
CW	continuous wave
DAC	digital-to-analog converter
DD	dipole–dipole (interaction or relaxation mechanism)
DEFT	driven-equilibrium Fourier transform
DNP	dynamic nuclear polarization
EHMO	extended Hückel molecular orbital
ENDOR	electron–nucleus double resonance
FC	Fermi contact
FET	field-effect transistor
FID	free induction decay
FPT	finite perturbation theory
FT	Fourier transform
GIAO	guage-invariant atomic orbitals
HR	high resolution
HSP	homogeneity-spoiling pulse
INDO	intermediate neglect of differential overlap
INDOR	internuclear double resonance
LCAO	linear combination of atomic orbitals
LIS	lanthanide-induced shift
LSR	lanthanide shift reagent
MO	molecular orbital
MOSFET	metal oxide silicon FET (q.v.)
NOE	nuclear Overhauser effect
NQCC	nuclear quadrupole coupling constant
OB	orbital (contribution to scalar coupling)
PRE	proton relaxation enhancement
PRFT	partially relaxed Fourier transform
QF	quadrupole moment/field gradient (interaction or relaxation mechanism)
QPD	quadrature phase detection
RPFT	repetitively pulsed Fourier transform
SC	scalar (interaction or relaxation mechanism)
SCF	self-consistent field
SD	spin-dipolar (contribution to scalar coupling)
SEFT	spin-echo Fourier transform
S/N	signal-to-noise ratio
SOS	sum over states
SPI	selective population inversion
SPT	selective population transfer
SR	spin-rotation (interaction or relaxation mechanism)
UCHF	uncoupled Hartree–Fock

Symbols and Abbreviations

VB	valence bond
WAHUHA	Waugh, Huber, and Haeberlen (cycle of pulses)
WEFT	water-eliminated Fourier transform

(b) *Chemical species**

acac	acetyl acetonato
ADP	adenosine diphosphate
AMP	adenosine monophosphate
ATP	adenosine triphosphate
bhmc	3-t-butylhydroxymethylene-d-camphorato
BSA	bovine serum albumin
CMP	cytidine monophosphate
cp	cyclopentadienyl
DAP	dodecylammonium propionate
DDT	1,1,1-trichloro-2,2-bis(p-chlorophenyl)ethane
DME	1,2-dimethoxyethane
DMF	dimethylformamide
DML	dimyristoyl lecithin
DMSO	dimethylsulphoxide
DNA	deoxyribonucleic acid
DPG	2,3-diphosphoglycerate
DPL	dipalmitoyl lecithin
DPPH	diphenyl picryl hydrazyl
DSS	2,2-dimethyl-2-silapentane-5-sulphonate (usually as the sodium salt)
EBBA	N-(p-ethoxybenzylidene)-p-butylaniline
EDTA	ethylenediaminetetra-acetic acid
facam	3-trifluoroacetyl-d-camphorato
fod	1,1,1,2,2,3,3-heptafluoro-7,7-dimethyl-4,6-octanedionato
HAB	4,4'-bis(heptyl)azoxybenzene
hfbc	3-heptafluorobutyryl-d-camphorato
HOAB	p-n-heptyloxyazoxybenzene
IHP	inositol hexaphosphate
KDP	potassium dihydrogen phosphate
MBBA	N-(p-methoxybenzylidene)-p-butylaniline
MMA	methyl methacrylate
NADH	nicotinamide adenine dinucleotide
OMTS	octamethyltetrasiloxane
PAA	p-azoxyanisole
PBA	pyrene butyric acid
PBLG	poly(α-benzyl L-glutamate)
PMAA	poly(methacrylic acid)
PMMA	poly(methyl methacrylate)
PSA	pyrene sulphonic acid
PTFE	polytetrafluoroethylene
PVC	poly(vinyl chloride)

* Lower case initials are used when the species is a ligand.

PVP	poly(vinyl pyrrolidone)
py	pyridine
RNA	ribonucleic acid
SDS	sodium dodecyl sulphate
TBBA	terephthal-bis-4-(4-n-butylaniline)
TCNQ	tetracyanoquinodimethane
TFA	trifluoroacetic acid
THF	tetrahydrofuran
tmhd†	2,2,6,6-tetramethylheptane-3,5-dionato
TMS	tetramethylsilane

† This ligand is sometimes referred to as dpm, dipivaloylmethanato.

N.M.R. Books and Reviews

This section lists all books and reviews with n.m.r. as the principle theme, published during 1974 and 1975 (following the list given in Volume 4), which were known to the Senior Reporter at the time of going to press. Some articles omitted from the lists in earlier Volumes are also included here. Titles and numbers of pages are given where appropriate, as are the number of references in each review and its Chemical Abstracts number. In general, texts of individual lectures from symposia are not listed; likewise references to technical bulletins, company or institute house journals, special reports *etc.*, are omitted, since they are judged to be of little value to the scientific community in view of their inaccessibility and ephemeral nature. N.m.r. sections of general physical chemistry or spectroscopy textbooks have not usually been included. The only foreign-language articles that are normally included are those in French, German, Russian, Spanish, or Portuguese; moreover, the coverage of Russian-language publications is not likely to be comprehensive. Usually, for such foreign-language reviews, the title is translated in the present list but the original language is indicated in brackets. The source of the references is in most cases the computer-based listing of n.m.r. publications provided by the U.K. Chemical Information Service Macroprofile 'N.M.R. – Chemical Aspects'; this may sometimes result in the list of authors being abbreviated. The compilation is sectionalized for the reader's convenience.

(a) *Books*

R1. E. Breitmaier and W. Voelter, '^{13}C N.M.R. Spectroscopy: Methods and Applications', (Monographs in Modern Chemistry Vol. 5), Verlag Chemie, Weinheim, Germany, 1974, 303 pages, CA 82:9909.

R2. N. F. Chamberlain, 'The Practice of N.M.R. Spectroscopy, with Spectra-Structure Correlations for ^{1}H', Plenum Press, New York, 1974, 424 pages, CA 82:132011.

R3. T. Clerc and E. Pretsch, 'Nuclear Resonance Spectroscopy, Part I: ^{1}H Resonance', (Textbook for undergraduate chemistry students, 2nd edn.), Akad. Verlagsges., Frankfurt, 1973, 182 pages, CA 82:148355.

R4. H. Friebolin, 'Pocket Texts Vol. 15. N.M.R. Spectroscopy. An Introduction with Exercises', Verlag Chemie, Weinheim, Germany, 1974, 210 pages (in German) CA 81:84296.

R5. F. A. Rushworth and D. P. Tunstall, 'Nuclear Magnetic Resonance', Gordon and Breach, New York, 1973, 253 pages, (120 refs.), CA 82:24248.

R6. W. W. Simons and M. Zanger, 'The Sadtler Guide to the N.M.R. Spectra of Polymers', Sadtler Research Labs. Inc., Philadelphia, 1973, 298 pages, CA 81:106 234.

(b) *Edited Books, Books of Partial Relevance, Chapters of Books, Symposia* etc.

R7. 'N.M.R. Spectroscopy (as applied to coatings)', L.C. Afremow, Amer. Soc. Test. Mater., Spec. Tech. Publ. 500, 1972, pp. 564—581 (7 refs.) CA 77: 141 547.

R8. 'Atlas of Spectral Data and Physical Constants for Organic Compounds', ed. J. G. Grasselli, Chemical Rubber Co., Cleveland, Ohio, 1973, CA 81: 31 729. Relevant sections are: (a) Sec. A III, pp. A-69 to A-92 'N.M.R. Spectroscopy: Introduction' and (b) Sec. C V, pp. C-455 to C-514 'N.M.R.: Index'.

R9. 'Dynamic N.M.R. Spectroscopy', ed. L. M. Jackman and F. A. Cotton, Academic Press, New York, 1975, 660 pages, CA 83:88 581.

R10. 'Methodicum Chimicum', Vol. 1A, ed. F. Korte, Academic Press, 1974. Relevant chapters are: (a) 'Principles of Magnetic Resonance', K.H. Hausser, pp. 318—322 (no refs.), CA 83:95 758, (b) 'Application of ^{1}H N.M.R. Methods to Structural and Stereochemical Problems', E. Glotter and D. Lavie, pp. 322—351 (109 refs.), (c) 'N.M.R. of Nuclei other than ^{1}H', H. Zimmer and D. C. Lankin, pp. 351—379 (262 refs.), CA 83:95 759, (d) 'Detection of Radicals', K. H. Hausser and K. Möbius, pp. 379—395 (98 refs.), CA 83:105 512, (e) 'Metal Complexes', H. J. Keller and K. E. Schwarzhans, pp. 395—403 (99 refs.), CA 83:123 331.

R11. 'Topics in ^{13}C N.M.R. Spectroscopy', Vol. 1. ed. G. C. Levy, Wiley-Interscience, New York, 1974, 292 pages, CA 82:148 358 and CA 83: 35 562.

R12. 'Metal Ions in Biological Systems', Vol. 4: Metal Ions as Probes. ed. H. Sigel, Marcel Dekker, New York, 1974. Relevant chapters are: (a) Ch. 2: 'Enzyme, Metal Ion, Substrate Complexes', J. J. Villafranca, pp. 29—59, (66 refs.), CA 83:39 205. (b) Ch. 3: 'Application of Paramagnetic Probes in Biochemical Systems', R. A. Dwek, R. J. P. Williams, and A. V. Xavier, pp. 61—210, (135 refs.), CA 83:39 577.

R13. 'Ions and Ion Pairs in Organic Reactions', Vol. 1. ed. M. Szwarc, Wiley-Interscience, New York, 1972. Relevant articles are: (a) 'N.M.R. Studies of Carbon-Lithium Bonding in Organolithium Compounds', L. D. McKeever, pp. 263—287 (71 refs.), CA 77:4400, (b) 'N.M.R. Studies of Alkali Radical Ion Pairs', E. de Boer and J. L. Sommerdijk, pp. 289—309 (42 refs.), CA 77: 26 573, (c) 'N.M.R. Studies of Solvation of Ions and Ion Pairs', M. Szwarc, pp. 311—319 (48 refs.), CA 77:10 163, and (d) 'E.S.R. and N.M.R. Studies of Ion Pairs – Quantitative Approach', J. L. Sommerdijk and E. de Boer, pp. 321—377 (135 refs.), CA 77:26 570.

R14. 'Instrumental Methods of Analysis', (fifth edition). H. H. Willard, L. L. Merritt, jun., and J. A. Dean, Van Nostrand, New York, 1974, CA 81: 32 951 and CA 81:44 992. Ch. 8 (pp. 203—235, 11 refs.) is on 'N.M.R. Spectroscopy'.

R15. 'On-Line Rechner in Der Chemie. Grundlagen und Anwendung in Der

Fourier-Spektroskopie', D. Ziessow, W. de Gruyter, Berlin, 1973, 376 pages, CA 79:131 272 (pages 252–303 are specifically on FT N.M.R. with 19 refs) (in German).
R16. 'Progress in N.M.R. Spectroscopy', M. G. Barlow in 'Fluorocarbon and Related Chemistry', ed. R. E. Banks and M. G. Barlow, (Specialist Periodical Reports), The Chemical Society, London, 1974, Vol. 2, pp. 456—469, (80 refs.), CA 81:83 719.
R17. 'N.M.R.', N. S. Bhacca, in 'Handbook of Spectroscopy' Vol. II, ed. J. W. Robinson, Chemical Rubber Co. Press, Cleveland, Ohio, 1974, pp. 330—485 (451 refs.).
R18. 'Use of N.M.R. for some Oceanological Measurements', A. I. Chernitsyn and S. A. Barinov, in 'Fiz. Metody Issled. Okeana', ed. V. P. Shevtsov, Akad. Nauk S.S.S.R., Vladivostok, 1973, pp. 44—47 (in Russian) (23 refs), CA 82:21 475.
R19. 'N.M.R. Investigations of the Interactions of Biomolecules', J. S. Cohen in 'Experimental Methods in Biophysical Chemistry', ed. C. Nicolau, Wiley, London, 1973, pp. 521—588, (364 refs.), CA 82:39 646.
R20. 'Some Recent Developments in N.M.R. Spectroscopy', J. A. Elvidge, in 'An Introduction to Spectroscopic Methods for the Identification of Organic Compounds', ed. F. Scheinmann, Pergamon Press, 1974, Vol. 2, pp. 211—231, (54 refs.), CA 81:168 727.
R21. 'Metal Ions as N.M.R. Probes in Biochemistry', S. J. Ferguson, in 'Techniques and Topics in Bioinorganic Chemistry', ed. C. A. McAuliffe, Wiley, New York, 1975, pp. 305—330, (43 refs.), CA 83:39 576.
R22. 'N.M.R. and Electronic Structure of the Actinides', F. Y. Fradin, Ch. 4 of 'The Actinides: Electronic Structure and Related Properties', ed. A. J. Freeman and J. B. Darby, Academic Press, New York, Vol. 1, pp. 181—235, (90 refs.), CA 83:17 787.
R23. 'Resonance Methods', D. Haneman, Ch. 14 of 'Characterization of Solid Surfaces,' ed. P. F. Kane and G. B. Larrabee, Plenum Press, New York, 1974, pp. 337–377, (100 refs.), CA 81:112 937.
R24. 'Absolute Configuration of Metal Complexes', C. J. Hawkins, Wiley-Interscience, New York, 1971. Ch. 6 (pp. 257—303, with 69 refs.) is on N.M.R., CA 74:149 051.
R25. 'Isomerization, Intramolecular Processes', R. D. W. Kemmitt and M. A. R. Smith, in 'Inorganic Reaction Mechanisms', ed. J. Burgess, (Specialist Periodical Reports), The Chemical Society, London, 1974, Vol. 3, pp. 460—484, (103 refs.), CA 82:72 249.
R26. 'N.M.R. Spectroscopy', B. E. Mann in 'Spectroscopic Properties of Inorganic and Organometallic Compounds', ed. N. N. Greenwood, (Specialist Periodical Reports), The Chemical Society, London, 1974, Vol. 7, pp. 1—166 (2213 refs.).
R27. 'N.M.R. Spectroscopy of Proteins', J. C. Metcalfe, N. J. M. Birdsall, and A. G. Lee, Ch. 3 of 'Companion to Biochemistry', ed. A. T. Bull, J. R. Lagnado, J. O. Thomas, and K. F. Tipton, 1974, Longmans, London, pp. 139—162, (34 refs.), CA 82:12 345.
R28. 'Magnetic Resonance Spectroscopy', J. A. S. Smith, in 'Modern Physical

Techniques in Materials Technology', ed. T. Mulvey and R. K. Webster, O.U.P., 1974, pp. 291—314, (14 refs.), CA 83:50 092.

R29. 'Physical Methods: N.M.R.', J. C. Tebby, in 'Organophosphorus Chemistry', ed. S. Trippett, (Specialist Periodical Reports), The Chemical Society, London, 1974, Vol. 5, pp. 247—269 (152 refs.).

R30. 'Physical Methods: N.M.R.', J. C. Tebby, in 'Organophosphorus Chemistry', ed. S. Trippett, (Specialist Periodical Reports), The Chemical Society, London, 1975, Vol. 6, pp. 221—238 (129 refs.).

R31. 'N.M.R. Spectrometry', D. S. Ware and R. S. Codrington, Ch. 13 of 'Systematic Materials Analysis', ed. J. H. Richardson and R. V. Peterson, Academic Press, New York, 1974, Vol. 2, pp. 73—118, (44 refs), CA 81: 179 148.

R32. 'N.M.R. Analysis of Silicones', D. E. Williams, in 'Analysis of Silicones', ed. A. Lee Smith, Wiley-Interscience, New York, 1974, Ch. 11 (pp. 287—323) (48 refs.).

R33. 'Applications of Magnetic Resonance in Biology: A Centenary Congress of the French Physics Society', May–June 1973. *J. Phys. (Paris)* 1973, **34**, Colloq. C–8, 63 pages. (Some contributions are in full, others are only abstracts) (In French, but abstracts also in English) (See, for example, CA 81:73 494, CA 81:73 689, CA 81:96 617).

R34. 'Proceedings of the Australian Polymer Symposium, May, 1974', ed. J. H. Bradbury, *J. Polymer. Sci., Polymer Symp.*, 1975, No. 49. Relevant articles are: (a) 'The Use of N.M.R. Techniques in the Study of Protein Hydration', L. J. Lynch and D. S. Webster, pp. 43—63 (45 refs.), CA 83:43 698, (b) 'Determination of the Sequence of Peptides Using Paramagnetic Probes with N.M.R. Spectroscopy', B. Warren and J. H. Bradbury, pp. 65—74 (18 refs.), CA 83:93 271.

R35. 'A Discussion of the Determination of Structures and Conformation of Molecules in Solution', Nov. 1974. Organized by R. E. Richards and R. J. P. Williams, *Proc. Roy. Soc.*, 1975, **A345**, No. 1640. Relevant articles are: B. A. Levine and R. J. P. Williams, 'The determination of the conformations of small molecules in solution by means of paramagnetic shift and relaxation perturbations of n.m.r. spectra', pp. 5—22 (14 refs.), CA 83:179 528; I. D. Campbell, C. M. Dobson, and R. J. P. Williams, 'Assignment of the ^{1}H n.m.r. spectra of proteins', pp. 23—40 (31 refs.), CA 83:109 885; I. D. Campbell, C. M. Dobson, and R. J. P. Williams, 'N.m.r. studies on the structure of lysozyme in solution', pp. 41—59 (18 refs.), CA 83:128 177; J. Feeney, 'The use of three-bond spin–spin coupling constants in the determination of conformations of molecules in solution', pp. 61—72 (45 refs.), CA 83:177 521.

R36. 'Magnetic Resonance – V. Selected specially invited and plenary lectures presented at the fifth international symposium on magnetic resonance', Bombay 1974, *Pure Appl. Chem.*, 1974, **40**, Issues 1 and 2 (275 pages).

R37. 'Critical Evaluation of Chemical and Physical Structural Information', (Proceedings of a conference, 1973) ed. D. R. Lide, jun., and M. A. Paul, National Academy of Sciences, Washington D.C., 1974, 628 pages. Relevant articles are: (a) 'Molecular Structure from N.M.R. in Liquid Crystalline

Solvents', L. C. Snyder and S. Meiboom, pp. 143—156 (17 refs.), CA 82: 30440, (b) 'Studies of Protein Structure by N.M.R.', F. A. Bovey, pp. 248—259 (27 refs.) CA 81:169831. (c) 'Internal Motions Studied by High-Resolution N.M.R.', A. A. Bothner-By, pp. 331—351 (31 refs.) CA 81: 179218. (d) 'Nuclear Quadrupole Coupling Constants', B. M. Fung, pp. 436—448 (64 refs.) CA 81:179215. (e) 'Magnetic Interactions and the Electronic Structure of Diamagnetic Molecules', W. H. Flygare, pp. 449—484 (44 refs.) CA 81:178969. (f) 'Anisotropies of Chemical Shifts and Their Relation to Magnetic Susceptibility', B. R. Appleman and B. P. Dailey, pp. 485—504 (81 refs.) CA 81:179210. (g) 'Studies of Molecular Properties Pertaining to Electronic Charge Distribution: A Comparison between Theory and Experiment', R. Ditchfield, pp. 565—590 (98 refs.) CA 82:21 857.

(c) *Regular Magnetic Resonance Review Series*

R38. J. L. Bjorkstam, 'N.M.R. Studies of Collective Atomic Motion near Ferroelectric Phase Transitions', *Adv. Magn. Resonance*, 1974, **7**, 1—70 (101 refs.) CA 81:179 133.

R39. D. Kivelson and K. Ogan, 'Spin Relaxation Theory in Terms of Mori's Formalism', *Adv. Magn. Resonance*, 1974, **7**, 71—155 (67 refs.) CA 81: 179 603.

R40. G. L. Closs, 'Chemically Induced Dynamic Nuclear Polarization', *Adv. Magn. Resonance*, 1974, **7**, 157—227 (66 refs.) CA 81:179 132.

R41. B. R. Appleman and B. P. Dailey, 'Magnetic Shielding and Susceptibility Anisotropies', *Adv. Magn. Resonance*, 1974, **7**, 231—320 (283 refs.) CA 81: 179 131.

R42. M. B. Hayes, J. S. Cohen, and M. L. McNeel, 'N.M.R. of Amino-acids, Peptides, and Proteins', *Magn. Resonance Rev.*, 1974, **3**, 1—62 (346 refs.) CA 83:43 696.

R43. J. D. Memory, 'High Resolution N.M.R. Theory', *Magn. Resonance Rev.*, 1974, **3**, 105—125 (200 refs.) CA 82:147 170.

R44. D. G. de Kowalewski and V. J. Kowalewski, 'High Resolution ^{1}H N.M.R.', *Magn. Resonance Rev.*, 1974, **3**, 171—206 (475 refs.) CA 82:161 906.

R45. R. H. Cox, 'N.M.R. of Other Nuclei', *Magn. Resonance Rev.*, 1974, **3**, 207—249 (570 refs.) CA 83:77 856.

R46. R. A. Komoroski and G. C. Levy, 'FT N.M.R. Spectroscopy', *Magn. Resonance Rev.*, 1974, **3**, 289—324 (188 refs.) CA 83:87 491.

R47. R. G. Lawler, 'CIDNP', *Progr. N.M.R. Spectroscopy*, 1973, **9**, 145—210 (259 refs.) CA 82:9911.

R48. P. D. Buckley, K. W. Jolley, and D. N. Pinder, 'Application of Density Matrix Theory to N.M.R. Lineshape Calculations', *Progr. N.M.R. Spectroscopy*, 1975, **10**, 1—26 (26 refs.) CA 83:123 341.

R49. J. Hilton and L. H. Sutcliffe, 'The "Through-Space" Mechanism in Spin–Spin Coupling', *Progr. N.M.R. Spectroscopy*, 1975, **10**, 27—39 (169 refs.), CA 83:123 342.

(d) *Review Articles in Regular Journals*

R50. J. L. Marshall, D. E. Müller, S. A. Conn, R. Seiwell, and A. M. Ihrig,

'Long-range (C,H) and (C,C) Spin–Spin Coupling Constants' *Accounts Chem. Res.*, 1974, **7**, 333—339 (60 refs.) CA 81:179 138.

R51. J. L. Markley, 'Observation of Histidine Residues in Proteins by N.M.R. Spectroscopy', *Accounts Chem. Res.*, 1975, **8**, 70—80 (112 refs.) CA 82:120 384.

R52. L. D. Hall, 'Solutions to the Hidden-Resonance Problem in ^{1}H N.M.R. Spectroscopy', *Adv. Carbohydrate Chem. Biochem.*, 1974, **29**, 11—40 (56 refs.) CA 81:37 739.

R53. A. S. Mildvan, 'Nuclear Relaxation Studies of the Role of Metals in Enzyme-Catalyzed Enolization and Elimination Reactions', in 'Bioinorganic Chemistry', ed. R. F. Gould, Adv. Chem. Series, 1971, No. 100, pp. 390—412 (46 refs.) CA 74:120 558.

R54. 'Carbohydrates in Solution' (American Chemical Society Symposium – Chairman, H. S. Isbell), Adv. Chem. Series, 1973, No. 117. Some relevant contributions are: (a) '^{13}C and Hydroxyl ^{1}H N.M.R. of Ketoses. A Conformational and Compositional Description of Ketohexoses in Solution', A. S. Perlin, P. H. du Penhoat, and H. S. Isbell, pp. 39–50 (28 refs.) CA 80:83 424, and (b) 'Conformational Equilibria of Acylated Aldopentopyranose Derivatives and Favoured Conformations of Acyclic Sugar Derivatives', pp. 147–176 (49 refs.) CA 80:83 418.

R55. 'Molecular Sieves', (Proc. 3rd. Internat. Conf., Zurich, Sept. 1973), ed. W. M. Meier and J. B. Uytterhoeven, Adv. Chem. Series, 1973, No. 121. Relevant articles are: (a) 'N.M.R. Investigations of the Framework Cations of Various Faujasite-Type Zeolites and Their Interpretation by Model Calculations', H. Lechert, pp. 74—86 (42 refs.), CA 79:85 305; (b) 'N.M.R. Relaxation and Molecular Motion in Zeolites', H. A. Resing and J. S. Murday, pp. 414—429 (53 refs.), CA 79:83 832; (c) 'N.M.R. Studies of Molecules Adsorbed on Zeolites A, X, and Y', H. Pfeifer, W. Schirmer, and H. Winkler, pp. 430—440 (25 refs.), CA 79:85 306.

R56. M. T. Melchior, L. P. Klemann, and A. W. Langer, jun., 'Magnetic Resonance Studies of Polytertiary Amine Chelated Alkali Metal Compounds', in 'Polyamine-Chelated Alkali Metal Compounds', ed. A. W. Langer, Adv. Chem. Series, 1974, No. 130, pp. 113—130 (18 refs.) CA 80:132 261.

R57. C. W. Beck, C. A. Fellows, and E. MacKennan, 'N.M.R. Spectrometry in Archaeology', in 'Archaeological Chemistry' ed. C. W. Beck, Adv. Chem. Series 1974, No. 138, pp. 226—235 (7 refs.) CA 83:42 209.

R58. M. J. Cook, A. R. Katritzky, and P. Linda, 'Aromaticity of Heterocycles', *Adv. Heterocyclic Chem.*, 1974, **17**, 256 (N.M.R. section pp. 276—283 with 63 refs.), CA 82:111 111.

R59. R. Lenk, 'Diffusion and Spin Relaxation', *Adv. Mol. Relaxation Processes*, 1975, **6**, 287—306 (77 refs.) CA 82:172 286.

R60. D. G. Farnum, 'Charge Density – N.M.R. Chemical Shift Correlations in Organic Ions', *Adv. Phys. Org. Chem.*, 1975, **11**, 123—175 (78 refs.).

R61. G. A. Gray, '^{13}C N.M.R. Spectroscopy', *Analyt. Chem.*, 1975, **47**, 546A—564A (4 refs. plus 26 general refs.) CA 82:177 308.

R62. E. Breitmaier, K.-H. Spohn, and S. Berger, '^{13}C Spin-Lattice Relaxation

Times and the Mobility of Organic Molecules in Solution', *Angew. Chem. Internat. Edn.*, 1975, **14**, 144—159 (114 refs.).
R63. I. H. Sadler, 'Physical Methods: N.M.R.', *Ann. Reports* (B), 1973, **70**, 22—45 (139 refs.) CA 82:111 127.
R64. R. W. Vaughan, 'Application of N.M.R. to Solids. High-Resolution Techniques', *Ann. Rev. Mater. Sci.*, 1974, **4**, 21—42 (93 refs.) CA 81:129 041.
R65. J. Homer, 'Solvent Effects on N.M.R. Chemical Shifts', *Appl. Spectroscopy Rev.*, 1975, **9**, 1—132 (137 refs.) CA 81:143 542.
R66. R. E. Walstedt, 'N.M.R. and Relaxation in Metals Containing Magnetic Impurities', *Arch. Sci.*, 1974, **27**, 329—342 (32 refs.) CA 82:131 154.
R67. N. J. Stone, 'Principles of Low-Temperature Nuclear Orientation and Nuclear Orientation N.M.R.', *Atomic Energy Rev.*, 1974, **12**, 585—604 (35 refs.) CA 83:50 070.
R68. W. Zinn, 'N.M.R. in Magnetically Ordered Solids', *Atomic Energy Rev.*, 1974, **12**, 709—725 (20 refs.) CA 83:34 841.
R69. P. Mansfield, 'Pulsed N.M.R. in Static Solids', *Atomic Energy Rev.*, 1974, **12**, 727—741 (42 refs.) CA 83:17 821.
R70. K. Möbius, 'Double Resonance Methods', *Ber. Bunsengesellschaft phys. Chem.*, 1974, **78**, 1116–1125 (75 refs., in German) CA 82:147 155. Also includes other forms of spectroscopy (E.S.R. *etc.*).
R71. H. Dreeskamp, 'Progress in N.M.R. Spectroscopy', *Ber. Bunsengesellschaft phys. Chem.*, 1974, **78**, 1160—1168 (51 refs., in German) CA 82:169 524.
R72. M. Che, J. Fraissard, and J. C. Vedrine, 'Application of E.P.R. and N.M.R. to the Study of Silicates and Clays', *Bull. Groupe Fr. Argiles*, 1974, **26**, 1—53 (N.M.R. specifically pp. 4—17, 25 refs., in French) CA 81:93 972.
R73. J. Kowalewski, 'Nonempirical Calculations of Nuclear Spin–Spin Coupling Constants', *Chem. Comm. Univ. Stockholm*, 1974, 29 pages (49 refs.) CA 82:131 120.
R74. P. C. Taylor, J. F. Baugher, and H. M. Kriz, 'Magnetic Resonance Spectra in Polycrystalline Solids', *Chem. Rev.*, 1975, **75**, 203—240 (284 refs., incl. E.S.R.) CA 82:177 297.
R75. W. B. Jennings, 'Chemical Shift Nonequivalence in Prochiral Groups', *Chem. Rev.*, 1975, **75**, 307—322 (149 refs.) CA 83:27 045.
R76. H. Günther, 'Physical Methods in Chemistry. ^{13}C N.M.R. Spectroscopy. II', *Chem. Unserer Z.* 1974, **8**, 84—94 (6 refs., In German) CA 81:97 066.
R77. R. R. Ernst, 'Die Anwendung von Digitalrechnern zur Steuerung und Datenverarbeitung im Spektroskopischen Laboratorium', *Chimia*, 1972, **26**, 53—64, (31 refs., in German) CA 77:11 521.
R78. R. R. Ernst, 'Two-Dimensional Spectroscopy', *Chimia*, 1975, **29**, 179—183 (14 refs.) CA 83:17 809.
R79. A. Lösche, 'N.M.R. Studies of Liquid Crystals', *Comments Solid State Phys.*, 1973, **5**, 119—131 (12 refs.) CA 80:100 811.
R80. H. M. El-Fatatry, 'Quantitative Aspects of N.M.R. Spectrometry. I. Fundamentals', *Egypt. Pharm. J.*, 1974, **56**, 45—52 (20 refs.) CA 82:147 201.
R81. J. Sandström, 'Dynamic N.M.R. Spectroscopy', *Endeavour*, 1974, **33**, 111—118 (25 refs.) CA 81:179 612.

R82. K. Wüthrich, 'N.M.R. in Protein Research', *Experientia*, 1974, **30**, 577—585 (28 refs.) CA 81:87259.

R83. H. Dutz and W. Poch, 'N.M.R. and E.S.R. Spectroscopy of Glass', *Fachausschussber, Dtsch. Glastech. Ges.*, 1974, **70**, 219—251 (79 refs., In German) CA 82: 49 336.

R84. J. A. Glasel, 'Impact of FT Techniques on High Resolution N.M.R. Spectroscopy', *Fed. Proc. (Fed. Amer. Soc. Exp. Biol)*, 1974, **33**, 1973—1977 (12 refs.) CA 81:70306.

R85. Y. Kyogoku, 'High-Resolution N.M.R. of Biochemical Substances', *Han'guk Saenghwahakhoe Chi.*, 1973, **6**, 51—61 (9 refs., in English) CA 81:146954.

R86. T. Yamane, 'Characterization of Proteins by N.M.R.', *J. Agric. Food Chem.*, 1971, **19**, 683—691 (73 refs.) CA 75:58694.

R87. C. L. Khetrapal, 'N.M.R. Spectroscopy of Molecules Dissolved in Liquid Crystal Solvents', *J. Indian Inst. Sci.*, 1974, **56**, 151-171 (40 refs.) CA 83: 42429.

R88. D. Shaw, 'Recent Advances in Experimental Techniques in High Resolution N.M.R. Spectroscopy', *J. Phys. (E)*, 1974, **7**, 689—697 (42 refs.) CA 81: 112934.

R89. F. A. Bovey, 'N.M.R. Observations of Polypeptide Conformations', *J. Polymer Sci., Macromol. Rev.*, 1974, **9**, 1—81 (233 refs.) CA 82:140467.

R90. H. Regel and W. Regel, 'N.M.R. Spectroscopy with Superconducting Magnets', *Messtechnik (Braunschweig)*, 1972, **80**, 328—333 (9 refs., In German) CA 78: 90739.

R91. J. J. Fischer, 'N.M.R. as Applied to Pharmacology', *Methods. Pharmacol.*, 1971, **1**, 431—453 (36 refs.) CA 76:54190, ed. A. Schwartz, Appleton-Century-Crofts, New York.

R92. J. J. Fischer, 'N.M.R. Studies of the Binding of Drugs to Macromolecules and Cellular Structures', *Mod. Pharmacol.*, 1973, **1**, 583—600 (63 refs.) CA 81:145499.

R93. G. J. Martin, M. L. Martin, and S. Odiot, 'Theoretical and Empirical Calculations of the Carbon Chemical Shift in Terms of the Electronic Distribution in Molecules', *Org. Magn. Resonance*, 1975, **7**, 2—17 (215 refs.).

R94. A. H. Lewin and M. Frucht, 'Restricted Rotation in Amides. VII. Methods of Resonance Assignment in Tertiary Amides – An Evaluation', *Org. Magn. Resonance*, 1975, **7**, 206—225 (46 refs.).

R95. G. Heinisch, 'Spectroscopic Methods in Pharmaceutical Analysis', *Österr. Apoth. Ztg.*, 1974, **28**, 275—284, 297—302, 361—368 (24 refs., in German) CA 81:82415.

R96. J. Seelig, 'Structure and Dynamics of Membranes. Application of N.M.R. and E.S.R.', *Pathol. Microbiol.*, 1974, **41**, 151—163 (27 refs., in German) CA 82:39641.

R97. W. Th. Wenckebach, T. J. B. Swanenburg, and N. J. Poulis, 'Thermodynamics of Spin Systems in Paramagnetic Crystals', *Phys. Letters (C)*, 1974, **14**, 181—255 (77 refs.) CA 82:79582.

R98. V. J. McBrierty, 'N.M.R. of Solid Polymers: A Review', *Polymer*, 1974, **15**, 503—520 (132 refs.) CA 82:43750.

R99. D. M. Rackham, 'Pharmaceutical Aspects of ^{1}H N.M.R. Spectroscopy', *Proc. Soc. Analyt Chem.*, 1972, **9**, 20—21 (3 refs.) CA 82:166782.

R100. D. M. Rackham, 'Recent Analytical Applications of N.M.R. Spectrometry', *Proc. Soc. Analyt. Chem.*, 1974, **11**, 335—337 (6 refs.) CA 83:21309.

R101. E. Sambuc and M. Naudet, 'Solid Content Determination of Fats and Margarines', *Rev. Franc. Corps Gras*, 1974, **21**, 309—312 (9 refs., in French) CA 81:167885.

R102. E. Sambuc, 'Broad-Band N.M.R. and Pulsed Broad-Band N.M.R.', *Rev. Franc. Corps Gras*, 1974, **21**, 689—698 (26 refs., in French) CA 83:17767.

R103. E. J. Vincent, 'Carbon-13 N.M.R.', *Rev. Franc. Corps Gras*, 1974, **21**, 699—703 (36 refs., In French) CA 83:17768.

R104. V. K. Voronov, 'Paramagnetic Reagents for the Investigation of the Structures of Organic Ligands', *Russ. Chem. Rev.*, 1974, **43**, 171—183 (*Uspekhi Khim.*, 1974, **43**, 432—454) (109 refs.) CA 80:144895.

R105. G. N. Boiku, Yu. I. Malko, and K. N. Semenenko, 'Use of the N.M.R. Method to Study Simple and Complex Hydrides of Light Metals', *Russ. Chem. Rev.*, 1975, **44**, 1—8 (*Uspekhi Khim.*, 1975, **44**, 3—16) (101 refs.) CA 82:147160.

R106. U. Sequin and A. I. Scott, '^{13}C as a Label in Biosynthetic Studies', *Science*, 1974, **186**, 101—107 (71 refs.) CA 82:13346.

R107. W. E. E. Stone, 'Application of N.M.R. to Certain Surface Problems', *Silic. Ind.*, 1974, **39**, 255—261 (9 refs., in French) CA 82:77370.

R108. J. D. Roberts, G. E. Hawkes, J. Husar, A. W. Roberts, and D. W. Roberts, 'N.M.R. Spectroscopy. Lanthanide Shift Reagents. Useful Tools for Study of Conformational Equilibria in Solution', *Tetrahedron*, 1974, **30**, 1833—1844, (17 refs.) CA 82:56892.

R109. B. Focher and A. Jus, 'Introduction of N.M.R. and its Application to the Research and Analysis of Textiles', *Tinctoria*, 1975, **72**, 11—20 (29 refs., in Italian) CA 82:157624.

R110. B. Schnabel, 'Untersuchungen zur Hochauflösenden Magnetischen Kern-Resonanz von Festkörpern. mit Hilfe des Verfahrens der Roetierden Probe', *Wiss. Ztschr., Friedrich-Schiller Univ. Jena, Math.-Nat. R*, 1973, **22**, 335—360 (62 refs., in German).

R111. H. Ernst, 'Nuclear Double Resonance Spectroscopy of Solids', *Wiss. Z. Karl-Marx-Univ. (Leipzig), Math.- Naturwiss. Reihe*, 1974, **23**, 449—465 (21 refs.) CA 82:147224.

1
Nuclear Shielding

BY R. DITCHFIELD

1 Introduction

Since the phenomenon of n.m.r. was first detected in bulk matter some thirty years ago,[1,2] the subject has expanded remarkably, with the result that high-resolution n.m.r. spectroscopy is now one of the most important branches of chemical spectroscopy. In measurements of n.m.r. spectra, nuclei are essentially being used to investigate local magnetic effects in a molecular system. The local magnetic field near a particular nucleus depends on the electronic environment of the nucleus and is determined by many factors. These include the electronic polarization of remote parts of the sample, magnetic moments (nuclear and electronic) of neighbouring molecules, and intramolecular effects due to other nuclei and electrons in the same molecule. Consequently, not only has n.m.r. spectroscopy become a powerful tool for elucidating molecular structure, but it also provides a sensitive probe into various aspects of the electronic structure of molecules.

As is well-known, information about molecular electronic structure is extracted from n.m.r. spectra in terms of nuclear magnetic shielding constants, σ_i, and nuclear spin–spin coupling constants, J_{ij}. Early n.m.r. studies were mainly devoted to protons, and theories were proposed to explain proton shieldings in many types of molecules. On the whole, although such theories were highly approximate, they were successful in explaining the gross trends in the values of proton magnetic shielding constants.

In the early 1960's, theoretical methods were developed to interpret shielding constants for first-row atoms. Again the theories were rather approximate but of qualitative value in rationalizing the main trends observed in the then limited experimental data. Since the late 1960's, developments in instrumentation and experimental techniques have meant that magnetic shielding data are now readily obtained for many nuclei.[3] Although advances have been made, developments in theoretical methods have not kept pace with experimental progress. Thus experimental data are still largely rationalized using empirical relationships developed some ten or fifteen years ago. For example, changes in shielding have been correlated with the changes in charge density, the changes in electric field effects, and the changes in magnetic anisotropy which occur when a substituent is varied. Although such empirical relationships can be valuable, the fact that there is a growing body of experimental data which cannot be explained adequately in this way suggests

---

[1] E. M. Purcell, H. C. Torrey, and R. V. Pound, *Phys. Rev.*, 1946, **69**, 37.
[2] F. Bloch, W. W. Hansen, and M. E. Packard, *Phys. Rev.*, 1946, **69**, 127.
[3] C. S. Peters, R. Codrington, H. C. Walsh, and P. D. Ellis, *J. Magn. Resonance*, 1973, **11**, 431.

that an accurate theoretical understanding of the factors which contribute to nuclear magnetic shielding is still lacking.

This chapter deals with articles on nuclear shielding that were published during the twelve months to the end of May 1975. As in previous Reports in this series,[4-7] the emphasis is on papers containing results which either do, or may, lead to a better understanding of the phenomenon of nuclear shielding in isolated molecules. Therefore, discussion of the following topics has been excluded: experimental methods of chemical-shift measurement, the details of methods for the quantum-mechanical calculation of shielding constants, and the mechanisms by which inter-molecular effects alter shielding constants. Solution phenomena, including the study of contact and pseudocontact shifts and of complex formation, are discussed in Chapter 11. During the period of writing a few journals were unavailable in the Reporter's library because they were being bound; apologies are offered to any author whose work is thereby overlooked.

2 Basic Aspects of Nuclear Shielding

A. General Theory.—It has become the practice in this series of Reports[4-7] to attempt to make them as self-contained as possible. Since readers have probably become accustomed to this kind of approach, the traditions so ably established by Drs. Raynes and Mallion will be followed here. The equations derived by Ramsey[8] relating nuclear magnetic shielding constants to electronic structure are appropriate to the case where the origin of the vector potential describing the uniform external magnetic field and the origin of the co-ordinate system are taken at the nucleus whose shielding is of interest. In this Report, the generalization presented by Raynes[6] in which the origin of co-ordinates, the gauge origin, and the nucleus of interest are located at different points will be followed.

The approach of Raynes[6] is conveniently discussed with reference to Figure 1. Here O is the co-ordinate origin and G is the origin with respect to which the vector potential is referred; μ is a point magnetic dipole placed at the position where the shielding is required. R, S, and r_k are vectors representing the position of G, the position of μ, and the position of electron k relative to O, respectively. Using Rayleigh–Schrödinger perturbation theory[9] and the clamped-nuclei approximation the following eight-term expression for the $\alpha\beta$ component of the shielding tensor is obtained:

$$\sigma_{\alpha\beta} = \sigma_{\alpha\beta}^{d} + \sigma_{\alpha\beta}^{dg} + \sigma_{\alpha\beta}^{d\mu} + \sigma_{\alpha\beta}^{dg\mu} + \sigma_{\alpha\beta}^{p} + \sigma_{\alpha\beta}^{pg} + \sigma_{\alpha\beta}^{p\mu} + \sigma_{\alpha\beta}^{pg\mu} \qquad (1)$$

In equation (1), the superscripts 'd' and 'p' denote diamagnetic and paramagnetic contributions, respectively. A superscript 'g' indicates those contributions which

---

[4] W. T. Raynes, in 'Nuclear Magnetic Resonance,' ed. R. K. Harris (Specialist Periodical Reports), The Chemical Society, London, 1972, Vol. 1, p. 1.
[5] W. T. Raynes, in 'Nuclear Magnetic Resonance,' ed. R. K. Harris (Specialist Periodical Reports), The Chemical Society, London, 1973, Vol. 2, p. 1.
[6] W. T. Raynes, in 'Nuclear Magnetic Resonance,' ed. R. K. Harris (Specialist Periodical Reports), The Chemical Society, London, 1974, Vol. 3, p. 1.
[7] R. B. Mallion, in 'Nuclear Magnetic Resonance,' ed. R. K. Harris (Specialist Periodical Reports), The Chemical Society, London, 1975, Vol. 4, p. 1.
[8] N. F. Ramsey, *Phys. Rev.*, 1950, **77**, 567; *ibid.*, 1950, **78**, 699; *ibid.*, 1952, **86**, 243.
[9] For a discussion see J. O. Hirschfelder, W. Byers-Brown, and S. T. Epstein, *Adv. Quantum Chem.*, 1964, **1**, 255.

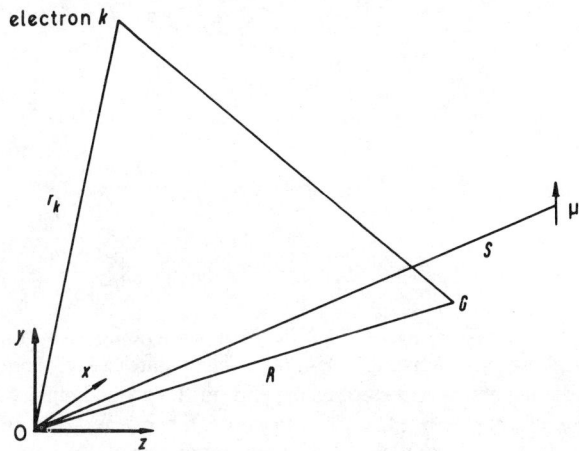

Figure 1 Co-ordinates used to define the vector potential describing the external magnetic field and the position of the point magnetic dipole μ

depend on the choice of gauge origin; such terms clearly vanish when $R = 0$. A superscript 'μ' denotes the point magnetic dipole, and contributions labelled in this way will be zero when $S = 0$. The expressions derived by Raynes for the contributions presented in equation (1) are given, in SI form, in equations (2) to (9).

$$\sigma_{\alpha\beta}^{d} = \frac{\mu_0}{4\pi}\frac{e^2}{2m}\langle 0|\sum_k \frac{r_k^2\delta_{\alpha\beta}-r_{k\alpha}r_{k\beta}}{|r_k-S|^3}|0\rangle \quad (2)$$

$$\sigma_{\alpha\beta}^{dg} = \frac{\mu_0}{4\pi}\frac{e^2}{2m}\langle 0|\sum_k \frac{R_\alpha r_{k\beta}-R_\gamma r_{k\gamma}\delta_{\alpha\beta}}{|r_k-S|^3}|0\rangle \quad (3)$$

$$\sigma_{\alpha\beta}^{d\mu} = \frac{\mu_0}{4\pi}\frac{e^2}{2m}\langle 0|\sum_k \frac{r_{k\alpha}S_\beta-r_{k\gamma}S_\gamma\delta_{\alpha\beta}}{|r_k-S|^3}|0\rangle \quad (4)$$

$$\sigma_{\alpha\beta}^{dg\mu} = \frac{\mu_0}{4\pi}\frac{e^2}{2m}\langle 0|\sum_k \frac{R_\gamma S_\gamma\delta_{\alpha\beta}-R_\alpha S_\beta}{|r_k-S|^3}|0\rangle \quad (5)$$

$$\sigma_{\alpha\beta}^{p} = \frac{\mu_0}{4\pi}\frac{e^2}{2m^2}\sum_n{}'(W_0-W_n)^{-1}\left\{\langle 0|\sum_k \frac{l_{k\alpha}}{|r_k-S|^3}|n\rangle\langle n|\sum_k l_{k\beta}|0\rangle + \right.$$
$$\left. \langle 0|\sum_k l_{k\beta}|n\rangle\langle n|\sum_k \frac{l_{k\alpha}}{|r_k-S|^3}|0\rangle\right\} \quad (6)$$

$$\sigma_{\alpha\beta}^{pg} = \frac{\mu_0}{4\pi}\frac{e^2}{2m^2}\varepsilon_{\beta\gamma\delta}R_\gamma\sum_n{}'(W_n-W_0)^{-1}\left\{\langle 0|\sum_k \frac{l_{k\alpha}}{|r_k-S|^3}|n\rangle\langle n|\sum_k p_{k\delta}|0\rangle + \right.$$
$$\left. \langle 0|\sum_k p_{k\delta}|n\rangle\langle n|\sum_k \frac{l_{k\alpha}}{|r_k-S|^3}|0\rangle\right\} \quad (7)$$

$$\sigma_{\alpha\beta}^{p\mu} = \frac{\mu_0}{4\pi} \frac{e^2}{2m^2} \varepsilon_{\alpha\omega\psi} S_\omega \sum_n{}'(W_n - W_0)^{-1}\left\{\langle 0|\sum_k \frac{p_{k\psi}}{|r_k - S|^3}|n\rangle\langle n|\sum_k l_{k\beta}|0\rangle + \right.$$

$$\left.\langle 0|\sum_k l_{k\beta}|n\rangle\langle n|\sum_k \frac{p_{k\psi}}{|r_k - S|^3}|0\rangle\right\} \quad (8)$$

$$\sigma_{\alpha\beta}^{pg\mu} = \frac{\mu_0}{4\pi}\frac{e^2}{2m^2}\varepsilon_{\alpha\omega\psi}\varepsilon_{\beta\gamma\delta}R_\omega S_\gamma \sum_n{}'(W_0 - W_n)^{-1}\left\{\langle 0|\sum_k \frac{p_{k\psi}}{|r_k-S|^3}|n\rangle \times \right.$$

$$\left.\langle n|\sum_k p_{k\delta}|0\rangle + \langle 0|\sum_k p_{k\delta}|n\rangle\langle n|\sum_k \frac{p_{k\psi}}{|r_k-S|^3}|0\rangle\right\} \quad (9)$$

From these equations, it is clear that the paramagnetic contributions require a knowledge of both ground-state $|0\rangle$ and excited-state wavefunctions $|n\rangle$. In contrast, the diamagnetic contributions depend on the ground-state wavefunction, $|0\rangle$, only. In equations (2)—(9), μ_0, e, and m are the permeability of free space, the electronic charge, and the electronic mass, respectively. The linear momentum for electron k is denoted by p_k, and $l_k(= r_k \wedge p_k)$ is the orbital angular momentum for this electron. W_0 and W_n are the energies of the ground and nth excited states, respectively, and the prime on the summations in equations (6)—(9) denotes a summation over all values of n except $n = 0$. It should be noted that integration over the continuum of excited states is implicitly included. The Greek subscripts denote Cartesian components x, y, and z; $\delta_{\alpha\beta} = 1$ if $\alpha = \beta$, and 0 if $\alpha \neq \beta$; $\varepsilon_{\beta\gamma\delta}$ is the alternating tensor and $= 1$ if $(\beta\gamma\delta)$ is an even permutation of (xyz), $= -1$ if $(\beta\gamma\delta)$ is an odd permutation of (xyz), and $= 0$ if any two of $(\beta\gamma\delta)$ are identical. If R and S are both null vectors then equation (1) reduces to the two-term expression (10) obtained by Ramsey.[8]

$$\sigma_{\alpha\beta} = \sigma_{\alpha\beta}^d + \sigma_{\alpha\beta}^p \quad (10)$$

The only paper in the review period which deals with the theory of nuclear magnetic shielding, as distinct from calculations of magnetic shielding constants, is the comment by Moss[10] in which he criticizes the approach of Weisenthal and de Graaf[11] to the theory of diamagnetism in neutral molecules. Weisenthal and de Graaf have expanded the molecular charge and current densities about a point R in the molecule, thereby obtaining a molecular Hamiltonian that involves the magnetic flux density B and the electric field strength E, rather than the external vector and scalar potentials. Use of this Hamiltonian gave an expression for the diamagnetic susceptibility that does not depend on the vector potential. This led Weisenthal and de Graaf[11] to claim that they had solved the problem of the most suitable gauge for calculations of diamagnetic susceptibilities when only approximate wavefunctions are available. Moss has pointed out that this is incorrect, their choice of R merely being the selection of a particular gauge. In fact, the molecular Hamiltonian may always be written so that it does not contain the potentials explicitly, by first selecting a gauge and then expressing the potentials in terms of B and E. Moss also states that the Weisenthal–de Graaf method is equivalent to the

---

[10] R. E. Moss, *J. Chem. Phys.*, 1974, **61**, 439.
[11] L. Weisenthal and A. M. de Graaf, *J. Chem. Phys.*, 1973, **58**, 249.

set of gauge transformations $A \to A - \nabla f$ with $f = (1/2c)B \wedge R \cdot \mu$, where μ is the molecular electric dipole moment. These conclusions are similar to those reached by Woolley and Cordle[12] (see Vol. 4, p. 5).

B. Basic Physical Aspects.—Following tradition, this sub-section contains the results of a number of calculations and experiments which add to the more fundamental knowledge of nuclear shielding phenomena.

Jameson et al.[13] have extended their studies of the temperature and density dependence of the ^{129}Xe chemical shift in rare-gas mixtures to the 3—28 amagat density range. As in previous work,[14,15] they write a virial expansion of shielding in ascending powers of density

$$\sigma(\rho) = \sigma_0 + \sigma_1 \rho + \sigma_2 \rho^2 + \ldots \quad (11)$$

The ultra-low densities used allowed the elimination of three-body and higher-order interactions, with the result that the present data give more accurate values for σ_1 than did the earlier measurements.[14,15] The main result of these studies is the precise determination of $\sigma_1(T)$ for ^{129}Xe in xenon gas and in mixtures with krypton and argon.

A knowledge of an accurate value of the proton shielding constant in molecular hydrogen would be particularly useful for the establishment of an absolute scale for proton magnetic shielding. In this regard, the theoretical studies of Reid,[16] who has attempted to obtain a precise value for the proton shielding constant in H_2, are especially important. Using Ramsey's theory in the adiabatic approximation, and with the origin of the vector potential and the origin of the co-ordinates taken at the nucleus of interest, the shielding in a state characterized by vibrational and rotational quantum numbers v and J is

$$\sigma_{vJ} = \sigma_{vJ}^d + \sigma_{vJ}^p \quad (12)$$

where

$$\sigma_{vJ}^d = \frac{1}{3}\frac{\mu_0}{4\pi}\frac{e^2}{m}\langle 0|\sum_k \frac{1}{r_k}|0\rangle_{vJ} \quad (13)$$

$$\sigma_{vJ}^p = \frac{1}{3}\frac{\mu_0}{4\pi}\frac{e^2}{m^2}\left[\sum_n{}'(W_0 - W_n)^{-1}\left\{\langle 0|\sum_k \frac{l_{kx}}{|r_k|^3}|n\rangle\langle n|\sum_k l_{kx}|0\rangle + \langle 0|\sum_k l_{kx}|n\rangle\langle n|\sum_k \frac{l_{kx}}{|r_k|^3}|0\rangle\right\}\right]_{vJ} \quad (14)$$

The diamagnetic contribution, σ^d, to the shielding was calculated using the 87-term James–Coolidge-type electronic wavefunction used by Reid and Vaida[17] previously.

---

[12] R. G. Woolley and J. E. Cordle, *Chem. Phys. Letters*, 1973, **22**, 411.
[13] C. J. Jameson, A. K. Jameson, and S. M. Cohen, *J. Chem. Phys.*, 1975, **62**, 4224.
[14] C. J. Jameson, A. K. Jameson, and S. M. Cohen, *J. Chem. Phys.*, 1973, **59**, 4540.
[15] A. K. Jameson, C. J. Jameson, and H. S. Gutowsky, *J. Chem. Phys.*, 1970, **53**, 2310.
[16] R. V. Reid, jun., *Phys. Rev. (A)*, 1975, **11**, 403.
[17] R. V. Reid, jun., and M. L. Vaida, *Phys. Rev. Letters*, 1972, **29**, 494; *Phys. Rev. (A)*, 1973, **7**, 1841.

The required averages over vibrational motion were performed using wavefunctions obtained numerically from the adiabatic potential of Kolos and Wolniewicz.[18] For the ground vibrational state of H_2 Reid has calculated values of 32.022 ($J = 0$) and 31.999 ($J = 1$) p.p.m. The corresponding values in HD are 32.072 ($J = 0$) p.p.m. and 32.055 ($J = 1$) p.p.m. Reid has used these calculated values of σ^d, together with experimental spin-rotation data and the experimental value of the isotope shift $\sigma(HD) - \sigma$ (ortho-H_2) at 295 K, to determine σ^p. The values of σ^p thus obtained were -5.622, -5.637, -5.654, and -5.650 p.p.m., for H_2 ($J = 1$), HD ($J = 1$), D_2 ($J = 1$), and D_2 ($J = 2$), respectively. These values are in good agreement with the results reported by Cade and Ramsey[19] discussed previously.[5]

The total shielding constants at 295 K obtained by Reid are presented in Table 1, where they are compared with the values obtained by other workers. One standard-deviation error in σ (ortho-H_2) at 295 K was found to be 0.014 p.p.m. Reid,[16] however, has reported an error of ±0.070 p.p.m. in order to allow for the contributions which relativistic and non-adiabatic effects make to σ (ortho-H_2). The calculated value is in good agreement with the experimental value quoted in Table 1

Table 1 Calculated magnetic shielding constants[a] for ortho-H_2 at 295 K

σ(ortho-H_2)	Reference
27.56	b
26.734	c
26.51 ± 0.30	d
26.297	e
26.366 ± 0.070	f
26.23 ± 0.18	g

[a] Units are p.p.m. [b] E. Ishiguro and S. Koide, *Phys. Rev.*, 1954, **94**, 350. [c] B. Mangeot, J. Guy, and F. Cabaret, *Compt. rend.*, 1963, **257**, 3134. [d] A. Saika and H. Narumi, *Canad. J. Phys.*, 1964, **42**, 1481. [e] Ref. 20. [f] Ref. 16. [g] Unpublished experimental data.

and with the experimental value estimated by Raynes[5] recently. Reid has also calculated the temperature dependence of the shielding in H_2 and in various isotopomers, and has found that the gross features of such temperature dependence are similar to those obtained by Raynes *et al.*[20]

New experimental data for the AlF molecule have been presented during the review period by Honerjaeger and Tischer.[21] The Zeeman effect in the microwave rotational spectrum of $^{27}Al^{19}F$ has been measured in the ground vibrational state. The spectrum obtained in the magnetic field exhibited resolvable hyperfine structure from the ^{27}Al nucleus, which enabled the determination of a shielding anisotropy value of -672 (±71) p.p.m. This value was compared with an anisotropy of -333 (±22) p.p.m. obtained from an interrelation with the spin-rotation interoction constant. No explanation of the discrepancy between these two values was affered.

Several experimental investigations have been published during the review period establishing absolute scales of shielding constants for cadmium, zinc, and potassium.

[18] W. Kolos and L. Wolniewicz, *J. Chem. Phys.*, 1968, **49**, 404.
[19] R. F. Cade and N. F. Ramsey, *Phys. Rev. (A)*, 1971, **4**, 1945.
[20] W. T. Raynes, A. M. Davies, and D. B. Cook, *Mol. Phys.*, 1971, **21**, 123.
[21] R. Honerjaeger and R. Tischer, *Z. Naturforsch.*, 1974, **29a**, 342.

Krueger et al.[22] have measured ^{111}Cd chemical shifts in aqueous solutions of CdCl$_2$, Cd(NO$_3$)$_2$, CdSO$_4$, and Cd(ClO$_4$)$_2$ as a function of concentration. From such data they have obtained a magnetogyric ratio for the cadmium ion in solution. A comparison of this ratio with the magnetogyric ratio for the free cadmium atom gives:

$$\sigma(^{111}\text{Cd}^{2+} \text{ in H}_2\text{O}) - \sigma(^{111}\text{Cd, free atom}) = -1106\ (\pm 4)\ \text{p.p.m.}$$

Using a similar approach, Lutz and co-workers[23] have reported FT n.m.r. studies of ^{67}Zn. The shielding constant for the hydrated zinc ion relative to the shielding constant for the free zinc atom was evaluated to be

$$\sigma(^{67}\text{Zn}^{2+} \text{ in H}_2\text{O}) - \sigma(^{67}\text{Zn, free atom}) = -690\ (\pm 10)\ \text{p.p.m.}$$

The trend of increasing magnitude for such shielding differences for Group IIB elements has been further substantiated by measurements on ^{199}Hg.[24]

Sahm and Schwenk[25] have reported n.m.r. studies of ^{39}K, ^{40}K, and ^{41}K. From the results of studies of n.m.r. lines of ^{39}K and ^{41}K in solutions of many potassium salts in H$_2$O and D$_2$O and the results of atomic beam magnetic resonance experiments they find

$$\sigma(\text{K}^+ \text{ in H}_2\text{O}) - \sigma(\text{K, free atom}) = -105.2\ (\pm 0.8)\ \text{p.p.m.}$$

This result implicitly indicates that the experimental data revealed no significant primary isotope effects.

All the studies referred to above have examined the effect of the isotopic composition of the solvent on the measured magnetic shielding constants. A discussion of such effects is deferred to Section 4H.

In 1970, Basch[26] published a theoretical paper in which he developed a relationship between the diamagnetic contribution to the shielding constant and the electrostatic potential which describes the chemical shift in ESCA. Since that work appeared there have been many attempts at establishing correlations between n.m.r. and ESCA chemical shifts with varying degrees of success. Thus the work of Lindberg,[27] who has addressed the question 'Can we expect any meaningful correlation between n.m.r.- and ESCA-shifts?', is quite timely. In an admittedly pragmatic approach, Lindberg[27] has discussed the correlations that have been found to date and has concluded his study with a set of rules to guide future searches for correlations between n.m.r. and ESCA shifts. Lindberg has suggested that a general correlation for the same atom in a varying chemical environment is unlikely. In particular, comparisons should preferably be made for molecules in which the substituent groups exert magnetically isotropic effects on the nucleus of interest. Furthermore, changes in hybridization of the atom of interest and of the first atom of a primary substituent should be kept as constant as possible. Since the success of such correlations depends on σ^p contributions remaining fairly constant or on the

---

[22] H. Krueger, O. Lutz, A. Schwenk, and G. Stricker, *Z. Physik*, 1974, **266**, 233.
[23] B. W. Epperlein, H. Krueger, O. Lutz, and A. Schwenk, *Z. Naturforsch.*, 1974, **29a**, 660; *ibid.*, p. 1553.
[24] O. Lutz, A. Nolle, A. Schwenk, and A. Uhl, *Verhandl. Deutsch. Physik, Gesellsch.*, 1974, 400.
[25] W. Sahm and A. Schwenk, *Z. Naturforsch.*, 1974, **29a**, 1754.
[26] H. Basch, *Chem. Phys. Letters*, 1970, **5**, 337.
[27] B. J. Lindberg, *J. Electron Spectroscopy Related Phenomena*, 1974, **5**, 149.

cancellation of those contributions to magnetic shielding which cannot be related to the electrostatic potential, this Reporter agrees that a general correlation between n.m.r. and ESCA shifts is unlikely.

Cohen and Brown[28] have continued their investigations into the relationship between σ^p and the temperature coefficient $(d\sigma/dT)$ for the total shielding. Measurements of the temperature sensitivity of ^{195}Pt chemical shifts were made to probe the large paramagnetic shieldings associated with transition-metal atoms. This work shows that the intrinsic temperature dependence of ^{195}Pt chemical shifts in complexes is described adequately by an equation developed by Benedek et al.[29] provided that the resonant nucleus lies in a site of effective octahedral or planar symmetry. Although the data on such complexes are sparse, the numerical results reported suggest that the proportionality between $(d\sigma/dT)$ and σ^p is, at best, rather approximate.

3 Calculations of Nuclear Shielding

This section reviews general methods of calculating magnetic shielding constants or actual calculations of shielding constants for isolated molecules. The discussion of papers concerned with the calculation of various types of contribution to magnetic shielding, such as ring-current contributions and shielding contributions from magnetically anisotropic groups, will be deferred to the next section.

One of the main difficulties in implementing Ramsey's theory involves finding an adequate representation of the unperturbed wavefunction, $\Psi^{(0)}$, and the first-order wavefunction, $\Psi^{(1)}$. In general, not only does one have little knowledge of the wavefunctions describing the high-energy discrete states but the summations in equations (6)—(9) include integration over continuum states, and little is known about these also for most molecules. For example, as Snyder and Parr[30] have demonstrated, hydrogenic expansion functions are unsatisfactory because the continuum contributions may be at least as important as those from the discrete states for the hydrogen atom. In the past few years, attention has turned to variational approaches to magnetic shielding, since such methods avoid some of the above difficulties;[31] as a result, there has been little research on expansion functions in recent years. Thus the current investigations of Emanuel[32] into the use of expansion methods in the calculation of the chemical shift and the diamagnetic susceptibility of the hydrogen atom are particularly important.

Emanuel[32] has used the normalized ground-state hydrogenic function for $\Psi^{(0)}$. For $\Psi^{(1)}$, a set of expansion functions whose radial parts are single exponent associated-Laguerre functions have been employed. Shull and Löwdin[33] have pointed out that this type of expansion set should be superior to the hydrogenic set since it is complete without continuum functions. Taking the magnetic dipole and the origin of the external magnetic field vector potential noncoincident with the hydrogen nucleus, Emanuel has employed such expansion functions centred on

[28] S. M. Cohen and T. H. Brown, J. Chem. Phys., 1974, **61**, 2985.
[29] G. B. Benedek, R. Englman, and J. A. Armstrong, J. Chem. Phys., 1963, **39**, 3349.
[30] L. C. Snyder and R. G. Parr, J. Chem. Phys., 1961, **34**, 827.
[31] R. Ditchfield, in 'Molecular Structure and Properties,' ed. G. Allen (M.T.P. International Review of Science), Physical Chemistry Series 1, Butterworths, London, 1972, Vol. 2, p. 91.
[32] R. V. Emanuel, Internat. J. Quantum Chem., 1974, **8**, 745.
[33] H. Shull and P.-O. Löwdin, J. Chem. Phys., 1955, **23**, 1362.

the hydrogen nucleus to calculate the magnetic shielding constant for the hydrogenic atom. This approach results in rapid convergence of the second-order energy summations of equations (6)—(9) to their exact values. In addition, the rate of convergence is fairly insensitive to the choice of the exponential coefficient of the expansion function. This work leads to the exciting conclusion that these functions may be suitable for the calculation of accurate values of magnetic shielding constants and diamagnetic susceptibilities in atoms and molecules. The results of such magnetic shielding calculations on small molecules are awaited with interest.

There have been relatively few *ab initio* calculations of magnetic shielding constants reported in the present review period. The first paper considered is that of Dixon and Palke,[34] who have calculated the diamagnetic contributions to the shielding constants for all the nuclei in the staggered and eclipsed forms of the BH_3NH_3 molecule. These contributions were calculated with the origin of the vector potential describing the magnetic field taken at the nucleus of interest in each case. The main point of interest in this work is sensitivity of σ^d to variations of molecular geometry. Table 2 presents a summary of the diamagnetic shielding data obtained by Dixon and Palke. The following points are worthy of note:

Table 2 *Diamagnetic shielding for nuclei in* $BH_3,NH_3{}^a$

Property	Eclipsed conformation				Staggered conformation			
	H_B	H_N	N	B	H_B	H_N	N	B
σ	111.01	125.70	393.77	283.26	110.94	125.66	393.72	283.25
$\dfrac{\partial\sigma}{\partial R_{BN}}$	−7.99	−5.96	−10.76	−17.84	−7.86	−5.89	−10.68	−17.79
$\dfrac{\partial\sigma}{\partial R_{BH}}$	−15.45	−1.06	−1.50	−7.22	−15.46	−1.06	−1.50	−7.19
$\dfrac{\partial\sigma}{\partial \widehat{HBN}}$	−3.45	−1.98	−2.50	0.57	−3.38	−1.93	−2.44	0.59
$\dfrac{\partial\sigma}{\partial R_{NH}}$	−0.47	−24.20	−8.03	−1.14	−0.50	−24.22	−8.04	−1.15
$\dfrac{\partial\sigma}{\partial \widehat{BNH}}$	−1.33	−2.37	−0.44	−1.96	−1.28	−2.30	−0.37	−1.98

a Units are p.p.m.

(1) The diamagnetic shieldings of the nuclei steadily decline as R_{BN} is increased, perhaps indicating that separating the fragments reduces their ability to screen a nucleus in the other fragment. It would be interesting to examine how the full tensor changes as R_{BN} is varied.

(2) Increasing R_{BH} decreases the diamagnetic shielding of H_B but causes little change in the shielding of the other nuclei.

(3) There is a rather large decrease in the diamagnetic shielding of H_N as R_{NH} is increased.

(4) All the changes in shielding produced by varying bond angles are small. It is also clear from Table 2 that the calculated values of σ^d are insensitive to internal rotation.

[34] M. Dixon and W. E. Palke, *J. Chem. Phys.*, 1974, **61**, 2250.

Raynes and Stanney[35] have considered the total magnetic shielding constants for the nuclei in carbon monoxide. The main aim of this work was to predict the ^{13}C and ^{17}O shielding constants at various temperatures for each of the isotopomers ^{13}C^{16}O, ^{13}C^{17}O, ^{13}C^{18}O, and ^{12}C^{17}O. From the coupled Hartree–Fock calculations of Stevens and Karplus,[36] it is clear that the bond-length dependence of the shielding of either nucleus is far greater than that found for hydrogen. The data presented in Table 3 indicate that there are substantial changes in the shieldings

Table 3 *Carbon and oxygen nuclear magnetic shielding constants (in p.p.m.) for CO at several values of the internuclear distance r (in atomic units)*

r	1.898	2.132	2.366	2.600
σ (C)	58.10	11.48	−47.79	−130.17
σ (O)	114.92	64.22	−1.62	−76.46

for quite small displacements from equilibrium ($r_e = 2.132$ a.u.), which for both nuclei involve a change in sign. Expanding the shielding as a power series about the equilibrium internuclear separation r_e gives

$$\sigma = \sigma_e^{(0)} + \sigma_e^{(1)} \xi + \frac{1}{2} \sigma_e^{(2)} \xi^2 + \frac{1}{6} \sigma_e^{(3)} \xi^3 + \ldots \quad (15)$$

where

$$\xi = (r - r_e)/r_e \quad (16)$$

is the relative displacement from equilibrium and the $\sigma_e^{(i)}$ are molecular parameters, $\sigma_e^{(0)}$ being the shielding constant for the equilibrium bond length. Figure 2 shows the shielding function for the carbon nucleus together with the lower part of the potential-energy curve for nuclear motion. By averaging over the nuclear motion and then over the occupied rotational and vibrational states Raynes and Stanney[35] have calculated a ^{13}C shielding constant in ^{13}C^{16}O at 300 K of 9.25 p.p.m. This can be compared with a value of 5.0 p.p.m. obtained[37] from a combination of rotational magnetic moment data and the calculated diamagnetic shielding constant. These workers have also predicted that the carbon shielding in ^{13}C^{16}O gas falls by approximately 0.05 p.p.m. on raising the temperature from 150 to 450 K. The isotope shifts associated with the various isotopomers of CO will be discussed in Section 4H.

Although it is not solely concerned with calculations of magnetic shielding in an isolated molecule, the recent theoretical work of Sadlej[38] on proton magnetic shielding in the bifluoride ion deserves special mention in this section. This paper highlights important limitations of some of the current theoretical methods for calculating magnetic shielding constants. At the outset of the work it is stated that the uncoupled Hartree–Fock (UCHF) variation–perturbation scheme of Karplus and Kolker[39] leads to a completely unsatisfactory value of the proton shielding

---

[35] W. T. Raynes and G. Stanney, *J. Magn. Resonance*, 1974, **14**, 378.
[36] R. M. Stevens and M. Karplus, *J. Chem. Phys.*, 1968, **49**, 1094.
[37] I. Ozier, L. M. Crapo, and N. F. Ramsey, *J. Chem. Phys.*, 1968, **49**, 2314.
[38] A. J. Sadlej, *Chem. Phys. Letters*, 1975, **30**, 432.
[39] M. Karplus and H. J. Kolker, *J. Chem. Phys.*, 1963, **38**, 1263.

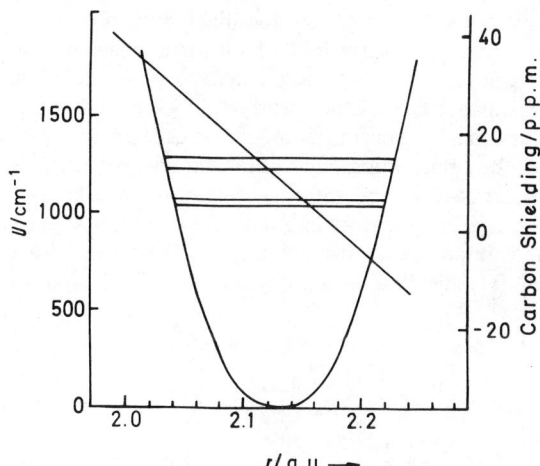

Figure 2 *Lower part of the potential curve, and the carbon nuclear shielding, for carbon monoxide. The horizontal lines denote (in ascending order) the levels $^{13}C^{16}O$ ($v = 0, J = 0$); $^{13}C^{18}O$ ($v = 0, J = 0$); $^{13}C^{16}O$ ($v = 0, J = 10$); $^{13}C^{18}O$ ($v = 0, J = 10$). The $v = 1, J = 0$ level of $^{13}C^{16}O$ is at about 3150 cm^{-1}*

constant in the bifluoride ion. Such uncoupled calculations performed with two different basis sets containing 33 and 52 contracted Gaussian functions gave a proton shielding constant in FHF$^-$ which was about 30 p.p.m. higher than the corresponding value for HF. Recent experimental studies[40,41] of dihalide ions have shown that the absolute proton shieldings are about 14 p.p.m. *lower* than in the corresponding hydrogen halides.

In order to study the effect of neglecting the coupling terms in the Hartree–Fock scheme, Sadlej[38] has performed coupled Hartree–Fock (CHF) calculations on FHF$^-$ and on HF, using basis sets of 33 and 22 contracted Gaussian functions, respectively.[42] The results are presented in Table 4. From these values, it is clear that the calculated proton magnetic shielding constant and the H-bond chemical shift for FHF$^-$ are even more disappointing than the results of the UCHF scheme of Karplus and Kolker. Various extensions of the basis set of Gaussian orbitals used in these calculations did not lead to significantly improved proton shielding

Table 4 *Calculated (CHF) proton magnetic shielding constants*[a]

Property	HF Gauge origin			FHF$^-$
	F	EC[b]	H	
σ_H	26.953	29.720	81.671	121.193

[a] Units are p.p.m. [b] EC denotes the electronic centroid.

[40] J. S. Martin and F. Y. Fujiwara, *Canad. J. Chem.*, 1971, **49**, 3071.
[41] F. Y. Fujiwara and J. S. Martin, *J. Chem. Phys.*, 1972, **56**, 4091.
[42] P. N. Noble and R. N. Kortzeborn, *J. Chem. Phys.*, 1970, **52**, 5375.

constants, and it appears that the required extension of the basis set goes far beyond any practical limit. Sadlej has concluded that the failure of both the uncoupled and the coupled Hartree–Fock calculations of magnetic properties of FHF$^-$ should be attributed to the inaccuracy of the representation of molecular excited states rather than to the inadequacy of the corresponding methods.

To improve the calculated results, Sadlej[38] has used an approximate method for the evaluation of the contribution due to the unknown part of the spectrum of the corresponding Hartree–Fock operator. The method proposed is based on the assumption that the average-energy approximation is applicable to the determination of the contribution due to the unknown SCF orbitals of a given molecule. Table 5 contains the shielding results obtained by Sadlej using this approximate

Table 5 Calculated[a] proton magnetic shielding constants[b]

Property	HF Gauge origin			FHF$^-$
	F	EC[c]	H	
σ_H	26.114	26.281	29.886	11.035

[a] Results of the approximate enforcement of the gauge origin independence of the magnetic susceptibility for each molecule. [b] Units are p.p.m. [c] EC denotes the electronic centroid.

scheme. The improvement in the agreement between the calculated and the observed H-bond shift is quite dramatic. Such improvement, however, is perhaps not too surprising, since it had been shown earlier[43] that methods that calculate magnetic shielding constants which depend on the gauge of the vector potential describing the magnetic field can predict proton chemical shifts which are in error by 1 to 60 p.p.m. Nonetheless, the method used is relatively simple and, if shown to be of comparable accuracy for other molecules, will be an important addition to the methods available for the calculation of magnetic shielding constants. The applications of this method to the calculations of magnetic shielding constants in larger molecules are awaited with interest.

In conclusion, Sadlej has suggested that the success of CHF schemes which use gauge-invariant atomic orbitals (GIAO) is mainly due to an inherent extension of the basis set. This led to the prediction that such gauge-invariant CHF schemes will suffer from the same deficiencies as the ordinary CHF method for the FHF$^-$ ion. In the light of earlier work[43] reporting the improvement in calculated proton shielding constants obtained by using GIAO in the CHF scheme, this Reporter feels that gauge-invariant CHF schemes may predict a downfield H-bond shift for the bifluoride ion even with minimal sets of basis functions. Clearly, the bifluoride ion represents a useful test case for the evaluation of theoretical methods for calculating magnetic shielding constants.

Ditchfield and Ellis[44] have reported CHF calculations of the magnetic shielding constants for all the carbon nuclei in the fluoromethanes, using the 4-31G set[45] of atomic functions appropriately modified by gauge factors. The calculated results are reported in Table 6, where they are compared with experimental data and with

[43] R. Ditchfield, *Chem. Phys. Letters*, 1972, **15**, 203.
[44] R. Ditchfield and P. D. Ellis, *Topics in Carbon-13 NMR*, 1974, **1**, 1.
[45] R. Ditchfield, W. J. Hehre, and J. A. Pople, *J. Chem. Phys.*, 1971, **54**, 724.

Nuclear Shielding

Table 6 Calculated and experimental ^{13}C chemical shifts for fluoromethanes[a,b]

Molecule	4-31G	4-31G+GIAO	Experimental
CH_3F	41.3	65.4	77.5
CH_2F_2	33.8	98.3	111.3
CHF_3	—	115.5	118.7
CF_4	−28.5	126.2	123.9

[a] Units are p.p.m. [b] All values are given relative to CH_4; a positive value indicates that the nucleus is less shielded than the carbon atom in methane.

the results of a previous theoretical study[46] which did not employ GIAO. The improvement obtained by using GIAO is particularly striking; the GIAO CHF values are in good agreement with the observed values.

These workers[44] have also presented some discussion of the diagonal elements of the ^{13}C shielding tensors in CH_3F and CH_2F_2. The calculated elements of the magnetic shielding tensor for the carbon atom in CH_2F_2 relative to the axis system shown in Figure 3 are 108.4 p.p.m. (σ_{xx}), 92.4 p.p.m. (σ_{yy}), 119.4 p.p.m. (σ_{zz}), and 15.6 p.p.m. (σ_{xz}). A comparison of these results with those obtained for CH_3F [112.4 p.p.m. (σ_{xx}) and 196.5 p.p.m. (σ_{zz})] indicates that the introduction of a second fluorine atom produces a large decrease in the σ_{zz} element and a moderate reduction in σ_{yy}. It was found that the cancellation of a large decrease in local atomic contributions with a comparable increase in non-local atomic cross-terms is responsible for the small overall change in σ_{xx}. A significant multiple-bond contribution to the shielding in the y direction was also reported. Furthermore, it was concluded that this result is consistent with the suggestion that there is partial double-bond character in the C–F bonds of CH_2F_2 resulting from electron donation of the type illustrated in Figure 3. In addition, it was noted that at least one

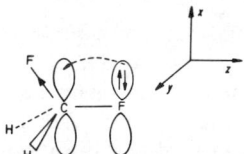

Figure 3 *Electron donation from a fluorie lonen-pair into a C—F bond*

pair of fluorine substituents is required before this multiple-bond contribution to shielding becomes significant. Similar conclusions have been reached from an analysis of the carbon shielding in CHF_3 and CF_4. As Ditchfield and Ellis have noted, Litchman and Grant[47] were only able to fit ^{13}C chemical shift data for the higher halogenomethanes by an empirical equation that included a term describing the combined effect of pairs of halogens.

The idea of interpreting molecular properties in terms of contributions associated with the constituent parts of the molecule has always appealed to chemists. For example, the concepts of bond energy, bond dipole, and bond polarizability all

---

[46] R. Ditchfield, D. P. Miller, and J. A. Pople, *J. Chem. Phys.*, 1971, **54**, 4186.
[47] W. M. Litchman and D. M. Grant, *J. Amer. Chem. Soc.*, 1968, **90**, 1400.

attest to attempts to associate contributions with various types of bonds, with the implicit understanding that such bond contributions can be transferred from molecule to molecule. As far as this Reporter is aware, such an approach has not been attempted for magnetic shielding constants. In this regard, the study of Maestro and Moccia[48] is particularly interesting. These workers have analysed CHF calculations of magnetic shielding constants in terms of localized orbital contributions. The problems of (1) ambiguities introduced by gauge transformations and (2) the lack of gauge invariance associated with the limited-basis-set results were examined in detail. Conditions were stated for resolving the above ambiguities, and a method was proposed for overcoming the effect of the limitation of the basis. By assuming that the localized orbital contributions associated with inner shells, lone pairs, and bonding pairs can be transferred from molecule to molecule, Maestro and Moccia[48] have estimated proton magnetic shielding constants in CH_4, CH_3F, CH_2F_2, and CHF_3. The shielding constants resulting from such an approach are presented in Table 7. Maestro and Moccia conclude that the agreement of these estimates

Table 7 *Proton magnetic shielding constants^a estimated from localized orbital contributions*

		CH_4	CH_3F	CH_2F_2	CHF_3
	Rb^b	29.66	27.37	26.77	25.39
σ_{Av}^H	RmP^c	29.98	26.92	24.64	23.57
	RmM^d	29.95	26.64	23.77	22.48
σ_{exp}^H		30.63 ± 0.6^e	26.63 ± 0.6^e	25.44 ± 0.6^f	24.64 ± 0.6^f

^{a} Units are p.p.m. ^{b} Rb denotes results determined with the gauge origin taken at the centre of charge. ^{c} RmP denotes results determined with the gauge origin corresponding to the minimum of the individual paramagnetic part of χ. ^{d} RmM denotes results determined with the gauge origin corresponding to the minimum modulus of the perturbed orbital. ^{e} G. P. Arrighini, M. Maestro, and R. Moccia, *J. Chem. Phys.*, 1970, **52**, 6411. ^{f} L. M. Jackman and S. Sternhell, 'Applications of N.M.R. Spectroscopy in Organic Chemistry,' 1969, Pergamon, 2nd edn., p. 160.

with the available experimental data probably is better than would have been expected.

There have been a number of papers involving the use of semi-empirical methods for calculating magnetic shielding constants. Witanowski *et al.*[49] have continued their earlier studies of nitrogen chemical shifts in some six-membered heterocycles containing nitrogen atoms. These workers have performed INDO calculations of magnetic shielding constants using the average-energy approximation to eliminate the evaluation of the energy denominators in equations (6)—(9). Within such an approach, the local diamagnetic contribution, σ_{loc}^d, and the local paramagnetic contribution, σ_{loc}^p, to the total shielding constant are given by equations (17) and (18), respectively;[50-52]

$$\sigma_{loc}^d = \frac{\mu_0}{4\pi}\frac{e^2}{3m}\langle 0|\sum_k r_k^{-1}|0\rangle \quad (17)$$

[48] M. Maestro and R. Moccia, *Mol. Phys.*, 1975, **29**, 81.
[49] M. Witanowski, L. Stefaniak, H. Januszewski, and G. A. Webb, *J. Magn. Resonance*, 1974, **16**, 69.
[50] J. A. Pople, *J. Chem. Phys.*, 1962, **37**, 60.
[51] J. A. Pople, *Mol. Phys.*, 1964, **7**, 301.
[52] M. Karplus and J. A. Pople, *J. Chem. Phys.*, 1963, **38**, 2803.

$$\sigma_{\text{loc}}^{\text{p}} = -\frac{\mu_0}{4\pi} \frac{h^2 e^2 \langle r^{-3} \rangle_{np}}{2m^2 \Delta E} \sum_{\text{B}} Q_{\text{AB}} \tag{18}$$

In these equations, r_k represents the distance of the kth electron in the local electron distribution from the nucleus, ΔE is an average excitation energy, $\sum_{\text{B}} Q_{\text{AB}}$ accounts for the amount of imbalance in the populations of the orbitals about the nucleus, and $\langle r^{-3} \rangle_{np}$ denotes the mean inverse cube of the distance of the valence-shell p-electrons from the nucleus. $\sum_{\text{B}} Q_{\text{AB}}$, in turn, may be expressed in terms of the ground-state bond-order and charge-density matrix elements.

The calculated changes in $\sigma_{\text{loc}}^{\text{d}}$ for the azines considered were much smaller than the experimental errors, and consequently were dismissed from further consideration in the discussion of nitrogen chemical-shift differences. By assuming that the average-energy denominator, ΔE, was constant for the series of azines investigated, Witanowski et al.[49] have demonstrated that the correlation between the measured shifts and the local paramagnetic term is highly linear. The relevant plot is shown in Figure 4. The correlation coefficient associated with this linear fit is 0.991 and the standard deviation from the least-squares fit is 7 p.p.m. These results were used as a strong argument in favour of use of the average-energy approximation for structurally related molecules. However, care was taken not to assign any physical significance

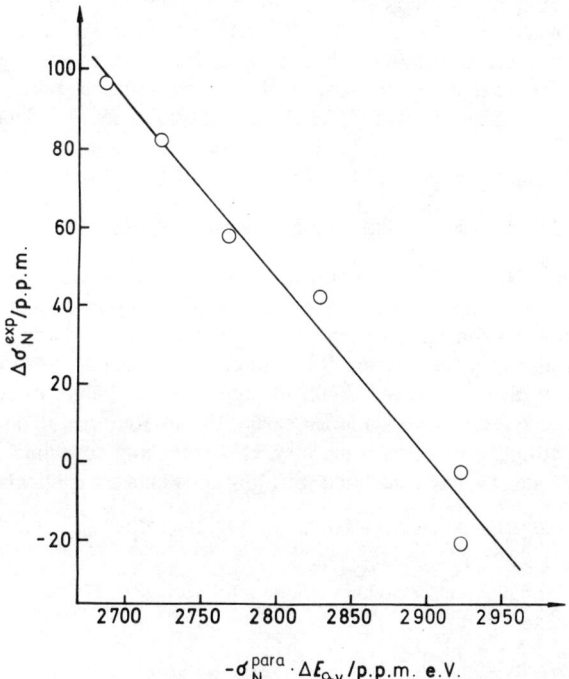

Figure 4 *Least-squares fit* $(\sigma_{N_1} - \sigma_{N_2}) = -0.4536 \, \Delta E_{\text{av}} \, (\sigma_{N_1}^{\text{p}} - \sigma_{N_2}^{\text{p}})$; *standard deviation* ± 7 p.p.m.; *correlation coefficient* 0.991; $\Delta E_{\text{av}} = 2.2$ eV

to the value of ΔE which gave the best agreement between calculated and observed values.

Aminova[53] has proposed a method for calculating magnetic shielding constants which is based on unperturbed LCAO MO wavefunctions expressed in terms of Gaussian basis functions. The results for the ^{13}C screening constants in CH_4 and CH_3F were compared with those obtained from other methods. Ando and co-workers[54-57] have published several papers dealing with the calculation of ^{13}C magnetic shielding constants. Using the approximations proposed by Pople[50,51] within a sum-over-virtual-states approach, Ando et al.[54] have calculated ^{13}C chemical shifts for some normal alkanes. The wavefunctions required were calculated using the original INDO method of Pople et al.[58] and the more recent MINDO/2 method proposed by Dewar and Haselbach.[59] The magnitudes of the paramagnetic contributions to the total shielding obtained with MINDO/2 wavefunctions were approximately twice as large as those obtained with INDO wavefunctions. This may be traced to the much smaller orbital energy separations predicted by the MINDO/2 approach. These in turn lead to excitation energies, $(W_n - W_0)$, which are about 5 eV smaller when evaluated within the MINDO/2 framework. The agreement between calculated and observed values is, however, quite poor. Ando et al.[54] have claimed that allowing for configurational mixing between the virtual states may be important. A discussion of other papers of Ando et al.[55-57] which deal mainly with shielding anisotropies is given in Section 4I.

Sterk and Steininger[60] have used expressions similar to those presented in equations (17) and (18), to calculate ^{13}C chemical shifts for E- and Z-oximes. Other semi-empirical calculations that have appeared in the review period are those of Schwind et al.[61] on ^{13}C chemical shifts of benzene, toluene, and cis-2-butene and those of Schroer[62] on ^{13}C chemical shifts in various ethylene derivatives. Finally, Martin et al.[63] have reviewed theoretical and empirical calculations of carbon chemical shifts in terms of the electronic distribution in molecules.

4 Transmission of Shielding Effects within Molecules

A. Introduction.—Many n.m.r. investigations have been devoted to the study of substituent effects. Such studies have not only played an important role in generating insights into the relationship between changes in magnetic shielding and electronic structure, but also they have given rise to several empirical correlations and additivity models which have proved useful in predicting the values of shielding constants in related systems. As mentioned earlier in this Report, the development of theoretical methods for calculating magnetic shielding constants has not yet progressed to a stage where calculated shielding constants are routinely in quantita-

---

[53] R. M. Aminova, *Doklady Akad. Nauk S.S.S.R.*, 1974, **219**, 625 (*Chem. Abs.*, 1975, **82**, 66 276).
[54] I. Ando, A. Nishioka, and M. Kondo, *Bull. Chem. Soc. Japan*, 1974, **47**, 1097.
[55] I. Ando and A. Nishioka, *Bull. Chem. Soc. Japan*, 1975, **48**, 841.
[56] I. Ando, A. Nishioka, and T. Asakura, *Makromol. Chem.*, 1975, **176**, 411.
[57] I. Ando, A. Nishioka, and M. Kondo, *Chem. Phys. Letters*, 1975, **25**, 212.
[58] J. A. Pople, D. L. Beveridge, and P. A. Dobosh, *J. Chem. Phys.*, 1967, **47**, 2026.
[59] M. J. S. Dewar and E. Haselbach, *J. Amer. Chem. Soc.*, 1970, **92**, 590.
[60] H. Sterk and H. Steininger, *Z. Naturforsch.*, 1974, **29a**, 1685.
[61] H. Schwind, D. Deininger, and D. Geschke, *Z. phys. Chem. (Leipzig)*, 1974, **255**, 149.
[62] W. Schroer, *Ber. Bunsengesellschaft phys. Chem.*, 1975, **79**, 36.
[63] G. J. Martin, M. L. Martin, and S. Odiot, *Org. Magn. Resonance*, 1975, **7**, 2.

tive agreement with experimental values. Thus, somewhat more approximate approaches have usually been used to explain the transmission of shielding effects revealed by the substituent-effect studies mentioned above.

Such schemes usually regard the shielding as being composed of three distinct types of contribution, each of which may be altered by a variety of 'through-bond' and 'through-space' mechanisms. In this approach, the total electron distribution in the molecule is partitioned into a 'local' part associated with the nucleus whose shielding is of interest and a 'non-local' part for the rest of the molecule. The three types of contribution are σ_{loc}^d due to the secondary screening magnetic fields arising from the induced local diamagnetic currents, σ_{loc}^p due to the magnetic fields of local paramagnetic currents, and σ_m arising from magnetic fields associated with currents induced in the 'non-local' electronic distribution. Both σ_{loc}^d and σ_{loc}^p may be altered by inductive and resonance effects, by intramolecular electric fields, by steric effects, and by van der Waals interactions. The value of σ_m is usually determined by the magnetic anisotropy of distant bonds or substituent groups but it may also be affected by the existence of ring currents elsewhere in the molecule. Approximate expressions for σ_{loc}^d and σ_{loc}^p have already been given in equations (17) and (18), respectively, of Section 3. In general, changes in σ_{loc}^p for proton shielding along a series of compounds are negligible due to the large values of ΔE involved. Thus proton chemical shifts are usually determined by changes in σ_{loc}^d and σ_m. For most other nuclei, however, changes in these contributions are very small relative to changes in σ_{loc}^p.

The sub-sections below are concerned with the interpretation of observed shielding data in terms of the three contributions outlined above. Inductive and resonance effects are treated in the same sub-section since many papers consider both types of effect in rationalizing changes in shielding. In the next sub-section proton and carbon shieldings are considered first, followed by discussions of substituent effects on boron, nitrogen, fluorine, silicon, phosphorus, and transition-metal shieldings.

B. Inductive, Resonance, and General Substituent Effects.—Since the early days of n.m.r., the relationship between proton magnetic shielding and the π-electron density at the carbon atom to which the hydrogen is bonded has been the subject of many studies. During the review period, there have been several studies of this type. Kloosterziel[64] has shown that a close relation exists between experimental proton chemical shifts of carbanions and π-electron densities as calculated by the ω-technique. The proportionality factor for phenyl protons was found to be 20 p.p.m. per unit charge. Mlochowski and Sliwa[65] have reported a relationship between ^1H chemical shifts and π-electron densities in diazaphenanthrenes and their protonated forms. The decrease in electron density at the carbon atoms due to protonation on nitrogen produces a deshielding relative to the free base. Nair and Gogte[66] have considered the ^1H chemical shifts in a number of condensed thiophens in terms of the π-electron density variations to be expected from different models for the bonding of sulphur. From such a study, these workers have concluded that sulphur appears to act as a π-donor.

[64] H. Kloosterziel, *Rec. Trav. chim.*, 1974, **93**, 215.
[65] J. Mlochowski and W. Sliwa, *Roczniki Chem.*, 1974, **48**, 1469 (*Chem. Abs.*, 1975, **82**, 85 524).
[66] P. M. Nair and V. N. Gogte, *Indian J. Chem.*, 1974, **12**, 589.

From studies of 2- and 2,7-substituted fluorenes, Mathieu et al.[67] have suggested that the proton chemical shifts found are related to the transmission of substituent effects across the biphenyl bridge. Condon[68] has rationalized the effects of methyl substitution on proton chemical shifts in alkyl-substituted hydrazines in terms of the electron-releasing inductive nature of the alkyl groups. Deady et al.[69] have examined chemical shifts for the ring protons in substituted methyl pyridinecarboxylates. The shifts for protons *meta* to the substituent suggest that substituent–nitrogen interactions may be important. Other studies of note are those of Englert et al.,[70] who have reported 261 substituent increments for ^{1}H n.m.r. chemical shifts of the 18- and 19-methyl protons of 9β,10α(retro)-steroids relative to 5β,9β,10α-androstane, and those of Blackwell et al.,[71] who have found correlations between ^{1}H substituent-induced chemical shifts for α- and β-methylene protons in, and rate constants for base-catalysed E2 elimination from, substituted phenethyl bromides and dimethyl-(phenethyl)sulphonium bromides.

By far the largest number of papers discussed in this section are concerned with the effects of substituents on carbon magnetic shielding constants. Wells et al.[72] have measured the ^{13}C n.m.r. spectra of a number of substituted naphthalenes and have derived substituent-induced chemical shifts from comparisons with the appropriate carbons of naphthalene. It appears that C-1 in substituted benzenes and C-2 in 2-substituted naphthalenes show the same response to the substituent (within experimental uncertainty). The 4α-interaction in naphthalene closely resembles the *para*-interaction in benzene, and both are well accounted for by the substituent parameters derived from reactivity and ^{19}F studies. The influence of substituents on the reactivity and the physical properties of functions in *ortho*, *meta*, *peri*, and related positions defied quantitative analysis. Wells et al.[72] have suggested that effects in the unsubstituted ring may arise more from small molecular distortions than from conventional changes in electron distribution. Schulman et al.[73] have determined the ^{13}C chemical shifts of 14 monosubstituted benzenes and fourteen 4-substituted biphenyls and have fitted the results to equation (19):

$$\Delta\delta = aAF^s + bBM^s + cCMF^s \qquad (19)$$

In this equation, $\Delta\delta$ is the chemical shift of the substituted compound relative to the parent; A, B, and C are the weighting factors chosen by Dewar et al.[74] to take account of the assumed molecular geometry. F^s, M^s, and MF^s are parameters representing field, mesomeric, and mesomeric-field effects, respectively, and a, b, and c are the regression parameters which measure the relative importance of Dewar's three terms to the ^{13}C chemical shift. The coefficients found are $a = -0.4833$, $b = -19.8484$, and $c - 1.9520$, with a correlation coefficient of 0.93. These results show that if the field effect is taken as 1.0, the mesomeric effect is 41 times

---

[67] A. Mathieu, J. C. Milano, and J. Douris, *Bull. Soc. chim. France*, 1974, 299.
[68] F. E. Condon, *Org. Magn. Resonance*, 1974, **6**, 517.
[69] L. W. Deady, P. M. Harrison, and R. D. Topsom, *Org. Magn. Resonance*, 1975, **7**, 41.
[70] G. Englert, W. Arnold, H. Els, A. Fuerst, A. Meier, and W. Meister, *Helv. Chim. Acta*, 1974, **57**, 1549.
[71] L. F. Blackwell, P. D. Buckley, and K. W. Jolley, *Austral. J. Chem.*, 1974, **27**, 2283.
[72] P. R. Wells, D. P. Arnold, and D. Doddrell, *J.C.S. Perkin II*, 1974, 1745.
[73] E. M. Schulman, K. A. Christensen, D. M. Grant, and C. Walling, *J. Org. Chem.*, 1974, **39**, 2686.
[74] M. J. S. Dewar, R. Golden, and J. M. Harris, *J. Amer. Chem. Soc.*, 1971, **93**, 4187.

greater, and the mesomeric field effects are 4 times greater and of opposite sign in their effect on the chemical shift. Thus the authors have concluded that electronic reorganization in these molecules through resonance structures is of greater significance than that due to electrostatic field effects. Schulman et al.[73] have pointed out that a correlation of both shifts and reaction rates with a single linear free energy relationship of the Hammett type would be fortuitous unless the appropriate dissection of the contributing components is first achieved. However, it is worth noting that carbon shift data can be used to characterize the important resonance ieature and thereby aid in the separation of these various effects in reaction or related equilibrium data.

Firl et al.[75, 76] have published interesting studies of ^{13}C n.m.r. chemical shifts and bonding in cumulated double bond systems. The aim of their work on ketens[75] and ketenimines[76] was to obtain a qualitative idea of the relative importance of the resonance structures (1a, b, and c) shown below:

$$\begin{array}{c}R^1\\ \\ R^2\end{array}\!\!\!\!C\overset{t}{=}C=X \quad \leftrightarrow \quad \begin{array}{c}R^1\\ \\ R^2\end{array}\!\!\!\!\overset{-t}{C}-C\equiv\overset{+}{X} \quad \leftrightarrow \quad \begin{array}{c}R^1\\ \\ R^2\end{array}\!\!\!\!\overset{t}{C}\overset{+}{=}C-\overline{X}$$

(1a)   (1b)   (1c)

With respect to the π-bonding situation, the most important ieature is the unusual shielding of the terminal carbon atom, C^+. Compared with the corresponding carbon atoms in analogously substituted olefins ($R^1R^2C=CH_2$) and allenes ($R^1R^2C=C=CH_2$), the terminal carbon nucleus in ketenimines is more shielded by more than 70 and 30 p.p.m., respectively. Firl et al.[76] have concluded that the stronger shielding found for C^t in ketenimines compared with C^t in allenes is mainly a consequence of the higher π-electron density at C^t and differences in the π-bond order. A comparison of the ketenimine results with those found in allenes and ketens suggests that the mesomeric interaction between the $C=C$ bond π-orbital and the group X orbitals of appropriate symmetry is more important when X = NR than when X = CH_2, but less significant than when X = O. These conclusions are at odds with those drawn from a study[77] of ^1H n.m.r. data on such molecules.

Hawkes et al.[78] have reported ^{13}C chemical shifts for a variety of perchlorocarbons, their hydrogen-substituted derivatives, and chlorocarbon ketones. Correlations between ^{13}C chemical shifts and structure were found for simple molecules, and these correlations appear to provide the possibility of reasonable structural assignments for complex perchlorocarbons. Substituent effects on ^{13}C chemical shifts of the pyrene molecule have been studied by Hansen et al.[79] Such effects on chemical shifts in molecules (2a and b) were found to alternate both in magnitude and sign along the periphery of the molecule. It was noted that such an alternation in sign is consistent with the variation in electron density at the carbon atoms, to be expected from a mesomeric withdrawal of electrons by the carbonyl group. The

---

[75] J. R. Firl and W. Runge, Z. Naturforsch, 1974, **29b**, 393.
[76] J. R. Firl, W. Runge, W. Hartmann, and H. P. Utikal, Chem. Letters, 1975, 51.
[77] J. L. Reilly and G. R. Krow, J. Org. Chem., 1972, **37**, 2364.
[78] G. E. Hawkes, R. A. Smith, and J. D. Roberts, J. Org. Chem., 1974, **39**, 1276.
[79] P. E. Hansen, O. K. Poulsen, and A. Berg, Org. Magn. Resonance, 1975, **7**, 23.

(2) a; R = $^{13}CO_2Et$
    b; R = $^{13}CO_2H$

effect on C-11, however, does not appear to be consistent with this explanation. It was suggested that the importance of electric field effects may be the reason for this. Adcock and co-workers[80] have performed ^{13}C n.m.r. studies for a number of model benzylic tin compounds in an attempt to evaluate the importance of hyperconjugation involving the carbon–tin bond in benzyltrimethylstannane. It was found that the shielding at C-4 was significantly higher in compounds (3a—e) and (4a) than in the appropriate parent system.

(3) a; X = $SnMe_3$
    b; X = $PbMe_3$
    c; X = $HgCH_2Ph$
    d; X = $HgCl$

(4) a; X = $SnMe_3$, Y = H
    b; X = $SnMe_3$, Y = F

(5) a; X = $SnMe_3$, Y = H
    b; X = $SnMe_3$, Y = F

(6) a; X = H
    b; X = $SnMe_3$

These data, together with the essential constancy of the shieldings at C-3 and C-5, and the zero value for the substituent-induced chemical shift for C-4 in (6b), are consistent with the established stereoelectronic requirements for a hyperconjugative mechanism for the electron-releasing metallomethyl substituents. This conclusion was based on experimental evidence[81] that for structurally similar compounds the chemical shift of C-4 is only affected by changes in local π-electron density.

Olah and co-workers have completed further studies of magnetic shielding in

[80] W. Adcock, B. D. Gupta, W. Kitching, D. Doddrell, and M. Geckle, *J. Amer. Chem. Soc.*, 1974, **96**, 7360.
[81] W. F. Reynolds, I. R. Peat, M. H. Freedman, and J. R. Lyerla, *Canad. J. Chem.*, 1973, **51**, 1857 and references therein.

various stable carbocations. Olah et al.[82] have measured ^{13}C n.m.r. chemical shifts of the σ complexes which result from the nitration and protonation of a series of hexasubstituted benzenes. Assuming that the shielding at the carbon nuclei reflects quite closely the local electron density, these results were used to determine details of the electron distribution in these ions. The ^{13}C n.m.r. spectra of benzenium (7a) and nitrobenzenium (7b) ions show distinct differences between the two types of

(7a) (7b)

σ-complex. C-4 in the nitrobenzenium ions is consistently deshielded relative to C-4 in the benzenium ions, while C-2 and C-6 are consistently shielded. In both types of ions, C-2, C-4, and C-6 are considerably deshielded relative to the *meta*-positions C-3 and C-5. The observed shielding differences have been interpreted as indicating a dissimilar distribution of the electron density in the cationic pentadienyl π-system of the nitrobenzenium and benzenium ions. The total change in magnetic shielding for the five carbon atoms in the cationic π-system relative to the five carbons of hexamethylbenzene is very nearly the same for ion (7b) [$\Delta\delta(^{13}C)$ = 192.1 p.p.m.] and ion (7a) [$\Delta\delta(^{13}C)$ = 197.0 p.p.m.] when X is a methyl substituent. This similarity in total shifts relative to hexamethylbenzene has led Olah et al.[82] to suggest that the same amount of positive charge has been introduced into the π-system of both ions. However, in nitrobenzenium ions, the electron distribution is apparently polarized towards the nitro-group, resulting in a gain in electron density with increased shielding at the *ortho*-positions and a loss of electron density with deshielding at the *para*-position, and to a lesser extent, at the *meta*-positions.

Olah and Liang[83] have studied ^{1}H and ^{13}C shielding in a series of 2-alkyl, 2-phenyl-, and 2-halogeno-substituted 2-adamantyl cations. A comparison of the carbenium carbon shifts in 2-methyl- and 2-ethyl-2-adamantyl cations with those observed in t-butyl and 1-methylcyclohexyl cations has indicated that the electron-releasing alkyl groups produce only slight shielding at the sp^2 centre. An examination of shielding at other positions has suggested that the strain produced by ionization may be an important contributing factor. The ^{13}C n.m.r. spectrum of the 2-phenyl-2-adamantyl cation has been interpreted as being indicative of extensive charge delocalization between the adamantyl nucleus and the phenyl ring. The extent of n–p conjugation between the halogen lone pairs of electrons and the empty p-orbital of the carbenium centre was also found to play an important role in determining the relative shielding in the 2-halogeno-2-adamantyl cations.

Olah and Grant[84] have measured ^{13}C chemical shifts in benzenediazonium ions in an attempt to provide further information about the relationship between such

---

[82] G. A. Olah, H. C. Lin, and D. A. Forsyth, *J. Amer. Chem. Soc.*, 1974, **96**, 6908.
[83] G. A. Olah and G. Liang, *J. Org. Chem.*, 1974, **39**, 3750.
[84] G. A. Olah and J. L. Grant, *J. Amer. Chem. Soc.*, 1975, **97**, 1546.

shifts and electron density in mono- and di-substituted aromatic systems. Previous work[85, 86] has suggested that the diazonium group (N_2^+) is the most powerful electron-withdrawing group known. It was also suggested that, in addition to the inductive effect, an extremely strong resonance effect is operative and is most important at the *para*-position. Olah and Grant have found that the *para*-^{13}C chemical shift and the total charge density at the C-4 position of the benzenediazonium ion fit a relationship proposed by Levy and Nelson[87] quite nicely. On the basis of ^{13}C chemical shift data for the benzenediazonium salts studied, Olah and Grant[84] have concluded that structures (8a—f) are significant resonance contributors. The

shift to higher frequency (relative to benzene) of 41.4 p.p.m. for 4-fluorobenzenediazonium ion, for example, was taken as providing strong support for the importance of the quinoidal-type resonance structures (8d).

Olah and Spear[88] have continued earlier studies of magnetic shielding in phenyl-, methyl-, and cyclopropyl-substituted alkenyl cations. Using ^{13}C chemical shifts as a measure of local electron density, the authors have found that the 1,3-substituted alkenyl cations exhibit strong charge delocalization between C-1 and C-3, whereas the charge in 1,1-substituted systems is substantially higher at the tertiary carbon atom. A neighbouring-group deshielding of carbenium ion centres by cyclopropyl groups was also detected. Olah *et al.*[89] have reported further results on vinyl cations in studies of carbon magnetic shielding in the carbocations (9a—e) and (10a—c). The marked deshielding of C_α and C_γ in the ions (9a—e) relative to their precursor alcohols was taken as a clear indication of a substantial contribution from the mesomeric cation form (11) in all cases. Similar comparisons for the ions (10a—c), however, suggest that the allenyl form (11) is not important for such ions.

Abraham *et al.*[90] have assigned the ^{13}C n.m.r. spectra of 55 pyrroles, and have found that the pyrrole ring-carbon chemical shifts can be accurately predicted on

---

[85] E. S. Lewis and M. D. Johnson, *J. Amer. Chem. Soc.*, 1959, **81**, 2070.
[86] J. D. Roberts, R. A. Clement, and J. J. Drysdale, *J. Amer. Chem. Soc.*, 1951, **73**, 2181.
[87] See G. C. Levy and G. L. Nelson, 'Carbon-13 Nuclear Magnetic Resonance for Organic Chemists,' Wiley-Interscience, New York, 1972.
[88] G. A. Olah and R. J. Spear, *J. Amer. Chem. Soc.*, 1975, **97**, 1539.
[89] G. A. Olah, R. J. Spear, P. W. Westerman, and J.-M. Denis, *J. Amer. Chem. Soc.*, 1974, **96**, 5855.
[90] R. J. Abraham, R. D. Lapper, K. M. Smith, and J. F. Unsworth, *J.C.S. Perkin II*, 1974, 1004.

$$R^1\!\!\diagdown\!\!\underset{R^2\diagup}{C_\alpha}\!\!=\!\!\overset{+}{C_\beta}\!\!\equiv\!\!C_\gamma\!-\!R^3 \qquad O\!\equiv\!\overset{+}{C_\alpha}\!\!=\!\!C_\beta\!\equiv\!C_\gamma\!-\!R$$

(9) a; $R^1 = R^2 = R^3 = $ Me
    b; $R^1 = R^2 = $ Ph, $R^3 = $ Me
    c; $R^1 = R^2 = $ Me, $R^3 = $ Ph
    d; $R^1 = R^3 = $ Ph, $R^2 = $ Me
    e; $R^1 = R^2 = R^3 = $ Ph

(10) a; $R = $ H
     b; $R = $ Me
     c; $R = $ Ph

the basis of additive substituent effects together with contributions due to adjacent substituents, and in some cases, a conjugation effect. The substituent chemical shift parameters are comparable to those found for monosubstituted thiophens,[91]

$$\underset{X\diagup}{\overset{Y\diagdown}{C_\alpha}}\!-\!C_\beta\!\equiv\!C_\gamma\!-\!Z \;\;\leftrightarrow\;\; \underset{X\diagup}{\overset{Y\diagdown}{C_\alpha}}\!=\!C_\beta\!=\!C_\gamma\!-\!Z$$

(11)

except for the substituted carbons. For C-2 substituents, C-4 is least affected, and for C-3 substituents both C-4 and C-5 are not very sensitive. This is in accord with the resonance and charge-delocalization effects expected between the substituent and the ring, [*e.g.* see (12) and (13)]. These qualitative conclusions were supported by

(12)   (13)

INDO calculations of charge densities. Miyajima *et al.*[92] have attempted to explain ^{13}C chemical shifts of 16 monosubstituted ethylenes in terms of σ- and π-electron densities calculated by the σ-included ω-HMO method.[93] The authors have claimed that a linear relationship between such calculated electron densities and ^{13}C chemical shifts exists. However, there is sufficient scatter in the plots of ^{13}C chemical shift *vs.* electron density to suggest that types of contribution other than inductive effects may be important. Yamamoto and Moritani[94] have measured ^{13}C n.m.r. spectra of alkenylboranes in an attempt to evaluate the importance of the π-type interaction between a π-electron system and a vacant orbital (of the appropriate symmetry) of a boron atom. They have concluded that the strong deshielding of the β-carbon atom found in $\alpha\beta$-unsaturated alkenylboranes supports the presence of π character in the B—C bond. However, since such conclusions were based on

---

[91] K. Takahashi, T. Sone, and K. Fujieda, *J. Phys. Chem.*, 1970, **74**, 2765.
[92] G. Miyajima and K. Takahashi, *Org. Magn. Resonance*, 1974, **6**, 413.
[93] A. Streitwieser, 'Molecular Orbital Theory for Organic Chemists,' Wiley, New York, 1961.
[94] Y. Yamamoto and I. Moritani, *Chem. Letters*, 1975, 57.

$$\underset{(14a)}{\diagup\!C\!\!=\!\!\overset{|}{C}\!-\!B\diagdown} \longleftrightarrow \underset{(14b)}{\diagup\!\overset{+}{C}\!-\!\overset{|}{C}\!=\!\overset{-}{B}\diagdown}$$

considerations of local carbon electron density, they should probably be viewed with caution. The loss of the C=C bond implied by the resonance structure (14b) may well play an important role in determining the value of the shielding at C_β.

Paulson et al.[95] have found that the ^{13}C chemical shift associated with the epoxy-carbon atom adjacent to the unsaturation in conjugated alicyclic epoxides is considerably more shielded than the other epoxy-carbon atom. A conjugative interaction between the epoxy-group and the olefinic π-bond has been tentatively proposed to explain these observations. Okuyama et al.[96] have treated the chemical shifts of side-chain carbons and protons in trans-1-phenyl-1,3-butadiene and phenylallene and their ring-substituted derivatives with a Hammett-type relationship. The carbon chemical shifts of the side-chain gave alternately opposite signs for ρ values while those of phenylallene showed little dependence on the ring substitution. Buchanan et al.[97] have reported a systematic investigation of the ^{13}C spectra of a series of 33 substituted methyl phenyl sulphides, sulphoxides, and sulphones. The observed trends were discussed in terms of the steric and electronic influences of the substituent groups. For example, the shielding effects (relative to benzene) observed for C-2, C-4, and C-6 in methyl phenyl sulphide were rationalized in terms of contributions from the resonance forms (15a—d). A possible alternative

suggested by the authors[97] is that the shielding at the *ortho*-carbons could be related to the fact that the favoured conformation is planar. In this case there would be a 'γ' steric interaction between the methyl group and the *ortho*-carbon which would be expected to shield the *ortho*-position.

Abronin et al.[98,99] have examined the redistribution of electron density which results from the protonation of benzene in an attempt to understand the changes which occur in carbon and proton shielding on protonation. Bjoergo et al.[100] have

---

[95] D. R. Paulson, F. Y. N. Tang, G. F. Moran, A. S. Murray, B. P. Pelka, and E. M. Vasquez, *J. Org. Chem.*, 1975, **40**, 184.

[96] T. Okuyama, K. Isawa, and T. Fueno, *Bull. Chem. Soc. Japan*, 1974, **47**, 410.

[97] G. W. Buchanan, C. Reyes-Zamora, and D. E. Clarke, *Canad. J. Chem.*, 1974, **52**, 3895.

[98] I. A. Abronin, S. G. Gagarin, and G. M. Zhidomirov, *Izvest. Akad. Nauk S.S.S.R., Ser. khim.*, 1974, 906 (*Chem. Abs.*, 1974, **81**, 37 089).

[99] I. A. Abronin, S. G. Gagarin, and G. M. Zhidomirov, *Izvest. Akad. Nauk S.S.S.R., Ser. khim.*, 1975, 922 (*Chem. Abs.*, 1975, **83**, 27 145).

[100] J. Bjoergo, D. R. Boyd, G. C. Watson, W. B. Jenning, and D. M. Jerina, *J.C.S. Perkin II*, 1974, 1081.

studied electronic and steric contributions to magnetic shielding in E- and Z-isomers of ortho-disubstituted C-arylaldimines. Bremser and Paust[101] have determined ^{13}C chemical shifts of β-carotene and have used these values to obtain information about how charge polarization of the double-bond changes as the distance of the double bond from the carbonyl function increases. Bundgaard et al.[102] have used ^{13}C n.m.r. spectroscopy to probe substituent effects in phosphorins. The trends observed are similar to those found for the analogous benzene and pyridine series. Geanangel[103] has found that the ^{13}C chemical-shift differences in the borane adducts Me_3N,BX_3 vary as $BF_3 < BCl_3 < BBr_3$, in agreement with the acceptor strengths of the BX_3 molecules. No correlations between ^{13}C and ^{11}B chemical shifts in the same molecule were found. Kalabin et al.[104] have obtained ^{13}C n.m.r. data for alkoxy- and alkylthio-ethylenes in order to determine the degree of p–π conjugation between the substituent group and the double bond. Kornprobst and Doucet[105] have examined the influence of alkyl substituents on the chemical shifts of sp carbons in alkynes $R^1C \equiv CR^2$ using ^{13}C n.m.r. Luedemann and Nimz[106] have determined the influence of methoxy-groups ortho to a phenolic group on the chemical shifts of the aromatic carbon atoms in methoxy-derivatives of 4-hydroxy-alkylbenzenes. Nesmeyanov et al.[107] have used ^{13}C chemical shifts to evaluate the stereochemical requirements for σ–π conjugation between C—Hg and C=O bonds in some oxo(alkyl)mercury compounds of the type shown in (16) and (17).

(16)   (17)

Okuyama and Fueno[108] have investigated the effect of substituents on ^1H and ^{13}C chemical shifts in the benzofuran system. The chemical shifts of H-2 and C-2 of 5- and 6-substituted derivatives were found to obey a modified Hammett equation. These authors have also concluded that the ground-state electronic structure is similar to that of styrene rather than phenyl vinyl ether. Peat[109] has investigated electronic effects in aromatic systems by n.m.r. spectroscopy and all-valence-electron MO calculations. ^{13}C Studies of benzofuran have also been completed by Platzer and Basselier,[110] who have investigated the effects of methyl substitution. They have proposed that a consideration of steric interactions is important to an understanding of the shielding changes produced by the introduction of methyl

---

[101] W. Bremser and J. Paust, Org. Magn. Resonance, 1974, 6, 433.
[102] T. Bundgaard, H. J. Jakobsen, and K. Dimroth, Tetrahedron Letters, 1974, 3179.
[103] R. A. Geanangel, Inorg. Chem., 1975, 14, 696.
[104] G. A. Kalabin, B. A. Trofimov, V. M. Bzhezovskii, D. F. Kushnarev, S. V. Amosova, N. K. Gusarova, and M. L. Al'pert, Izvest. Akad. Nauk S.S.S.R., Ser. khim., 1975, 576 (Chem. Abs., 1975, 82, 154 724).
[105] J. M. Kornprobst and J. P. Doucet, J. Chim. phys., 1974, 71, 1129.
[106] H. D. Luedemann and H. Nimz, Makromol. Chem., 1974, 175, 2393.
[107] A. N. Nesmeyanov, V. A. Blinova, and E. I. Fedin, Doklady Akad. Nauk S.S.S.R., 1975, 220, 1336 (Chem. Abs., 1975, 82, 138 769).
[108] T. Okuyama and T. Fueno, Bull. Chem. Soc. Japan, 1974, 47, 1263.
[109] I. R. Peat, Diss. Abs. (B), 1974, 35b, 162 (Chem. Abs., 1974, 81, 119 635).
[110] N. Platzer and J. J. Basselier, Bull. Soc. chim. France, 1974, 905.

groups. ^{13}C N.m.r. spectra of methyl derivatives of benzo[b]thiophen, indole, and indene were also reported. Repmann[111] has correlated the ^{13}C chemical shifts of several of the sterols and steroidal hormones measured by Reich et al.[112] with atomic charges calculated by the extended Hückel MO method. The largest discrepancies were found for carbons in the neighbourhood of methyl groups. Riand et al.[113] have determined the effect of Me and NH_2 substituents on the carbon chemical shifts of the pyrimidine system. Substituent parameters derived from measurements on disubstituted pyrimidines were used to predict ^{13}C chemical shifts for tri- and tetra-substituted pyrimidines.

Tori et al.[114] have reported data on ^{13}C chemical shifts of norbornane, norbornene, norbornadiene, benzonorbornene, and benzonorbornadiene. The extraordinary deshielding of the bridge methylene carbon in the latter molecule was explained in terms of electron delocalization between the bridge methylene and the unsaturated groups. Volz et al.[115] have measured ^{1}H, ^{13}C, and ^{19}F n.m.r. spectra of p-$FC_6H_4C^+MeR$ (R = Me,Ph,cyclopropyl) ions in order to evaluate the relative stabilization of carbonium ions by such substituents. The order found from ^{19}F data was cyclopropyl > Ph > Me.

A number of other ^{13}C and ^{1}H n.m.r. studies which have contributed to the understanding of general substituent effects will now be mentioned briefly. Wenkert et al.[116] have suggested that the marked γ shielding and δ deshielding effects observed on acylation of allyl alcohols are a consequence of through-bond polarization of the olefinic carbons. The influence of substituents at the 6-position on the ^{13}C chemical shifts of purine carbon atoms has been investigated by Breitmaier and Voelter.[117] These workers have found a linear correlation between such substituent effects on carbon shielding and the Pauling electronegativity value of the substituent. Davis et al.[118] have reported N-methyl chemical shifts in quaternized azoles, isazoles, diazoles, and their benzologues. The shifts in the 1,3-di-N-Me compounds were found to be affected by resonance effects associated with the heteroatom. In contrast, electronegativity effects are important for the 1,2-di-N-Me-substituted compounds. Sterk and Holzer[119] have shown that the chemical shifts of protons in aryl-ortho-diketo-compounds can be rationalized adequately in terms of charge densities calculated by the CNDO/2 MO method. Solkan and Sergeev[120] have examined ^{13}C magnetic shielding in monosubstituted benzenes and have proposed a general equation relating the magnitude of the ^{13}C chemical shift to the effective charge of the adjacent atoms in the molecule as evaluated by the INDO MO method. Torri and Azzaro[121] have investigated the effect of methyl substituents on

[111] H. Repmann, Z. Naturforsch., 1974, 29a, 1172.
[112] H. J. Reich, M. Jautelat, M. T. Messe, F. J. Weigert, and J. D. Roberts, J. Amer. Chem. Soc., 1969, 91, 7445.
[113] J. Riand, M. T. Chenon, and N. Lumbroso-Bader, Tetrahedron Letters, 1974, 3123.
[114] K. Tori, T. Tsushima, H. Tanida, K. Kushida, and S. Satoh, Org. Magn. Resonance, 1974, 6, 324.
[115] H. Volz, J. H. Shin, and H. J. Streicher, Tetrahedron Letters, 1975, 1297.
[116] E. Wenkert, M. J. Gasic, E. W. Hagaman, and L. D. Kwart, Org. Magn. Resonance, 1975, 7, 51.
[117] E. Breitmaier and W. Voelter, Tetrahedron, 1974, 30, 3941.
[118] M. Davis, L. W. Deady, and E. Homfeld, J. Heterocyclic Chem., 1974, 11, 1011.
[119] H. Sterk and H. Holzer, Z. Naturforsch., 1974, 29a, 974.
[120] V. N. Solkan and N. M. Sergeev, Vestnik Mosk. Univ., Khim., 1975, 16, 97 (Chem. Abs., 1975, 83, 35 271).
[121] J. Torri and M. Azzaro, Bull. Soc. chim. France, 1974, 1633.

the ^{13}C chemical shifts in cyclohex-2-en-1-ones. The shifts were related to the effect of substitution on the planarity of the conjugated system and to the inductive effect of the substituents.

Other studies of substituent effects which have appeared in the review period are those of Yamazaki and Niwa[122] on the effect of *meta-* and *para-*substituents on the ^1H chemical shifts of thioformanilides and those of Zeisberg and Bohlmann[123] on ^{13}C chemical shifts of poly-ynes. Imanari and Masahiro[124] have investigated the effects of *ortho-* and *para-*halogen substituents on the carbon chemical shifts of biphenyl derivatives. Nakajima *et al.*[125] have reported substituent effects on the ring-proton chemical shifts in iodomethylbenzenes. Niwa and Yamazaki[126] have studied substituent effects on the ^{13}C chemical shifts of carboxy-carbons in substituted benzoic acids. Pehk *et al.*[127] have compared carbon chemical shifts in bicyclo[3,3,1]nonanes and adamantanes substituted in the 1-position by OH, CH_2OH, and CO_2H. The studies of Szarek and Vyas[128] have indicated that the changes in ^{13}C shieldings resulting from the replacement of a hydroxyl-group by a chloro-group in chlorodeoxy-sugars are configuration dependent. Trofimov *et al.*[129] have examined ^{13}C chemical shifts in alkoxy- and alkylthio-ethylenes. Other studies worth noting here are those of Zheglova *et al.*[130] on the effect of substituents on ^1H n.m.r. spectra of benzoylacetates and those of Zanger and Simons[131] who have developed an additivity scheme for the determination of the chemical shifts of ring protons of substituted pyridines.

Sprecher *et al.*[132] have determined ^{11}B n.m.r. chemical shifts for all possible isomers of monochloro-, monobromo-, and monoiodo-dodecaborane(14). All shift trends were interpreted in terms of changes of the $2p$ orbital size, which in turn produce variations in the values of $\langle r^{-3} \rangle$ in the paramagnetic shielding term. It was suggested that a small value of σ^p accounts for the shielding of B_2 and B_4 relative to the rest of the cage.

Lichter and Roberts[133] have reported ^{15}N chemical shifts of aniline, the toluidines, xylidines, and several halogen- and oxygen-substituted anilines. The methyl-substituent parameters determined in this work contrast strikingly with those found in saturated systems, in which substantial deshielding effects are induced by substitution at all except the γ positions. The comparable increases in shielding induced by the *ortho-* and *para-*methyl groups are in accord with qualitative concepts of the electron-donating properties of the methyl substituent in a benzene ring. Adler and

[122] M. Yamazaki and J. Niwa, *Nippon Kagaku Kaishi*, 1974, **8**, 1501.
[123] R. Zeisberg and F. Bohlmann, *Chem. Ber.*, 1974, **107**, 3800.
[124] M. K. Imanari and O. Masahiro, *Bull. Chem. Soc. Japan*, 1974, **47**, 708.
[125] R. Nakajima, H. Motono, and T. Hara, *Bunseki Kagaku*, 1975, **24**, 68 (*Chem. Abs.*, 1975, **83**, 8588).
[126] J. Niwa and M. Yamrzaki, *Chem. Letters*, 1975, 765.
[127] T. Pehk, E. Lippmaa. V. F. Baklan, T. Utochka, and A. G. Yurchenko, *Eesti NSV Tead. Akad. Toim., Fuus. Mat.*, 1974, **23**, 425 (*Chem. Abs.*, 1975, **82**, 131694).
[128] W. A. Szarek and D. M. Vyas, *Canad. J. Chem.*, 1974, **52**, 3394.
[129] B. A. Trofimov, G. A. Kalabin, V. M. Bzhesovskii, N. K. Gusarova, D. F. Kushnarev, and S. V. Amossova, *Reakts. spos. org. Soedinenii*, 1974, **11**, 367 (*Chem. Abs.*, 1975, **83**, 8885).
[130] D. Kh. Zheglova, B. A. Ershov, and A. I. Kol'tsov, *Zhur. org. Khim.*, 1974, **10**, 2396.
[131] M. Zanger and W. W. Simons, *Analyt. Chem.*, 1974, **46**, 2042.
[132] R. F. Sprecher, B. E. Aufderheide, G. W. Luther, and J. C. Carter, *J. Amer. Chem. Soc.*, 1974, **96**, 4404.
[133] R. L. Lichter and J. D. Roberts, *Org. Magn. Resonance*, 1974, **6**, 636.

Lichter[134] have extended such studies of nitrogen magnetic shielding in aniline, the toluidines, and the xylidines. To gain further insight into the methyl electronic effect, and to determine whether nitrogen shielding can serve as a probe of electron distribution, Adler and Lichter have calculated electron densities using the INDO MO method. Figure 5 is an example of the correlations they find between electron

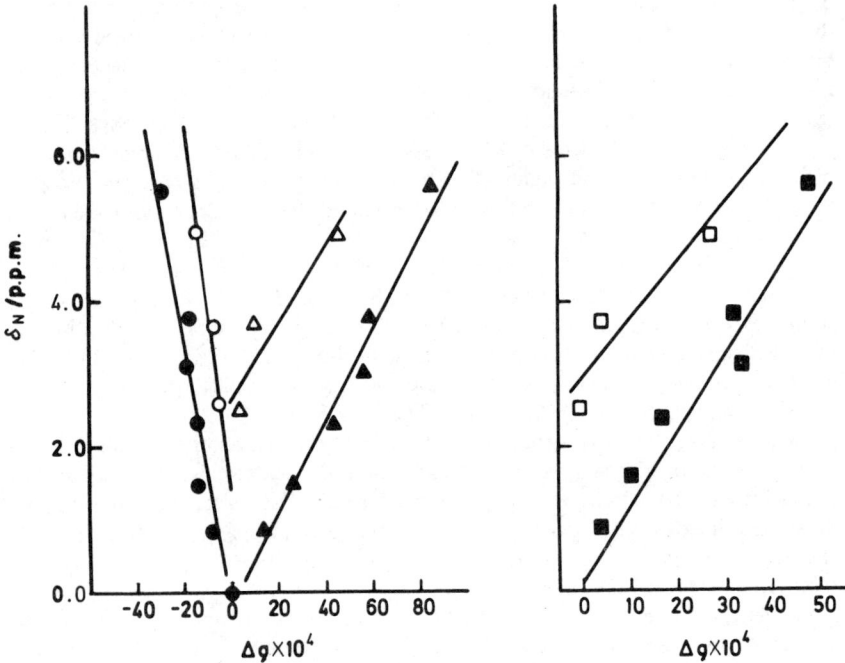

Figure 5 *Plot of methylaniline nitrogen chemical shifts* vs. *average excess nitrogen electron densities. The open symbols represent compounds with p-methyl substituents.* (●, ○) π *density*; (▲, △) σ *density*; (■, □) *total density*

densities and nitrogen chemical shifts. Although there is considerable scatter, it is clear that the shieldings at nitrogen increase with total and with σ-electron density. Changes in nitrogen shieldings in anilines substituted in the *para*-position with conjugatively interacting substituents have been interpreted[135] in terms of variations in nitrogen π-electron density. It is clear from Figure 5 that such an approach cannot explain the methylaniline results. Methyl substitution decreases the nitrogen π-electron density, and the increased shielding apparently arises because of larger increases in the nitrogen σ-electron density. This behaviour is consistent with the different correlation that the *o*-, *m*-, and *p*-methyl substituent chemical shifts display with Hammett σ values compared to the conjugatively electron-donating halogens and alkoxy-substituents.[136] In contrast to the latter cases the methyl group

[134] G. Adler and R. L. Lichter, *J. Org. Chem.*, 1974, **39**, 3547.
[135] T. Axenrod, P. S. Pregosin, M. J. Wieder, E. D. Becker, R. B. Bradley, and G. W. A. Milne, *J. Amer. Chem. Soc.*, 1971, **93**, 6536.
[136] For a discussion see ref. 133.

appears to enhance inductive electron withdrawal to the amine nitrogen. The parallel decrease in π-electron density at nitrogen may be due to the increase in σ-electron density; this type of $\sigma-\pi$ repulsion has been noted[137] for the *para*-carbon of monosubstituted benzenes and for the fluorine-bearing carbon of *para*-substituted fluorobenzenes. The authors note[134] that their inference that polarization effects and nitrogen electron densities dominate the nitrogen shieldings is at variance with other studies of fluorine,[137] carbon,[81, 138] and nitrogen[139] shieldings, where π-polarization and π-electron-density changes were found to be the most important effects. However, in all these cases, substituents have been capable of direct conjugative interaction with the nucleus of interest, which is not possible in the methylanilines.

Warren and Roberts[140] have reported ^{15}N chemical-shifts for some secondary alkylamines, amides, ureas, urethanes, heterocycles, and miscellaneous compounds. Such data were used to derive substituent chemical shift parameters for secondary alkylamine ^{15}N chemical shifts. A regression analysis was performed on the secondary alkylamine ^{15}N chemical shifts using the equation

$$\delta_N = n_\beta \beta + n_\gamma \gamma + C \quad (20)$$

Here n_β and n_γ are the numbers of carbon atoms two bonds and three bonds removed from the nitrogen, respectively; β and γ are the substituent parameters, and C is a constant. The final regression analysis gave alkyl-substituent parameter values of $\beta = 14.3 \pm 0.5$, $\gamma = -2.1 \pm 0.5$, and $C = -28.2$ p.p.m. Data on primary and secondary amines were compared to give the result that alkyl substitution on ^{15}N causes a shift of $+9$ p.p.m. With this and calculated values for the local diamagnetic nitrogen shielding term, σ_N^d, of methylamine and dimethylamine, α, β, and γ substituent parameters were derived for σ_N^p. The values obtained were $\alpha = 38$, $\beta = 14.3$, and $\gamma = -2.1$ p.p.m. In contrast to results reported by Mason[141] for ^{13}C chemical shifts of alkanes, the ratio of the α and β parameters for σ_N^p obtained by Warren and Roberts was not close to that expected for an inverse-cube law of the internuclear distances.

(18)

(19) (20)

[137] R. T. C. Brownlee and R. W. Taft, *J. Amer. Chem. Soc.*, 1970, **92**, 7007.
[138] P. Lazzeretti and F. Taddei, *Org. Magn. Resonance*, 1971, **3**, 283.
[139] P. Hampson, A. Mathias, and R. Westhead, *J. Chem. Soc. (B)*, 1971, 397.
[140] J. P. Warren and J. D. Roberts, *J. Phys. Chem.*, 1974, **78**, 2507.
[141] J. Mason, *J. Chem. Soc. (A)*, 1971, 1038.

In an attempt to determine the mechanisms by which a substituent X in (18) affects the ^{19}F shielding, Fukunaga and Taft[142] have examined ^{19}F shielding in ketones (19) and (20). Polar effects on such shielding will be little altered in (20) relative to (19) if transmission is by an electrostatic field mechanism, but will be substantially reduced if transmission is by the internal π-electron framework. These arguments are based on the assumption that the dihedral angle between the planes of the phenyl rings will be much larger in (20) than it is in (19). Such twisting of the phenyl rings has little effect on the X—F distance, but extension of the π-system beyond the phenyl ring should be quite sensitive to such steric influences. The π-electron delocalization effects are reduced by almost a factor of 2 in (20) compared to (19). Thus the authors[142] have concluded that the transmission of polar substituent effects upon ^{19}F n.m.r. shifts is largely carried internally by the π-bond framework and that any transmission through space is by comparison entirely minor for this series of fluoroaromatics.

Dawson and Reynolds[143] have continued their investigations of the relationship between chemical shifts and charge densities. Correlations of these parameters for 4-substituted phenylacetylenes with Hammett σ_I and σ_R^0 parameters indicate that the important mechanisms of transmission of substituent effects are field, resonance, and π-polarization effects. CNDO/2 calculations on a variety of aromatic derivatives suggest that the σ_I dependence of ^{19}F chemical shifts in these derivatives may be primarily due to π-polarization. Lichter and Wasylishen[144] have demonstrated that fluorine substituent effects in fluoropyridines are additive over a fairly large range of ring positions. The largest deviations were found where extensive conjugative interaction between fluorine and nitrogen is possible. For example, C-2 in (21a) shows the largest deviation, perhaps due to the importance of interactions of the

(21a) (21b) (21c)

type represented by (21b) [and similarly (21c)]. Mukhomorov[145] has calculated ^{19}F chemical shifts in polysubstituted derivatives of fluorobenzene as sums of electronic and intramolecular dispersion interaction contributions. The author has concluded that the shielding of ^{19}F atoms in such systems is determined mainly by the local π-electron distribution.

Ernst et al.[146] have reported ^{29}Si n.m.r. chemical shifts for a number of aryl-trifluoro-, aryltrichloro-, aryltriethoxy-, and aryltrihydro-silanes. With the exception of the chloro-compounds, such chemical shifts exhibited good linear correlations with Hammett σ constants. In contrast to chemical-shift trends in aryltrihydrosilanes

---

[142] J. Fukunaga and R. W. Taft, *J. Amer. Chem. Soc.*, 1975, **97**, 1612.
[143] D. A. Dawson and W. F. Reynolds, *Canad. J. Chem.*, 1975, **53**, 373.
[144] R. L. Lichter and R. E. Wasylishen, *J. Amer. Chem. Soc.*, 1975, **97**, 1808.
[145] V. K. Mukhomorov, *Teor. i eksp. Khim.*, 1974, **10**, 685 (*Chem. Abs.*, 1975, **82**, 42 502).
[146] C. R. Ernst, L. Spialter, G. R. Buell, and D. L. Wilhite, *J. Amer. Chem. Soc.*, 1974, **96**, 5375.

the fluoro- chloro- and ethoxy-compounds exhibited decreased shieldings with increasing electron donation by *meta* and *para* substituents on phenyl. CNDO/2 calculations of electron-density at silicon for aryltrifluoro-, aryltrihydro- and aryltrimethyl-silanes indicated no unusual reversal in the electron-density trends to account for these observations. An additive although non-linear relationship between silicon chemical shifts and the summed electronegativities of the groups bonded to silicon was found.

Fritz and Schäfer[147] have determined substituent effects on proton and phosphorus chemical shifts in silylphosphines. An empirical method for the calculation of phosphorus chemical shifts in terms of bond increments is reported. Radeglia and Teichmann[148] have provided an interpretation of the effect of substituents on the ^{31}P chemical shifts of some phosphonium ions of the type $[PR^1_{4-n}R^2_n]^+$. In a rather similar study, Radeglia and Englehardt[149] have found that the substituent dependence of the ^{119}Sn chemical shifts of stannanes of the type $Me_{4-n}SnX_n$ may be explained qualitatively in terms of changes in the local paramagnetic contribution to the total shielding.

Barnard *et al.*[150] have reported ^{119}Sn chemical shifts for some aryl- and substituted-aryl-tin compounds. A comparison of ^{119}Sn chemical-shift values of the three tributyltin derivatives Bu_3SnPh, $Bu_3SnC_6H_4CF_3$-*m*, and $Bu_3SnC_6H_4CF_3$-*o* indicated that the electron-withdrawing effect of the trifluoromethyl group causes a deshielding of the tin nucleus. These results and ^{119}Sn shifts for members of the related butylphenyltin series, $Bu_{4-n}SnPh_n$, were used to estimate a value for tetraphenyltin. A somewhat surprising result was that the tin nucleus in $PhSnCl_3$ is more shielded than the tin nucleus in tributylphenyltin by approximately 20 p.p.m. This result was rationalized by noting that σ^d_{loc} is increased by approximately 130 p.p.m. when the three butyl groups are replaced by three chlorine atoms. Harris *et al.*[151] have determined ^{119}Sn chemical shifts for 35 organotin compounds containing transition-metal atoms. The exceptionally low shielding values found for species with light transition-metal atoms were interpreted in terms of a large paramagnetic contribution arising from small values of ΔE associated with d_π–d_π bonding. From considerations of ^{59}Co magnetic shielding in the series $R_nX_{m-n}Sn\{Co(CO)_4\}_{4-m}$ ($m = 1,2,3$; $n<m$; R = alkyl, phenyl; X = Cl), Wuyts and van der Kelen[152] have shown that the ^{59}Co chemical shift is mainly sensitive to π-effects in the metal–metal bond.

Although equations (17) and (18) have proved to be very useful, it is nevertheless important to recognize that the simplicity of these equations is only achieved by the introduction of severe approximations. Thus, rationalizations of experimental data which are based on such equations must, of necessity, be mainly of qualitative value. In the opinion of this Reporter, studies which attempt to push these types of rationalization beyond such qualitative limits should be viewed with some caution. In this regard, the paper by Evans and Norton[153] provides a fitting conclusion to

---

[147] G. Fritz and H. Schäfer, *Z. anorg. Chem.*, 1974, **409**, 137 (*Chem. Abs.*, 1975, **82**, 56 990).
[148] R. Radeglia and H. Teichmann, *Z. Chem.*, 1974, **14**, 249 (*Chem. Abs.*, 1974, **81**, 90 693).
[149] R. Radeglia and G. Englehardt, *Z. Chem.*, 1974, **14**, 319 (*Chem. Abs.*, 1974, **81**, 161 752).
[150] M. Barnard, P. J. Smith, and R. F. M. White, *J. Organometallic Chem.*, 1974, **77**, 189.
[151] D. H. Harris, M. F. Lappert, J. S. Poland, and W. McFarlane, *J.C.S. Dalton*, 1975, 311.
[152] L. F. Wuyts and G. P. van der Kelen, *J. Mol. Structure*, 1974, **23**, 73.
[153] J. Evans and J. R. Norton, *Inorg. Chem.*, 1974, **13**, 3043.

this sub-section. These workers have pointed out that detailed rationalizations of ^{13}C chemical shifts for carbon atoms bound to transition metals have been both contradictory and misleading. Evans and Norton have also suggested that simple explanations of the shielding changes which occur for nuclei involved in the co-ordination sphere of a transition metal are unlikely to be valid. These statements were substantiated by discussion of the chemical shifts of the carbonyl carbon atom in metal carbonyls,[154] and the ^{13}C chemical shifts of π-bonded carbon atoms in olefin complexes.[155, 156] From a consideration of such n.m.r. data, Evans and Norton have concluded that current approximate shielding theories are unable to provide a convincing explanation of ^{13}C chemical shifts in transition-metal complexes.

C. **Electric Field Effects.**—An electric field distorts the electron distribution in a molecule and thereby changes the magnetic shielding at each of the nuclei. Buckingham[157] has shown that such a shielding change due to a uniform electric field, E, present in the vicinity of a conically symmetric bond containing a hydrogen or a fluorine atom (*e.g.* X—H or X—F) is given by

$$\sigma_E = -AE_z - BE^2 \tag{21}$$

In this equation, A and B are parameters characteristic of the X—H or X—F bond and E_z is the component of the field along the bond from X to H or F. For the case of a proton and a C—H bond, A and B have been calculated[157] to be 2×10^{-12} and 10^{-18} in e.s.u.,* respectively. Hence to a first approximation the second term in equation (21) is usually neglected. The difficulties involved in using this equation have been summarized by Raynes[5] in an earlier report. The papers by Ebraheem and Webb,[158] Ellis *et al.*,[159] Ando *et al.*,[56] and Garreau *et al.*[160] are noted in this section. It should be mentioned that many of the papers referred to in Section 4B consider electric field effects in their more general consideration of substituent effects on magnetic shielding.

Ebraheem and Webb[158] have investigated magnetic shielding associated with the fluorine nucleus in *para*-substituted fluorobenzenes. By choosing a co-ordinate system such that the molecular dipole moment, μ, is directed along the z-axis, which is also taken as the axis of the C—F bond, the electric field component, E_z, becomes

$$E_z = \frac{2\mu}{r^3} \tag{22}$$

In this equation, r is the separation between the ^{19}F nucleus and the weighted centre of mass of the molecule considered. The values of r, μ, E_z, and Δ, the observed ^{19}F

---

[154] G. M. Bodner and L. J. Todd, *Inorg. Chem.*, 1974, **13**, 1335 and references therein.
[155] D. Cooper, R. P. Hughes, and J. Powell, *J. Amer. Chem. Soc.*, 1972, **94**, 9244.
[156] M. H. Chisholm, H. C. Clark, L. E. Manzer, and J. B. Stothers, *J. Amer. Chem. Soc.*, 1972, **94**, 5087.
[157] A. D. Buckingham, *Canad. J. Chem.*, 1960, **38**, 300.
[158] K. A. K. Ebraheem and G. A. Webb, *J. Mol. Structure*, 1975, **25**, 387.
[159] G. E. Ellis, R. G. Jones, and M. G. Papadopoulos, *J.C.S. Perkin II*, 1974, 1381.
[160] M. Garreau, G. J. Martin, M. L. Martin, J. Morel, and C. Paulmier, *Org. Magn. Resonance*, 1974, **6**, 648.

* For conversion of e.s.u. into the appropriate SI units, multiply by 0.33×10^{-4}, after which the units are m V^{-1} (= C mJ^{-1}).

Table 8 *Parameters used to estimate substituent electric field effects, and ^{19}F chemical shifts (relative to p-difluorobenzene) for some p-substituted fluorobenzenes*

Substituent	μ/Debye	r/Å	$E_z \times 10^5$/e.s.u.	Δ/p.p.m.
F	0.0	2.750	0.0	0.0
NH$_2$	−3.373	2.755	−3.226	7.24
OH	−2.071	2.755	−2.596	4.72
CH$_3$	−1.931	3.017	−1.407	−1.43
H	−1.837	2.075	−0.356	−6.90
NO$_2$	2.908	3.071	2.007	−15.69

chemical shifts relative to p-difluorobenzene, are given in Table 8. The values of μ presented in this Table were calculated from INDO MO wavefunctions. Ebraheem and Webb have considered that such calculated values were reliable for use in equation (22), since INDO MO calculations of dipole moments usually give results which are in good agreement with experimental values.[161] The plot of E_z against Δ revealed a good correlation, and the value of A found from the slope was -46.0×10^{-12} in e.s.u. The negative sign indicates that fluorine forms the negative end of the dipole and the increased magnitude of A over values found for C—H bonds reflects the greater polarizability of C—F bonds.

A rather similar study has been conducted by Ellis et al.,[159] who were interested in understanding the effect of nitroso-substituents on the carbon shielding in piperidines and piperazines. In this work, CNDO/2 MO theory was used to calculate the charge distribution in N-nitrosodimethylamine. These authors emphasize the dangers associated with drawing conclusions about the importance of various shielding contributions from investigations of small numbers of molecules. Garreau et al.[160] have discussed proton chemical shifts in mono- and di-substituted selenophens in terms of electric field effects arising from the substituents. In addition, these authors note that magnetic anisotropy effects may also be important. Ando et al.[56] have calculated proton chemical shifts of diads, triads, and tetrads of poly(vinyl chloride) using Pople's method together with electric field corrections evaluated using Buckingham's equation.

D. Magnetic Anisotropy Effects.—A number of interesting developments concerning shielding by magnetically anisotropic groups have appeared in the review period. In several important papers, Stiles[162-164] has reconsidered the theory of nuclear shielding by magnetically anisotropic molecules and functional groups. Recently, Buckingham and Stiles[165] have presented a generalization of earlier work by McConnell[166] in which isotropic magnetic shieldings at points (R, θ, ϕ) *further* from a molecular origin than the most distant part of the molecular electron density are described by the multipolar expansion

---

[161] J. A. Pople and D. L. Beveridge, 'Approximate Molecular Orbital Theory,' McGraw-Hill, New York, 1970.
[162] P. J. Stiles, *Proc. Roy. Soc.*, 1974, **A336**, 251.
[163] P. J. Stiles, *Mol. Phys.*, 1975, **29**, 1271.
[164] P. J. Stiles, *Chem. Phys. Letters*, 1975, **30**, 259.
[165] A. D. Buckingham and P. J. Stiles, *Mol. Phys.*, 1972, **24**, 99.
[166] H. M. McConnell, *J. Chem. Phys.*, 1957, **27**, 226.

$$\sigma = \sum_{L=M}^{\infty} \sum_{M=0}^{L} (A_{LM} \cos M\phi + B_{LM} \sin M\phi) P_L^M(\cos\theta)/R^{L+1} \quad (23)$$

Stiles has extended this work even further,[162] and has shown that shieldings at points (r,θ,ϕ) *closer* to an arbitrary origin than the nearest region of molecular electron density are conveniently calculated using the expansion of equation (24)

$$\sigma = \sum_{L=0}^{\infty} \sum_{M=0}^{L} (F_{LM} \cos M\phi + G_{LM} \sin M\phi) r^L P_L^M(\cos\theta). \quad (24)$$

In both expansions, the angular dependence of the shielding is determined by the associated Legendre polynomials $P_L^M(\cos\theta)$, and the coefficients A_{LM}, B_{LM}, F_{LM}, and G_{LM} measure anisotropy in the multipolar magnetic susceptibilities of the functional group responsible for the shielding.

Stiles[163] has also shown that these expressions can be simplified for magnetic shielding associated with symmetric functional groups. By examining the symmetry restrictions on the coefficients in the expansions of equations (23) and (24), Stiles has obtained the shieldings associated with groups which have C_{2v}, C_{3v}, $C_{\infty v}$, D_{2h} and $D_{\infty h}$ symmetry. As Stiles has recognized,[164] there are difficulties in obtaining rigorous multipolar expansions in spherical harmonics for isotropic shielding near non-spherical groups. In an attempt to overcome such difficulties, it has been proposed that expansions of the isotropic shielding in ellipsoidal harmonics may provide a more general solution to this problem for groups which are ellipsoidal in shape. Using such an approach, Stiles has derived the dipolar contributions to isotropic shielding associated with an axially symmetric group. The final results are given in equations (25) and (26).

$$\sigma^{(N)} = \frac{\mu_0}{4\pi a_0^3} (\chi_{\parallel} - \chi_{\perp}) \left[\frac{1}{2} \ln\left(\frac{U+1}{U-1}\right) - \frac{U}{U^2 - V^2} \right] \begin{Bmatrix} \text{prolate} \\ \text{molecular} \\ \text{shape} \end{Bmatrix} \quad (25)$$

$$\sigma^{(N)} = -\frac{\mu_0}{4\pi a_0^3} (\chi_{\parallel} - \chi_{\perp}) \left[\tan^{-1}\left(\frac{1}{U}\right) - \frac{U}{U^2 + V^2} \right] \begin{Bmatrix} \text{oblate} \\ \text{molecular} \\ \text{shape} \end{Bmatrix} \quad (26)$$

In these equations, U and V are confocal spheroidal co-ordinates and $\chi_{\parallel}$ and $\chi_{\perp}$ are parallel and perpendicular components of the magnetic susceptibility tensor associated with the axially symmetric group. It is worth noting that equations (25) and (26) reduce to the McConnell equation[166]

$$\sigma^{(N)} = -\frac{1}{3} \frac{\mu_0}{4\pi} (\chi_{\parallel} - \chi_{\perp})(3\cos^2\theta - 1)/R^3 \quad (27)$$

as $U \to \infty$. Stiles has illustrated the advantages gained by using such ellipsoidal expansions with the aid of the Pauling ring-current model[167] of a benzene molecule. For this model the shielding at any point on the axis through the centre of the ring and perpendicular to the plane of the ring can be calculated exactly and compared

[167] L. Pauling, *J. Chem. Phys.*, 1936, **4**, 673.

with the results from the dipolar expressions (26) and (27). The considerable improvement in accuracy which results from using the ellipsoidal expression for the dipolar shielding is shown in Table 9.

Rae[168] has explored the effects of protonation on the anisotropic shielding associated with the amide and the thioamide groups. He concludes that anisotropic deshielding by $>$C=O and $>$C=S is sharply reduced when these groups become protonated, and in a number of cases shieldings were found for adjacent protons. Sprecher and Aufderheide[169] have examined the ^{11}B chemical shifts for all monochloro-, monobromo-, and monoiodo-decaborane isomers reported by Sprecher et al.[132] earlier. They have found that the magnetic anisotropy associated with the halogen substituent is a major factor in determining the shielding at the halogenated boron. The shieldings observed for borons in parts of the cage remote from the halogen can be understood in terms of the variation of the paramagnetic shielding term with the changing boron $2p$ atomic orbital size.

Table 9 Errors due to dipolar approximations to the magnetic shielding along the ring axis

Distance from centre of ring in units of ring-radii	Spheroidal approximation eq. (26)	Spherical approximation eq. (27)
10	0.4%	1.6%
6	0.8%	4.1%
2	6.8%	40%
1.5	11%	74%
1	21%	183%
0.5	48%	1020%

Scherr et al.[170] have measured the ^{7}Li chemical shifts of a variety of organolithium derivatives and have found a reasonable correlation between the changes in shielding at lithium nuclei and the magnetic anisotropy of the substituent group. From a study of proton shielding in oxides of 1,3-dithian and 1,3,5-trithian, Khan et al.[171] have found that the relative order of the proton resonances in such systems can be explained by a model in which the overall diamagnetic anisotropies of the C—S and C—(SO$_2$) bonds have the opposite sign to those of the C—C, C—O, and C—(SO) bonds. Apsimon and co-workers[172] have continued their attempts to evaluate bond magnetic susceptibility anisotropies from proton chemical-shift data. In their most recent studies, they have concluded that it is impossible to calculate reliable magnetic susceptibilities and susceptibility anisotropies of C—C and C—H bonds from the collection of chemical shifts reported by Pretsch et al.[173] However, reasonably reliable values for the magnetic anisotropies of the C=C and C=O bonds were obtained.

[168] I. D. Rae, Austral. J. Chem., 1974, **27**, 2621.
[169] R. F. Sprecher and B. E. Aufderheide, Inorg. Chem., 1974, **13**, 2287.
[170] P. A. Scherr, R. J. Hogan, and J. P. Oliver, J. Amer. Chem. Soc., 1974, **96**, 6055.
[171] S. A. Khan, J. B. Lambert, O. Hernandez, and F. A. Carey, J. Amer. Chem. Soc., 1975, **97**, 1468.
[172] J. W. Apsimon, J. Elguero, and A. Fruchier, Canad. J. Chem., 1974, **52**, 2296.
[173] E. Pretsch, H. Immer, C. Pascual, K. Schaffner, and W. Simon, Helv. Chim. Acta, 1967, **50**, 105.

Nakashima and co-workers[174-177] have published a number of papers in which they have shown that the proton shielding in cobalt(III) compounds may be correlated with the anisotropy of the second-order paramagnetism of the central cobalt-(III) ion. Other studies worth noting here are those of Weissberger,[178] who has estimated and compared the strengths of the shielding cones of ketonic and olefinic linkages as determined from ^{1}H n.m.r. measurements on polycyclic ketones and olefins, and those of Ellis et al.,[159] who have estimated the shielding anisotropy of the N-nitroso-group from ^{13}C n.m.r. studies of substituted piperidines and piperazines. Finally, Martin[179] has shown that there is a substantial contribution to the shielding of the hydrogen-bonded proton in the bichloride ion from currents associated with the magnetic anisotropy of the chlorine atoms.

E. **Ring-current Effects.**—Although it has received criticism[180, 181] from time to time, the concept of the 'ring current' has been widely accepted and used by chemists since Pople[182] introduced the idea to explain the low shielding values found for the protons of benzene nearly twenty years ago. During the period under review there have been a number of interesting contributions to this field. Roberts[183] has continued his studies in this area by commenting on the calculation of ring-current contributions to magnetic susceptibilities. He has shown that the finite-perturbation method proposed by Davies[184] should give identical results to those obtained from the uncoupled form of Hartree–Fock perturbation theory, all other things (e.g. parametrization in semi-empirical approaches) being equal. This work has thus resolved the discrepancies between the results of Davies and those of Corradi et al.[185] for the ring currents in thiophen, pyrrole, and furan. Roberts has also pointed out that uncoupled Hartree–Fock perturbation theory has important advantages over the finite-perturbation approach of Davies since it can be corrected, using the geometric approximation,[186] to give values which are very close to those which would be obtained by using the coupled form of Hartree–Fock perturbation theory directly.

Lazzeretti[187] has presented an uncoupled Hartree–Fock perturbation theory for the density matrix and has used such an approach to calculate second-order properties for some conjugated molecules. The uncoupled results were suitably corrected using the geometric approximation, and then compared with the results of coupled Hartree–Fock calculations. Such a comparison for the shielding contributions arising from the π-electrons is shown in Table 10. Long and Memory[188] have

[174] Y. Nakashima, U. Sakaguchi, and H. Yoneda, *Chem. Letters*, 1974, 503.
[175] H. Yoneda, U. Sakaguchi, and Y. Nakashima, *Bull. Chem. Soc. Japan*, 1975, **48**, 209.
[176] Y. Nakashima, U. Sakaguchi, and H. Yoneda, *Bull. Chem. Soc. Japan*, 1975, **48**, 762.
[177] Y. Nakashima, *Bull. Chem. Soc. Japan*, 1975, **48**, 766.
[178] E. Weissberger, *J. Org. Chem.*, 1974, **39**, 3701.
[179] J. S. Martin, *Chem. Phys. Letters*, 1974, **27**, 67.
[180] J. I. Musher, *J. Chem. Phys.*, 1965, **43**, 4081; ibid., 1967, **46**, 1219.
[181] J. I. Musher, *Adv. Magn. Resonance*, 1966, **2**, 177.
[182] J. A. Pople, *J. Chem. Phys.*, 1956, **24**, 1111.
[183] H. G. Ff. Roberts, *Mol. Phys.*, 1975, **29**, 1277.
[184] D. W. Davies, *Trans. Faraday Soc.*, 1961, **57**, 2081.
[185] E. Corradi, P. Lazzeretti, and F. Taddei, *Mol. Phys.*, 1973, **26**, 41.
[186] H. G. Ff. Roberts, *Theor. Chim. Acta*, 1974, **33**, 269.
[187] P. Lazzeretti, *Mol. Phys.*, 1974, **28**, 1389.
[188] E. R. Long and J. D. Memory, *J. Chem. Phys.*, 1974, **61**, 3865.

Table 10 *Calculated values of the π-electron contributions to proton magnetic shielding constants*[a]

Molecule	Proton	σ_{CHF}^{b}	σ_{GA}^{c}
Pyridine	2	2.384	2.378
	3	2.435	2.432
	4	2.429	2.424
Triazine	1	2.096	2.095
Pyrazine	2	2.487	2.473
Pyrimidine	2	2.256	2.243
	4	2.295	2.292
	5	2.342	2.333
Benzene	1	2.487	2.487

[a] Units are p.p.m. [b] Values calculated using CHF theory directly. [c] Values calculated using UCHF theory together with corrections evaluated by the geometric approximation.

reported self-consistent field calculations of the shielding contributions associated with the ring currents in some polycyclic hydrocarbons. Their results were compared to those obtained by Haigh and Mallion,[189] who used McWeeny's approach,[190] which in turn is based on Hückel MO theory. No improvement in the agreement between theory and experiment relative to the Haigh and Mallion calculations was observed if *ratios* of chemical shifts were considered. However, Long and Memory have found much better agreement between theory and experiment for the *absolute* values of chemical shifts. These findings essentially substantiate earlier conclusions of Amos and Roberts.[191]

Two other theoretical studies worth noting here are those of Ege and Vogler.[192,193] In the first of these, Ege and Vogler[192] have derived a simple formula for calculating non-local contributions to the magnetic susceptibilities and shieldings in systems with degenerate eigenvalues. Applications of this formula to the calculation of ¹H chemical shifts of the circumpolyacenes have been reported. Ege and Vogler have also presented[193] Hückel MO calculations of the non-local contributions of the π-electrons to the ¹H n.m.r. chemical shifts in macrocyclic annulenes and phenanthrene. By comparing the calculated values with the experimental shifts, estimates of the relative annulenoid character of such compounds were made.

With the goal of contributing to the understanding of the factors which affect ring-current contributions to proton shielding, Du Vernet *et al.*[194] have studied proton shielding in bridged [18]annulenes in order to determine the dependency of the ring-current contribution on the contour of the annulene perimeter. In particular, these workers have studied the two hexahydrocoronenes (22) and (23). An examination of molecular models of (22) and (23) suggests that the annulene perimeter of each should be essentially planar. The only obvious difference is in the contour of the annulene loop, one following a benzo[*ghi*]perylene perimeter and the other a

[189] C. W. Haigh and R. B. Mallion, *Mol. Phys.*, 1970, **18**, 737.
[190] R. McWeeny, *Mol. Phys.*, 1958, **1**, 311.
[191] A. T. Amos and H. G. Ff. Roberts, *Mol. Phys.*, 1971, **20**, 1073.
[192] G. Ege and H. Vogler, *Theor. Chim. Acta*, 1974, **35**, 189.
[193] G. Ege and H. Vogler, *Tetrahedron*, 1975, **31**, 569.
[194] R. Du Vernet, T. Otsubo, J. A. Lawson, and V. Boekelheide, *J. Amer. Chem. Soc.*, 1975, **97**, 1629.

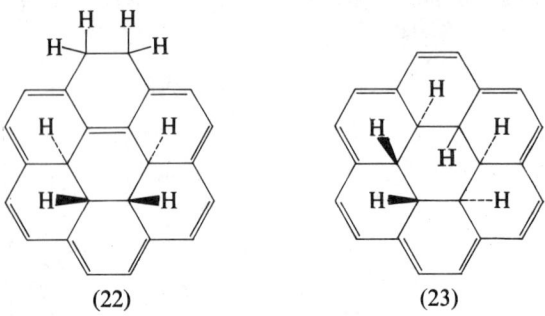

(22)   (23)

coronene perimeter. As might be expected from simple considerations, the ring-current contribution of (22) is smaller than that of (23). The surprising result, however, is the magnitude of the effect, the signals for the internal protons of the molecule with the coronene perimeter reflecting a 5 p.p.m. greater shielding than the corresponding internal protons where a benzo[*ghi*]perylene perimeter is present. Thus Du Vernet *et al.* have concluded that the contour of the annulene perimeter, aside from its planarity, plays an important role in determining the ring-current contribution to proton shielding in such systems.

As is illustrated by the above studies, the changes in the ring-current shielding contributions for protons are usually large compared to other factors affecting the

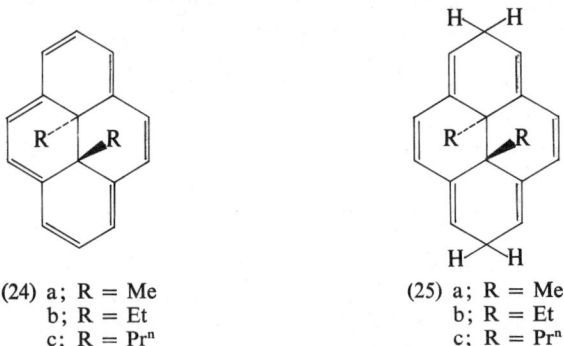

(24) a; R = Me
    b; R = Et
    c; R = Pr^n

(25) a; R = Me
    b; R = Et
    c; R = Pr^n

values of proton shielding constants. For heavier nuclei, however, the ring-current effects are relatively less important and are often completely overshadowed by other kinds of influence. In an important study, Du Vernet and Boekelheide[195] have examined ^{1}H and ^{13}C n.m.r. chemical shifts in compounds (24a—c) and (25a—c). These molecules appeared to be ideally suited for analysing ring-current contributions to carbon shielding since (i) the molecules (24a—c) are bridged [14]annulenes in which the carbons under consideration are internal substituents lying within the shielding region of the aromatic π-electron cloud; and (ii) an examination of

[195] R. Du Vernet and V. Boekelheide, *Proc. Nat. Acad. Sci., U.S.A.*, 1974, **71**, 2961.

molecular models clearly indicates that the comparison molecules (25a—c), although lacking an aromatic ring current, are essentially identical to (24a—c) with regard to the geometry of the internal framework. Thus, to a close approximation, the differences in shielding between these two series of compounds should be attributable directly to ring-current contributions in (24a—c). Du Vernet and Boekelheide have found[195] that for the same position in space relative to the mean plane of delocalization of the aromatic π-electron cloud, the magnitude of the ring-current effect on chemical shifts is essentially the same for carbon as for protons. The effects for both types of nuclei were found to follow the theoretical curve predicted by Johnson and Bovey.[196]

Ojima et al.[197] have examined the ^{1}H n.m.r. chemical shifts in dibenzo[f,l]-8,10-bisdehydro[17]annulenone (26), and have compared them to the values obtained by Howes et al.[198] for the octahydro-derivative (27). Such a comparison revealed

that the paratropic nature of the macrocyclic ring in dibenzo-fused compound (26) is smaller than that of the macrocyclic ring in octahydro-compound (27). Thus the authors have concluded[197] that the two benzene rings reduced the paramagnetic ring current arising from polarization of the carbonyl group in the inner ring. Honeybourne has suggested[199] that the n.m.r. and susceptibility data associated with (28) are compatible with the presence of a paramagnetic contribution to the

(28) R = H

[196] C. E. Johnson, jun. and F. A. Bovey, *J. Chem. Phys.*, 1958, **29**, 1012.
[197] J. Ojima, A. Kimura, and T. Tokoyama, *Chem. Letters*, 1975, 207.
[198] P. D. Howes, E. LeGoff, and F. Sondheimer, *Tetrahedron Letters*, 1972, 3691.
[199] C. L. Honeybourne, *Tetrahedron Letters*, 1974, 3075.

susceptibility associated with the conjugated $4n$ π-electron, $(4n-2)$-atom inner great ring of (28). Hirohashi et al.[200] have concluded that steric considerations play an important role in determining the shielding contributions of the ring current of the benzene ring to the shielding of H_α and H_β of the thiophen ring in benzoylthiophen, thienopyrimidine, and thienodiazepine derivatives.

Alscher et al.[201] have measured proton shieldings in bridged [14]annulenes with anthracene perimeters. From such a study, they have concluded that the deshielding of the bridge protons which occurs as the number of links in the bridge is increased is attributable to geometric factors. Turchin[202] has used chemical-shift measurements for the protons in 1H-benzo-1,5-diazepinium mono-cation (29) to demon-

(29)

strate the presence of a paramagnetic ring current in the diazepine ring. Mamaev and Smirnova[203] have calculated the distribution of the magnetic field due to two current loops of π-electrons in the benzene molecule. The results were presented in tabulated form and used to evaluate ring-current contributions to proton and carbon-13 shielding constants in some aromatic hydrocarbons. Glotsmann et al.[204] have examined the diamagnetic anisotropy of [2,2] metacyclophanes and have concluded that a reliable interpretation of proton shieldings in such systems based on ring-current contributions requires that bond-length and angle deformations due to the puckered topology of the molecule be taken into account. In a related study, Sakamoto and Oki[205] have measured ^{1}H n.m.r. spectra of a series of dioxadioxo-[n]paracyclophanes. These workers have found that the rigidity of the ansa chain is an important factor in determining the proton chemical shifts of the ansa-chain protons in addition to the ring-current contributions. Patel and Tonelli[206] have used ring-current calculations to evaluate the shielding of the G-N1H and T-N3H protons in Watson–Crick base pairs. The perturbations on these shielding ring-current contributions due to twisting and tilting a base pair adjacent to the ring NH were also examined.

Although not strictly concerned with nuclear shielding, the work of Davidson et al.[207] deserves mention here. These workers have measured the magnetic sus-

[200] T. Hirohashi, S. Inaba, and H. Yamamoto, *Bull. Chem. Soc. Japan*, 1975, **48**, 147.
[201] A. Alscher, W. Bremser, D. Cremer, H. Günther, H. Schmickler, W. Sturm, and E. Vogel, *Chem. Ber.*, 1975, **108**, 640.
[202] K. F. Turchin, *Khim. geterotsikl. Soedinenii*, 1974, **6**, 828 (*Chem. Abs.*, 1974, **81**, 119 384).
[203] V. M. Mamaev and F. M. Smirnova, *Vestnik Moskov. Univ., Khim.*, 1974, **15**, 621 (*Chem. Abs.*, 1975, **82**, 57 375).
[204] C. Glotsmann, E. Langer, and H. Lehner, *Monatsh.* 1974, **105**, 354.
[205] K. Sakamoto and M. Oki, *Bull Chem. Soc. Japan*, 1974, **47**, 2739.
[206] D. J. Patel and A. E. Tonelli, *Biopolymers*, 1974, **13**, 1943.
[207] J. R. Davidson, A. K. Burnham, B. M. Siegel, P. Beak, and W. M. Flygare, *J. Amer. Chem. Soc.*, 1974, **96**, 7394.

ceptibility anisotropies of oxazole and isoxazole and have found slightly lower non-local contributions to the out-of-plane susceptibility than those observed for furan. From this study, it appears that the replacement of C=C by C=N and the relative positions of the heteroatoms in the heteroatomic ring do not have a major effect on the electron delocalization as measured by the magnetic susceptibility. Using similar techniques, Czieslik and Sutter[208] have determined the magnetic susceptibility anisotropy of methylenecyclobutenone. This system shows a positive non-local contribution, in contrast to the large negative contributions associated with aromatic molecules. Thus, Czieslik and Sutter have concluded that the rather strong deshielding observed for protons in such four-membered rings[209, 210] should probably not be explained in terms of the molecular ring current induced in the π-system of the four-membered ring.

F. van der Waals and Steric Effects.—The applications of ^{13}C magnetic resonance to the determination of stereochemical assignment are in general based on empirical correlations of carbon-13 shieldings and molecular geometry.[211, 212] The larger shielding observed for the carbon which is *gauche* to another carbon[213, 214] or to a heteroatom[215-217] at the γ position (30a) relative to the resonance of an analogous carbon in the *anti*-conformation (30b) is particularly useful for stereochemical assignment. This well-known γ effect is attributed[218, 219] to a steric perturbation which is considered to polarize the ^{13}C—H bonds of the encumbered groups such that the carbon nucleus is shielded and the attached protons are

(30a)   (30b)

deshielded. However, the recent report[220] of deshielding δ effects in molecules beset with severe *syn*-axial interactions with atoms other than hydrogen suggests that additional mechanisms may be necessary to explain ^{13}C chemical shifts in such crowded situations.

[208] W. Czieslik and D. H. Sutter, *Z. Naturforsch*, 1974, **29a**, 1820.
[209] W. S. Trahanovsky and M.-G. Park, *J. Amer. Chem. Soc.*, 1973, **95**, 5412.
[210] W. D. Hunstman and H. J. Wristers, *J. Amer. Chem. Soc.*, 1967, **89**, 342.
[211] J. B. Stothers, 'Carbon-13 NMR Spectroscopy,' Academic Press, New York, 1972.
[212] N. K. Wilson and J. B. Stothers, *Topics Sterochem.*, 1974, **8**, 1.
[213] D. K. Dalling and D. M. Grant, *J. Amer. Chem. Soc.*, 1972, **94**, 5318.
[214] D. K. Dalling, D. M. Grant, and E. G. Paul, *J. Amer. Chem. Soc.*, 1973, **95**, 3718.
[215] J. B. Grutzner, M. Jautelat, J. B. Dence, R. A. Smith, and J. D. Roberts, *J. Amer. Chem. Soc.*, 1970, **92**, 7107.
[216] T. Pehk and E. Lippmaa, *Org. Magn. Resonance*, 1971, **3**, 679.
[217] H. Eggert and C. Djerassi, *J. Amer. Chem. Soc.*, 1973, **95**, 3710.
[218] D. M. Grant and B. V. Cheney, *J. Amer. Chem. Soc.*, 1967, **89**, 5315.
[219] B. V. Cheney and D. M. Grant, *J. Amer. Chem. Soc.*, 1967, **89**, 5319.
[220] S. H. Grover, J. P. Guthrie, J. B. Stothers, and C. T. Tan, *J. Magn. Resonance*, 1973, **10**, 227.

Thus the recent study of Eliel et al.[221] on the shieldings caused by nitrogen, oxygen, and fluorine atoms located at γ positions and which are anti-periplanar to the nucleus of interest is especially important. The results of this study have demonstrated clearly that:

(1) a carbon atom located *anti* to a first-row heteroatom in the γ position (30b; X = N,O,F) is generally more shielded than an analogous nucleus *anti* to a methyl or methylene group;

(2) a gauche heteroatom (30a; X = N,O,F,S,Cl) produces greater shielding at a carbon atom than that arising from a methyl or methylene group;

(3) the shielding created at the *gauche* carbon atom by X = N,O,F,S,Cl (30a), as compared with X = Me or -CH$_2$-, is generally less than the corresponding shielding at the *anti*-carbon atom.

These results clearly highlight the dangers involved in associating increased shielding with the presence of *gauche* conformations for systems containing first-row heteroatoms. These authors[221] have suggested that a hyperconjugative-type interaction of free electron pairs centred on the N, O, and F atoms with the C_α—C_β bond, accompanied by a subsequent alternation of the electron density at the γ antiperiplanar carbon, can provide a plausible explanation of the observed n.m.r. data. This mechanism will be enhanced by coplanarity of the *p*-orbital contribution in the four orbitals indicated in Figure 6.

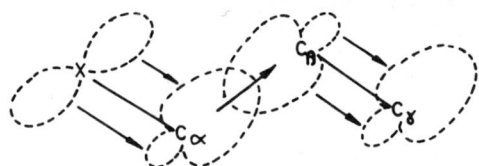

Figure 6 *Hyperconjugative transfer of charge from free-electron pairs on an* X *atom to a* trans γ *carbon. (This mechanism will be enhanced by coplanarity of the p-orbital contributions in the four orbitals indicated)*

Grant[222] has summarized his group's extensive investigations of shieldings in cyclic alkanes. As is well known, the ^{13}C chemical shifts in such hydrocarbons can be accurately determined using the additive parameters presented in Table 11. Ernst[223] has reinvestigated the ^{13}C chemical shifts of branched alkylbenzenes and has found major discrepancies with earlier values.[211, 224] The results of Ernst are more in keeping with what would be expected from shift predictions based on additive parameters of the type listed in Table 11. Kiewiet et al.[225] have found that the behaviour of neighbouring bulky substituents in furans can be determined from ^{13}C n.m.r. measurements. These workers have argued that the hybridization of the carbon atoms in the aromatic part of the molecule does not change in such systems,

---

[221] E. L. Eliel, W. F. Bailey, L. D. Kopp, R. L. Willer, D. M. Grant, R. Bertrand, K. A. Christensen, D. K. Dalling, M. W. Duch, E. Wenkert, F. M. Schell, and D. W. Cochran, *J. Amer. Chem. Soc.*, 1975, **97**, 322.
[222] D. M. Grant, *Pure Appl. Chem.*, 1974, **37**, 61.
[223] L. Ernst, *Tetrahedron Letters*, 1974, 3079.
[224] W. R. Woolfenden, Ph.D. Thesis, University of Utah, 1965.
[225] A. Kiewiet, J. De Wit, and W. D. Weringa, *Org. Magn. Resonance*, 1974, **6**, 461.

Table 11 *Linear additive parameters for chemical shifts in cyclic alkanes*

Structural feature		Parametric value/p.p.m.
α-carbon	^{13}C—C	−9.94
β-carbon	^{13}C(C)(C)	−8.49
Trigonal	^{13}C with three C substituents and one C	2.91
Quaternary	C—^{13}C(C)(C)—C	9.04
Vicinal	C—C—^{13}C—C—C	3.50
γ-gauche	H,H,^{13}C gauche arrangement	4.56
γ_2-gauche	^{13}C with two γ-gauche H interactions	8.51
β-gauche-γ-trans	^{13}C—C(C)—C—C	−1.91

and thus deviations from values predicted from additivity rules are interpreted as arising from normal steric shifts. Zimmermann et al.[226] have compared ^{13}C chemical shifts in monospiro- and dispiro-derivatives of cyclopentane and cyclohexane and have thus obtained information about the 1,5- or δ-effect, which is deshielding. Vdovin et al.[227] have found that the ^{19}F n.m.r. of poly(trifluoromethyl)-benzene reflects steric interaction between *ortho*-groups. Stephens et al.[228] have measured ^{13}C n.m.r. spectra of several methyl-substituted bicyclo[2,2,2]-octenones and -octanones and have found that increased steric crowding of the vicinal carbon atoms consistently produces increased shielding.

---

[226] D. Zimmermann, R. Ottinger, and J. Reisse, *Org. Magn. Resonance*, 1974, **6**, 346.
[227] G. P. Vdovin, Y. P. Egorov, E. V. Konovalov, A. P. Krasnoshchek, V. G. Lukmanov, E. P. Soenko, and L. M. Yagupol'skii, *Teor. i eksp. Khim.*, 1975, **11**, 44.
[228] K. R. Stephens, J. B. Stothers, and C. T. Tan, *Mass Spectrom. N.M.R. Spectrosc. Pestic. Chem. Proc. Symp.*, 1973, 179 (*Chem. Abs.*, 1974, **81**, 119 331).

Quin et al.[229] have investigated the effects of phosphorus substituents on the ^{13}C chemical shifts of alkyl chains. The results were interpreted on the basis of the usual α, β-deshielding, and γ-shielding effects. The γ-effects were small (0.2—1.6 p.p.m.(but of significance in suggesting reduced steric compression relative to that found for methyl and most of the common organic functional groups. The authors have attributed this to the greater length of the C—P bond. In a similar study, Casanova et al.[230] have measured ^{13}C chemical shifts for some symmetrical dialkylmercurials in order to determine empirical substituent parameters for the HgR group. A somewhat surprising result of this work was the finding of a deshielding γ-effect for the HgR substituent.

Adcock et al.[231] have found that the C_α shieldings in tetralin, indane, and benzocyclobutene are more responsive to strain than are the C_β shieldings. Recognition of this trend has led these workers to suggest a reversal of the assignments given by Motell et al.[232] for benzocyclopropene. Lindberg et al.[233] have estimated steric contributions to the chemical shifts of methyl-group protons in ring-chlorinated methylbenzenes. This has been achieved by using substituent constants derived from systems in which there are no unusual steric effects to determine proton shieldings in ring-chlorinated methylbenzenes. Using the assumption that such substituent constants are only indicative of differences in inductive, anisotropy, and resonance effects, the difference between proton shieldings estimated in this way and experimental values was taken as a measure of steric contributions.

Several other papers deserve mention in this section. Diegnan[234] has used ^{13}C n.m.r. to investigate the steric requirements associated with a sulphur lone pair; Evans and Ternay[235] have examined the conformational preferences of mesoalkylated thioxanthenes using proton magnetic resonance spectroscopy. Lemieux and Koto[236] have measured ^{13}C chemical shifts in some glycosidic systems in order to determine the relative steric compression at the anomeric centre. Pehk et al.[237] have used ^{13}C chemical shifts to investigate the conformations of cis- and trans-decalins and of 1- and 2-methyldecalins. Somin and Gindin[238] have utilized the chemical shifts of the azomethine protons to study the conformational features of $E(syn)$-α-oxoaldoximes.

G. Intramolecular Hydrogen Bonding.—The marked conformational preference for the *gauche* form of 2-fluoroethanol has been attributed to the formation of a strong intramolecular hydrogen bond between the electronegative fluorine atom and the hydroxy-proton. Recently, Abraham and Monasterios[239] have found that 2-fluoroethyl acetate and 2-fluoroethyl trichloroacetate exist predominantly in the

---

[229] L. D. Quin, M. D. Gordon, and S. O. Lee, *Org. Magn. Resonance*, 1974, **6**, 503.
[230] J. Casanova, H. R. Rogers, and K. L. Servis, *Org. Magn. Resonance*, 1975, **7**, 57.
[231] W. Adcock, B. D. Gupta, T. C. Khor, D. Doddrell, D. Jordan, and W. Kitching, *J. Amer. Chem. Soc.*, 1974, **96**, 1595.
[232] E. L. Motell, D. Lauer, and G. E. Maciel, *J. Phys. Chem.*, 1973, **77**, 1865.
[233] J. G. Lindberg, G. Y. Sugiyama, and R. L. Mellgren, *J. Magn. Resonance*, 1975, **17**, 112.
[234] G. A. Diegnan, *Diss. Abs.* (*B*), 1974, **35**, 127 (*Chem. Abs.*, 1974, **81**, 120 087).
[235] S. A. Evans and A. L. Ternay, *J. Org. Chem.*, 1974, **39**, 2941.
[236] R. U. Lemieux and S. Kato, *Tetrahedron*, 1974, **30**, 1933.
[237] T. Pehk, A. Laht, I. A. Musaev, E. K. Kurashova, and P. I. Sanin, *Neftekhimiya*, 1974, **14**, 541 (*Chem. Abs.*, 1975, **82**, 42 898).
[238] I. N. Somin and V. A. Gindin, *Zhur. org. Khim.*, 1974, **10**, 414 (*Chem. Abs.*, 1974, **81**, 107 839).
[239] R. J. Abraham and J. R. Monasterios, *Org. Magn. Resonance*, 1973, **5**, 305.

Nuclear Shielding

gauche forms even though there is no possibility for an intramolecular hydrogen bond in such cases.

In view of such findings, Griffith and Roberts[240] have re-examined the question of intramolecular hydrogen bonding in 2-fluoroethanol. These workers have studied the concentration dependence of the hydrogen bonding in 2-fluoroethanol by following the OH chemical shift in CCl_4. They have found that the chemical shift of the methylene protons is essentially invariant with dilution, and that the OH resonance remains a sharp singlet. The monomeric species is found to be approximately 1.33 p.p.m. less shielded than TMS. This value compares quite closely with the value of 1.04 p.p.m. for ethanol itself. A proton engaged in a strong intramolecular hydrogen bond should be markedly deshielded with respect to the OH proton of ethanol. The authors have thus concluded that this result and the 3.42 p.p.m. decrease in shielding found on going from an infinitely dilute to a neat solution of 2-fluoroethanol are indicative of, at most, only weak intramolecular association. Thus it would appear that other explanations are required to explain the observed preference for the *gauche* conformation. Other studies worth noting in this section are those of Rowbotham and Schaefer,[241] who have reported a proton magnetic resonance study of intramolecular hydrogen bonding in halogenophenols, and those of Robillard and Shulman[242] on the hydrogen-bonded protons of the 'charge relay' system of chymotrypsin.

H. Isotope Effects.—When one isotopic species of a nucleus in a molecule is replaced by another isotope there is usually a change of shielding not only at the site of substitution but also at other nuclei in the molecule. Such changes of shielding are usually referred to as isotope shifts, the former type being described as primary shifts while the latter are called secondary isotope shifts. In the absence of other effects such as hydrogen bonding and strong intermolecular effects in solution, the substitution of a nucleus by one of its heavier isotopes in general produces greater shielding for all the nuclei involved. For example, the shielding of a fluorine nucleus is increased when deuterium is substituted for hydrogen either geminally (HCF → DCF) or vicinally (HCCF → DCCF).[243]

Theoretical treatments of such isotope effects have focused on the way in which vibrational changes due to isotopic substitution affect nuclear shielding.[244] Batiz-Hernandez and Bernheim[245] have explained geminal isotope effects in terms of the slight changes in bond lengths and bond angles associated with the lower zero-point vibrational energy in an anharmonic potential well when the heavier isotope is present. Implicit in this approach is the assumption that such changes permit the electrons to provide slightly greater nuclear shielding. Gutowsky,[246] on the other hand, has suggested that different vibrational amplitudes of C-H and C-D systems, even within the harmonic approximation, are the cause of increased shielding in the -CHD relative to the -CH_2 fragment. In this model, isotope effects are attributed to

[240] R. C. Griffith and J. D. Roberts, *Tetrahedron Letters*, 1974, 3499.
[241] J. B. Rowbotham and T. Schaefer, *Canad. J. Chem.*, 1974, **52**, 3037.
[242] G. Robillard and R. G. Shulman, *J. Mol. Biol.*, 1974, **86**, 519, 541.
[243] J. B. Lambert and L. G. Greifenstein, *J. Amer. Chem. Soc.*, 1973, **95**, 6150.
[244] T. W. Marshall, *Mol. Phys.*, 1961, **4**, 61.
[245] H. Batiz-Hernandez and R. A. Bernheim, *Progr. N.M.R. Spectroscopy*, 1967, **3**, 63.
[246] H. S. Gutowsky, *J. Chem. Phys.*, 1959, **31**, 1683.

the change in the intramolecular electric field experienced by the resonating nucleus. Since C—D has a smaller vibrational amplitude, the average H—D distance in the CHD fragment is slightly larger than the average H—H distance in the CH_2 fragment. Since such electric field deshielding effects are inversely proportional to the distance from the source of the field, this approach provides a plausible rationalization of the geminal isotope effect. Pendlebury and Phillips[247] have offered yet a different type of explanation for vicinal H(D)CCF isotope effects. Vicinal delocalization of σ-electrons[248] plays an important role in this approach. These authors have proposed that the reduced vibrational amplitudes associated with deuterium substitution produce a lowering of the long-range bond order between H and F in such systems, resulting in an increased shielding of the vicinal fluorine.

There have been several interesting contributions to this area during the period under review. In his study of nuclear shielding in H_2, Reid[16] not only considered the temperature dependence of the observed shielding constant, but also the isotope shifts of the various isotopic derivatives of molecular hydrogen. The nuclear shielding constant σ_{vJ} associated with a particular rotational level J of the vibrational state v can be calculated from equation (28);

$$\sigma_{vJ} = \sigma_e^{(0)} + \sigma_e^{(1)}\langle\xi\rangle_{vJ} + \frac{1}{2}\sigma_e^{(2)}\langle\xi^2\rangle_{vJ} + \frac{1}{6}\sigma_e^{(3)}\langle\xi^3\rangle_{vJ} + \ldots \quad (28)$$

Here, $\langle\xi^n\rangle_{vJ}$ are the expectation values of ξ defined in equation (16). With the required values of $\langle\xi^n\rangle_{vJ}$ determined using wavefunctions obtained numerically from the adiabatic potential of Kolos and Wolniewicz,[18] Reid has performed the necessary averaging over a Boltzmann distribution to give the isotope shifts at 295 K shown in Table 12; these calculated values are substantial and in good agreement with the available experimental data.

Table 12 *Calculated isotope shifts at 295 K compared with experiment[a]*

Isotope shift	Calculated value	Experiment
$\sigma(HD) - \sigma(\text{ortho-}H_2)$	0.036	0.036 ± 0.002[b]
$\sigma(D_2) - \sigma(HD)$	0.042	0.048 ± 0.032[b]
$\sigma(D_2) - (\text{ortho-}H_2)$	0.078	0.065 ± 0.059[c]

[a] Units are p.p.m. [b] D. F. Evans, *Chem. and Ind.*, 1961, 1960. [c] T. F. Wimett, *Phys. Rev.*, 1953, **91**, 476.

As is well known, isotopic substitution in molecular hydrogen produces relatively large effects on the positions and spacings of the rotational and vibrational levels. Raynes and Stanney[35] have recently considered the rather different situation which prevails for the effects of isotopic substitution in carbon monoxide. Although, in this case, isotopic substitution has only a small effect on the rotational and vibrational levels, the bond-length dependence of the shielding of either nucleus in carbon monoxide is far greater than that in molecular hydrogen as has already been illustrated in Table 3. To determine the values of σ_{vJ}, Raynes and Stanney have obtained the quantities $\langle\xi^n\rangle_{vJ}$ for carbon monoxide using the method of Herman

[247] M. H. Pendlebury and L. Phillips. *Org. Magn. Resonance*, 1972, **4**, 529.
[248] J. A. Pople and D. P. Santry, *Mol. Phys.*, 1963, **7**, 269.

Table 13 *Calculated carbon and oxygen nuclear magnetic shielding constants (in p.p.m.) of some isotopomers of carbon monoxide at several temperatures*

T/K	$\sigma(^{13}C)$			$\sigma(^{17}O)$	
	$^{13}C^{16}O$	$^{13}C^{17}O$	$^{13}C^{18}O$	$^{12}C^{17}O$	$^{13}C^{17}O$
0	9.35	9.38	9.41	61.74	61.80
100	9.30	9.33	9.35	61.68	61.74
200	9.27	9.30	9.32	61.64	61.70
300	9.25	9.28	9.31	61.62	61.68
400	9.24	9.27	9.30	61.61	61.67
500	9.22	9.25	9.28	61.59	61.65

and Short.[249] The results for σ_{vJ} were then used with averaging over a Boltzmann distribution to give the nuclear shielding constants presented in Table 13. From this work it is clear that the effects of both isotopic substitution and temperature variations are small but nonetheless significant. From such results, Raynes and Stanney have determined that an enriched mixture of $^{13}C^{16}O$ and $^{13}C^{18}O$ has a carbon isotope shift of about 0.06 p.p.m., which should be detectable.

Another study deserving mention here is that of Lambert and Greifenstein.[250] In an interesting and important paper, these workers have attempted to distinguish between various explanations of vicinal isotope effects by examining the stereochemical dependence of the H(D)CCF chemical-shift isotope effect. The electric field model of Gutowsky[246] requires that the effect diminish as the separation of the H(D) and F nuclei increases. In contrast, the Pendlebury–Phillips approach,[247] in which vicinal delocalization is important, predicts a maximum effect when the separation of the nuclei has its greatest value (*anti*-periplanar conformation).

With this goal in mind, Lambert and Greifenstein[250] have measured the ^{19}F chemical shifts of compounds (31) and (32) and compared the results with the values obtained in the corresponding undeuteriated compounds. The isotope shifts determined in this way are plotted as a function of the H(D)CCF dihedral

(31a) (31b) (31c)

(31d) (32a) (32b)

[249] R. M. Herman and S. Short, *J. Chem. Phys.*, 1968, **48**, 1266; *ibid.*, 1969, **50**, 272.
[250] J. B. Lambert and L. G. Greifenstein, *J. Amer. Chem. Soc.*, 1974, **96**, 5120.

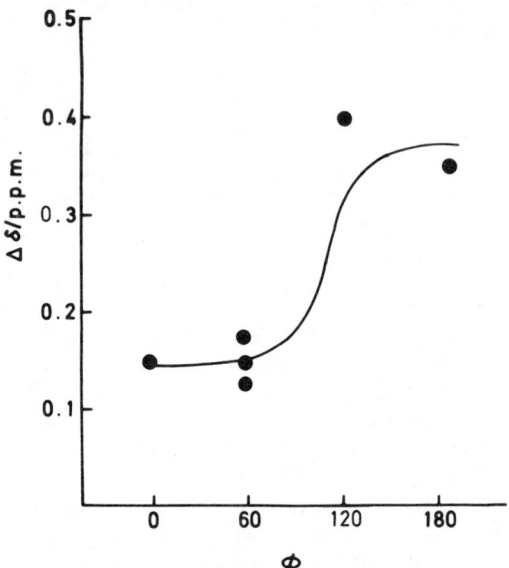

Figure 7 *The H(D)CCF chemical shift isotope effect ($\Delta\delta$) as a function of dihedral angle (Φ)*

angle in Figure 7. Clearly, the largest isotope shifts occur when the distance between the resonating (^{19}F) and the perturbing (^{2}H) nuclei is the greatest. On the basis of these results, Lambert and Greifenstein have concluded that the electric field effect cannot be the only contributor to the isotope shift. Furthermore, since the largest long-range bond order occurs when the H(D) and F atoms are anti-periplanar, these workers have argued that the isotope-shift observations are in accord with an angular-dependent inductive effect, analogous to that invoked to explain vicinal substituent effects on chemical shifts[251] and coupling constants.[252,253] Another interesting paper in this regard is the study by Canuel and St-Jacques[254] in which they have discovered a novel long-distance (four-bond) transannular deuterium isotope effect in a partially deuteriated derivative of 5-oxabenzocycloheptene.

Other studies worth noting here are those of Linde and Jakobsen,[255] who have reported a 0.009 p.p.m. ^{13}C isotope shift induced by ^{34}S in ^{13}C^{34}S(^{32}S), and those of Al-Rawi et al.[256] who have examined the primary and secondary isotope shifts which occur when ^{1}H is replaced by ^{3}H. The authors of this latter study have concluded that both primary and secondary isotope shifts of this type are negligibly small. These results are rather puzzling in view of earlier theoretical studies[257] on

---

[251] L. Phillips and V. Wray, *J.C.S. Perkin II*, 1972, 223.
[252] D. H. Williams and N. S. Bhacca, *J. Amer. Chem. Soc.*, 1964, **86**, 2742.
[253] H. Booth, *Tetrahedron Letters*, 1965, 411.
[254] L. Canuel and M. St.-Jacques, *Canad. J. Chem.*, 1974, **52**, 3581.
[255] S. A. Linde and H. J. Jakobsen, *J. Magn. Resonance*, 1975, **17**, 411.
[256] J. M. A. Al-Rawi, J. P. Bloxsidge, C. O'Brien, D. E. Caddy, J. A. Elvidge, J. R. Jones, and E. A. Evans, *J.C.S. Perkin II*, 1974, 1635.
[257] W. T. Raynes, A. M. Davies, and D. B. Cook, *Mol. Phys.*, 1971, **21**, 113.

nuclear shielding in isotopomers of molecular hydrogen which have demonstrated that such isotope shifts are substantial.

Although not strictly relevant to the considerations of this chapter, two studies on solvent isotope effects are also included here. The shielding at a nucleus of an ion in aqueous solution can depend on the isotopic composition of the solvent. For example, the difference in shielding at ^{207}Pb in H_2O and D_2O solutions of lead nitrate has been found[258] to be -31 ± 3 p.p.m. Such shielding differences are usually attributed[259] to the different vibronic states of the H_2O and D_2O molecules surrounding the ions. During the review period, Lutz and co-workers have reported such solvent isotope effects for ^{111}Cd [22] and ^{67}Zn.[23]

I. Chemical Shift Anisotropy.—Recently, Buckingham and Malm[260] have pointed out that the nuclear shielding tensor is, in general, non-symmetric. These workers divided the total shielding into the sum of an isotropic part, $\bar{\sigma}$, a traceless symmetric part, $\sigma^{(s)}$, and an antisymmetric part, $\sigma^{(a)}$:

$$\sigma_{\alpha\beta} = \bar{\sigma}\delta_{\alpha\beta} + \sigma^{(s)}_{\alpha\beta} + \sigma^{(a)}_{\alpha\beta} \tag{29}$$

The only nuclear-shielding information that can be obtained from conventional high-resolution n.m.r. measurements on liquids is the value of the isotropic part of the tensor. Recent developments in experimental resonance techniques, however, have led to the availability of much more information about such tensors. In some methods[261] all the elements of the symmetric part of the magnetic shielding tensor can be measured, while in others, certain differences of diagonal elements (shielding anisotropies) are obtained.[262-264]

During the period under review, there has been increasing theoretical and experimental interest in magnetic shielding anisotropies. Ando and co-workers[55, 57] have calculated ^{13}C chemical-shift anisotropies in PhH, PhMe, C_6Me_6, Me_2CO, $MeCO_2H$, and $(CH_3CO_2H)_2$ using Pople's approach,[50] with the required wavefunctions determined within the INDO or the MINDO/2 framework. These workers[55, 57] have compared calculated anisotropies for the paramagnetic contribution to the shielding tensor with experimental anisotropies of the total shielding tensor. The adequacy of such an approach clearly depends on the diamagnetic contribution to the shielding tensor being symmetric. It seems unlikely that this would be true in general, and so the calculated results of Ando et al. should probably be regarded with some caution. Appleman and Dailey[265, 266] have considered the relationship between magnetic shielding and magnetic susceptibility tensors in some detail, and have shown that there is some correlation between the magnitudes of their respective anisotropies for many small systems. It was concluded, however, that the occurrence

---

[258] O. Lutz and G. Stricker, Phys. Letters (A), 1971, 35, 397.
[259] C. Deverell, K. Schaumburg, and H. J. Bernstein, J. Chem. Phys., 1968, 49, 1276.
[260] A. D. Buckingham and S. M. Malm, Mol. Phys., 1971, 22, 1127.
[261] See e.g., A. Pines, M. G. Gibby, and J. S. Waugh, Chem. Phys. Letters, 1962, 15, 373.
[262] A. D. Buckingham and K. A. McLaughlan, Progr. N.M.R. Spectroscopy, 1967, 2, 63.
[263] G. Englert and A. Saupe, Mol. Cryst., 1966, 1, 503.
[264] S. C. Wofsy, J. S. Muenter, and W. Klemperer, J. Chem. Phys., 1971, 55, 2014.
[265] B. R. Appleman and B. P. Dailey, Adv. Magn. Resonance, 1974, 7, 231.
[266] B. R. Appleman and B. P. Dailey, in 'Critical Evaluation of Chemical and Physical Structural Information', ed. D. R. Lide and M. A. Paul, Washington Academy of Sciences, 1974, p. 485.

of a $1/r^3$ factor in the shielding results in this property being a predominantly local quantity. The susceptibility, on the other hand, is a property of the entire molecule and increases in magnitude roughly with the size of the molecule. Another theoretical study worth noting is that of Schmiedl,[267] who has used a method similar to that presented by Stephens[268] to calculate the proton shielding anisotropy in molecular hydrogen.

Pausak et al.[269] have used high-resolution double-resonance methods to determine experimental values for the anisotropic ^{13}C chemical shifts in single crystals of polyalkylbenzenes. It was found that methyl and methylene carbons have small shielding anisotropies and that shielding of the ring carbons is a maximum when the magnetic field is normal to the ring and a minimum when it lies along the bond to the substituent. Pines and Abramson[270] have reported shielding tensors for the methyl and carbonyl carbons in solid dimethyl oxalate. In a rather similar study, Ackerman et al.[271] have determined ^{13}C shielding tensors in solid calcium formate. Kempf et al.[272] have measured ^{13}C chemical-shift anisotropies associated with the carboxy- and carbonyl-carbons in benzoic acid derivatives, benzophenone, and thiobenzophenone. The mean shielding anisotropy $[\Delta\sigma = \sigma_{zz} - \frac{1}{2}(\sigma_{xx} + \sigma_{yy})]$ for the central carbon atom was found to increase from 151 p.p.m. in benzophenone to 280 p.p.m. in thiobenzophenone. Mahnke et al.[273] have analysed the ^{13}C chemical-shift tensors in CO, Ni(CO)$_4$, and Fe(CO)$_5$, and have discussed the results in relation to the presence of of π back-bonding in such carbonyl systems.

Three studies concerned with the shielding anisotropies in acetonitrile have been published in the review period. Kaplan et al.[274] have obtained high-resolution ^{13}C and ^{15}N n.m.r. spectra for powdered CH$_3$CN. From such results, these workers have found $\sigma_\parallel = 20 \pm 3$ and $\sigma_\perp = -184 \pm 3$ p.p.m. for the nitrile carbon relative to the methyl carbon in solid neopentane and $\sigma_\parallel = -366 \pm 5$ and $\sigma_\perp = 122 \pm 8$ p.p.m. for nitrogen relative to (NH$_4$)$_2$SO$_4$. Kennedy and McFarlane[275] have performed heteronuclear double resonance experiments on isotopically enriched samples of CH$_3$CN dissolved in the nematic mesophase of a liquid-crystal solvent to determine the chemical-shift anisotropies associated with the nuclei in the C≡N group. The values of $(\sigma_\parallel - \sigma_\perp)$ found were 307 and 452 p.p.m. for ^{13}C and ^{15}N, respectively. In a similar study, Bhattacharyya and Dailey[276] have determined the ^{13}C shielding anisotropy in CH$_3$^{13}CN to be 302 ± 8 p.p.m. Clearly the liquid-crystal results are in good agreement with each other, but such values differ significantly from the powder results. It is also worth noting that the powder studies give a value for the isotropic shift which is in disagreement with the standard value for the liquid state. These large discrepancies between the liquid and solid-state values are rather puzzling. Another carbon study worth noting is that of Stoll et al.,[277] who

---

[267] H. Schmiedl, *Phys. Status Solidi*, 1975, **67B**, K27.
[268] M. J. Stephens, *Proc. Roy. Soc.*, 1957, **A243**, 264.
[269] S. Pausak, J. Tegenfeldt, and J. S. Waugh, *J. Chem. Phys.*, 1974, **61**, 1338.
[270] A. Pines and E. Abramson, *J. Chem. Phys.*, 1974, **60**, 5130.
[271] J. L. Ackerman, J. Tegenfeldt, and J. S. Waugh, *J. Amer. Chem. Soc.*, 1974, **96**, 6843.
[272] J. Kempf, H. W. Spiess, U. Haeberlen, and H. Zimmermann, *Chem. Phys.*, 1974, **4**, 269.
[273] H. Mahnke, R. K. Sheline, and H. W. Spiess, *J. Chem. Phys.*, 1974, **61**, 55.
[274] S. Kaplan, A. Pines, R. G. Griffin, and J. S. Waugh, *Chem. Phys. Letters*, 1974, **25**, 78.
[275] J. D. Kennedy and W. McFarlane, *Mol. Phys.*, 1975, **29**, 593.
[276] P. K. Bhattacharyya and B. P. Dailey, *Chem. Phys. Letters*, 1975, **32**, 305.
[277] M. E. Stoll, R. W. Vaughan, R. B. Saillant, and T. Cole, *J. Chem. Phys.*, 1974, **61**, 2896.

have measured the ^{13}C chemical-shift tensor in potassium tetracyanoplatinate(II) bromide.

Sagnowski[278] has found that the shielding tensor for the ^{19}F nucleus in solid BF_3 is axially symmetric and has principal components $\sigma_{11} = 215 \pm 3$ and $\sigma_{33} = 554 \pm 6$ p.p.m. Bhattacharyya and Dailey[279] have determined ^{19}F and ^{31}P magnetic shielding anisotropies in phosphoryl fluoride from studies of the temperature dependence of the ^{19}F and ^{31}P chemical shifts in nematic solutions. Appleman and Dailey[280] have found that the temperature dependence of the isotropic-phase ^{19}F shift for $CFCl_3$ is quite large. Thus the method of plotting $\sigma_{nem}(T)$ vs. the ordering parameter S used in the POF_3 work was not suitable for determining the ^{19}F chemical-shift anisotropy in $CFCl_3$. Consequently an alternative approach in which the ^{13}C satellites of the fluorine resonance are observed in nematic liquid-crystal solvents has been used. Appleman and Dailey[280] have found that the spinning and concentration dependences of the ordering parameter S provide suitable means of determining an anisotropy which is independent of phase-change chemical shifts and temperature effects. These authors have also pointed out that the resulting anisotropy $(\sigma_{\parallel} - \sigma_{\perp}) = 51 \pm 7$ p.p.m. is suitable for obtaining reliable fluorine spin-rotation constants.

Schweitzer and Spiess[281, 282] have determined the principal elements of the nitrogen chemical shift tensor in pyridine[281] and in nitrobenzene from Fourier transform ^{15}N n.m.r. powder spectra. Schreiber and Vaughan[283] have determined the elements of the chemical-shift tensor for the hydroxy-proton in $Ca(OH)_2$. Finally, Kennedy and McFarlane[284] have studied the magnetic shielding anisotropy of ^{199}Hg in methyl mercuric bromide dissolved in a nematic solvent. The value of $(\sigma_{\parallel} - \sigma_{\perp}) = 5345 \pm 25$ p.p.m. found is an order of magnitude larger than values so far observed for other nuclei in nematic-phase n.m.r. investigations.

5 Shieldings of Particular Nuclear Species

A. Introduction.—The studies discussed in the previous sections have been mainly concerned with calculating, interpreting, correlating, and, in general, understanding nuclear shielding phenomena. There are a number of papers which are not conveniently discussed under the headings of Sections 2—4 but which nonetheless contain important contributions to considerations of nuclear shielding. Although attention is given to such studies in the present section, limitations of space prevent detailed discussions of such papers from being made.

Many ^{13}C shielding studies have already been discussed in Section 4. In subsection 5B a number of additional papers which report magnetic shielding data for carbon nuclei are presented. Sub-sections 5C—E consider shielding information from n.m.r. studies of fluorine, phosphorus, and other nuclei.

B. Carbon Chemical Shifts.—Carbon-13 shieldings have been used in investigations

[278] S. Sagnowski, *Acta Phys. Polonica.*, 1974, **A45**, 397.
[279] P. K. Bhattacharyya and B. P. Dailey, *Mol. Phys.*, 1974, **28**, 209.
[280] B. R. Appleman and B. P. Dailey, *J. Magn. Resonance*, 1974, **16**, 265.
[281] D. Schweitzer and H. W. Spiess, *J. Magn. Resonance*, 1974, **15**, 529.
[282] D. Schweitzer and H. W. Spiess, *J. Magn. Resonance*, 1974, **16**, 243.
[283] L. B. Schreiber and R. W. Vaughan, *Chem. Phys. Letters*, 1974, **28**, 586.
[284] J. D. Kennedy and W. McFarlane, *J.C.S. Chem. Comm.*, 1974, 595.

of the conformations of retinal isomers,[285] polypeptide chains,[286] 2-cyclohexanol,[287] and the 18-norsteroidal structure of fukujusonorones.[288] Such shieldings have also been used to study anomeric configurations of furanoside derivatives[289] and di-d-fructose dianhydrides[290] and to investigate *trans*-fused hexapyranoside derivatives.[291] Torchia et al.[292] have examined the effects of the carbonyl substituent and *syn–anti* geometries on ^{13}C chemical shifts of amides and imino-acid residues; the stereochemical dependence of ^{13}C chemical shifts in olean-12-enes has been discussed,[293] and Cotton and Hunter[294] have given a structural explanation of the ^{13}C n.m.r. spectra of tri-iron dodecacarbonyl. Correlations have been found between ^{13}C n.m.r. chemical shifts of carbonyl and thiocarbonyl groups,[295] and between the ^{13}C chemical shifts of carbenium ions and the ^{11}B chemical shifts of their boron analogues.[296] Radeglia[297] has examined the ring-size effect of cyclic amino-groups on the ^{13}C n.m.r. chemical shifts of polymethine carbon atoms of simple cyanines.

^{13}C Chemical shifts have been reported for propane,[298] phosphorinans,[299] bicyclofulvenes,[300] fulvenes,[301] ureas,[302] methylquinolizidines,[303] monosubstituted adamantanes,[304] methylene carbons in phenol novolac resins,[305] oxepin,[306] 1,3,5-cycloheptatriene,[306] and α-picolines.[307] Olah and co-workers have continued their studies of shielding in carbonium ions and have reported ^{13}C chemical shifts for homocyclopropenyl,[308] alkyl- and aryl-hydridohalonium,[309] and 8,9-dehydro-2-adamantyl[310] cations. ^{13}C Chemical shifts have also been reported for 9-pentacyclo-[4,3,0,0^{2,4},0^{3,8},0^{5,7}]nonyl cations,[311] and Skvortsov and Lizina[312] have studied the protonation of ketones by ^{13}C n.m.r.

[285] R. S. Becker, S. Berger, D. K. Dalling, D. M. Grant, and R. J. Pugmire, *J. Amer. Chem. Soc.*, 1974, **96**, 7008.
[286] F. A. Bovey, *Macromol. Rev.*, 1974, **9**, 1.
[287] Y. Senda and S. Imaizumi, *Tetrahedron*, 1974, **30**, 3813.
[288] P. H. Solomon and K. Nakanishi, *Chem. and Pharm. Bull. (Japan)*, 1974, **22**, 1671.
[289] T. T. Usui, and Y. N. Shigeru, *Agric. and Biol. Chem. (Japan)*, 1974, **38**, 1409.
[290] R. W. Binkley, W. W. Binkley, and B. Wickberg, *Carbohydrate Res.*, 1974, **36**, 196.
[291] E. Conway, D. R. Guthrie, S. D. Gero, G. Lukacs, and A. M. Sepulchre, *J.C.S. Perkin II*, 1974, 542.
[292] D. A. Torchia, J. R. Lyerla, and C. M. Deber, *J. Amer. Chem. Soc.*, 1974, **96**, 5009.
[293] D. M. Doddrell, P. W. Khong, and K. G. Lewis, *Tetrahedron Letters*, 1974, **27**, 2381.
[294] F. A. Cotton and D. L. Hunter, *Inorg. Chim. Acta*, 1974, **11**, L9.
[295] H. O. Kalinowski, and H. Kessler, *Angew. Chem.*, 1974, **86**, 43.
[296] B. F. Spielvogel, W. R. Nutt, and R. A. Izydore, *J. Amer. Chem. Soc.*, 1975, **97**, 1609.
[297] R. Radeglia, *J. prakt. Chem.*, 1974, **316**, 344.
[298] R. E. Wasylishen and T. Schaefer, *Canad. J. Chem.*, 1974, **52**, 3247.
[299] S. I. Featherman and S. O. Lee, *J. Org. Chem.*, 1974, **39**, 2899.
[300] R. W. Hoffmann and H. Kurz, *Chem. Ber.*, 1975, **108**, 119.
[301] R. Hollenstein and A. Mooser, *Angew. Chem.*, 1974, **86**, 595.
[302] H. O. Kalinowski and H. Kessler, *Org. Magn. Resonance*, 1974, **6**, 305.
[303] R. T. LaLonde and T. N. Donvito, *Canad. J. Chem.*, 1974, **52**, 3778.
[304] G. E. Maciel, H. C. Dorn, R. L. Greene, W. A. Kleschick, M. R. Peterson, jun., and G. H. Wahl, jun., *Org. Magn. Resonance*, 1974, **6**, 178.
[305] M. Tsuge, T. Miyabayashi, and S. Tanaka, *Bunseki Kagaku.*, 1974, **23**, 520 (*Chem. Abs.*, 1974, **81**, 136 852).
[306] R. Wehner and H. Guenther, *Chem. Ber.*, 1974, **107**, 3149, 3152.
[307] W. M. Litchman, A. E. Zune, and U. Hollstein, *J. Magn. Resonance*, 1975, **17**, 242.
[308] G. A. Olah, J. S. Staral, and G. Liang, *J. Amer. Chem. Soc.*, 1974, **96**, 6233.
[309] G. A. Olah and Y. Yamada, *J. Amer. Chem. Soc.*, 1975, **97**, 680.
[310] G. A. Olah, G. Liang, and K. A. Babiak, *J. Amer. Chem. Soc.*, 1974, **96**, 6794.
[311] R. M. Coates and R. E. Fretz, *J. Amer. Chem. Soc.*, 1975, **97**, 2538.
[312] N. K. Skvortsov and V. P. Lizina, *Zhur. obshchei Khim.*, 1974, **44**, 2293 (*Chem. Abs.*, 1975, **82**, 30 823).

Several studies of carbon magnetic shielding in organometallic compounds have been reported. These include investigations of additivity parameters for organotin ^{13}C chemical shifts,[313] characteristic ^{13}C chemical shifts for organolithium compounds,[314] and the review of ^{13}C n.m.r. data for organometallic compounds presented by Mann.[315] Cronin et al.[316] have presented a ^{13}C n.m.r. study of isocyanide transition-metal complexes.

Some ^{13}C n.m.r. studies of biochemical interest are: those of Moon and Richards,[317] who have investigated the binding of CO to various haemoglobins; those of Boxer et al.,[318] who have examined the effect of magnesium co-ordination on the ^{13}C and ^{15}N n.m.r. spectra of chlorophyll A, and those of Harris et al.,[319] who have studied the spatial relation of the metal- and anion-binding sites of human transferrin. A ^{13}C magnetic resonance study evaluating polypeptide secondary structure has been presented by Urry et al.[320] ^{13}C n.m.r. data have also been reported for phenobarbital,[321] pentobarbital,[321] dilantin,[321] flavone and deuteriated analogues,[322] and for phenalenone.[323]

C. **Fluorine Chemical Shifts.**—Sears[324] has measured ^{19}F chemical shifts for polycrystalline samples of the alkali-metal fluorides and has found that the observed trends can be correlated with changes in σ_{loc}^{p}. Using the results of measurements of ^{19}F and ^{129}Xe n.m.r. data, Rupp and Seppett[325] have concluded that the structure of xenon hexafluoride in solution is Xe_4F_{24}; in a similar study, LeGeyt and Paddock[326] have determined the structure of tris(p-fluorophenyl)triphosphorus trisulphide from ^{19}F and ^{31}P n.m.r. considerations. Alexandre and Rigny[327] have determined the ^{19}F chemical shift between the non-equivalent fluorine atoms in $ClOF_3$.

Martin and Fujiwara[328] have continued their investigations of magnetic shieldings in dihalide ions by measuring ^{19}F n.m.r. spectra for HF in solution and for the difluoride ion. Daneshrad et al.[329] have investigated the electronic effects associated with the silatrane group by measuring ^{19}F n.m.r. spectra for the [1-m(or p)-fluorobenzyl]silatranes. ^{19}F Chemical shifts have been measured for trifluoro-oxirans,[330] tantalum ethoxyfluoro-complexes,[331] polyfluorinated α-fluorodiphenylmethyl[332]

[313] D. E. Axelson, S. A. Kandil, and C. E. Holloway, *Canad. J. Chem.*, 1974, **52**, 2968.
[314] J. P. C. M. van Dongen, H. W. D. van Dijkman, and M. J. A. De Bie, *Rec. Trav. chim.*, 1974, **93**, 29.
[315] B. E. Mann, *Adv. Organometallic Chem.*, 1974, **12**, 135.
[316] D. L. Cronin, J. R. Wilkinson, and L. J. Todd, *J. Magn. Resonance*, 1975, **17**, 353.
[317] R. B. Moon and J. H. Richards, *Biochemistry*, 1974, **13**, 3437.
[318] S. G. Boxer, G. L. Closs, and J. J. Katz, *J. Amer. Chem. Soc.*, 1974, **96**, 7058.
[319] D. C. Harris, G. A. Gray, and P. Aisen, *J. Biol. Chem.*, 1974, **249**, 5261.
[320] D. W. Urry, L. W. Mitchell, and T. Ohnishi, *Proc. Nat. Acad. Sci., U.S.A.*, 1974, **71**, 3265.
[321] R. C. Long, jun. and J. H. Goldstein, *J. Magn. Resonance*, 1974, **16**, 228.
[322] P. Joseph-Nathan, J. Mares, Ma. C. Hernandez, and J. N. Shoolery, *J. Magn. Resonance*, 1974, **16**, 447.
[323] R. J. Highet and J. M. Edwards, *J. Magn. Resonance*, 1975, **17**, 336.
[324] R. E. J. Sears, *J. Chem. Phys.*, 1974, **61**, 4368.
[325] H. H. Rupp and K. Seppelt, *Angew. Chem.*, 1974, **86**, 669.
[326] M. R. LeGeyt and N. L. Paddock, *J.C.S. Chem. Comm.*, 1975, 20.
[327] M. Alexandre and P. Rigny, *Canad. J. Chem.*, 1974, **52**, 3676.
[328] J. S. Martin and F. Y. Fujiwara, *J. Amer. Chem. ., Soc* 1974, **96**, 7632.
[329] A. Daneshrad, C. Eaborn, and D. R. M. Walton, *J. Organometallic Chem.*, 1975, **85**, 35.
[330] K. W. Jolley, L. H. Sutcliffe, and K. L. Williamson, *Spectrochim. Acta*, 1974, **30A**, 1455.
[331] Yu. A. Buslaev, Yu. V. Kokunov, V. D. Kopanev, and M. P. Gustyakova, *J. Inorg. Nuclear Chem.*, 1974, **36**, 1569.
[332] Yu. V. Pozdnyakovich and V. D. Shteingarts, *J. Fluorine Chem.*, 1974, **4**, 283, 296.

and polyfluorinated benzyl cations,[332] cyanine dyes containing fluorine,[333] 2,5-difluorophenyl methyl sulphide,[334] (methylthio)tetrafluorophosphorane,[335] cis-1,2-dichlorohexafluorocyclobutane,[336] and complexes of phosphorus pentafluoride with amines.[337] Others studies include the determination of ^{19}F chemical shifts for E- and Z-isomers of p- and m-FC$_6$H$_4$NHCHO,[338] 10-substituted 9-fluoroanthracenes,[339] 4-substituted 3,5-dimethylfluorobenzenes,[339] derivatives of CF$_3$SeH,[340] phosphorofluoridates derived from 7-cholestane,[341] perfluorocyclobutanone,[342] 3-chloro-4-bromo-3,4,4-trifluoro-1-butene,[343] 2-chloro-1,4-dibromo-1,2,2-trifluorobutane,[344] 9,9,10-trifluoro-9-anthracenonium ions,[345] and fluoro-derivatives of 1,2-dithioles[346] and Xe^{II} compounds.[347] Nixon et al.[348] have obtained ^{19}F n.m.r. spectra of (1—3-η-allyl)tris(trifluorophosphine)rhodium(I) complexes, Jarvis and Kemmitt[349] have reported ^{19}F n.m.r. studies of fluoro-olefin complexes of rhodium(I), and Zemskov et al.[350] have found anomalous screening of ^{19}F nuclei in rhodium(III) hexafluoro-complexes.

D. Phosphorus Chemical Shifts.—In an important paper, Gorenstein[351] has proposed an empirical correlation between ^{31}P chemical shifts and O—P—O bond angles in phosphate esters. Such conclusions were drawn from a study of phosphorus shielding in phosphates for which X-ray crystallographic data are available. The author suggests that such ^{31}P shifts may be useful in determining solution structures of complex phosphate esters such as those found in biochemically important nucleotides. From a study of ^{1}H and ^{31}P n.m.r. spectra of substituted phospholan-3-one 1-oxides Purdum and Berlin[352] have concluded that an enol–keto tautomerism exists both in the solid state and in TFA solutions of such compounds. Chemical-shift considerations were also used to investigate the stereochemistry of the substituent relative to the P→O group of the phospholan-3-one ring. Buchner and Wolfsberger[353] have studied the influence of alkyl substituents on the ^{31}P chemical shifts of a series of N-trimethylsilyl thio-organophosphine imides and have concluded that such substituent effects can be explained in terms of a mechanism involving intramolecular co-ordination of the methyl group to phosphorus using

[333] V. L. Troitskaya, V. I. Rudyk, E. V. Konovalov, and L. M. Yagulpol'skii, Zhur. org. Khim., 1974, 10, 1524.
[334] G. Haegele, J. Richter, and M. Peach, Z. Naturforsch., 1974, 29b, 619.
[335] R. B. Johannesen, S. C. Peake, and R. Schmutzler, Z. Naturforsch., 1974, 29b, 699.
[336] V. J. Gazzard and R. K. Harris, Org. Magn. Resonance, 1974, 6, 404.
[337] K. P. John and R. Schmutzler, Z. Naturforsch., 1974, 29b, 730.
[338] Z. Arnold, V. Krchnak, and P. Trska, Tetrahedron Letters, 1975, 347.
[339] W. Adcock, M. J. S. Dewar, R. Golden, and M. A. Zeb, J. Amer. Chem. Soc., 1975, 97, 2198.
[340] C. J. Marsden, J. Fluorine Chem., 1975, 5, 401.
[341] G. H. Cooper and R. A. Chittenden, Org. Magn. Resonance, 1974, 6, 563.
[342] R. G. Green, L. H. Sutcliffe, B. Taylor, and S. M. Walker, Spectrochim. Acta, 1974, 30A, 703.
[343] J. F. Hinton and L. W. Jaques, J. Magn. Resonance, 1974, 15, 564.
[344] J. F. Hinton and L. W. Jaques, J. Magn. Resonance, 1975, 17, 95.
[345] B. G. Oksenenko and V. D. Shteingarts, Zhur. org. Khim., 1974, 10, 1190 (Chem. Abs., 1974, 81, 77 719).
[346] B. J. Lindberg, R. Pinel, and Y. Mollier, Tetrahedron, 1974, 30, 2537.
[347] R. J. Gillespie, A. Netzer, and G. J. Schrobilgen, Inorg. Chem., 1974, 13, 1455.
[348] J. F. Nixon, B. Wilkins, and A. D. Clement, J.C.S. Dalton, 1974, 1993.
[349] A. C. Jarvis and R. D. W. Kemmitt, J. Organometallic Chem., 1974, 81, 415.
[350] S. V. Zemskov, S. P. Gabuda, Yu. I Nikonorov, E. D. Pastukhova, and V. A. Selezneva, Doklady Akad. Nauk. S.S.S.R., 1974, 216, 123 (Chem. Abs., 1974, 81, 43 697).
[351] D. G. Gorenstein, J. Amer. Chem. Soc., 1975, 97, 898.
[352] W. R. Purdum and K. D. Berlin, J. Org. Chem., 1974, 39, 2904.
[353] W. Buchner and W. Wolfsberger, Z. Naturforsch., 1974, 29b, 328.

d-orbitals. Radeglia and Teichmann[354] have provided a semi-quantitative interpretation of ^{31}P chemical-shift differences between equally substituted phosphorus and phosphoryl compounds.

^{31}P Chemical shifts have also been reported for tetramethylmethoxyphosphoranes (together with ^{13}C n.m.r. data),[355] Group VI metal carbonyl derivatives of diphenylphosphinomethyl,[356] diphenylphosphinomethyl phosphinites,[356] Group VIB metal unsymmetrical phosphine complexes,[357] phosphoryl compounds,[358] hexachlorophosphates,[359] adducts formed from tributylphosphines and isothiocyanates,[360] cis-Pt(CNS)$_2$[P(OPh)$_3$]$_2$,[361] phosphorimidates and related compounds,[362] o- and p-substituted primary phenylphosphines,[363] mixed trihalides of phosphorus(III),[364] phenylchlorophosphonium hexachloroantimonates,[365] and quasi-phosphonium intermediates.[366] Other studies include those of Featherman,[367] who has used ^{31}P, ^{13}C, and ^1H n.m.r. to investigate the conformations of phosphorinans and the structure of isomeric 4-phosphorinanols, and those of Gordon and Quin,[368] who have employed ^{31}P n.m.r. in a study of the conformations of cis-(4-methylcyclohexyl)phosphine and its P,P-di-Me- and P,P-di-Cl-derivatives. Rueger et al.[369] have reported a ^{31}P n.m.r. study of rhodium carbonyl halogenophosphine derivatives, and Taylor et al.[370] have presented a similar study of tungsten carbonyl complexes containing unsaturated organophosphorus cationic ligands.

E. Chemical Shifts of Other Nuclei.—A long-range shielding effect has been observed in icosahedral carboranes and their metalloborane derivatives by Siedle et al.[371] This perturbation of the chemical shift upon substitution occurs at a position antipodal to the point of substitution and leads to a net shielding of endo-polyhedral ^{11}B, ^{13}C, and ^{31}P n.m.r. resonances. Hermanek and Plesek[372] have correlated some ^{11}B n.m.r. chemical shifts with structure factors in nido-boranes. Other ^{11}B n.m.r. studies of carboranes are those of Janousek et al.[373] and those of Akitt and Savory[374] on small nido-carbaborane systems. Noeth and Wrackmeyer[375]

[354] R. Radeglia and H. Teichmann, Z. Chem., 1974, **14**, 282.
[355] M. Schmidbaur, W. Buchner, and F. H. Koehler, J. Amer. Chem. Soc., 1974, **96**, 6208.
[356] S. O. Grim, W. L. Briggs, R. C. Barth, C. A. Tolman, and J. P. Jesson, Inorg. Chem., 1974, **13**, 1095.
[357] S. O. Grim, J. Del Gaudio, R. P. Molenda, C. A. Tolman, and J. P. Jesson, J. Amer. Chem. Soc., 1974, **96**, 3416.
[358] V. E. Bel'skii, L. A. Kudryavtseva, E. I. Gol'dfarb, and B. E. Ivanov, Zhur. obshchei Khim., 1974, **44**, 2657 (Chem. Abs., 1975, **82**, 138 697).
[359] K. B. Dillon, R. J. Lynch, R. N. Reeve, and T. C. Waddington, J. Inorg. Nuclear Chem., 1974, **36**, 815.
[360] K. Akiba, T. Yoneyama, and N. Inamoto, Chem. Letters, 1974, 561.
[361] A. J. Carty and E. S. Jacobsen, J.C.S. Chem. Comm., 1975, 175.
[362] H. Goldwhite, P. Gysegen, S. Schow, and C. Swyke, J.C.S. Dalton, 1975, 12.
[363] L. Maier, Phosphorus, 1974, **4**, 41.
[364] K. B. Dillon, T. C. Waddington, and D. Younger, Inorg. Nuclear Chem. Letters, 1974, **10**, 777.
[365] W. Buder and A. Schmidt, Spectrochim. Acta, 1974, **30A**, 1311.
[366] H. R. Hudson and J. C. Roberts, J.C.S. Perkin II, 1974, 1575.
[367] S. I. Featherman, Diss. Abs. (B), 1974, **35**, 129.
[368] M. C. Gordon and L. D. Quin, J.C.S. Chem. Comm., 1975, 35.
[369] C. Rueger, A. Mehlhorn, and K. Schwetlick, Z. Chem., 1974, **14**, 196.
[370] R. C. Taylor, R. L. Keiter, and L. W. Cary, Inorg. Chem., 1974, **13**, 1928.
[371] A. R. Siedle, G. M. Bodner, and A. K. Garber, Inorg. Chem., 1974, **13**, 2321.
[372] S. Hermanek and J. Plesek, Z. anorg. Chem., 1974, **409**, 115.
[373] Z. Janousek, S. Hermanek, J. Plesek, and B. Stibr, Coll. Czech. Chem. Comm., 1974, **39**, 2363.
[374] J. W. Akitt and C. G. Savory, J. Magn. Resonance, 1975, **17**, 122.
[375] H. Noeth and B. Wrackmeyer, Chem. Ber., 1974, **107**, 3070.

have reported ^{11}B and ^{14}N n.m.r. studies of adducts formed from trialkylboranes and various amines. In a similar study, Drake and Rapp[376] have investigated boron-11 shieldings in mixed boron trihalide adducts of phosphine and methylphosphine. ^{11}B N.m.r. data have also been given for triallylborane,[377] tetrachloroaminoborane,[378] bis(dihalogenboron) compounds,[379] and sterically over-crowded aminoboranes.[380]

Several nitrogen magnetic resonance studies have been published during the review period. These include the ^{15}N n.m.r. investigations of liquid nitrogen (^{14}N–^{15}N) by Bradley et al.,[381] ^{15}N n.m.r. studies of nitroalkanes by Witanowski et al.,[382] measurements of ^{14}N n.m.r. isotropic shifts in complexes of the nitrate ion with cobalt(II) in acetone,[383] ^{14}N n.m.r. studies of monosubstituted pyridine N-oxides,[384] and the work of Barboiu[385] on the ^{14}N n.m.r. of formamide, using double-resonance techniques. Several ^{15}N n.m.r. studies of systems of biochemical interest are those of Richards and Thomas,[386] who have reported ^{15}N n.m.r. investigations of amino-acids, peptides, and other biologically important molecules, Suzuki et al.,[387] who have studied the pH dependence of the ^{15}N chemical shifts in L-arginine, Leipert and Noggle,[388] who have presented a ^{15}N n.m.r. study of glycine and Coxon,[389] who has completed a ^{15}N n.m.r. study of amino-sugars of biomedical interest. Also noted here is the brief review of nitrogen n.m.r. spectroscopy presented by Witanowski.[390]

Harris and Kimber[391] have reported ^{29}Si and ^{13}C chemical shifts for trimethylsilyl compounds. In this work, a suggestion, arising from earlier work by Engelhardt et al.,[392] that shielding at silicon nuclei depends on a power series in the substituent electronegativity has been investigated. Although a general relationship is found, the quantitative correlation between ^{29}Si chemical shifts and substituent electronegativity is poor. Lippmaa et al.[393] have measured ^{29}Si chemical shifts for various types of organosilicon compound. These workers have concluded that the changes in such shieldings produced by varying substituents are not a directly additive

[376] J. E. Drake and B. Rapp, J. Inorg. Nuclear Chem., 1974, 36, 2613.
[377] B. M. Mikhailov, V. V. Negrebetskii, V. S. Bogdanov, A. V. Kessenikh, Yu. N. Bubnov, T. K. Baryshnikova, and V. N. Smirnov, Zhur. obshchei Khim., 1974, 44, 1878 (Chem. Abs., 1974, 81, 151 270).
[378] J. C. Haasnoot and W. L. Groeneveld, Z. Naturforsch., 1974, 29b, 52.
[379] W. Haubold and J. Weidlein, Z. anorg. Chem., 1974, 406, 171.
[380] J. Casanova and M. Geisel, Inorg. Chem., 1974, 13, 2783.
[381] C. H. Bradley, G. E. Hawkes, E. W. Randall, and J. D. Roberts, J. Amer. Chem. Soc., 1975, 97, 1958.
[382] M. Witanowski, L. Stefaniak, H. Januszewski, and H. Piotrowska, Bull. Acad. polon. Sci., Sér. Sci. chim., 1975, 23, 333.
[383] V. V. Matveev, Yu. G. Gladkii, G. I. Skubnevskaya, and Yu. N. Molin, Zhur. strukt. Khim., 1974, 15, 931 (Chem. Abs., 1975, 82, 9713).
[384] L. Stefaniak and A. Grabowska, Bull. Acad. polon. Sci., Sér. Sci. chim., 1974, 22, 267.
[385] V. Barboiu, Mol. Phys., 1974, 28, 707.
[386] R. E. Richards and N. A. Thomas, J.C.S. Perkin II, 1974, 368.
[387] T. Suzuki, T. Yamaguchi, and M. Imanari, Tetrahedron Letters, 1974, 1809.
[388] T. K. Leipert and J. H. Noggle, J. Amer. Chem. Soc., 1975, 97, 269.
[389] B. Coxon, Carbohydrate Res., 1974, 35, C1.
[390] M. Witanowski, Pure Appl. Chem., 1974, 37, 225.
[391] R. K. Harris and B. J. Kimber, J. Magn. Resonance, 1975, 17, 174.
[392] G. Engelhardt, R. Radeglia, H. Jancke, E. Lippmaa, and M. Mägi, Org. Magn. Resonance, 1973, 5, 561.
[393] E. Lippmaa, M. Mägi, G. Engelhardt, H. Jancke, V. Chvalovsky, and J. Schraml, Coll. Czech. Chem. Comm., 1974, 39, 1041.

property of the substituents. Such conclusions, which are illustrated by the shielding effects associated with the chloromethyl substituent, are in contrast to those reported in earlier work by Scholl et al.[394] Other ^{29}Si n.m.r. studies include those of Harris and Kimber[395] on polymeric silicones, and those of Gould et al.,[396] who have investigated aqueous sodium metasilicate solutions.

^{119}Sn Chemical shifts have been given for polysubstituted aryltrimethyltin derivatives and related organotin compounds,[397] tin derivatives of cyclopentadiene,[398] and halogen derivatives of diethylstannane.[399] Kim and Bray[400] have measured ^{207}Pb chemical shifts in polycrystalline samples of $Pb(NO_3)_2$, $PbCl_2$, Pb_3O_4, PbO, and PbO_2. The results of an analysis of the chemical shifts have been correlated with the bonding character and site symmetry associated with the lead atoms. Seppelt and Rupp[401] have reported ^{129}Xe n.m.r. spectra for XeF_2, XeF_4, $XeOF_4$, XeO_3, $Xe(OSeF_5)_2$, $Xe(OTeF_5)_2$, $FXeOSeF_5$, $FXeOTeF_5$, and $F_5SeOXeOTeF_5$. ^{113}Cd N.m.r. studies have been presented by Cardin et al.[402] The chemical-shift range was found to exceed 640 p.p.m. and was considered to be consistent with large paramagnetic contributions to the shielding constant. Large substituent effects on ^{113}Cd chemical shifts were also reported.

Other nuclei and associated compounds for which new chemical shift data have been presented are as follows: ^{55}Mn (aqueous permanganate solutions[403, 404]), ^{183}W(WF_6, WO_4^{2-}, and WCl_6),[405] ^{121}Sb [Sb^V compounds],[406] ^{7}Li and ^{35}Cl (perchlorate, chloride, bromide, iodide, tri-iodide and tetraphenylborate salts of lithium),[407] ^{27}Al (octahedral and tetrahedral solvates of the aluminum cation),[408] ^{17}O (aqueous Na_2MoO_4),[409] ^{45}Sc [solution of Sc^{III} salts],[410, 411] ^{195}Pt (Pt–cyclopropane complexes),[412] and ^{59}Co [Co^{III} complexes[413] and alkyl Co^{III} derivatives[414]].

It is a pleasure to thank Monique C. Cleland, Physical Sciences Librarian at Dartmouth College, for her help with computer searches of the chemical literature. I am also grateful to Wendy Odell-Berryman for her help and patience in preparing this manuscript.

[394] R. L. Scholl, G. E. Maciel, and W. K. Musker, *J. Amer. Chem. Soc.*, 1972, **94**, 6376.
[395] R. K. Harris and B. J. Kimber, *J.C.S. Chem. Comm.*, 1974, 559.
[396] R. O. Gould, B. M. Lowe, and N. A. MacGilp, *J.C.S. Chem. Comm.*, 1974, 720.
[397] H. J. Kroth, H. Schumann, H. G. Kuivila, C. D. Schaeffer, jun., and J. J. Zuckerman, *J. Amer. Chem. Soc.*, 1975, **97**, 1754.
[398] V. N. Torocheshnikov and A. Tupciauskas, *J. Organometallic Chem.*, 1974, **81**, 351.
[399] Y. Limouzin and J. C. Maire, *J. Organometallic Chem.*, 1974, **82**, 99.
[400] K. S. Kim and P. J. Bray, *J. Magn. Resonance*, 1974, **16**, 334.
[401] K. Seppelt and H. H. Rupp, *Z. anorg. Chem.*, 1974, **409**, 331.
[402] A. D. Cardin, P. D. Ellis, and J. D. Odom, *J. Amer. Chem. Soc.*, 1975, **97**, 1672.
[403] O. Lutz and W. Steinkilberg, *Z. Naturforsch.*, 1974, **29a**, 1467.
[404] D. Gudlin and H. Schneider, *J. Magn. Resonance*, 1975, **17**, 268.
[405] J. Banck and A. Schwenk, *Z. Physik.*, 1975, **20b**, 75.
[406] R. G. Kidd and R. W. Matthews, *J. Inorg. Nuclear Chem.*, 1975, **37**, 661.
[407] Y. M. Cahen, P. R. Handy, E. T. Roach, and A. I. Popov, *J. Phys. Chem.*, 1975, **79**, 80.
[408] J. J. Delpuech, M. R. Khaddar, A. Peguy, and P. Rubin, *J.C.S. Chem. Comm.*, 1974, 154.
[409] R. R. Vold and R. L. Vold, *J. Chem. Phys.*, 1974, **61**, 4360.
[410] G. A. Melson, D. J. Olszanski, and E. T. Roach, *J.C.S. Chem. Comm.*, 1974, 229.
[411] Yu. A. Buslaev, S. P. Petrosyants, V. P. Tarasov, and U. I. Chagin, *Zhur. neorg. Khim.*, 1974, **19**, 1790 (*Chem. Abs.*, 1974, **81**, 129 491).
[412] P. W. Hall, R. J. Puddephatt, and C. F. H. Tipper, *J. Organometallic Chem.*, 1974, **71**, 145.
[413] F. Yajima, Y. Koike, A. Yamasaki, and S. Fujiwara, *Bull. Chem. Soc. Japan*, 1974, **47**, 1442.
[414] B. E. Reichert and B. O. West, *J.C.S. Chem. Comm.*, 1974, 177.

2
Nuclear Spin–Spin Coupling

BY R. GRINTER

1 Introduction

The form of this chapter of the Report follows that of earlier Volumes, particularly Volume 4, very closely. In the section on π-electron systems, however, an attempt has been made to relate the material more closely to the species of the coupled nuclei than hitherto.

It appears to the Reporter that there has been noticeably less activity in the coupling constant sphere in the year under review than in the immediately preceding ones. This is probably understandable in terms of the developments in n.m.r. instrumentation during the past few years and the new aspects of the subject which they have opened up. A conscious effort has therefore been made to prevent this chapter expanding to fill the space available. It is hoped that readers, even those whose primary interest is in coupling constants, will feel that, in view of the extensive literature which any practising scientist must attempt to cover, it is right that reviewers should always be prepared to see the length of their products reduced.

Readers are referred to the beginning of this volume for a list of abbreviations, symbols, *etc.*, which are extensively used throughout the chapter.

2 Theoretical Work

A. *Ab Initio* Calculations. Hoarau and Paviot[1,2] have continued their work on the detailed perturbation variation calculation of the coupling in HD. They have examined the double perturbation formalism, paying particular attention to the form of the Fermi contact operator and to the convergence of the total second-order energy, *i.e.* the self-coupling energy plus the coupling energy. They show that if a variational function of the form (1) is used, then the following consequences ensue: (i) the minimum in the total second-order energy does not coincide with a minimum in the self-coupling energy, or with a stationary state in the coupling energy when the basis functions ϕ_k and ϕ'_l are identical. On the contrary, the theory is less satisfying for this

$$\psi = \psi_{00} + \mu\psi_{10} + \nu\psi_{01} \qquad (1)$$

$$\psi_{10} = \sum_k a_k \phi_k \quad \psi_{01} = \sum_l b_l \phi'_l$$

(the basis functions ϕ_k and ϕ'_l are not necessarily the same)

[1] J. Hoarau and J. Paviot, *Theor. Chim. Acta*, 1974, **35**, 243.
[2] J. Hoarau and J. Paviot, *Theor. Chim. Acta*, 1974, **35**, 251.

$$\mu = \frac{4}{3}\beta\hbar\gamma_D I_{Dz} \quad \nu = \frac{4}{3}\beta\hbar\gamma_H I_{Hz}$$

situation; (ii) the coupling energy does not converge monotonically as the basis is extended; (iii) it is necessary to develop the functions ψ_{10} and ψ_{01} in terms of bases well adapted to the corresponding perturbations $\mathscr{H}_{10}$ and $\mathscr{H}_{01}$, *i.e.* the Fermi operators, and this is where the major difficulty lies. The authors review many of the previous calculations of $^1J(DH)$ in the light of these remarks.

In their second paper[2] Hoarau and Paviot give some results for calculations of $^1J(DH)$ using a variety of wavefunctions and the Gregson–Hall–Rees form of the FC operator.[3] The results, though somewhat higher than the experimental value, do appear to show good convergence characteristics both for the self-coupling and for $^1J(DH)$. In fact, the rather high numerical result may well be simply a consequence of the choice of effective nuclear charge, a parameter to which the calculated result is very sensitive. It is encouraging to note that the coupling is very insensitive to the value of the parameter K in the Gregson–Hall–Rees operator even though the self-coupling energy is very sensitive to it.

Maestro, Moccia, and Zandomeneghi[4] have calculated $^2J(HXH)$ and $^1J(XH)$ for water, ammonia, and methane using extended bases of Slater-type orbitals to construct SCF wavefunctions and, in the case of water, localized orbitals. Like Hoarau and Paviot, they have adopted double perturbation theory and have used both the Rayleigh–Schrödinger and Epstein–Nesbet formulations. The results are disappointing. However, signs are always correctly predicted, with the exception of the localized orbital result for water which leads to incorrect signs for both $^1J(OH)$ and $^2J(HOH)$. This bad failure of the localized orbital calculations is particularly disappointing, but the authors suggest that the use of such orbitals may still be worthwhile with small basis sets. As the size of the basis is increased the agreement with experiment deteriorates due to the presence at particular nuclei of the tails of orbitals which are localized elsewhere. It is worth noting that the Epstein–Nesbet results show a better agreement with experiment than do the Rayleigh–Schrödinger.

Hiroike[5] has considered the effect of electron correlation on $^1J(CH)$ in saturated hydrocarbons. No numerical results are available but her conclusion that the effect of carbon $1s$ electrons on $^1J(CH)$ is small is important. The work also suggests that reasonably accurate values of $^1J(CH)$ may be obtained even if all electron correlation is neglected, and that in such cases the absolute value of the coupling will be somewhat smaller than the true value.

B. Semi-Empirical Calculations (Excluding Those on π-Electron Systems).—In view of the fact that several years have now elapsed since Blizzard and Santry[6] and Hirao and co-workers[7] gave detailed descriptions of MO methods for the calculation of spin-dipolar (SD) and orbital (OB) contributions to nuclear spin–spin coupling, it is somewhat surprising that very few calculations of these contributions to coupling have been published.

[3] M. J. Gregson, G. G. Hall, and D. Rees, *J. Phys. (B)*, 1970, **3**, 1195.
[4] M. Maestro, R. Moccia, and M. Zandomeneghi, *Chem. Phys. Letters*, 1974, **28**, 373.
[5] E. Hiroike, *J. Phys. Soc. Japan*, 1974, **37**, 856.
[6] A. C. Blizzard and D.P. Santry, *J. Chem. Phys.*, 1971, **55**, 950; 1973, **58**, 4714.
[7] K. Hirao, H. Nakatsuji, H. Kato, and T. Yonezawa, *J. Amer. Chem. Soc.*, 1972, **94**, 4078.

Schulman and Newton[8] have used Blizzard and Santry's[6] programme to calculate FC, SD, and OB contributions to $^1J(CC)$ for 17 (mostly saturated) hydrocarbons. On the whole the results are in good agreement with experiment. For three-membered rings it is found that the FC term is much reduced so that OB and SD terms become significant. Equation (2), in which α_C^2 is the percentage carbon-2s character in the C—C bonding hybrid[9] (as calculated by INDO), gives quite good agreement with experiment except where non-FC terms are important.

$$^1J(C_1 C_2) = 0.0621\alpha_{C_1}^2 \alpha_{C_2}^2 - 10.2 \qquad (2)$$

Safiullin, Aminova, and Samitov[9-11] have also reported calculations involving OB and SD terms; they have been particularly interested in phosphorus compounds. According to their calculations, (P,C), (P,O), (P,P), and (P,Cl) couplings are all FC dominated.[9,10] This is a very important result and translations of their papers are awaited with interest. The same workers have also investigated the dependence of $^3J(HPCH)$ in $MePH_2$ on dihedral angle and find results in agreement with experimental observations.[11]

Hall et al.[12] have measured $^1J(CB)$ values for 11 organoboranes and compared their results with INDO FPT calculations of the FC term only. The rather good correlation between theory and experiment which is obtained suggests either a dominant FC term or smoothly varying other terms. However, there is no correlation of the experimental coupling with the C_{2s}—B_{2s} bond order so there appears to be no relationship between $^1J(CB)$ and s-character of the C—B bond. This is very strange if the FC term is dominant. Purcell and Martin[13] have studied the influence of hyperconjugation on borane adducts theoretically, calculating, among other things, some one- and two-bond coupling constants.

Aminova and Samitov[14] have investigated the dependence of one-, two- and three-bond couplings on the orientation of adjacent nitrogen and oxygen lone pairs. They have used a simple MO method with a sum-over-states procedure of the Pople–Santry type.[15] In general, the calculations of $^2J(H,H)$ were rather unsatisfactory. General trends were reproduced but the changes in coupling with lone-pair orientation were not as great as those observed experimentally. Better results were obtained for $^3J(HNCH)$ and $^3J(HOCH)$ in CH_3NH_2 and CH_3OH respectively. For methylamine $^3J(HNCH)$ shows maxima of 3 and 8 Hz at dihedral angles, ϕ, of 0 and 180°, respectively. The minimum coupling lies at $\phi \simeq 80°$ with a value of -0.4 Hz or $+0.5$ Hz depending upon the orientation of the lone pair. The negative figure corresponds to the case in which the lone pair is perpendicular to the C—H bond, the positive to the case in which it is approximately parallel to the bond. The plot of $^3J(HOCH)$ against ϕ has a similar form though the range of coupling is small, -0.4 to $+4.0$ Hz. The effect of lone-pair orientation on $^1J(CH)$ in methylamine and methanol was also calculated. These results are described in Section 3A(ii), where

---

[8] J.M. Schulman and M.D. Newton, *J. Amer. Chem. Soc.*, 1974, **96**, 6295.
[9] R. K. Safiullin, R.M. Aminova, and Yu. Yu. Samitov, *Zhur. strukt. Khim.*, 1975, **16**, 42.
[10] R. K. Safiullin, Yu. Yu. Samitov, and R. M. Aminova, *Teor. i eksp. Khim.*, 1974, **10**, 829.
[11] R. K. Safiullin, R. M. Aminova, and Yu. Yu. Samitov, *Zhur. strukt. Khim.*, 1974, **15**, 907.
[12] L. W. Hall, D. W. Lowman, P. D. Ellis, and J. D. Odom, *Inorg. Chem.*, 1975, **14**, 580.
[13] K. F. Purcell and R. L. Martin, *Theor. Chim. Acta*, 1974, **35**, 141.
[14] R. M. Aminova and Yu. Yu. Samitov, *Zhur. strukt. Khim.*, 1974, **15**, 607.
[15] J. A. Pople and D. P. Santry, *Mol. Phys.*, 1964, **8**, 1.

the calculations of Sergeyev and Solkan[16] on the dependence of ^1J(CH) on bond length are also reported.

Solkan and Bystrov[17] have made a detailed theoretical study of the variation of ^3J(CNCC), ^3J(NCCH), ^3J(HNCH), ^3J(CNCH), and ^3J(CCHN) in peptide fragments with the H—N—C—H dihedral angle. The results of this work, which used the INDO FPT method, are described in Section 5. Ostlund and Pruniski[18] have also used INDO and CNDO FPT methods to examine three- and four-bond (H,H) coupling in compounds related to peptides. The INDO and CNDO calculations were used to estimate the σ and π contributions to the couplings. The calculated values of ^3J(HNCH) show a conformational dependence describable by a Karplus relationship and appear to be in good qualitative agreement with experiment. There seems to be little contribution to this coupling from π mechanisms. ^4J(HNCCH) couplings are predicted to be small in general, due to the rotational averaging of positive and negative terms. However, larger couplings should be observable for structures held in favourable positions, e.g. $+2$ Hz for 'W' and -0.5 Hz for 'sickle' ⌊⌋ paths. The π mechanism is unimportant in cis-peptides but dominates in trans-peptides. However, even in the latter it is small and essentially opposite in character to the relatively large four-bond couplings in analogous compounds with full double bonds.

Barfield et al.[19] have made an extremely detailed investigation of ^4J(HCCCH) including both theoretical and experimental work. They have considered, in particular, the ^4J(H,H) coupling in propanic (1) and allylic (2) fragments, and also

couplings in which a methyl group is involved. They have used the INDO FPT technique to separate the contributions to the coupling into two parts, the direct or through-bond contribution and the indirect or through-space contribution which arises when the coupled protons are close to one another in space. The former contribution is positive and takes its maximum value for the well-known 'W' configuration while the latter contribution is negative. The results of the INDO FPT calculations are complemented by other calculations based upon extended Hückel theory and valence bond theory.

For coupling involving a methyl group the various theoretical methods still leave much to be desired. But a combination of the theoretical and experimental evidence is rather convincing in predicting that significant coupling across four single bonds and involving a methyl group is diagnostic of an anti-periplanar arrangement.

For the allylic couplings the INDO FPT results have been compared with the sum of σ-electron effects and π-electron effects calculated using EHMO theory and VB

---

[16] N. M. Sergeyev and V. N. Solkan, J.C.S. Chem. Comm., 1975, 12.
[17] V. N. Solkan and V. F. Bystrov, Izvest. Akad. Nauk S.S.S.R., Ser. khim., 1974, 1308.
[18] N. S. Ostlund and M. J. Pruniski, J. Magn. Resonance, 1974, 15, 549.
[19] M. Barfield, A. M. Dean, C. J. Fallick, R. J. Spear, S. Sternhell, and P. W. Westerman, J. Amer. Chem. Soc., 1975, 97, 1482.

theory, respectively. Either of these methods reproduces the variation of *transoid* allylic coupling with dihedral angle well, but for *cisoid* allylic coupling the INDO FPT gives definitely inferior results as far as the magnitudes of the couplings are concerned. The general form of the graph of J against ϕ is, however, the same for both theoretical methods.

INDO FPT calculations have also been reported for $^3J(\text{HNCH})$ in 3-isoquinuclidone[20] and (H,H) and (C,H) in oxetans.[21]

C. **Semi-empirical Calculations on π-Electron Systems.**—Reynolds, Peat, and Hamer[22] have reported practical and theoretical investigations of the effect of substituents on coupling to vinyl protons in styrenes. This work is described in Section 10A. An FPT study of long-range (H,H) coupling in styrene, benzaldehyde, and naphthalene has been published by the same laboratory.[23] In this work σ- and π-electron contributions to the coupling were estimated using conventional INDO and CNDO FPT calculations, and also by means of INDO FPT calculations in which appropriate one-centre exchange integrals had been set equal to zero giving $J(\text{INDO},\sigma)$. This procedure was compared with the CNDO method and found to give a good interpretation of σ-electron coupling, though the CNDO method may be preferable where strongly stereospecific σ contributions are present. The π-electron contributions obtained by subtracting the σ-electron couplings from the results of the full INDO FPT treatment are reasonable but generally overestimated. The best estimate of π contributions appears to be given by $J(\text{INDO})-J(\text{INDO},\sigma)$ rather than $J(\text{INDO})-J(\text{CNDO})$.

Rowbotham and Schaefer and their co-workers have reported studies of the relationship between long-range (H,H) couplings and conformation in *p*-methylbenzyl bromide[24] and pyridine aldehydes.[25] They have also compared the experimental couplings with theoretical (INDO/CNDO FPT) values. INDO FPT calculations of (H,H) coupling in selenium heterocycles have been described.[26]

Radeglia *et al.*[27,28] have continued their work on coupling constant calculations focusing particular attention upon the prediction of (H,H) couplings in conjugated systems by means of the simpler MO methods of McConnell[29] and Pople and Santry.[15] They define[27] 'corrected' bond orders, P'_{AB}, and atom–atom polarizabilities, π'_{AB}, according to equations (3) and (4), respectively. In these equations q_A is the

$$P'^2_{AB} = (1.3-0.3q_A)^3(1.3-0.3q_B)^3\, P^2_{AB} \qquad (3)$$

$$\pi'_{AB} = -4(1.3-0.3q_A)^3(1.3-0.3q_B)^3 \times \sum_i \sum_j (E_j-E_i-\gamma_{ij})^{-1} c_{Ai}c_{Bi}c_{Aj}c_{Bj} \qquad (4)$$

[20] A. Aubry, C. Giessner-Prettre, M. T. Chung, M. Marraud, and J. Neel, *Bipolymers*, 1974, **13**, 523.
[21] J. Jokisaari, *Z. Naturforsch.*, 1974, **29a**, 1907.
[22] W. F. Reynolds, I. R. Peat, and G. K. Hamer, *Canad. J. Chem.*, 1974, **52**, 3415.
[23] I. R. Peat and W. F. Reynolds, *Canad. J. Chem.*, 1974, **52**, 2403.
[24] K. Chum, J. B. Rowbotham, and T. Schaefer, *Canad. J. Chem.*, 1974, **52**, 3489.
[25] W. Danchura, T. Schaefer, J. B. Rowbotham, and D. J. Wood, *Canad. J. Chem.*, 1974, **52**, 3986.
[26] V. Galasso and A. Bigotto, *Org. Magn. Resonance*, 1974, **6**, 475.
[27] R. Radeglia, E. Gey, H. Gey, H. Hennig, W. Kühnel, D. Labes, and H.-G. Sieg, *Z. phys. Chem. (Leipzig)*, 1974, **255**, 202.
[28] R. Radeglia, E. Gey, T. Steiger, S. Kulpe, R. Lück, M. Ruthenberg, M. Stierl, and S. Dähne, *J. prakt. Chem.*, 1974, **316**, 766.
[29] H. M. McConnell, *J. Mol. Spectroscopy*, 1957, **1**, 11; *J. Chem. Phys.*, 1959, **30**, 126.

charge on atom A, c_{Ai} is the coefficient of the valence s-orbital on atom A in the MO i which has energy E_i, and γ_{ij} is a parameter which allows for the effect of electron repulsion in estimating the energy of the excited state obtained by promoting an electron from MO i to MO j. It can be seen from equations (3) and (4) that P'_{AB} and π'_{AB} are bond orders and mutual polarizabilities corrected for the charge on the atoms concerned and for interelectronic repulsion. When the quantities P'^2_{AB} and π'_{AB} are calculated using CNDO/2 MO's it is found[27] that both give a very good linear correlation with experimental (H,H) couplings. Thus either the McConnell (P'^2_{AB}) or Pople and Santry (π'_{AB}) method will predict (H,H) couplings with equal success. These methods have been applied to couplings in polymethines.[28]

Steiger, Gey, and Radeglia[30] have also used the INDO FPT method to calculate (C,H) and (H,H) coupling in vinyl compounds. The general trends are given well but there are some notable difficulties. ^2J(HCH) and trans ^3J(HCCH) are usually too positive, often giving rise to a change of sign in the former. The effects of changing substituents on individual couplings are not well reproduced and the authors suggest that this may reflect unsuitable parameterization of the INDO MO scheme rather than a failure of the method itself. Finally, the slopes of the graphs of J_{expt} vs. J_{calc} are rather low, showing a minimum of 0.82 and a maximum of 0.95, i.e. the experimental coupling changes more slowly than the theoretical estimate.

With reference to their earlier work[27] described above, the authors find no significant difference in the empirical correlation of experimental (H,H) couplings with theoretical values obtained from the INDO FPT method, the Pople–Santry or the McConnell methods (using corrected mutual polarizabilities and bond orders, respectively, in the last two procedures).

Ritschl and Dorn[31] have also considered the simpler MO methods and have calculated vicinal (H,H) coupling constants for hydroxypyrazoles and Δ^4-pyrazolinones using the Hückel ω technique to calculate the required bond orders and energy differences. The authors found it necessary to include the electronegativities of the neighbouring atoms as independent variables in the calculation. Quite good agreement of theory and experiment was then obtained, and the range of validity of this type of calculation of coupling constants is discussed. Günther et al.[32] have reported prediction of bond orders from ^3J(HCCH) values in benzo[n]annulenes.

Lichter and Wasylishen[33] have made a study of fluorine–carbon coupling in fluoropyridines and fluoropyridinium ions using the INDO FPT method, including the FC term only. The results are poor, though they do seem to reflect the experimental trends. There is no direct relationship between ^1J(FC) and π-bond order. ^1J(FC) is negative, but a positive contribution to the coupling appears to arise from the proximity of the N lone pair, and this reduces the absolute value of the coupling in some cases; e.g. the coupling has values of -236.2 and -255.1 Hz for 2- and 4-fluoropyridine, respectively. Calculations seem to confirm this observation.

Abraham et al.[34] have made an interesting study of ^4J(FCCCF) in 1-substituted 2,6-difluorobenzenes. If the benzene ring is placed in the xz plane then there is

---

[30] T. Steiger, E. Gey, and R. Radeglia, Z. phys. Chem. (Leipzig), 1974, **255**, 1102.
[31] F. Ritschl and H. Dorn, J. Prakt. Chem., 1974, **316**, 705.
[32] H. Günther, A. Shyoukh, D. Cremer, and K. H. Frisch, Tetrahedron Letters, 1974, 781.
[33] R. L. Lichter and R. E. Wasylishen, J. Amer. Chem. Soc., 1975, **97**, 1808.
[34] R. J. Abraham, J. F. Busby, M. A. Cooper, and A. M. Roe, Org. Magn. Resonance, 1974, **6**, 612.

a good correlation of the coupling with the F_{2p_y}–F_{2p_y} bond order as calculated from INDO MO's. The authors suggest that this shows that the substituent effect on the coupling operates through the π-electron framework, and, furthermore, that the interaction is directly through the π-electrons, and is not a σ–π effect, since there is no correlation of $^4J(F,F)$ with the bond orders between other F orbitals. It is also interesting to note that a correlation of the coupling with the C_{2p_y}–C_{2p_y} bond orders is found, tracing the coupling along the path, so-to-speak.

D. Calculations of Through-space Coupling.—Buckingham and Cordle[35] have described a semi-empirical approach to the problem of through-space coupling in the form of equation (5) in which $S_{s_A s_B}$ is the overlap between the valence s-orbitals

$$K_{AB} = \text{constant} \times S_A^2(0) S_B^2(0) \, (S_{s_A s_B})^2 \qquad (5)$$

of atoms A and B, and the other symbols have their usual meanings. The constant is determined by comparison with experiment. Not only it is found to be suitable for a particular (A,B) pair over a range of separations, but the value found for alkali-metal halides is also appropriate for (F,F) coupling. Equation (5) is applied particularly to the case of (F,F) coupling, for which the largest quantity of data is available, and a table of $J(F,F)$ *versus* internuclear distance shows an excellent correlation with experiment for values of the separation in excess of 0.22 nm. Rather good agreement with alkali-metal halide spin–spin coupling constants is also obtained using equation (5). However, the equation implies that the sign of the coupling will be given by the constant and hence by the product of the magnetogyric ratios of the coupling nuclei and this appears to be at variance with some recent results of Schaefer and Rowbotham.[36] They have reported negative signs for the sizeable spin–spin coupling constants between hydroxyl protons and fluorine nuclei in derivatives of 2-fluorophenol and 2-hydroxybenzotrifluoride. Their result[36] is, however, in accord with INDO and CNDO/2 FPT calculations[36] and with INDO SOS results.[37] But there may be other important effects, *e.g.* core polarization, at work here, and more sign determinations are clearly required before any final verdict can be reached with regard to the application of Buckingham and Cordle's equation.

E. Other Topics of Theoretical Interest.—Emsley and Tabony[38] have made a theoretical study of the variation of (F,H) and (F,F) coupling with temperature. This work is described in Sections 5C(iv) and 5F(i).

3 Coupling of Directly-bonded Nuclei

A. Coupling to Hydrogen.—(*i*) *Elements of Group IIIB*. Tarasov and Bakum[39] have measured $^1J(^{27}Al,H)$, $^1J(^{69}Ga,H)$ and $^1J(^{71}Ga,H)$ for the MH_4^- ions. Together with the result for BH_4^- these couplings fit the curve given by equation (6). This result is discussed in terms of Pople–Santry theory.

---

[35] A. D. Buckingham and J. E. Cordle, *J.C.S. Faraday II*, 1974, **70**, 994.
[36] T. Schaefer and J. B. Rowbotham, *Chem. Phys. Letters*, 1974, **29**, 633.
[37] K. Hirao, H. Nakatsuji, and H. Kato, *J. Amer. Chem. Soc.*, 1973, **95**, 31.
[38] J. W. Emsley and J. M. Tabony, *Mol. Phys.*, 1974, **28**, 423.
[39] V. P. Tarasov and S. I. Bakum, *J. Magn. Resonance*, 1975, **18**, 64.

$$\{J(MH)/\gamma_H\gamma_M\}^{\frac{1}{2}} = 0.205\, Z_M + 1.87 \tag{6}$$

Jouany et al.[40] have made $^1J(BH)$ measurements for the series $Bu_{3-n}^tF_nP \rightarrow BH_3$. Values of $^1J(BH)$ for carbaboranes continue to appear,[41,42] and Beall, Elvin, and Bushweller[41] have drawn attention to the effect of ^{10}B and ^{11}B nuclear spin relaxation on measurements of this coupling. They correlate the effect with the molecular volume of the carbaborane.

(ii) Elements of Group IVB. Theoretical investigations of the effect of lone-electron pairs,[14] C—H bond length[16] and hybridization at carbon[43] have been published. The dependence of $^1J(CH)$ on the orientation of the lone pair(s) of an adjacent nitrogen or oxygen atom has been studied by Aminova and Samitov.[14] They find that for an adjacent nitrogen lone pair the coupling constant is a minimum when the directions of the C—H bond and the lone pair are coincident. When oxygen is the adjacent atom $^1J(CH)$ is a minimum when the direction of the C—H bond and the bisector of the angle between the two lone pairs are coincident.

Sergeyev and Solkan[16] have used INDO FPT calculations to show that $^1J(CH)$ in methane increases if the C—H bond length is increased. The value of $^1J(CH)$ for the other bonds, whose lengths have not been increased, also increases substantially. From these results an isotope effect on the coupling of -1.0 Hz is predicted, in agreement with experimental estimates of -1.0 to -1.5 Hz. Pomerantz and Hildenbrand[43] have continued their investigations of 1-cyanobicyclo[1,1,0]butane. They find that the $^1J(CH)$ values are in good agreement with carbon hybridizations calculated by a variety of methods. However, they note that hybridizations at carbon calculated from (C,H) and (C,C) couplings do not always sum to unity. They point out that errors of 10% in this sum are quite common in the literature.

A solvent dependence of $^1J(CH)$ in *p*-substituted phenyltrimethylsilanes has been noted,[44] and Coxon[45] has discussed the almost linear dependence of $^1J(CH)$ upon ^{13}C chemical shift in carbohydrates. Inamoto and co-workers[46] have found a correlation of $^1J(CH)$ with Hammett σ constants in 4- and 4'-substituted *N*-benzylineanilines.

Thorpe, Coburn, and Montgomery[47] found that $^1J(CH)$ values in substituted purines (3) were sufficiently characteristic to be useful for making assignments. For

(3)

---

[40] C. Jouany, G. Jugie, J.-P. Laurent, R. Schmutzler, and O. Stelzer, *J. Chim. phys.*, 1974, **71**, 395.
[41] H. Beall, A. T. Elvin, and C. H. Bushweller, *Inorg. Chem.*, 1974, **13**, 2031.
[42] J. W. Akitt and C. G. Savory, *J. Magn. Resonance*, 1975, **17**, 122.
[43] M. Pomerantz and D. F. Hillenbrand, *Tetrahedron*, 1975, **31**, 217.
[44] P. Brouant, Y. Limouzin, and J. C. Maire, *Helv. Chim. Acta*, 1973, **56**, 2057.
[45] B. Coxon, *Ann. New York Acad. Sci.*, 1973, **222**, 953.
[46] N. Inamoto, K. Kushida, S. Masuda, H. Ohta, S. Satoh, Y. Tamura, K. Tokumaru, K. Tori, and M. Yoshida, *Tetrahedron Letters*, 1974, 3617.
[47] M. C. Thorpe, W. C. Cobrun, and J. A. Montgomery, *J. Magn. Resonance*, 1974, **15**, 98.

11 compounds $^1J[C(2)H(2)]$ had a mean value of 203.9 ± 3.8 Hz and a total range of 11.6 Hz. Corresponding figures for $^1J[C(6)H(6)]$ (four compounds) and $^1J[C(8)H(8)]$ (23 compounds) were 185.1 ± 3.4 Hz, 10 Hz; 211.1 ± 1.2 Hz, 6.3 Hz. Values of $^1J(CH)$ have also been reported for organolithium compounds[48] and a variety of cyclopentane derivatives.[49, 50]

Fritz and Schäfer[51] have measured $^1J(SiH)$ values in silyl- and silylmethyl-phosphines. The coupling decreases rapidly and uniformly as the number of methyl groups bound to silicon increases. For polysilyl anions of the form $[SiH_{3-n}(SiH_3)_n]^-$ a good linear correlation between $^1J(SiH)$ and the force constant of the Si—H bond is found.[52] This would appear to support the view that both of these properties depend upon the silicon s-character of the Si—H bond.

Schumann and Kroth[53] have measured $^1J(SnH)$ in Me_3SnH for the tin isotopes of mass numbers 119, 117, and 115. There is good agreement with the ratios of the magnetogyric ratios.

(*iii*) *Elements of Group VB.* Ahlbrecht and Papke[54] have discussed the effect of hybridization at N on $^1J(NH)$ in diphenylamine and isobutenylaniline.

$^1J(PD)$ in PD_3 has been measured and the coupling shows no isotope effect.[55] In contrast to $^1J(SiH)$, $^1J(PH)$ in silyl- and silylmethyl-phosphines shows no great variation or uniform trend as the number of groups bonded to silicon is changed.[51] A very high value (898 ± 3 Hz) of $^1J(PH)$ has been observed in the ion $HP^+(OCH_2)_3$-CMe by Vande Griend and Verkade.[56] The authors propose that this very high coupling results from the constraint introduced by the alkoxy-groups which enhances either the phosphorus positive charge and/or the s-character of the P—H bond.

(*iv*) *Elements of Group VIB.* Burnett and Zeltmann[57] have made a detailed investigation of $^1J(OH)$ in water. The coupling is not a function of pH within the range 4.5—9.2, nor of temperature in the range 4—46 °C. Its value is 89.8 ± 2.3 Hz.

(*v*) *Elements of Group VIIB.* Fujiwara and Martin[58] have discussed the coupling $^1J(FH)$ in hydrogen-bonded systems for which they have observed the following values: FHF^-, 121 Hz; HF, 476 Hz; $FHCl^-$, 403 Hz; $FHBr^-$, 427 Hz; FHI^-, 437 Hz.[59] They find that $^1J(FH)$ correlates strongly with measures of H-bond strength, and advance simple MO arguments to explain why $^1J(FH)$ is so sensitive to H-bonding while $^1J(CH)$ and $^1J(NH)$ are not. Their analysis suggests that the coupling in polar bonds should be more sensitive to H-bonding than the coupling in non-polar bonds, that H-bonding should decrease coupling in highly polar bonds, and

---

[48] J. P. C. M. van Dongen, H. W. D. van Dijkman, and M. J. A. de Bie, *Rec. Trav. chim.*, 1974, **93**, 29.
[49] R. L. Lipnick, *J. Mol. Structure*, 1974, **21**, 411.
[50] R. L. Lipnick, *J. Mol. Structure*, 1974, **21**, 423.
[51] G. Fritz and H. Schäfer, *Z. anorg. Chem.*, 1974, **409**, 137.
[52] H. Bürger, R. Eujen, and H. C. Marsmann, *Z. Naturforsch.*, 1974, **29b**, 149.
[53] H. Schumann and H. J. Kroth, *Z. Naturforsch.*, 1974, **29b**, 573.
[54] H. Albrecht and G. Papke, *Chem. Letters*, 1975, 91.
[55] N. Zumbulyadis and B. P. Dailey, *J. Chem. Phys.*, 1974, **60**, 4223.
[56] L. J. Vande Griend and J. G. Verkade, *Phosphorus*, 1973, **3**, 13.
[57] L. J. Burnett and A. H. Zeltmann, *J. Chem. Phys.*, 1974, **60**, 4636.
[58] J. S. Martin and F. Y. Fujiwara, *J. Amer. Chem. Soc.*, 1974, **96**, 7632.
[59] F. Y. Fujiwara and J. S. Martin, *J. Amer. Chem. Soc.*, 1974, **96**, 7625.

that the coupling in a symmetrical complex such as FHF⁻ should be no greater than one half of that in the related molecule. All available experimental data are in accord with these predictions.

Heteronuclear ^{1}H–{^{35}Cl} double resonance has been used to find a value of 9.3 ± 0.3 Hz for ^1J(ClH) in SiHCl$_3$.[60]

Further reports of one-bond coupling to hydrogen may be found in the following papers: ^1J(CH) (refs. 5, 22, and 61—85), ^1J(SiH) (refs. 81, 86, and 87), ^1J(NH) (refs. 88 and 89) and ^1J(CdH) (ref. 90).

B. Coupling to Elements of Groups IIIB and IVB.—(i) *Coupling to Boron and Thallium.* Hall *et al.*[12] have determined ^1J(CB) values for 11 organoboranes and have compared their experimental results with theoretical calculations of the INDO FPT type using the FC term only. The rather good correlation of theory with experiment suggests a dominant FC term and/or a smooth variation of the other terms in the series of compounds considered. However, there is no correlation of ^1J(CB) with the s_Cs_B bond order, *i.e.* with the *s*-character of the C—B bond. This is very strange if the FC term is dominant, but a similar result has been found for ^1J(CC) by Marshall and Miiller.[91] In this context too the results of Negrebetskii and his colleagues,[92] who

[60] A. Briguet, J. C. Duplan, D. Graveron-Demilly, and J. Delman, *Mol. Phys.*, 1974, **28**, 177.
[61] Y. Takeuchi and N. Dennis, *J. Amer. Chem. Soc.*, 1974, **96**, 3657.
[62] G. E. Hawkes, R. A. Smith and J. D. Roberts, *J. Org. Chem.*, 1974, **39**, 1276.
[63] R.-M. Lequan, M.-J. Pouet, and M.-P. Simonnin, *J.C.S. Chem. Comm.*, 1974, 475.
[64] H. A. Brune, H. Hanebeck, G. Horlbeck, and H. Hüther, *Z. Naturforsch.*, 1974, **29b**, 219.
[65] H. Schmidbaur, W. Buchner, and F. H. Köhler, *J. Amer. Chem. Soc.*, 1974, **96**, 6208.
[66] K. Tori, T. Tsuchima, H. Tanida, K. Kushida, and S. Satoh, *Org. Magn. Resonance*, 1974, **6**, 324.
[67] G. Miyajima, K. Takahashi, and K. Nishimoto, *Org. Magn. Resonance*, 1974, **6**, 413.
[68] V. Nair, *Org. Magn. Resonance*, 1974, **6**, 483.
[69] R. E. Wasylishen and T. Schaefer, *Canad. J. Chem.*, 1974, **52**, 3247.
[70] S. Sternhell and P. W. Westerman, *J. Org. Chem.*, 1974, **39**, 3794.
[71] E. Haloui and D. Canet, *J. Mol. Structure*, 1975, **24**, 85.
[72] J. F. Hinton and L. W. Jaques, *J. Magn. Resonance*, 1974, **15**, 564.
[73] R. D. Singh and S. N. Singh, *J. Magn. Resonance*, 1974, **16**, 110.
[74] C. Th. Pedersen and K. Schaumburg, *Org. Magn. Resonance*, 1974, **6**, 586.
[75] H. U. Schwering, J. Weidlein, and P. Fischer, *J. Organometallic Chem.*, 1975, **84**, 17.
[76] M. Begtrup, *J.C.S. Chem. Comm.*, 1974, 702.
[77] A. S. Perlin, N. Cyr, R. George, S. Ritchie, and A. Parfondry, *Carbohydrate Res.*, 1974, **37**, C1.
[78] M. Begtrup, *Acta Chem. Scand. (B)*, 1974, **28**, 61.
[79] F. Fringuelli, S. Gronowitz, A.-B. Hörnfeldt, J. Johnson, and A. Taticchi, *Acta Chem. Scand. (B).*, 1974, **28**, 184.
[80] J. F. Hinton and L. W. Jaques, *J. Magn. Resonance*, 1975, **17**, 95.
[81] R. K. Harris and B. K. Kimber, *J. Magn. Resonance*, 1975, **17**, 174.
[82] M. Garreau, G. J. Martin, M. L. Martin, J. Morel, and C. Paulmier, *Org. Magn. Resonance*, 1974, **6**, 648.
[83] T. N. Mitchell, *Org. Magn. Resonance*, 1975, **7**, 59.
[84] Y. Takeuchi, *J.C.S. Perkin II*, 1974, 1927.
[85] J. L. Marshall, D. E. Miiller, H. C. Dorn, and G. E. Maciel, *J. Amer. Chem. Soc.*, 1975, **97**, 460.
[86] F. Fehér and R. Freund, *Inorg. Nuclear Chem. Letters*, 1974, **10**, 561.
[87] G. Fritz and H. Schäfer, *Z. anorg. Chem.*, 1974, **406**, 167.
[88] L. J. Turbini and R. F. Porter, *Org. Magn. Resonance*, 1974, **6**, 456.
[89] H. Ahlbrecht and G. Papke, *Tetrahedron*, 1974, **30**, 2571.
[90] A. D. Cardin, P. D. Ellis, J. D. Odom, and J. W. Howard, *J. Amer. Chem. Soc.*, 1975, **97**, 1672.
[91] J. L. Marshall and D. E. Miiller, *Org. Magn. Resonance*, 1974, **6**, 395.
[92] V. V. Negrebetskii, V. S. Bogadanov, P. V. Petrovskii, N. Yu. Bubnov, and B. M. Mikhailov, *Zhur. obshchei Khim.*, 1974, **44**, 1882.

found that ^1J(CB) depends upon hybridization at B and C, are interesting. It would appear that more theoretical work would be helpful here, particularly calculations of orbital and spin-dipolar contributions to ^1J(CB); significant non-FC contributions to ^1J(CC) have been found for three-membered rings.[8] Kitching et al.[93] have made a study of (Tl,C) coupling in arylthallium trifluoroacetates. They find that methyl substitution, particularly *ortho* or *para* to thallium, decreases ^1J(TlC) significantly.

^1J(PB) in the series $Bu^t_{3-n}F_nP \rightarrow BH_3$ has been found to be approximately linearly related to the first ionization potentials of the fluorophosphines $Bu^t_{3-n}F_nP$.[40] As the ionization potential increases the coupling decreases. ^1J(FB) in boron trihalide adducts can be determined by means of a pair-wise additivity relationship[94] of the type proposed by Malinowsky.[95]

(ii) Elements of Group IVB. The theoretical work of Schulman and Newton[8] on ^1J(CC) has been described in Section 2B. Pomerantz and Hillenbrand[43] have noted that hybridization at a particular carbon atom calculated from its (C,C) and (C,H) couplings often deviates from unity by as much as 10%, while Marshall and Miiller[91] find that ^1J(CC) is not proportional to the product of $S^2(0)$ for sp^2–sp^2 bonds. However, they find that the sum of three ^1J(CC) values involving the same carbon atom is constant (156—159 Hz) for a carbon having one double- and two single-bonded neighbours. Marshall and Miiller[91] also suggest that steric effects may explain some of the observed variations in ^1J(CC).

^1J(CC) in amino-acids has been the subject of several investigations. Sogn et al.[96] found little variation among 30 $^1J(C_0C_\alpha)$ values for nine amino-acids, the couplings all lying between 59.1 and 59.9 Hz. They also noted a rough correlation of the (C_α, C_β) coupling with the Hammett σ_I parameter of the β substituent. The dependence of $^1J(C_0C_\alpha)$ upon pH has been investigated.[97]

^1J(SiC) in trimethylsilyl compounds depends upon nearest-neighbour substituent electronegativity in a way consistent with Bent's rules, which suggests a dominant FC term.[81] Substituent effects on the coupling parallel those found for ^1J(CC) in t-butyl systems and there is a good linear correlation between the two types of coupling.

Kennedy and McFarlane[98] have measured a value of 155 ± 5 Hz for ^1J(SnC) in Me_3SnLi. This leads to a reduced coupling constant of negative sign, the first such between any pair of Group IV elements. They note that there is a clear linear relationship between ^1J(SnC) and ^2J(SnCH), but it does not pass through the origin, indicating that it is unwise to use either of these couplings to estimate the hybridization of tin. Interpreting these observations in terms of the core polarization theory of Jameson and Gutowsky, the authors suggest that the change of sign in ^1J(SnC) and ^2J(SnCH) can be seen as analogous to the corresponding changes of sign in ^1J(PC) and ^2J(PCH) when P goes from P^{III} to P^V, and also to certain coupling constants involving selenium.[98] Mitchell[83] has observed that the linear relationship between

---

[93] W. Kitching, D. Praeger, C. J. Moore, D. Doddrell, and W. Adcock, *J. Organometallic Chem.*, 1974, **70**, 339.
[94] J. S. Hartman and J. M. Miller, *Inorg. Chem.*, 1974, **13**, 1467.
[95] T. Vladimiroff and E. R. Malinowsky, *J. Chem. Phys.*, 1967, **46**, 1830.
[96] J. A. Sogn, L. C. Craig, and W. A. Gibbons, *J. Amer. Chem. Soc.*, 1974, **96**, 4694.
[97] Tran Dinh Son and S. Fermandjian, *J. Phys. (Paris)*, 1973, 45.
[98] J. D. Kennedy and W. McFarlane, *J.C.S. Chem. Comm.*, 1974, 983.

^1J(SnC) and ^2J(SnCH) does not hold for couplings in the $SnCH_nCL_{3-n}$ group of mono- and di-chloromethyltins. $^1J(^{117}SnC)$ and $^1J(^{119}SnC)$ values have been used to predict the value of $^1J(^{115}SnC)$, but the coupling has not yet been observed experimentally.[53]

(*iii*) *Elements of Group VB*. Berger and Roberts[99] have discussed the effect of the nitrogen lone-pair on ^1J(NC).

The sign of ^1J(PC) has been the subject of much interest. Gray and Cremer[100] have examined this coupling in a number of phosphorus(III) heterocycles carrying a phenyl group on the phosphorus. The coupling of P to the adjacent phenyl carbon varies smoothly from -12.5 to -35.4 in the series, and the coupling to the adjacent carbon in the heterocyclic ring lies between -10 and -15 Hz in all cases but one. The exception involves a four-membered ring with $^1J_{PC} = +0.6$ Hz. Lequan *et al.*[63] have also measured (P,C) couplings in systems of general form (4). For the two phosphines, $R^1 = C{\equiv}CH$, $R^2 = NPh_2$

$$\begin{array}{c} R^1 \\ {}{\searrow} \\ R^2 \end{array} P(X){-}C{\equiv}C{-}H$$

(4) X = O or lone pair

and $R^1 = R^2 = OEt$, ^1J(PC) is negative, with values of -9.8 and -50 Hz, respectively. For three phosphine oxides ^1J(PC) is positive with values as follows: $R^1 = R^2 = Ph$, $+164.6$ Hz; $R^1 = R^2 = NMe_2$, $+224$ Hz; $R^1 = R^2 = OEt$, $+294$ Hz. For these compounds, as can be seen, ^1J(PC) increases with increasing electronegativity of the phosphorus substituents, suggesting a dominant FC term. Albright, Freeman, and Schweizer[101] have noted the marked changes in ^1J(PC) across the ylide bond for $Ph_3P{=}CH_2$ (51.9 Hz), $Me_3P{=}CH_2$ (90.5 Hz) and $Ph_3P{=}CMe_2$ (121.5 Hz). They compare these with the relatively constant values in the phosphonium salts, $Me_3P^+MeI^-$ (56 Hz), $Ph_3P^+MeI^-$ (57 Hz), and $Ph_3P^+CHMe_2$-Br^- (47 Hz). They interpret these results in terms of hybridization at carbon and charge on that atom.

Fritz and Schäfer[51] have examined the effect of the substituents H, Me, H_3Si, and Me_3Si on coupling constants in silyl- and silymethyl-phosphines. They find that ^1J(PSi) can be calculated to within 2 Hz, and usually within 1 Hz, by means of simple additivity relationships.

(*iv*) *Elements of Group VIIB*. Schuster[102] has measured ^1J(FC) in *ortho-*, *meta-*, and *para-*substituted benzotrifluorides and similar species. He finds that ^1J(FC) correlates well with Taft's dual substituent parameter according to equation (7), where J_X is the coupling constant for substituent X, J_H the corresponding coupling with X replaced by H and ρ_I and ρ_R^+ measure the susceptibility of J to the polar and

$$J_X = J_H + \rho_I \sigma_I + \rho_R^+ \sigma_R^+ \tag{7}$$

resonance effects of substituents, respectively.

[99] S. Berger and J. D. Roberts, *J. Amer. Chem. Soc.*, 1974, **96**, 6757.
[100] (*a*) G. A. Gray and S. E. Cremer, *J.C.S. Chem. Comm.*, 1974, 451; (*b*) *ibid.*, 1975, p. 304.
[101] T. A. Albright, W. J. Freeman, and E. E. Schweizer, *J. Amer. Chem. Soc.*, 1975, **97**, 940.
[102] I. I. M. Schuster, *J. Magn. Resonance*, 1975, **17**, 104.

(v) *Metals*. Köhler et al.[103] have measured ^nJ(W,C) values in an attempt to relate the magnitude of this coupling to s-character or hybridization at carbon. They find that ^1J(WC) to the adjacent carbon atom in the organic chain goes from 29.5 Hz in $(CO)_3W(C_5H_5)CH_2CH{=}CH_2$ to 185.5 Hz in $Br(CO)_4WCC{\equiv}CPh$. Although the general trend is clear the detailed correlation of ^1J(WC) with hybridization at C is poor. The relationships between the coupling of Pt to H and C in a variety of cyclo-octa-1,5-dieneplatinum(II) complexes have been discussed.[104]

Schumann, Dreeskamp, and Hildenbrand[105] have measured ^1J(HgC) and ^2J(CHgC) in dimethylmercury in nematic solvents. They find that the anisotropies of these two couplings are both about one half of their isotropic values. This is much larger than the theoretical estimates for (C,C) coupling and the authors suggest that this is due to the presence of the heavy Hg atom. It is noteworthy, however, that even in these situations the isotropic FC term clearly dominates the coupling.

Other one-bond couplings to elements of Groups IIIB and IVB may be found in the following papers: ^1J(BB) (ref. 42), ^1J(CB) (ref. 106), ^1J(CC) (refs. 85 and 107—111), ^1J(SiC) (ref. 112), ^1J(PC) (refs. 65 and 113—116), ^1J(PSi) (ref. 87), ^1J(SeC) (ref. 82), ^1J(FC) (refs. 72, 73, 80, and 117—119), ^1J(SiF) (ref. 81), ^1J(CdC) (ref. 90), ^1J(HgC) (ref. 120), and ^1J(PtC) (refs. 121—124).

C. Coupling to Elements of Group VB.—(*i*) ^1J(SnP). McFarlane and Rycroft[125] have noted that the reduced coupling ^1K(SnP) has substantial negative values in $(Me_3\text{-}Sn)_{3-n}Ph_nP$ and increases algebraically to values close to zero in transition-metal carbonyl complexes. This behaviour parallels that of ^1K(PC) in analogous compounds but there is a substantial negative contribution to the (Sn,P) coupling which is attributed to the larger energy gap between the P and Sn valence atomic orbitals. ^1J(SnP) has been measured for ^{119}Sn, ^{117}Sn, and ^{115}Sn in tris(trimethylstannyl)-phosphine.[53] There is no isotope effect.

[103] F. H. Köhler, H. J. Kalder, and E. O. Fischer, *J. Organometallic Chem.*, 1975, **85**, C19.
[104] M. H. Chisholm, H. C. Clark, L. E. Mauzer, J. B. Stothers, and J. E. H. Ward, *J. Amer. Chem. Soc.*, 1975, **97**, 721.
[105] C. Schumann, H. Dreeskamp, and K. Hildenbrand, *J. Magn. Resonance*, 1975, **18**, 97.
[106] J. D. Odom, L. W. Hall, and P. D. Ellis, *Org. Magn. Resonance*, 1974, **6**, 360.
[107] M. Tanabe and K. T. Susuki, *Tetrahedron Letters*, 1974, 4417.
[108] R. C. Long, jun. and J. H. Goldstein, *J. Magn. Resonance*, 1974, **16**, 228.
[109] J. L. Marshall, A. M. Ihrig, and D. E. Miiller, *J. Magn. Resonance*, 1974, **16**, 439.
[110] Tran Dinh Son, S. Fermandjian, E. Sala, R. Mermet-Bouvier, and P. Fromageot, *J. Amer. Chem. Soc.*, 1975, **97**, 1267.
[111] P. E. Hansen, O. K. Poulsen, and A. Berg, *Org. Magn. Resonance*, 1975, **7**, 23.
[112] C. D. Schaeffer and J. J. Zuckerman, *J. Amer. Chem. Soc.*, 1974, **96**, 7160.
[113] W. Buchner and W. Wolfsberger, *Z. Naturforsch.*, 1974, **29b**, 328.
[114] S. Aime, R. K. Harris, E. M. McVicker, and M. Fild, *J.C.S. Chem. Comm.*, 1974, 426.
[115] S. I. Featherman, S. O. Lee, and L. D. Quin, *J. Org. Chem.*, 1974, **39**, 2899.
[116] H. Manke, R. J. Clark, R. Rosanske, and R. K. Sheline, *J. Chem. Phys.*, 1974, **60**, 2997.
[117] D. D. Giannini, P. A. Kollman, N. S. Bhacca, and M. E. Wolff, *J. Amer. Chem. Soc.*, 1974, **96**, 5462.
[118] J. W. Emsley and J. C. Lindon, *Chem. Phys. Letters*, 1974, **26**, 361.
[119] O. A. Subbotin and N. M. Sergeyev, *J. Amer. Chem. Soc.*, 1975, **97**, 1080.
[120] J. Casanova, H. R. Rogers, and K. L. Servis, *Org. Magn. Resonance*, 1975, **7**, 57.
[121] P. W. Hall, R. J. Puddephatt, and C. F. H. Tipper, *J. Organometallic Chem.*, 1974, **71**, 145.
[122] H. C. Clark, L. E. Manzer, and J. E. H. Ward, *Canad. J. Chem.*, 1974, **52**, 1973.
[123] B. E. Reichert, *J. Organometallic Chem.*, 1974, **72**, 305.
[124] H. C. Clark, J. E. H. Ward, and K. Yasufuku, *Canad. J. Chem.*, 1975, **53**, 186.
[125] W. McFarlane and D. S. Rycroft, *J.C.S. Dalton*, 1974, 1977.

(ii) *Elements of Group VB.* Schweiger et al.[126] have used double resonance techniques to determine the signs of a number of directly bonded (P,N) coupling constants. They find that ^1K(PN) changes from negative to positive and then becomes progressively larger as the percentage of P(3s)-character in the P—N bond increases. The authors suggest an explanation in terms of Jameson and Gutowsky's analysis in which increasing the valence s-character at phosphorus causes a change-over in the dominant coupling mechanism from an indirect negative (core polarization) interaction to a direct positive (FC) interaction.

Albrand and Robert[127] have noted some remarkable effects in (P,P) couplings of (5a) and (5b). The couplings in (5b) (See Table 1) show the same trends as were

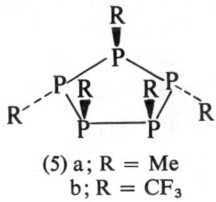

(5) a; R = Me
b; R = CF$_3$

observed with (5a).[128] However, comparison of the results for the two molecules shows that changes in the corresponding couplings are not in the same direction for J_{cis} and J_{trans}. If ^1J(PP) is assumed to be negative then J_{cis} is more negative in

Table 1 (P,P) *couplings* (Hz) *in* (5a) *and* (5b)

	$^1J_{2,3}$	$^1J_{3,4}$	$^1J_{1,2}$	$^2J_{1,3}$	$^2J_{2,5}$	$^2J_{2,4}$
(5a)	−236.4	−310.3	−248.6	+23.3	−3.6	−8.1^a
(5b)	∓223.3	∓343.2	∓214.5	±36.9	±0.8	±5.3^b

^{a} Values from J.P. Albrand, D. Gagnaire, and J.B. Robert, *J. Amer. Chem. Soc.*, 1973, **95**, 6498.
^{b} The upper sign is preferred.

(PCF$_3$)$_5$ than in (PMe)$_5$, while J_{trans} exhibits the reverse relationship. These observations show that stereochemical factors must be very important in determining these couplings.

(iii) *Elements of Group VIB.* ^1J(SeP) and ^1J(TeP) values have been reported by Loginova et al.[129] Both reduced coupling constants are negative, and the former depends sensitively on the nature of the substituents at phosphorus. This dependence could be explained quite well in terms of the Jameson–Gutowsky model.

(iv) ^1J(PH). Grosse and Schmutzler[130] have discussed phosphorus–fluorine coupling

---

[126] J. R. Schweiger, A. H. Cowley, E. A. Cohen, P. A. Kroon, and S. L. Manatt, *J. Amer. Chem. Soc.*, 1974, **96**, 7122.
[127] J. P. Albrand and J. B. Robert, *J.C.S. Chem. Comm.*, 1974, 644.
[128] J. P. Albrand, D. Gagnaire, and J. B. Robert, *J. Amer. Chem. Soc.*, 1973, **95**, 6498.
[129] E. I. Loginova, I. A. Nuretdinov, and Yu. A. Petrov, *Teor. i eksp. Khim.*, 1974, **10**, 75.
[130] J. Grosse and R. Schmutzler, *Phosphorus*, 1974, **4**, 49.

in difluorophosphoranes (R_3PF_2) in detail. They find that $^1J(PF)$ is correlated linearly with the sum of the substituent group electronegativities in the series $(RS)_nMe_{3-n}PF_2$ (R = alkyl, n = 0—2) and $Ph_nMe_{3-n}PF_2$ (n = 0—3). For the latter series there is also a linear correlation of the coupling with σ_R^*, the Taft inductive constant. A similar correlation of $^1J(PF)$ with σ^+ was found by Szafraniec for phenylphosphonic difluorides.[131] Grosse and Schmutzler[130] also found that the $^1J(PF)$ values depended on solvent, concentration, and temperature, and attributed these effects to intermolecular interactions.

Harris, Schmutzler, and their co-workers have continued their work on diazadiphosphetidines (6) and have discussed $^1J(PF)$ in detail.[132] The effect of molecular geometry on the coupling is surprisingly large and values for *gauche-* and *trans-*isomers can differ by as much as 40 Hz. At low temperatures the difference between couplings to axial and equatorial fluorines is of the order of 150 Hz. $^1J(PF)$ decreases, *i.e.* becomes more negative, with increasing Cl substitution in $[Cl_nH_{3-n}F_2PNMe]_2$.[133]

$$R_{3-n'}F_{n'}P\underset{\underset{Me}{|}}{\overset{\overset{Me}{|}}{\underset{N}{\overset{N}{\diamond}}}}PF_nR_{3-n}$$

(6) R = Me or OMe

Buslaev *et al.*[134] have reported measurements of $^1J(PF)$ for octahedral mixed fluorochlorides of phosphorus(v) anions, and Bhattacharyya and Dailey[135] have found evidence of some anisotropy in this coupling in POF_3.

(v) *Metals.* Vande Griend and Verkade[56] have observed that the point for L = $P(OCH_2)_3CMe$ lies on the graph of $^1J(WP)$ for $LW(CO)_5$ complexes *versus* $^1J(PH)$ for the ligand L, supporting the hypothesis that there is no π-electron contribution to $^1J(WP)$. The unusually low $^1J(PtP)$ value of 1491.8 Hz found in *cis*-[PtPh$_2$(Ph$_2$-PCH$_2$PPH$_2$)] may be due to a distortion of the valency angles at the P and Pt atoms.[136]

Other values of one-bond couplings to elements of Group VB may be found in the following papers: $^1J(NN)$ (ref. 137), $^1J(PP)$ (ref. 138), $^1J(PF)$ (refs. 40, 46, 126, and

[131] L. L. Szafraniec, *Org. Magn. Resonance*, 1974, **6**, 565.
[132] R. K. Harris, M. I. M. Wazeer, O. Schlak, and R. Schmutzler, *J.C.S. Dalton*, 1974, 1921.
[133] R. K. Harris, M. Lewellyn, M. I. M. Wazeer, R. E. Dunmur, M. J. C. Hewson, and R. Schmutzler, *J.C.S. Dalton*, 1975, 61.
[134] Yu. A. Buslaev, E. G. Ilin, and M. N. Shcherbakova, *Doklady Akad. Nauk S.S.S.R.*, 1974, **217**, 337.
[135] P. K. Bjattacharyya and B. P. Dailey, *Mol. Phys.*, 1974, **28**, 209.
[136] P. S. Braterman, R. J. Cross, L. Manojlovic-Muir, K. W. Muir, and G. B. Young, *J. Organometallic Chem.*, 1975, **84**, C40.
[137] R. K. Harris and N. C. Pyper, *Mol. Phys.*, 1975, **29**, 205.
[138] G. Hägele, W. Kuchen, and H. Steinberger, *Z. Naturforsch.*, 1974, **29b**, 349.

Nuclear Spin–Spin Coupling 73

139—147), $^1J(\text{HgP})$ (ref. 148), $^1J(\text{PtP})$ (refs. 148—150), and $^1J(\text{WP})$ (refs. 125, 148 and 151—153).

D. Coupling to Elements of Group VIB.—Vold and Vold have found a value for $^1J(^{95}\text{Mo}^{17}\text{O})$ of 40.3 Hz in aqueous sodium molybdate solution.[154] $^1J(\text{PtSe})$ and $^1J(\text{PtTe})$ for the ions $[\text{PtX}_3\text{SeMe}_2]^-$, $[\text{PtX}_5\text{SeMe}_2]^-$, and $[\text{PtX}_3\text{TeMe}]^-$ (X = Cl, Br, or I) decrease markedly in the series Cl > Br > I, and they are much smaller than $^1J(\text{PtTe})$ in $[(\text{PtX}_3)_2\text{TeMe}_2]^{2-}$ (X = Cl or Br). The sign reverses[155] between Br and I in the series $[\text{PtX}_5\text{SeMe}_2]^-$, behaviour which is said[155] to parallel that of $^1J(\text{PP})$.

E. Coupling to Elements of Group VIIB.—The good correlation of $^1J(\text{XeF})$ with the fluorine chemical shift has been established for a variety of xenon compounds by Gillespie and Schrobilgen.[156, 157] As the XeF group becomes more ionic the coupling and the chemical shift increase. This is interpreted as the effect on both phenomena of changing ΔE.[157] There are also good correlations between $^1J(\text{XeF})$ and the Xe—F bond length and i.r. stretching frequency.[156] Seppelt and Rupp[158, 159] have also reported $^1J(\text{XeF})$ values and have observed that the coupling decreases with higher oxidation and fluorination, *e.g.* XeF_2, 5550 Hz; $(\text{XeF}_6)_4$, 330 Hz.

$^1J(^{45}\text{Sc}^{19}\text{F})$ in ScF_6^{3-} has been found to be 180 ± 10 Hz.[160]

Other reports of one-bond couplings to fluorine may be found in the following references: $^1J(\text{TeF})$ (ref. 161), $^1J(\text{XeF})$ (refs. 162 and 163), and $^1J(\text{WF})$ (ref. 164).

F. Miscellaneous One-bond Couplings.—Kennedy and McFarlane[165] have reported the surprisingly small value of $+290$ Hz for $^1J(\text{PbPb})$ in hexamethyldilead. Any

[139] M. Fild and T. Stahkiewicz, *Z. anorg. Chem.*, 1974, **406**, 115.
[140] T. Kruck, H. Jung, M. Höfler, and H. Blume, *Chem. Ber.*, 1974, **107**, 2156.
[141] L. F. Doty and R. G. Cavell, *Inorg. Chem.*, 1974, **13**, 2722.
[142] R. B. Johannesen, S. C. Peake, and R. Schmutzler, *Z. Naturforsch.*, 1974, **29b**, 699.
[143] A. H. Cowley, R. W. Braun, and J. W. Gilje, *J. Amer. Chem. Soc.*, 1975, **97**, 434.
[144] K. P. John and R. Schmutzler, *Z. Naturforsch.*, 1974, **29b**, 730.
[145] M. A. Busch and R. J. Clark, *Inorg. Chem.*, 1975, **14**, 226.
[146] J. F. Nixon, B. Wilkins, and D. A. Clement, *J.C.S. Dalton*, 1974, 1993.
[147] O. Schlak, R. Schmutzler, H. M. Schiebel, M. I. M. Wazeer, and R. K. Harris, *J.C.S. Dalton*, 1974, 2153.
[148] R. L. Keiter, K. M. Fasig, and L. W. Cary, *Inorg. Chem.*, 1975, **14**, 201.
[149] A. J. Carty and S. E. Jacobson, *J.C.S. Chem. Comm.*, 1975, 175.
[150] N. Sonoda, S. Araki, T. Onishi, and T. Tanaka, *J. Inorg. Nuclear Chem.*, 1974, **36**, 1985.
[151] S. O. Grim, J. Del Gaudio, R. P. Molenda, C. A. Tolman, and J. P. Jesson, *J. Amer. Chem. Soc.*, 1974, **96**, 3416.
[152] R. C. Taylor, R. L. Keiter, and L. W. Cary, *Inorg. Chem.*, 1974, **13**, 1928.
[153] R. J. Mynott, P. S. Pregosin, and L. M. Venanzi, *J. Coordination Chem.*, 1973, **3**, 145.
[154] R. R. Vold and R. L. Vold, *J. Chem. Phys.*, 1974, **61**, 4360.
[155] P. L. Goggin, R. J. Goodfellow, and S. R. Haddock, *J.C.S. Chem. Comm.*, 1975, 176.
[156] R. J. Gillespie, A. Netzer, and G. J. Schrobilgen, *Inorg. Chem.* 1974, **13**, 1455.
[157] R. J. Gillespie and G. J. Schrobilgen, *Inorg. Chem.*, 1974, **13**, 2370.
[158] K. Seppelt and H. H. Rupp, *Z. anorg. Chem.*, 1974, **409**, 331.
[159] K. Seppelt and H. H. Rupp, *Z. anorg. Chem.*, 1974, **409**, 338.
[160] Yu. A. Buslaev, S. P. Petrosyants, V. P. Tarasov, and U. I. Chagin, *Zhur. neorg. Khim.*, 1974, **19**, 1790.
[161] U. Elgad and H. Selig, *Inorg. Chem.*, 1975, **14**, 140.
[162] R. J. Gillespie and G. J. Schrobilgen, *Inorg. Chem.*, 1974, **13**, 1694.
[163] J. H. Holloway, G. J. Schrobilgen, and P. Taylor, *J.C.S. Chem. Comm.*, 1975, 40.
[164] J. Banck and A. Schwenk, *Z. Phys.*, 1975, **20B**, 75.
[165] J. D. Kennedy and W. McFarlane, *J. Organometallic Chem.*, 1974, **80**, C47.

reasonable assumption concerning the *s*-character of the orbitals used to form the Pb—Pb bond cannot, they suggest, lead to such a low value of the coupling if used in conjunction with the average energy approximation. The coupling must be interpreted in terms of mutual polarizabilities, and relationships between *J* and hybridization *etc.* must be used with great care.

^1J(ClF) in ClOF$_3$ is found to be 192 Hz for coupling to the axial fluorine but <20 Hz for the equatorial atoms.[166] A very elegant analysis of the nuclear hyperfine structure of the $R(127)$ line of the 11-5 band in the $B \leftarrow X$ electronic absorption band of molecular iodine has given a value of $-36\,300 \pm 8000$ Hz for the (I,I) coupling constant in the *excited* state.[167] The authors point out that an investigation of the change of coupling constant with vibrational band would give information about the mixing of excited electronic states. Such information would be of great interest to theoretical chemists and in the theory of nuclear spin–spin coupling.

4 Coupling between Nuclei Separated by Two Chemical Bonds, ^2J

A. Geminal Proton–Proton Coupling.—The detailed calculations of the effect of lone-pair orientation on ^2J(HCH) made by Aminova and Samitov[14] have been discussed in Section 2B.

Hinton and Jaques[72] have analysed the ^{19}F and ^{1}H spectra of 4-bromo-3-chloro-3,4,4-trifluorobut-1-ene. They find that the sign of the geminal (H,H) coupling depends both upon solvent and temperature. It changes from approximately -0.5 Hz at 190 K to about $+0.1$ Hz at 350 K but the reasons for this behaviour are unclear. Geminal (H,H) coupling in three-membered rings has been reported by two laboratories. Ivin *et al.*[168] find that in mono-substituted thiirans (7a), ^2J(HCH) is much smaller and ^3J(HCCH) much larger than in the related oxirans (7b). This they interpret as due to the lower electronegativity of S relative to

R\C——C/H
H/   \X/   \H

(7) R = Me, Bu^n, Ph, CH$_2$Cl, or MeOCH$_2$
a; X = S
b; X = O

O. Nakanishi and Yamamoto[169, 170] have made a very detailed study of the (H,H) couplings in ethyleneimine (10) (see Section 5B). Values of ^2J(HCH) have also been reported for thietans,[171] alkylcyclohexanones,[172, 173] and —SCH$_2$—groups.[174]

[166] M. Alexandre and P. Rigny, *Canad. J. Chem.*, 1974, **52**, 3676.
[167] P. R. Bunker and G. R. Hanes, *Chem. Phys. Letters*, 1974, **28**, 377.
[168] K. J. Ivin, E. D. Lillie, and I. H. Petersen, *Internat. J. Sulfur Chem.*, 1973, **8**, 411.
[169] H. Nakanishi and O. Yamamoto, *Tetrahedron*, 1974, **30**, 2115.
[170] H. Nakanishi and O. Yamamoto, *Chem. Letters*, 1973, 465.
[171] C. Cistaro, G. Fronza, R. Mondelli, S. Bradamante, and G. A. Pagani, *J. Magn. Resonance*, 1974, **15**, 367.
[172] K. L. Servis, D. J. Bowler, and C. Ishii, *J. Amer. Chem. Soc.*, 1975, **97**, 73.
[173] K. L. Servis and D. J. Bowler, *J. Amer. Chem. Soc.*, 1975, **97**, 80.
[174] J. T. Wróbel and K. Kabzińska, *Bull. Acad. Polon. Sci., Ser. Sci. chim.*, 1974, **22**, 173.

Other values of geminal proton–proton coupling *via* a carbon atom may be found in refs. 22, 50, 80, and 175—185.

B. Two-bond Coupling of Other Nuclei to Protons.—(i) *Elements of Group IIIB.* Negrebetskii and his co-workers[92] have measured $^2J(BCH)$ in a number of organoboron compounds. They find that the coupling depends upon hybridization at B and at C. Weibel and Oliver[186] have given a detailed discussion of metal–proton coupling in trimethyl-tin, -aluminium, -gallium, -indium, and -thallium organometallic compounds. The correlation of $^2J(TlCH)$ with hybridization at Tl is good, supporting an FC mechanism for the coupling. There are also linear relationships between $^2J(SnCH)$ and $^2J(TlCH)$; $^3J(SnTlCH)$ and $^2J(SnCH)$; and $^3J(TlSnCH)$ and $^2J(TlCH)$ in the series $Li[(Me_3Sn)_nTlMe_{4-n}]$.

(*ii*) *Elements of Group IVB.* Takeuchi and Dennis[61] have used ^{13}C FT spectra to determine (C,H) couplings in 2-, 3-, and 4-cyanopyridine. $^2J(CCH)$ normally lies between 1 and 4 Hz but the coupling to a carbon adjacent to nitrogen is much enhanced reaching values of 7—9 Hz. The calculations of Aminova and Samitov[14] suggest that $^2J(COH)$, $^2J(CNH)$ and $^2J(NCH)$ should not, however, be very much affected by the orientation of adjacent lone pairs.

Wasylishen and Schaefer[69] have analysed the ^{13}C spectrum of propane and [2,2-$^2H_2$]propane to give the (C,H) and (H,H) couplings in high accuracy. Their results suggest that some earlier reports of $^2J(CCH)$ in isopropyl derivatives may actually refer to $^3J(CCCH)$.

A detailed account of (C,H) coupling in cyclobutadienyltricarbonyl-iron derivatives has been given[64] and Fritz and Sauter[187] have shown how the $^2J(CCH)$ values for diethylfumarate (-2.8 Hz) and maleic anhydride ($+3.6$ Hz) can be obtained from an off-resonance coherently decoupled spectrum.

$^2J(SnCH)$ has been the subject of a number of studies. Schumann and Kroth[53] found no isotope effect for the coupling. Gielen and Topart[188] have reported $^2J(SnCH)$ values for stannacycloalkanes for which they find that the $^2J(Sn-CH_3)$ coupling in $CH_3SnR^1R^2R^3$ with R^1, R^2, R^3 = alkyl groups, phenyl, $(CH_2)_4$, $(CH_2)_5$, and $(CH_2)_6$ shows a very good additivity relationship of the form (8) in which the $J(R)$ *etc.* are simply additive parameters associated with the groups R. For

$$^2J(SnCH) = J(R^1) + J(R^2) + J(R^3) \qquad (8)$$

---

[175] J.-J. Barieux, J. Gore, and J.-C. Richer, *Bull. Soc. chim. France*, 1974, 1020.
[176] J. P. Albrand, J. P. Dutasta, and J. B. Robert, *J. Amer. Chem. Soc.*, 1974, **96**, 4584.
[177] J. D. Remijnse, H. van Bekkum, and B. M. Wepster, *Rec. Trav. chim.*, 1974, **93**, 93.
[178] D. G. Streefkerk, M. J. A. de Bie, and J. F. G. Vliegenthart, *Carbohydrate Res.*, 1974, **33**, 249.
[179] J. L. Marshall and S. R. Walter, *J. Amer. Chem. Soc.*, 1974, **96**, 6358.
[180] H. Günther, M. Görlitz, and H. Meisenheimer, *Org. Magn. Resonance*, 1974, **6**, 388.
[181] E. Diez, A. L. Esteban, and M. Rico, *J. Magn. Resonance*, 1974, **16**, 137.
[182] R. A. Newmark, R. E. Watson, and S. Croft, *Tetrahedron*, 1974, **30**, 3253.
[183] T. A. Crabb and M. J. Hall, *J.C.S. Perkin II*, 1974, 1419.
[184] W. G. Bentrude, H.-W. Tan, and K. C. Yee, *J. Amer. Chem. Soc.*, 1975, **97**, 573.
[185] C. Cistaro, G. Fronza, R. Mondelli, S. Bradamante, and G. A. Pagani, *J. Magn. Resonance*, 1975, **17**, 219.
[186] A. T. Weibel and J. P. Oliver, *J. Organometallic Chem.*, 1974, **74**, 155.
[187] H. Fritz and H. Sauter, *J. Magn. Resonance*, 1974, **15**, 177.
[188] M. Gielen and J. Topart, *J. Organometallic Chem.*, 1974, **81**, 357.

compounds of the form $(CH_2)_5SnMeR$ (R = alkyl) there is also a good correlation of $^2J(SnCH)$ with $\sigma^*(R)$. For dimethylstannacycloalkanes the coupling constants are related to the angles CH_2—Sn—CH_2 and CH_3—Sn—CH_3.[188]

McFarlane and Kennedy's observations[98] on the relationship between $^1J(SnC)$ and $^2J(SnCH)$ are mentioned in Section 3B(ii), and the linear relationships observed by Weibel and Oliver[186] in Section 4B. The latter authors have also studied $^2J(SnCH)$ in alkali-metal trimethyltin derivatives as a function of temperature.[189] The concentration dependence of $^2J(SnCH)$[190] and $^2J(HgCH)$[191] has been investigated by Petrosyan et al.

Barbieri and Taddei[192] have found that $^2J(SnCH)$ and $^2J(PbCH)$ in oxiran derivatives of tin and lead are not linearly correlated with the corresponding (H,H) couplings, in contrast to the situation with furan and thiophen derivatives.

(iii) Elements of Group VB. Theoretical calculations suggest that $^2J(NCH)$ in CH_3NH_2 is only slightly dependent on the orientation of the CH_3 group.[14]

Schweiger et al.[126] have found that $^2K(PNH)$ can change sign; it is positive in $F_3P(NH_2)_2$ and negative in $(CF_3)_2PNH_2$. It is a rather remarkable coincidence that the moduli of these two couplings (+2.98 and −2.92) should be so nearly equal. $|^2J(PCH)|$ is much larger in phosphine oxides than in the corresponding phosphines[193] but in silyl and silylmethyl phosphines it is appreciably smaller than $|^2J(PSiH)|$.[51]

Two Russian laboratories have reported on correlations of $^2J(PCH)$ with solvent parameters[194] and substituent constants.[195]

(iv) Elements of Group VIIB. Phillips, Wray, and Pendlebury have published some very detailed investigations of $^2J(FCH)$ and $^2J(FSiH)$.[196, 197] For four series of molecules of the type XYCHF, where X = F, Cl, Me, or H and the group Y was variable, analysis of the dependence of $^2J(FCH)$ upon the Huggins electronegativity of X and Y gave equation (9). This equation predicts $^2J(FCH)$ to within ±2.1 Hz.

$$^2J(FCH) = 78.76 + 8.45(\chi_X\chi_Y) - 16.73(\chi_X + \chi_Y) \qquad (9)$$

A similar analysis of $^2J(FSiH)$ in molecules of the form XYSiHF gave equation (10) which predicts the coupling

$$^2J(FSiH) = 49.08 + 7.85(\chi_X\chi_Y) - 9.03(\chi_X + \chi_Y) \qquad (10)$$

[189] A. T. Weibel and J. P. Oliver, *J. Organometallic Chem.*, 1974, **82**, 281.
[190] V. S. Petrosyan, N. S. Yashina, V. I. Bakhmutov, A. B. Permin, and O. A. Reutov, *J. Organometallic Chem.*, 1974, **72**, 71.
[191] V. S. Petrosyan, V. I. Bakhmutov, and O. A. Reutov, *J. Organometallic Chem.*, 1974, **72**, 79.
[192] G. Barbieri and F. Taddei, *J. Organometallic Chem.*, 1974, **70**, 383.
[193] P. W. Clark, J. L. S. Curtis, P. E. Garrou, and G. E. Hartwell, *Canad. J. Chem.*, 1974, **52**, 1714.
[194] L. I. Vinogradov, Yu. Yu. Samitov, I. Ya. Kuramshin, A. A. Muratova, and A. N. Pudovik, *Zhur. obshchei Khim.*, 1974, **44**, 37.
[195] B. N. Laskorin, V. V. Yakshin, and L. I. Sokal'skaya, *Doklady Akad. Nauk S.S.S.R.*, 1974, **215**, 595.
[196] M. H. Pendlebury, L. Phillips, and V. Wray, *J.C.S. Perkin II*, 1974, 787.
[197] L. Phillips and V. Wray, *J.C.S. Perkin II*, 1974, 928.

to within ±2.4 Hz.[196] An analysis of ^2J(FCH) in fluorocarbohydrates, polyfluoro-1,4-dioxans and polyfluoro-1,4-oxathians produced a simple additive method of calculating these couplings. The root-mean-square deviation of calculated from experimental couplings is ±1 Hz.[197]

Further two-bond couplings to hydrogen may be found in the following publications: ^2J(CCH) (refs. 43, 63, 68, 71, 72, 75, 77—80, 84, 85, 198—201), ^2J(SiCH) (refs. 81 and 202), ^2J(SnCH) (refs. 125, 150, 203, and 204), ^2J(PBH) (ref. 40), ^2J(PCH) (refs. 65, 87, 138, 176, 202, and 205), ^2J(PSiH) (ref. 87), ^2J(PMnH) (ref. 206), ^2J(TeCH) and ^2J(SeCH) (ref. 79), ^2J(FCH) (ref. 207), ^2J(SnCH) (ref. 83), ^2J(CdCH) (ref. 90), and ^2J(PtCH) (refs. 104 and 122—124).

C. **Two-bond Coupling to Carbon [Excluding $^2J(^{13}CX^1H)$].**—(i) *Elements of Group IIIB.* Abraham, Hawkes, and Smith[208] have amplified their earlier reports of ^2J(TlCC) values and Kitching *et al.*[93] have noted that in arylthallium trifluoroacetates ^3J(TlCCC) is larger in magnitude than ^2J(TlCC).

(ii) *Elements of Group IVB.* Marshall and Müller[91] have reported ^2J(CCC) values for olefins. The coupling is small when a C=C bond is involved but much larger for the situation in which both carbon–carbon bonds are single. In contrast ^3J(CCCC) seems to be little affected by the presence of one double bond. Comparisons of couplings in ^{13}C-labelled tetrolic ester (12) and propyne (13) have shown that ^2J(CCC) values compare closely with geometrically equivalent ^2J(CCH) values, adding further support to the view that ^{13}C behave very much like ^{1}H where spin–spin coupling is concerned [see Section 5C(ii)].[85] Some observations concerning the anisotropy of ^2J(CHgC)[105] are reported in Section 3B(v).

(iii) *Elements of Group VB.* Berger and Roberts[99] have discussed the effect of the nitrogen lone pair on ^2J(NCC). Buchner and Wolfsberger[113] have measured ^2J(PCC) for a series of *N*-trimethylsilyl-triorganophosphine imines of general formula $R^1R^2R^3P$=NSiMe$_3$. For $R^1 = R^2 = R^3$ they find a steady change of ^2J(PCC) from -4.7 Hz for R = Et, through -3.8 Hz for R = Pr^n, Bu^i, and Bu^n, -2.4 Hz for R = Pr^i, and finally $+1.1$ Hz for R = Bu^t. Lequan, Pouet, and Sinonnin[63] have reported ^2J(PCC) values for phosphines and phosphine oxides. The couplings are very much larger in the latter.

(iv) *Elements of Group VIIB.* A rough correlation of ^2J(FCC) with F_{2s}—C_{2s} bond order has been found for a 9α steroid.[117]

[198] N. Platzer, J.-J. Basselier, and P. Demerseman, *Bull. Soc. chim. France*, 1974, 905.
[199] U. Ewers, H. Günther, and L. Jaenicke, *Chem. Ber.*, 1974, **107**, 3275.
[200] J. L. Marshall and R. Seiwell, *J. Magn. Resonance*, 1974, **15**, 150.
[201] P. E. Hansen, J. Feeney, and G. C. K. Roberts, *J. Magn. Resonance*, 1975, **17**, 249.
[202] R. D. Bertrand, J. W. Rathke, and J. G. Verkade, *Phosphorus*, 1973, **3**, 1.
[203] D. Hänssgen and W. Roelle, *J. Organometallic Chem.*, 1974, **71**, 231.
[204] D. H. Harris, M. F. Lappert, J. S. Poland, and W. McFarlane, *J.C.S. Dalton*, 1975, 311.
[205] H. R. Hudson, R. G. Rees, and J. E. Weekes, *J.C.S. Perkin I*, 1974, 982.
[206] A. G. Ginzburg, P. O. Okulevich, V. N. Setkina, G. A. Panosyan, and D. N. Kursanov, *J. Organometallic Chem.*, 1974, **81**, 201.
[207] R. J. Abraham, M. A. Cooper, T. M. Siverns, P. F. Swinton, H. G. Weder, and L. Cavalli, *Org. Magn. Resonance*, 1974, **6**, 331.
[208] R. J. Abraham, G. E. Hawkes, and K. M. Smith, *J.C.S. Perkin II*, 1974, 627.

(v) *Metals.* Two values of $^2J(WCC)$ (34.0 and 53.5 Hz) for some tungsten complexes have been reported,[103] and it has been observed that $^3J(PtNCC)$ is frequently greater than $^2J(PtNC)$.[209]

Other two-bond couplings to carbon may be found in the following references: $^2J(CCB)$ (ref. 106), $^2J(TlCC)$ (ref. 210), $^2J(CCC)$ (refs. 108, 110, and 111), $^2J(CNC)$ (ref. 108), $^2J(SnNSi)$ (ref. 112), $^2J(PCC)$ (refs. 100, 101, and 115), $^2J(PNC)$ (refs. 101, 143, and 211), $^2J(PPC)$ (ref. 114), $^2J(POC)$ (ref. 65), $^2J(FCC)$ (refs. 72, 80, 118, 119, and 212), $^2J(CdCC)$ (ref. 90), $^2J(SnCC)$ (ref. 212), $^2J(HgCC)$ (refs. 120 and 212), $^2J(PbCC)$ (ref. 212), $^2J(PtCC)$ (refs. 104, 121, and 124), and $^2J(PtAsC)$ (refs. 104 and 124).

D. Two-bond Coupling of Group VB Atoms.—Harris and his colleagues have reported many $^2J(PNP)$ values and have discussed them in some detail. For bis(phosphinoyl)amines (8a) and bis(phosphinothioyl)amines (8b) they find that $^2J(PNP)$ values are positive and in the range 11—44 Hz.[211] The sulphides show larger

(8) R^x = Cl, Me, NMe_2, Ph in various combinations
a; X = O
b; X = S

couplings than the oxides and the coupling is particularly sensitive to the group $PClNMe_2$. $^2J(PNP)$ is also found to be positive in the diazadiphosphetidines (6),[132] where it is very geometry-dependent and can be reasonably well reproduced by equation (11) in which $\lambda(P)$ and $\lambda(P')$ are group parameters for the PF_nR_{3-n} and

$$^2J(PNP') = \lambda(P)\lambda(P') \quad (11)$$

$PF_{n'}R_{3-n'}$ groups, respectively. It is important to note that these parameters do not contain additive substituent contributions. A large value of $^2J(PNP)$, +107.5 Hz, has been recorded for the zwitterion (9a).[147] The much smaller value of ±54.4 Hz

(9a)   (9b)

[209] S. T. Chow and R. B. Martin, *Inorg. Nuclear Chem. Letters*, 1974, **10**, 1131.
[210] L. Ernst, *J. Organometallic Chem.*, 1974, **82**, 319.
[211] G. Hägele, R. K. Harris, M. I. M. Wazeer, and R. Keat, *J.C.S. Dalton*, 1974, 1985.
[212] W. Adcock, B. D. Gupta, W. Kitching, D. Doddrell, and M. Geckle, *J. Amer. Chem. Soc.*, 1974, **96**, 7360.

is found for (9b) and this may be due to the fact that the ring is not planar. ^2J(PPP) values of different sign and widely different magnitudes are found in $(PCF_3)_5$ (5b).[127]

^2J(PSP) in the cage molecule P_4S_3 has been found to have a value of $+71 \pm 1.5$ Hz.[213]

Goodfellow and Taylor[214] have made an extensive study of ^2J(PMP) in complexes of Pd, Pt, Ir, Hg, Au, and Rh using $^1H-\{^{31}P\}$ INDOR experiments. They obtained both the signs and magnitudes of many of the couplings, and discuss their results in detail. All the *trans* couplings are positive and >250 Hz, whereas the *cis* couplings are much smaller and negative, with the exception of *cis*-[PdX$_2${P-(OMe)$_3$}$_2$] (X = Cl, Br, or I). Carty and Jacobsen[149] have reported measurements of ^2J(NPtN), ^2J(PPtN), and ^2J(PPtP) for square-planar platinum(II) thiocyanate complexes. The dependence of ^2J(PPtN) on geometry is very marked, the *cis* couplings lying between 0 and 7 Hz while the *trans* couplings have values of 63 to 95 Hz.

Further values of ^2J(PXP) may be found in the following references: ^2J(PNP) (ref. 133), ^2J(POP) (ref. 215), ^2J(PSP) (ref. 141) and ^2J(PMP) where M is a transition metal, (refs. 145, 149—151, and 153).

E. $^2J(^{31}PC^{19}F)$.—Cavell and his co-workers[141, 216] have discussed the values of ^2J(PCF) in phosphoranes. Axial CF_3 groups show very low (~ 50 Hz) couplings while equatorial CF_3 groups have couplings of the order of 150 Hz.

F. Two-bond ($^{19}F,^{19}F$) Couplings.—^2J(FCF) values for trifluoro-oxirans have been published[217] as have measurements of ^2J(FPF) in octahedral mixed fluorochlorides of phosphorus (V) anions.[134] Harris *et al.*[132] have reported that ^2J(FPF) in some diazadiphosphetidines (6) is negative.

Buslaev and Ilyin[218] have measured ^2J(FTaF) in tantalum fluorobromo and fluorochloro anion complexes, $[TaF_nX_{6-n}]^-$ (X = Cl or Br). The coupling is 36 Hz and is unchanged when the number of fluorine atoms in the complex is changed.

Other two-bond (F,F) couplings may be found in the following references: ^2J(FCF) (refs. 72, 80, 182, 219, and 220), ^2J(FPF) (refs. 139, 142, and 144), ^2J(FSF) (ref. 221), ^2J(FTeF) (ref. 161), ^2J(FXeF) (refs. 156, 157, and 163), ^2J(FTaF) (ref. 222), and ^2J(FWF) (ref. 163).

G. Miscellaneous Two-bond Couplings.—The dependence on concentration and temperature of ^2J(HgCF) in CF_3HgX [X = Cl, I, or $OC(O)CF_3$] has been investigated.[223]

---

[213] N. Zumbulyadis and B. P. Dailey, *Chem. Phys. Letters*, 1974, **26**, 273.
[214] R. J. Goodfellow and B. F. Taylor, *J.C.S. Dalton*, 1974, 1676.
[215] W. E. Morgan, T. Glonek, and J. R. van Wazer, *Inorg. Chem.*, 1974, **13**, 1832.
[216] D. D. Poulin and R. G. Cavell, *Inorg. Chem.*, 1974, **13**, 2324.
[217] K. W. Jolley, L. H. Sutcliffe, and K. L. Williamson, *Spectrochim. Acta*, 1974, **30A**, 1455.
[218] Yu. A. Buslaev and E. G. Ilyin, *J. Fluorine Chem.*, 1974, **4**, 271.
[219] V. J. Gazzard and R. K. Harris, *Org. Magn. Resonance*, 1974, **6**, 404.
[220] A. C. Jarvis and R. D. W. Kemmitt, *J. Organometallic Chem.*, 1974, **81**, 415.
[221] P. Meakin, D. W. Ovenall, W. A. Sheppard, and J. P. Jesson, *J. Amer. Chem. Soc.*, 1975, **97**, 522.
[222] Yu. A. Buslaev, Yu. V. Kokunov, V. D. Kopanev, and M. P. Gustyakova, *J. Inorg. Nuclear Chem.*, 1974, **36**, 1569.
[223] V. S. Petrosyan, V. I. Bakhmutov, and O. A. Reutov, *J. Organometallic Chem.*, 1974, **72**, 87.

Values of $^2J(\text{SiOSi})$,[81] $^2J(\text{PNSi})$,[113] $^2J(\text{RhCF})$,[220] $^2J(\text{RhPF})$,[146] and $^2J(\text{XeOTe})$[159] have been published.

5 Coupling between Nuclei Separated by Three Chemical Bonds, ^3J

A. Vicinal Proton–Proton Coupling via Carbon, $^3J(^1\text{HCC}^1\text{H})$.—The prediction of π bond order in the benzene ring of benzo[n]annulenes from measurements of $^3J(\text{HCCH})$ has been found useful;[32] The equation (12) was used.

$$P_{\mu v} = 0.104\ {}^3J(\text{H,H}) - 0.120 \qquad (12)$$

The measurement of $^3J(\text{HCCH})$ in ring systems has been the subject of several papers and some very detailed results have been published. Nakanishi and Yamamoto's[169,170] results for ethyleneimine are described in Section 5B. Lipnick[49,50] has reported values for a variety of cyclopentane derivatives, and Günther and his co-workers[180,224] have studied cyclohepta-1,3,5-trienes. They have used spin-tickling to determine signs, and have compared their results with theoretical discussions in the literature and also with EHMO calculations using Pople and Santry's formulation of the SOS theory. The agreement of theory and experiment is satisfactory for $^3J(\text{H,H})$ and $^5J(\text{H,H})$ but not for $^4J(\text{H,H})$. The dependence of the couplings on molecular geometry is discussed.

Cistaro et al.[171] have compared the $^3J(\text{HCCH})$ couplings in thietans with similar couplings in four-membered rings, and Ivin, Lillie, and Petersen[168] have compared $^3J(\text{HCCH})$ in thiirans and oxirans; the coupling is much larger in the former. Values of $^3J(\text{HCCH})$ have also been reported for alkylcyclohexanones[172,173] and oxathiolanes.[225]

Forrest[226] has investigated the relationship between gauche $^3J(\text{HCCH})$ couplings and substituent electronegativities. He deduces that equation (13) can be used to predict such couplings. In this equation $\Delta\chi_i$ is the difference between the Huggins

$$J^g(\text{H}_1,\text{H}_2) = (4.1 + 0.63 \sum_i \Delta\chi_i)(1 - 0.462\ \Delta\chi_1) \times (1 - 0.462\ \Delta\chi_2) \qquad (13)$$

electronegativity of H and that of the substituent characterized by i, $\sum_i \Delta\chi_i$ is the sum of these values for all substituents, and $\Delta\chi_1$ refers to the substituent trans to H_1 etc. In general the predictive power of equation (13) is good, but problems arise because of the dihedral angle dependence of the coupling. Forrest proposes that the equation should first be used to obtain the substituent electronegativity dependence of the coupling after which the dihedral angle dependence can be evaluated with a Karplus-type relationship. In a subsequent paper[227] Forrest applies this idea to the problem of $^3J(\text{HCCH})$ and dihedral angle in sugars. The results indicate that the modified approach has more general applicability than a Karplus equation in which the constants have been adjusted to fit the couplings observed for some members of a series of compounds.

Anteunis et al.[228] have expressed some doubt about the applicability of the

---

[224] H. Günther and M. Görlitz, *Org. Magn. Resonance*, 1974, **6**, 384.
[225] R. Keskineu, A. Nikkilä, K. Pihlaja, and F. G. Riddell, *J.C.S. Perkin II*, 1974, 466.
[226] T. P. Forrest, *Org. Magn. Resonance*, 1974, **6**, 355.
[227] T. P. Forrest, *Canad. J. Chem.*, 1974, **52**, 4095.
[228] M. Anteunis, J. Gelan, and R. Van Gauwenberghe, *Org. Magn. Resonance*, 1974, **6**, 362.

Karplus–Conroy equation to ring systems in some acetals, but Lambert and his co-workers[229] have suggested that the Karplus method is better than the *R*-value method for relating ^3J(HCCH) to geometry in five-membered rings. Their reason is that these rings do not possess the nominal three-fold symmetry of the Newman projection as six-membered rings do. Various methods of calculating dihedral angles in six-membered rings from vicinal (H,H) couplings have been discussed by Cazaux and Navech.[230]

The Karplus relationship between ^3J(HCCH) and dihedral angle ϕ has been strongly criticized by Vorontsova and Bochkov.[231] They suggest that the alternative equation (14) gives much improved results. In this equation ψ is the smaller

$$^3J(\text{HCCH}) = (4.37 + 4.39 \cos \psi + 3.32 \cos 2\psi) \times (1.28 - 0.041\ S) \quad (14)$$

angle between C—H and the line *ab* drawn parallel to C′—H′ as in Figure 1. *S* is the sum of the chemical shifts of the coupled protons in p.p.m. It remains to be seen if

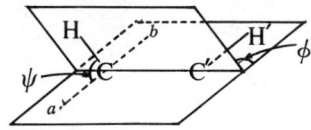

Figure 1

this equation will have the range of applicability which the authors forecast.

Elguero and Fuchier[232] have compared the conformational analysis of 2-pyrazoline derivatives by means of (i) the differences and (ii) the ratios of the vicinal coupling constants. The latter method had the better results although both were only approximate. The effect of the electronegativities of the nitrogen atoms and of their lone pairs is discussed.

Culvenor and Ham[233] have considered the relationship of the n.m.r. parameters *N* and *L* of the [AB]$_2$ system to the conformational populations of 1,2-disubstituted ethanes.

Other values of ^3J(HCCH) may be found in the following references: 22, 69—72, 80, 175, 177—179, 181—185, 200, 201, 207, and 234—240.

B. Three-bond Coupling of Protons *via* Other Atoms.—Theoretical work on ^3J-(HNCH),[14,18] ^3J(HOCH),[14] and ^3J(HPCH)[11] is described in Section 2B.

Nakanishi and Yamamoto[169, 170] have made a very accurate study of (H,H) coupling in ethyleneimine (10). They find the coupling constants given in Table 2, where

---

[229] J. B. Lambert, J. J. Papay, S. A. Khan, K. A. Kappauf, and E. S. Magyar, *J. Amer. Chem. Soc.*, 1974, **96**, 6112.
[230] L. Cazaux and J. Navech, *Org. Magn. Resonance*, 1975, **7**, 26.
[231] L. G. Vorontsova and A. F. Bochkov, *Org. Magn. Resonance*, 1974, **6**, 654.
[232] J. Elguero and A. Fruchier, *An. Quim.*, 1974, **70**, 141.
[233] C. C. J. Culvenor and N. S. Ham, *Austral. J. Chem.*, 1974, **27**, 2191.
[234] J. P. Zahra, B. Waegell, and H. Bodot, *Bull. Soc. chim. France*, 1974, 1107.
[235] G. Hägele, J. Richter, and M. Peach, *Z. Naturforsch.*, 1974, **29b**, 619.
[236] B. D. Lavrukhin, T. M. Filippova, and E. I. Fedin, *Org. Magn. Resonance*, 1974, **6**, 368.
[237] T. Ibusuki and Y. Saito, *Org. Magn. Resonance*, 1974, **6**, 436.
[238] F. G. Riddell and H. Labaziewicz, *Org. Magn. Resonance*, 1974, **6**, 599.
[239] R. J. Abraham, L. J. Kricka, and A. Ledwith, *J.C.S. Perkin II*, 1974, 1648.
[240] A. Schouteeten and M. Julia, *Tetrahedron Letters*, 1975, 607.

$$\begin{array}{c} H^1 \diagdown \diagup H^2 \\ H^4 \diagup \diagdown H^3 \\ | \\ N \\ | \\ H^5 \end{array}$$

(10)

they are compared with the corresponding values for N-methylaziridine, *i.e.* H-5 replaced by Me.

Zahra, Waegell, and Bodot[234] have discussed the dependence of ^3J(HOCH) upon geometry, and Solkan and Bystrov[17] have done the same for ^3J(HNCH) [see Section 5C(ii)].

Table 2 (H,H) *Coupling in ethyleneimine and N-methylaziridine. See structure* (10) *for numbering of protons.*

	$^2J_{1,4}$	$^3J_{1,2}$	$^3J_{3,4}$	$^3J_{1,3}$	$^3J_{1,5}$	$^3J_{3,5}$
Ethyleneimine	1.06	5.21	6.34	3.84	7.63	9.67
N-methylaziridine[a]	1.0	5.3	7.0	3.8		

[a] Couplings from T. Yonezawa and I. Morishima, *J. Mol. Spectroscopy*, 1968, **27**, 210.

Ader and Lowenstein[241] have reported values of ^3J(DCSiH), ^3J(DCGeH), ^3J(DSiCH), and ^3J(DGeCH). Their results correct an earlier error.[242]

Other values of three-bond (H,H) coupling may be found in the following references: ^3J(HSiCH) (ref. 87), ^3J(HSiSiH) (ref. 86), and ^3J(HNCH) (refs. 20 and 124).

C. Three-bond Coupling of Other Nuclei to Protons.—(*i*) *Elements of Group IIIB.* Negrebetskii *et al.*[92] have measured ^3J(BCCH) in organoboron compounds. They find that when boron is bound to a cyclopropyl ring ^3J(BCCH) is smaller when B is *trans* to the cyclopropyl H than when the two atoms are in *cis* relationship. Weibel and Oliver[186] have observed correlations between ^3J(TiSnCH) and ^2J(TlCH).

(*ii*) *Elements of Group IVB.* Solkan and Bystrov[17] have used INDO FPT methods to investigate ^3J(CNCH) and ^3J(CCNH) in peptide fragments (11) as a function of

$$\begin{array}{c} C^\beta \\ | \\ -C-N-C^\alpha-C- \\ \| | | \| \\ O H H O \end{array}$$

(11)

the HNCH dihedral angle, ϕ. Coefficients in equation (15) are given for the couplings ^3J(HNCH), ^3J(CNCH), ^3J(CCNH), ^3J(CNCC), and ^3J(NCCH). Hansen,

$$^3J = A \cos^2 \phi - B \cos \phi + C \sin^2 \phi \qquad (15)$$

[241] R. Ader and A. Lowenstein, *J. Amer. Chem. Soc.*, 1974, **96**, 5336.
[242] R. Ader and A. Lowenstein, *Mol. Phys.*, 1974, **27**, 1113.

Feeney, and Roberts[201, 243] have also investigated the use of (C,H) couplings for conformational analysis in peptides and aminoacids. They show that side-chain conformational information may be obtained from the coupling between the α-CO_2^- and the β-protons of the amino-acids (11). The ^3J(HCCH) couplings alone do not give this information. They conclude that (C,H) coupling constants will be very useful tools in this field, but a well-established Karplus-type curve for ^3J(CC'NH) is first required.

Marshall and his co-workers have made some very interesting comparisons of (C,C) and (C,H) coupling in tri-^{13}C-labelled tetrolic ester (12) and propyne (13),[85] and in crotonic (14) and isocrotonic (15) acids[200] labelled at the carboxy-group with ^{13}C. In the case of the former compounds they find that for nuclei in corresponding

$$H_3C^*-C^*\equiv C-C^*O_2CH_3$$
$$\quad\ 4\ \ \ \ 3\ \ \ 2\ \ \ \ \ 1$$
(12)

$$H_3C-C\equiv C-H$$
(13)

positions J(C,C)/J(C,H) and J(C,H)/J(H,H) both lie in the range 0.4 to 0.7. Thus, J(C,C)/J(H,H) = {J(C,C)/J(C,H)} × {J(C,H)/J(H,H)} = $(0.4)^2$ to $(0.7)^2$ = 0.1 to 0.5. This prediction seems to be in accord with the value of $J_{1,4}$ for (12) (1.84 Hz) and J(H,H) for acetylene which is +9.54 Hz. For the crotonic acids[200] extensive

tickling experiments gave unambiguous sign determination for all (C,H) couplings—assuming the (H,H) coupling signs – and, using these results and others from the literature, the authors show that there is a remarkably good correlation between geometrically equivalent J(C,H) and J(H,H) over three, four, and five bonds. The graph passes through the origin which, the authors suggest, indicates identical coupling mechanisms. These interesting relationships between J(H,H) and J(C,H) and between J(C,H) and J(C,C) have also been noted by Solkan and Bystrov,[17] who also find a proportionality between ^3J(H,H) and ^3J(N,H). But for oxiran derivatives of tin and lead Barbieri and Taddei[192] found no linear relationship between ^3J-(SnCCH), ^3J(PbCCH), and the corresponding (H,H) couplings, which contrasts strongly with the situation in furan and thiophen derivatives.

Takeuchi and Dennis[61] have used ^{13}C FT spectra to determine all (C,H) couplings in 2-, 3-, and 4-cyanopyridines. They find that ^3J(C,H) lies normally in the range 5—8 Hz but the coupling through nitrogen is considerably enhanced (12—13 Hz) while that through the substituted carbon is reduced to 3—5 Hz.

Further work on the ^{13}C spectrum of 1-cyanobicyclo[1,1,0]butane has lead Pomerantz and Hillenbrand[43] to the conclusion that the ^3J(CCCH) values show a dependence on factors other than dihedral angle. They propose that one such factor

[243] J. Feeney, P. E. Hansen, and G. C. K. Roberts, *J.C.S. Chem. Comm.*, 1974, 465.

might be the rear lobe of the C—H bond orbital interacting with the coupling carbon. A detailed account of (C,H) coupling in cyclobutadienyltricarbonyliron derivatives has been given.[64]

Linear relationships between two- and three-bond (Sn,H) and (Tl,H) couplings in the series $Li[(Me_3Sn)_nTlMe_{4-n}]$ have been reported.[186]

(*iii*) *Elements of Group VB*. Benassi et al.[244] have used the extended Hückel method with the Pople–Santry formalism to calculate values of $^3J(PCCH)$ in *p*-substituted derivatives of triphenylphosphine. The calculations show a qualitative agreement with experiment, and the effects of phosphorus *d*-orbital participation, change in the *s*-character of the P—C bond, and molecular conformation are discussed.

The geometry dependence of (P,H) and (N,H) couplings is a theme which has attracted the interest of several laboratories. The work of Solkan and Bystrov[17] on $^3J(NCCH)$ in peptide fragments has been noted in Section 5C(ii) above. A dependence of $^3J(PCCH)$ upon geometry has been observed in diphenyl-butylphosphine complexes of transition metals,[245] and for methylplatinum(IV) tertiary phosphine complexes the stereochemical dependence of the absolute sign of $^3J(PPtCH)$ has been determined.[246] It is found to be positive for coupling of phosphorus to a *cis* methyl proton and negative for the *trans* coupling. Clearly, these couplings are of considerable importance in structural studies but it should not be assumed that they will be governed by the same stereochemical factors as (H,H) coupling. Measurements on 2-phenyl-1,3,2-dithiaphosphorinane (16),[247] for example, suggest that $^3J(PSCH)$ may not follow a Karplus-type curve as $^3J(POCH)$ does.

$$\left\langle \begin{matrix} S \\ S \end{matrix} \right\rangle P\!-\!Ph$$

(16)

Hägele et al.[211] have reported values of $^3J(PNCH)$ for bis(phosphinoyl)amines (7a) and bis(phosphinothioyl)amines (7b). They find that the electronegativity of the phosphorus substituents has a very large effect on this coupling, which increases with substituent electronegativity. Russian workers have noted good correlations of $^3J(POCH)$ and $^3J(PNCH)$ in phosphoryl compounds with substituent constants[195] and have investigated the effect of solvent and complex formation on $^3J(POCH)$.[194]

Lequan, Pouet, and Simonnin[63] have measured $^3J(PCCH)$ in a series of acetylenic phosphines and phosphine oxides (4). They find that the coupling increases as the substituent groups become more electron-attracting, *cf.* the work of Hägele et al.[211] cited above, and that it is negative for phosphines and positive for phosphine oxides. In contrast to this, Clark and his colleagues[193] found that $^3J(PCCH)$ in diphenylalkenylphosphines and the related oxides did not change much in magnitude on going from P^{III} to P^V. Could there have been a change of sign?

---

[244] R. Benassi, M. L. Schenetti, F. Taddei, P. Vivarelli, and P. Dembech, *J.C.S. Perkin II*, 1974, 1338.
[245] C. H. Bushweller and M. Z. Lourandos, *Inorg. Chem.*, 1974, **13**, 2514.
[246] R. Bramley, J. R. Hall, G. A. Swile, and I. B. Tomkins, *Austral. J. Chem.*, 1974, **27**, 2491.
[247] J. Martin and J. B. Robert, *Org. Magn. Resonance*, 1975, **7**, 76.

(iv) **Elements of Group VIIB.** Emsley and Tabony[38] have made a theoretical study of the effect of temperature on (F,F) and (F,H) coupling. For ethyl fluoride ^3J-(FCCH) is predicted to be independent of temperature but the other couplings are expected to show appreciable changes over a temperature range of 200 K. This predicted temperature-independence of ^3J(FCCH) parallels the behaviour of ^3J-(HCCH) found experimentally, and it is suggested that temperature insensitivity may be a feature of any coupling constant which varies with dihedral angle in a way describable by a Karplus equation. However, an experimental study of 4-bromo-3-chloro-3,4,4-trifluorobut-1-ene (17) by Hinton and Jaques[72] shows a marked

$$\begin{array}{cccc} F^5 & F^4 & H^3 & H^1 \\ | & | & | & | \\ Br\!-\!C^4\!-\!C^3\!-\!C^2\!=\!C^1 \\ | & | & & | \\ F^6 & Cl & & H^2 \end{array}$$

(17)

temperature dependence for $^3J(F_4,H_3)$ in this molecule. Furthermore, the theoretically predicted[38] variation of the vicinal (F,F) coupling with temperature was not observed – see Section 5F(i). ^3J(FCCH) in chelate complexes of di(tertiary arsines) may be interpreted on a conformational basis.[248]

Other reports of three-bond coupling to hydrogen may be found in the following references: ^3J(TlCCH) (refs. 210 and 249), ^3J(CCCH) (refs. 68, 69, 72, 77, 79, 108, 198, and 199), ^3J(CSiCH) (ref. 81), ^3J(CNCH) (refs. 75 and 84), ^3J(SiOCH) (ref. 202), ^3J(SnNCH) (ref. 203), ^3J(NBNH) (ref. 88), ^3J(NCCH) (ref. 89), ^3J(PCCH) (refs. 40, 138, and 205), ^3J(PSiCH) (ref. 87), ^3J(PSnCH) (ref. 125), ^3J(PNCH) (refs. 133, 143, 216, 250, and 251), ^3J(POCH) (refs. 56, 65, 176, 184, 205, 250, 251, and 252), ^3J(PSCH) (ref. 142), ^3J(PRhCH) (ref. 146), ^3J(SeCCH) (ref. 79), ^3J(TeCCH) (ref. 79), ^3J(FCCH) (refs. 80, 182, 207, 234, and 235), ^3J(FPCH) (ref. 130), ^3J-(FPNH) (ref. 126), ^3J(CdCCH) (ref. 90), ^3J(HgCCH) (ref. 237), ^3J(PtCCH) (ref. 124), ^3J(PtPCH) (ref. 150), and ^3J(PtAsCH) (ref. 124).

D. Three-bond Coupling to Carbon.—(i) *Elements of Group IIIB.* The coupling ^3J(TlCCC) has attracted some attention[93, 208] and Kitching *et al.*[93] have found that in arylthallium trifluoroacetates $^3J(Tl,C_{meta}) > {}^2J(Tl,C_{ortho}) > {}^4J(Tl,C_{para})$.

(ii) *Elements of Group IVB.* The work of Solkan and Bystrov[17] on ^3J(CNCC) in peptide fragments has been referred to in Section 5C(ii), as has the comparison of ^nJ(C,C) and ^nJ(C,H) by Marshall and his colleagues.[85] Marshall and Miiller[91] have also made a study of (C,C) coupling in olefins. In contrast to ^2J(CCC), ^3J(CCCC) seems to be little affected by the presence of *one* C=C bond, but if the system is conjugated right through as in butadiene then ^3J(C,C) is much increased. Tran-Dinh

---
[248] W. R. Cullen, L. D. Hall, and J. E. H. Ward, *J. Amer. Chem. Soc.*, 1974, **96**, 3431.
[249] H. C. Bell, J. R. Kalman, J. T. Pinhey, and S. Sternhell, *Tetrahedron Letters*, 1974, 3391.
[250] D. B. Cooper, J. M. Harrison, and T. D. Inch, *Tetrahedron Letters*, 1974, 2697.
[251] G. H. Cooper and R. A. Chittenden, *Org. Magn. Resonance*, 1974, **6**, 563.
[252] A. Cogne, A. G. Guimaraes, J. Martin, R. Nardin, J.-B. Robert, and W. J. Stec, *Org. Magn. Resonance*, 1974, **6**, 629.

et al.[110] find that amino-acids with more than four carbon atoms show large values of $^3J(C_\alpha,C_\gamma)$, while $^3J(C_oC_\gamma)$ is <1.5 Hz.

(*iii*) *Elements of Group VB*. Di Blasi and Kopple[253] have examined $^3J(NCCC)$ in peptides and they confirm that the coupling is indeed very small. Three such couplings, for which the dihedral angle is constrained to 110° or 160°, are smaller than 1 Hz. This small value does not therefore appear to be due to rotational averaging and the use of these couplings in conformational studies will be much smaller than might have been hoped. The effect of the nitrogen lone pair on $^3J(NCCC)$ has been discussed by Berger and Roberts.[99]

Bertrand *et al.*[202] have determined the magnitudes and relative signs of the couplings in the cyclic phosphine (18) and have compared them with those of other bicyclo[2,2,2]octane systems.

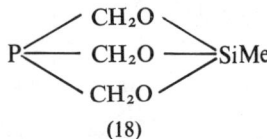

(18)

(*iv*) *Elements of Group VIIB*. Giannini *et al.*[117] have noted that in 9α-steroids, $^3J(FCCC)$ to the carbon *trans* to fluorine is much greater than the coupling to the *gauche* carbon atom. Subbotin and Sergeyev[119] find that $^3J(FCCC)$ for axial and equatorial conformers of fluorocyclohexane is ∼1 and 11.5 Hz, respectively. Inspection of appropriate models shows that these observations are effectively the same. A rough correlation of $^3J(FCCC)$ with F_{2s}—C_{2s} bond order was also observed.[117]

(*v*) *Metals*. $^3J(HgCCC)$ shows a very marked dependence on dihedral angle, with a maximum at 180° (265—285 Hz) and a minimum (< 10 Hz) at *ca*. 90°.[254] It appears that $^3J(M,C)$ values for M = Sn, Hg, Pb, and Tl all show a similar dependence on geometry.[254] The fact that metal–carbon couplings frequently do not fall off uniformly with increasing number of intervening bonds has been noted for (Hg,C) by Casanova *et al.*[120] and for (Pt,C) by Chow and Martin.[209] Values of $^2J(WCC)$ (34 and 53.5 Hz) and of $^3J(WCCC)$ (100 Hz) show that this is not invariably so.[103] It has been suggested[90] that in the case of $^nJ(Cd,C)$ there may be an alternation of sign in the couplings, n = 1 being negative, but more evidence on this point is required.

Other values of three-bond coupling to carbon may be found in the following papers: $^3J(CCCB)$ (ref. 106), $^3J(TlCCC)$ (ref. 210), $^3J(CCCC)$ (ref. 111), $^3J(CNCC)$ (ref. 108), $^3J(SnNSiC)$ (ref. 112), $^3J(PCCC)$ (refs. 100, 101, 113, and 115), $^3J(PNCC)$ (ref. 101), $^3J(PNNC)$ (ref. 101), $^3J(PNSiC)$ (ref. 113), $^3J(FCCC)$ (refs. 80 and 212), $^3J(SnCCC)$ (ref. 212), $^3J(HgCCC)$ (ref. 212), $^3J(PbCCC)$ (ref. 212), and $^3J(PtCNC)$ (ref. 124).

E. **Three-bond Coupling to Phosphorus.**—(*i*) *Phosphorus*. Taylor, Keiter, and Cary[152]

---

[253] R. Di Blasi and K. D. Kopple, *J.C.S. Chem. Comm.*, 1975, 33.
[254] W. Kitching, D. Praeger, D. Doddrell, F. A. L. Anet, and J. Krane, *Tetrahedron Letters*, 1975, 759.

have measured ^3J(PCCP) in ligands of the form $Ph_2PCH=CHPPh_2$ and in tungsten carbonyl complexes of those ligands. The magnitudes of the observed couplings cannot be accounted for by simple qualitative arguments and the signs of the couplings appear to vary from compound to compound.

(ii) Fluorine. Harris, Schmutzler, and their co-workers have reported further investigations of diazadiphosphetidines (6).[132,133] ^3J(PNPF) is greatly affected by geometry, and *gauche-* and *trans-*isomers can show opposite signs for this coupling.[132] There are marked similarities in the behaviour of ^1J(PF) and ^3J(PNPF),[132,133] and it is suggested that the factors controlling these two couplings are the same. The FC term may be the dominant influence, but if it is the differing signs of ^3J(P,F) require that changes in mutual polarizability rather than *s*-electron density predominate.

Values of ^3J(PNiPF) in complexes of general form $Ni(PF_3)_{4-n}(PF_2NR_2)_n$ (R = alkyl group or H) are reported to lie in the range 30—45 Hz.[140]

Other three-bond couplings to phosphorus can be found in the following references: ^3J(PCCP) (refs. 148 and 153), ^3J(PNPF) (ref. 147), ^3J(PSPF) (ref. 141), and ^3J(PFePF) (ref. 145).

F. Three-bond Coupling to Fluorine.—*(i) Fluorine.* Emsley and Tabony[38] have predicted a temperature dependence for ^3J(FCCF) in CF_3CH_2F and CF_3CFO but this is not observed experimentally in 4-bromo-3-chloro-3,4,4-trifluorobut-1-ene (17)[72] although the coupling is found to be solvent-dependent, a fact which may prove useful in sign determination.[72] However, the predicted[38] change in ^3J(FCCF) over 200 °C is rather small, $+0.55$ Hz for CF_3CFH_2 and 0.1—0.2 Hz for CF_3CFO, and bearing in mind the many doubts concerning the calculation of (F,F) couplings by MO methods[37] and the effect of substituents on this couplings, no firm conclusions can be drawn from this comparison.

Measurements on *cis-*difluoroethylene in the nematic phase[255] have shown that ^3J(FCCF) is negative. Calculations of this coupling constant give values between $+47.6$[7] and -16.6,[256] which illustrates rather well the point made above. The experimentally determined magnitude of the coupling lies between 19 and 20 Hz, depending on solvent *etc.*

Values of ^3J(FCCF) from the ^{19}F spectrum of *cis-*1,2-dichlorohexafluorocyclobutane are in accord with the suggestion that *cis* ^3J(FCCF) are negative while the *trans* couplings are probably positive.[219] Jolley *et al.*[217] have discussed (F,F) coupling in trifluoro-oxirans.

Fild and Stankiewicz[139] have noted that in compounds of the form $R(C_6F_5)PF_3$ (R = Bu^t, Ph, or C_6F_5), the coupling ^3J(FPCF) to the *ortho* fluorine atom is only observed for the axial fluorine on phosphorus and not for the equatorial fluorine.

(ii) Xenon. Seppelt and Rupp[159] have measured ^3J(Xe,F) values in systems such as $Xe(OTeF_5)_2$ and $Xe(OSeF_5)_2$. The coupling of xenon to the equatorial fluorine on tellurium or selenium is observed to have a value of 30—37 Hz, but the coupling to the axial fluorine is not seen. The authors suggest that this is due to the large

[255] J. W. Emsley and J. C. Lindon, *Mol. Phys.*, 1974, **28**, 1253.
[256] K. Schaumburg, private communication to the authors of reference 255.

distance which separates the axial, as opposed to the equatorial, fluorine from the xenon atom.

(*iii*) *Metals*. Dixon, Moss, and Smith[257] have reported an excellent linear correlation (16) of ^3J(PtSCF) *cis* with ^3J(PtSCF) *trans* for some platinum(II) complexes of SCF_3.

$$^3J_{cis} = -0.25\ ^3J_{trans} + 92.0 \qquad (16)$$

Other reports of three-bond coupling to fluorine may be found in the following papers: ^3J(FCCF) (refs. 182, 207, 220, and 221), ^3J(FNNF) (ref. 137), ^3J(SnCCF) (ref. 258), and ^3J(PtSCF) (ref. 259).

6 Coupling between Nuclei Separated by Four Chemical Bonds, ^4J

A. Proton–Proton Coupling *via* Carbon, ^4J(^{1}HCCC^{1}H).—The detailed experimental and theoretical work on this coupling by Barfield *et al.*[19] is described in Section 2B.

Marshall and his co-workers[85] have made a comparison of ^nJ(C,C) with ^nJ(C,H) in tri-^{13}C-labelled tetrolic ester (12) and propyne (13), and have also made a careful study of norcamphor.[179] The observed values of ^4J(HCCCH) in the last compound have been compared with theoretical analyses of the coupling, and it is found that only in the case of the large 'W' couplings are theory and experiment in good agreement. This conclusion may need to be modified in the light of the most recent theoretical work.[19] The authors suggest that strain may be responsible for the differences between theory and experiment.[179]

Günther, Görlitz, and Meisenheimer[180, 224] have made extensive investigations of (H,H) coupling in cycloheptatrienes using spin tickling to determine relative signs. They have compared their results with theoretical work from the literature and also with EHMO calculations using the Pople–Santry approach. The agreement between theory and experiment is satisfactory for ^3J and ^5J but not for ^4J. The dependence of the couplings on molecular geometry is discussed.

Chalmers and Hall[260] have found that for O-acetylated D-glycals and related molecules, ^4J[H(2),H(4)] is the coupling most sensitive to conformational change. For 6-oxo-7-unsaturated steroids the configuration at C-5 can be obtained from the four-bond coupling between the protons at C-7 and C-5.[261] If the C-7 proton resonance is split into a doublet then the configuration at C-5 is α; if the resonance is not split then the configuration is β. Some unusual values of ^4J(HCCH) have been reported for some dioxan derivatives,[262] and Barieux *et al.*[175] have measured ^4J(HCCCH) in cyclohexanones at 220 MHz.

Morris and Murray[263] have found a good correlation between ^4J(H,H) and σ_I in 4-substituted boranones, and the influence of solvent on ^4J(HCCCH) in pyrrole carbaldehydes has been discussed.[264]

[257] K. R. Dixon, K. C. Moss, and M. A. R. Smith, *Inorg. Nuclear Chem. Letters*, 1974, **10**, 373.
[258] H.-J. Kroth, H. Schumann, H. G. Kuivala, C. D. Schaeffer, and J. J. Zuckerman, *J. Amer. Chem. Soc.*, 1975, **97**, 1754.
[259] K. R. Dixon, K. C. Moss, and M. A. R. Smith, *J.C.S. Dalton*, 1974, 971.
[260] A. A. Chalmers and R. H. Hall, *J.C.S. Perkin II*, 1974, 728.
[261] W. B. Smith and G. P. Newsoroff, *Steroids*, 1974, **23**, 579.
[262] E. Bernaert, M. Anteunis, and R. de Waele, *Bull. Soc. chim. belges*, 1973, **82**, 795.
[263] D. G. Morris and A. M. Murray, *Org. Magn. Resonance*, 1974, **6**, 510.
[264] M. Farnier and T. Drakenberg, *J.C.S. Perkin II*, 1975, 333.

Other reports of ^4J(HCCCH) may be found in references 69, 181, 183, 185, 200, 235, 238, and 240.

B. Four-bond Coupling of Protons via Other Atoms.—Cistaro et al.[171] have reported a study of proton–proton coupling in thietan 1-oxide and thietan 1,1-dioxide at 300 and 100 MHz. The ^4J(H,H) values in systems of the form (19), which have effective C_{2v} symmetry either because the molecule is planar or because of rapid intercon-

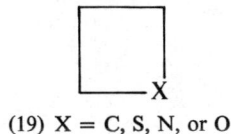

(19) X = C, S, N, or O

version between non-planar conformers, behave according to theoretical prediction.[265] Thus, $^4J_{cis}$ becomes more positive as the electronegativity of X is increased while $^4J_{trans}$ becomes more negative.[171] There is a strong steric dependence of substituent effects in the bent structures.[171]

The theoretical work of Ostlund and Pruniski[18] on ^4J(HNCCH) in molecules related to peptides is described in Section 2B.

Other reports of four-bond proton–proton coupling may be found in the following papers: ^4J(HCOCH) (ref. 181) and ^4J(HSiPCH) (ref. 87).

C. Four-bond Coupling of Other Nuclei to Protons.—(i) *Elements of Group IVB*. Takeuchi and Dennis[61] have measured (C,H) coupling in cyanopyridines. They set an upper limit of 2 Hz on ^4J(C,H). The work of Marshall and his colleagues comparing (C,C), (C,H), and (H,H) coupling[85, 200] is described in Section 5C(ii).

(ii) *Elements of Group VB*. Some INDOR experiments have shown that ^4J(PCCOH) is positive for the *RS,RS* diastereoisomers of β-hydroxyphosphonate esters and a 'W' geometry makes the coupling more positive.[266] No correlation was found between experimental values of ^4J(PCCCH) for *p*-substituted derivatives of triphenylphosphine and calculated values using EHMO's and Pople–Santry theory.[244]

Other reports of four-bond coupling to protons may be found in the following papers: ^4J(TlCCCH) (refs. 210 and 249), ^4J(SnNSiCH) (ref. 112), ^4J(PCCCH) (ref. 205), ^4J(POCCH) (ref. 184), ^4J(PSCCH) (ref. 247), ^4J(POPCH) (ref. 138), ^4J(PSPCH) (ref. 138), ^4J(FCCCH) (refs. 72 and 235) and ^4J(FPSCH) (ref. 142).

D. Four-bond Coupling to Carbon.—Reports of four-bond coupling to carbon may be found in the following papers: ^4J(TlCCCC) (refs. 93 and 210), ^4J(CCCCB) (ref. 106), ^4J(CCCCC) (ref. 111), ^4J(PCCCC) (refs. 100, 101, and 113), ^4J(POCCC) (ref. 267), ^4J(SnCCCC) (ref. 212), ^4J(PbCCCC) (ref. 212), ^4J(HgCCCC) (refs. 120 and 212), and ^4J(FCCCC) (refs. 119 and 212).

---

[265] M. Barfield, R. J. Spear, and S. Sternhell, *J. Amer. Chem. Soc.*, 1971, **93**, 5322.
[266] T. Bottin-Strzalko, M. J. Pouet, and M. P. Simonnin, *Org. Magn. Resonance*, 1974, **6**, 419.
[267] R. H. Sarma, C. H. Lee, F. E. Evans, N. Yathindra, and M. Sundaralingam, *J. Amer. Chem. Soc.*, 1974, **96**, 7337.

E. **Four-bond Coupling to Fluorine.**—The theoretical work of Abraham et al.[34] is described in Section 2C. Emsley and Tabony[38] have predicted that ^4J(FCCCF) in $CF_3CF=CF_2$ will show an appreciable change over a temperature range of 200 K [see also Section 5F(i)].

The (F,F) coupling in (20) has been investigated by Meakin, Ovenall, and Shep-

(20)

pard.[221] The coupling between the *meta* fluorines is linearly related to the chemical shift of the *para* fluorine, as others have also noticed. At -66 °C in toluene $^4J(F_b,F_o') = \mp 6.8$ Hz and $^4J(F_b,F_o) = \pm 31.2$ Hz. For *cis*-1,2-dichlorohexafluorocyclobutane ^4J(FCCCF) is negative with the *trans* coupling greater in magnitude than the *cis*.[219] The (F,F) couplings in trifluoro-oxirans have been discussed.[217] For complexes of the general form $Ni(PF_3)_{4-n}(PF_2NR_2)_n$ (R = alkyl or H), ^4J(FPNiPF) lies between 4.3 and 6.4 Hz.[140]

The work of Schumann and Cavagna[268] is reported in Section 11.

Other reports of four-bond coupling to fluorine may be found in the following papers: ^4J(FCCCF) (ref. 220), ^4J(FCOCF) (ref. 269), ^4J(FPNPF) (refs. 132 and 147), ^4J(FPSPF) (ref. 141), ^4J(PPtSCF) (ref. 259), ^4J(PPdSCF) (ref. 259), and ^4J(SnCCCF) (refs. 258 and 270).

7 Coupling between Nuclei Separated by Five Chemical Bonds, ^5J

A. **Proton–Proton Coupling *via* Carbon, $^5J(^1HCCCC^1H)$.**—The work of Günther et al.[180, 224] on (H,H) coupling in cycloheptatrienes is described in Sections 5A and 6A.

Riddell and Labaziewicz[238] have reported a study of (H,H) coupling in the 3,6-dihydro-1,2-oxazine ring (21). The changes of ^5J(HCCCCH) with substituent,

(21)

R, suggest that there may be some coupling through the O—N path.

Other reports of ^5J(HCCCCH) can be found in refs. 235 and 264.

B. **Five-bond Proton–Proton Coupling *via* Other Atoms.**—Rowbotham and Schaefer

---

[268] C. Schumann and F. Cavagna, *J. Magn. Resonance*, 1975, **18**, 172.
[269] F. A. Hohorst, J. V. Paukstelis, and D. D. Des Marteau, *J. Org. Chem.*, 1974, **39**, 1298.
[270] M. Barnard, P. J. Smith, and R. F. M. White, *J. Organometallic Chem.*, 1974, **77**, 189.

have discussed[271] ^5J(HOCCCH) in halogenophenols in connection with intramolecular H-bonding in those molecules. The influence of solvent on ^5J(HCCCCH) and ^5J(HCNCCH) in pyrrolecarbaldehydes has been investigated.[264]

C. **Five-bond Coupling of Other Nuclei to Protons.**—(*i*) *Elements of Group IVB.* The work of Zeisberg and Bohlmann[272] on ^nJ(C,H) in polyynes is described in Section 10D.

(*ii*) *Elements of Group VB.* Tseng and Mihailovski[273] have reported values of ^5J(P,H) for iminothiazolines and imino-oxazolines (22). They find that the magnitudes of the

H\\ /S\\   X
               ||
   C=C—N—P—R²
H/ N   R³
    |
    R¹

(22) X = O, S, or lone pair
R¹ = alkyl group
R², R³ = SMe, OMe, SEt, OEt, Ph, Et

couplings decrease in the following series; (i) $|\,^5J(\text{PNCNCH})| > |\,^5J(\text{PNCSCH})| > |\,^5J(\text{PNCOCH})|$; (ii) for $|\,^5J(\text{PNCNCH})|$ P > PS > PO; (iii) for $|\,^5J(\text{PNCSCH})|$ PO ≃ PS > P. There are insufficient data for ^5J(PNCOCH). The differences in the orders (ii) and (iii) might be explained by a difference of sign but unfortunately the signs are unknown. However, the magnitudes of the couplings, 0.46—4.89 Hz, are sufficiently small to make changes of sign a reasonable possibility.
^5J(PCOSiCH) in (18) = −0.32 Hz.[202]

Other reports of five-bond coupling to protons may be found in the following papers: ^5J(TlCCCCH) (refs. 210 and 249), ^5J(FCCCCH) (ref. 72), ^5J(FCCSiCH) (ref. 274), ^5J(FCCCNH) (ref. 275), ^5J(FCPNCH) (ref. 216), and ^5J(FCOOCH) (ref. 269).

D. **Five-bond Coupling to Carbon.**—Adcock *et al.*[93, 212] have continued their active investigations of the coupling of Tl, Sn, Pb, Hg, and F to carbon in organometallic systems. They have recently published an investigation of the contribution of a σ–π hyperconjugative mechanism to five-bond tin–carbon couplings.[212] They find that for X = Sn, Pb, or Hg, $|\,^5J(X,C)|$ is always greater than $|\,^4J(X,C)|$ for benzylmetallic systems, and that for planar systems the coupling is much less than for non-planar molecules. They propose that the ^5J coupling is being determined predominantly by a π-electron mechanism which is enhanced when direct access to the metal *s*-orbitals is possible *via* σ–π interactions.[212] This proposal might well be examined theoretically as well as experimentally.

[271] J. B. Rowbotham and T. Schaefer, *Canad. J. Chem.*, 1974, **52**, 3037.
[272] R. Zeisberg and F. Bohlmann, *Chem. Ber.*, 1974, **107**, 3800.
[273] C. K. Tseng and A. Mihailovski, *Org. Magn. Resonance*, 1974, **6**, 494.
[274] G. Hägele and V. Lueg, *Z. Naturforsch.*, 1975, **30a**, 175.
[275] H. Fritz and T. Winkler, *Helv. Chim. Acta*, 1974, **57**, 836.

The work of Ernst[210] on (Tl,C) coupling and that of Hansen et al.[111] on ^nJ(C,C) is described in Section 10A. Casanova, Rogers, and Servis[120] have reported values of ^5J(HgCCCCC) for symmetrical dialkylmercury compounds.

E. Five-bond Coupling to Fluorine.—Barnard, Smith, and White[270] have compiled a table of ^nJ(Sn,F) values for $n = 1$—5 and have discussed the trends in these couplings. The work of Schumann and Cavagna[268] on ^5J(F,F) is described in Section 11.

Other values of five-bond couplings to fluorine may be found in the following papers: ^5J(TlCCCCF) (ref. 93), ^5J(SnCCCCF) (ref. 258), ^5J(FCCCCF) (refs. 221 and 235), ^5J(FPSPCF) (ref. 141), and ^5J(FCOOCF) (ref. 269).

8 Coupling between Nuclei Separated by Six Chemical Bonds, ^6J

The systematic study of long-range couplings has attracted rather little attention in the past year. The important work in this subject has been concerned with π-electron systems and is described in Section 10.

Hägele and Lueg[274] have described a method of bandshape analysis based upon Lorentzian lineshapes. They have illustrated their technique with a determination

(23)

(24) R^1 = H or Me
R^2 = H, Me, p-ClC$_6$H$_4$, or o-ClC$_6$H$_4$

of ^5J(FCCSiCH) and ^6J(FCCSiSiCH) in (23). The value for ^6J(F,H) of 1.1 Hz recorded by Fritz and Winkler[275] for (24) with $R^1 = R^2$ = Me may well be a through-space coupling. Long-range (F,F) couplings have been reported for (CF$_3$)$_2$P(S)SP(CF$_3$)$_2$[141] and C$_6$F$_5$SF$_3$ (20).[221]

The work of Hansen et al.[111] on (C,C) coupling in pyrene and its derivatives, and the study of (C,H) coupling in polyynes by Zeisberg and Bohlmann[272] are reported in Section 10.

Ernst[210] has noted a dependence of ^6J(Tl,C) in alkyl-substituted arylthallium trifluoroacetates on alkyl-group conformation. He attributes the effect to changes in the π-electron contribution to the coupling.

9 Coupling between Nuclei Separated by Seven Chemical Bonds, ^7J

The ^{1}H n.m.r. spectrum of 1,2,4,5-hexatetraene (25) in a nematic solvent reveals values of 3.5 Hz for both seven-bond (H,H) couplings.[276]

---

[276] B. Pederson, J. Schaug, and H. Hopf, *Acta Chem. Scand.* (A), 1974, **28**, 846.

(25)   (26)

10 Coupling in π-Electron Systems

A. Homoaromatic Molecules.—(i) *Proton–Proton coupling.* Mathieu, Milano, and Douris[277] have measured three-, four-, and five-bond (H,H) couplings in fluorene derivatives (26). They find good correlations of $^3J(3,4)$ and $^4J(1,3)$ with substituent electronegativity, but $^5J(1,4)$ shows no such relationship.

Reynolds and co-workers[22] have studied the effect of substituents on coupling to vinyl protons in styrenes. Theory and experiment are in agreement concerning the fact that couplings between ring and vinyl protons are independent of substituent. For benzaldehyde derivatives, where the experimental couplings do change with substituent, the calculations suggest that this is due to substituent-induced changes in the C-1—C_α bond length which effects both the σ- and the π-electron contributions to the coupling. Within the vinyl fragment geminal (H,H) couplings show a substituent dependence whereas the *cis* and *trans* (H,H) couplings vary only slightly. The authors conclude that the substituents affect these couplings by a through-space rather than through-bond mechanism.

The benzylic coupling constants in α-tetralone show quite good agreement with the theoretical predictions of Wasylishen and Schaefer[278] concerning the effect on the coupling of the angle between the C—H bond and the benzene ring.[279]

Other homoaromatic systems for which (H,H) couplings have been reported include benzonitrile,[280] acetylated [3]ferrocenophanes,[281] a variety of methylnaphthalenes,[282, 283] and *p*-methylbenzyl bromide.[24]

(ii) *Coupling of Other Nuclei to Protons.* Investigations of the effect of substituents on $^1J(CH)$ in aromatic systems have been reported from two laboratories. In the vinyl group of substituted styrenes the coupling shows strongly stereospecific substituent effects, and theory indicates that this is due to a through-space interaction.[22] In substituted fluorobenzenes $^1J(CH)$ for a particular substituent orientation increases linearly with the electron-withdrawing power of the substituent.[73] For a particular substituent the $^1J(CH)$ values are largest for the *para* and smallest for the *meta* isomer.[73]

Indirect (C,H) couplings for 1,4-dibromo[^{13}C]benzene partially orientated in a nematic phase have been determined from the spectra of samples with three different

---

[277] A. Mathieu, J. C. Milano, and J. Douris, *Bull. Soc. chim. France*, 1974, 299.
[278] R. Wasylishen and T. Schaefer, *Canad. J. Chem.*, 1972, **50**, 1852.
[279] K. Gatto, J. D. Reinheimer, K. Shafer, and J. T. Gerig, *Org. Magn. Resonance*, 1974, **6**, 577.
[280] J. P. Jacobsen and K. Schaumburg, *Mol. Phys.*, 1974, **28**, 1505.
[281] R. R. McGuire, R. E. Cochoy, and J. A. Winstead, *J. Organometallic Chem.*, 1975, **84**, 269.
[282] V. Lucchini and P. R. Wells, *J.C.S. Perkin II*, 134, 1434.
[283] N. K. Wilson and J. B. Stothers, *J. Magn. Resonance*, 1974, **15**, 31.

degrees of orientation.[284] (C,H) couplings have also been obtained for benzonitrile from measurements in isotropic and nematic phases.[280]

With the help of shifts induced by solution in [^{2}H$_6$]benzene Sorensen and Jakobsen[285] have analysed completely the 100 MHz spectrum of triphenylphosphine. The magnitudes of the ^nJ(P,H) couplings are very similar to those of the corresponding ^nJ(H,H) in benzene, and the phosphorus substituent effect on the (H,H) coupling is negligible. The (P,H) couplings show the expected dependence upon the angle of twist of the benzene rings.

The influence of substituents on (F,F) and (F,H) coupling in polyfluorinated α-fluorodiphenylmethyl cations has been studied,[286] and the ^{19}F spectra of some substituted fluoronaphthalenes have been discussed.[287]

(*iii*) *Coupling to Carbon.* Ernst[210, 280] has reported some detailed investigations of (Tl,C) coupling over two to six bonds in arylthallium(III) bis(trifluoroacetates)

$$\begin{array}{c} \text{F}_3\text{CCO} \\ \diagdown \\ \text{Tl}\text{—}\text{C}_6\text{H}_5 \\ \diagup \\ \text{F}_3\text{CCO} \end{array}$$

(27)

(27). The couplings show characteristic ranges. For the ring carbons; ^2J(Tl,*ortho* C) +460 to +570 Hz, ^3J(Tl,*meta* C) +990 to +1100 Hz, ^4J(Tl,*para* C) −170 to −220 Hz. For couplings to the α-carbons of substituent alkyl groups we have ^3J(Tl,*ortho* αC) +430 to +460 Hz, ^4J(Tl,*meta* αC) +75 to +95 Hz, ^5J(Tl,*para* αC) +75 to +115 Hz. The (Tl,C) couplings over two, three, and four bonds parallel the (C,C) couplings in sodium benzoate, which suggests a similar mechanism for the two types of coupling.[288] ^6J(Tl,C) was observed and shows a dependence on alkyl group conformation which is probably due to the contribution of π-coupling.[210] ^1J(T1C) was not seen.

Marshall, Ihrig, and Müller[109] have measured (C,C) couplings over one to four bonds in anthracene derivatives. ^3J(*cisoid*) is approximately twice as large as ^3J(*transoid*), in marked contrast to the case of (H,H) coupling. However, Weigert and Roberts[289] have proposed that two mechanisms contribute to the *cisoid* coupling, and if these are assumed to contribute equally then the experimental observation may be explained. If the ^3J(*cisoid*) values are halved then ^2J(C,C) and ^3J(C,C) show linear correlations with the sums of the π bond orders of the bonds in the coupling paths. The large ^4J(C,C) values observed (1.72 and 1.92 Hz) may also be thought to have effectively two contributing paths. If they are divided by two then the results fall within the range of previously determined ^4J(C,C) values.

[284] E. E. Burnell and M. A. J. Sweeney, *Canad. J. Chem.*, 1974, **52**, 3565.
[285] S. Sørensen and H. J. Jakobsen, *Acta Chem. Scand.* (*A*), 1974, **28**, 249.
[286] Yu. V. Pozdnyakovich and V. D. Shteingarts, *J. Fluorine Chem.*, 1974, **4**, 296.
[287] L. V. Vlasova, L. S. Kobrina, and N. V. Popkova, *Izvest. Sib. Otdel Akad. Nauk S.S.S.R., Ser. khim. Nauk*, 1974, 112.
[288] L. Ernst, *Org. Magn. Resonance*, 1974, **6**, 540.
[289] F. J. Weigert and J. D. Roberts, *J. Amer. Chem. Soc.*, 1972, **94**, 6021.

Hansen et al.[111] have determined (C,C) coupling over one to six bonds in pyrene and pyrene derivatives, and have reported an unusually small cisoid ^3J(CCCC) of 0.97 Hz. There is some evidence of an alternation in the magnitude of ^nJ with n.[111]

Singh and Singh[73] have reported that ^1J(FC) in substituted fluorobenzenes does depend both on substituent and on position, but the variations are not as regular as they are in the case of ^1J(CH) [see Section 10A(ii)].

(iv) (^{19}F,^{19}F) Coupling. Lustig, Hansen, and Lincoln[290] have given a complete analysis of the $[AX]_2Z$ spectra of the pentafluorobenzene moiety. They show that if it is assumed that ortho couplings are negative and greater in magnitude than meta or para couplings, then two tickling experiments allow the spectrum to be completely analysed and all couplings to be assigned an absolute sign. They illustrate the method with analyses of the ^{19}F spectra of pentafluorophenylthiocyanate, pentafluorobenzene sulphenyl chloride, and decafluoroazoxybenzene.[290]

Shteingarts and co-workers have investigated the influence of substituents on ^{19}F coupling constants in polyfluorinated α-fluorophenylmethyl cations,[286] and have also reported (F,F) couplings for carbocations of the form (28) substituted with fluorine at various positions.[291]

(28)

B. Heterocyclic Molecules.—(i) *Proton–Proton Coupling.* Jokisaari[21, 292] has reported INDO FPT calculations of (H,H) and (C,H) coupling in oxetans. Danchura et al.[25] have examined the relationship between the ^{1}H n.m.r. spectra of pyridine aldehydes and their conformational equilibria in solution. The long-range coupling constants are found to be a precise measure of this property. The experimental couplings are compared with theoretical values calculated using INDO/CNDO FPT methods. Deady, Harrison, and Topsom[293] have also measured (H,H) couplings in pyridine derivatives.

Turbini and Porter[88] have compared ^3J(HNBH) and ^4J(HNBNH) in a series of ^{15}N-labelled borazine derivatives with the corresponding couplings in benzene derivatives. They find no obvious correlation between the two sets of values. Further (H,H) couplings have been reported for strychnine derivatives,[294, 295] thiophen complexes of chromium,[296] substituted quinazoline diones,[297] and coumarin partially orientated in a nematic phase.[298]

---

[290] E. Lustig, E. A. Hansen, and D. N. Lincoln, *J. Magn. Resonance*, 1974, **15**, 1.
[291] B. G. Oksenenko and V. D. Shteingarts, *Zhur. org. Khim.*, 1974, **10**, 1190.
[292] J. Jokisaari, *Z. Naturforsch*, 1974, **29a**, 1902.
[293] L. W. Deady, P. M. Harrison, and R. D. Topsom, *Org. Magn. Resonance*, 1975, **7**, 41.
[294] J. C. Carter, G. W. Luther, and T. C. Long, *J. Magn. Resonance*, 1974, **15**, 122.
[295] G. W. Luther, J. Valentini, and J. C. Carter, *J. Magn. Resonance*, 1974, **15**, 132.
[296] C. Segard, B. P. Roques, C. Pommier, and G. Guiochon, *J. Organometallic Chem.*, 1974, **77**, 59.
[297] M. Khalife el Saleh, G. Pastor, C. Montginoul, E. Torreilles, L. Giral, and A. Texier, *Bull. Soc. chim. France*, 1974, 1667.
[298] E. Cappelli, A. Di Nola, and A. L. Segre, *Mol. Phys.*, 1974, **27**, 1385.

(ii) *Coupling of Other Nuclei to Protons.* The work of Thorpe et al.[47] on $^1J(CH)$ in substituted purines is described in Section 3A(ii). Linear correlations between (C,H), (Se,H), and (Te,H) coupling constants and heteroatom electronegativity have been observed in some 2-substituted furans, thiophens, selenophens, and tellurophens.[79] Chalmers, Pachler, and Wessels[299] have used difference selective population inversion spectra to assign and determine the relative signs of the (C,H) couplings in 2,3-dibromothiophen.

(iii) *Coupling to Carbon.* Bundgaard et al.[300] have measured (P,C) couplings over one to seven bonds in phosphabenzenes (29). The relationship between (P,C) and

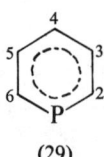

(29)

(P,H) couplings in these systems shows similarities to that between (N,C) and (N,H) coupling in pyridine. In particular, the larger magnitude of $^3J[P,C(4)]$ relative to $^2J[P,C(3)]$ is comparable to the situation observed with $^3J[N,C(4)]$ and $^2J[N,C(3)]$ in pyridine and pyridinium ion. Also, $^2J[P,C(3)]$ and $^3J[P,H(3)]$, like $^2J[N,C(3)]$ and $^3J[N,H(3)]$ in pyridine, have opposite signs.

The theoretical work on (C,F) coupling by Lichter and Wasylishen[33] has been described in Section 2C. $^nJ(C,F)$ values for substituted purines have been reported.[47]

C. **Annulenes.**—Günther and co-workers[301] have discussed the factors which are responsible for the small variations of $^3J(HCCH)$ in bridged [14]annulenes. They find that a consideration of C—C bond length, HCC angle and the dihedral angle is sufficient to explain the experimentally observed couplings.

D. **Linear Conjugated Systems.**—All the (H,H) couplings in 1,2,4,5-hexatetraene (25) have been measured.[276] (H,H) couplings in 9-*cis*- and 13-*cis*-retinal[302] and in the η-allyl fragment of some η-allyl tris(trifluorophosphine)rhodium(II) complexes[146] have been reported.

Zeisberg and Bohlmann[272] have made a detailed study of the ^{13}C n.m.r. spectra of polyynes with particular reference to the (C,H) coupling constants of acetylenic carbon atoms. Their results are summarized in Figures 2, 3, and 4, in which the couplings are given in Hz.

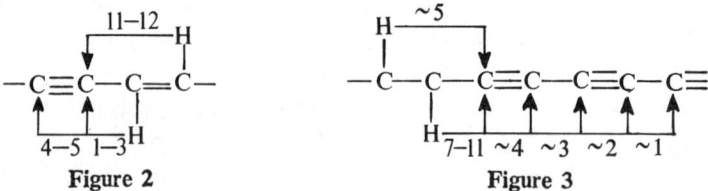

Figure 2   Figure 3

[299] A. A. Chalmers, K. G. R. Pachler, and P. L. Wessels, *J. Magn. Resonance*, 1974, **15**, 415.
[300] T. Bundgaard, H. J. Jakobsen, K. Dimroth, and H. H. Pohl, *Tetrahedron Letters*, 1974, 3179.
[301] A. Alscher, W. Bremser, D. Cremer, H. Günther, H. Schmickler, W. Sturm, and E. Vogel, *Chem. Ber.*, 1975, **108**, 640.
[302] R. Rowan and B. D. Sykes, *J. Amer. Chem. Soc.*, 1975, **97**, 1023.

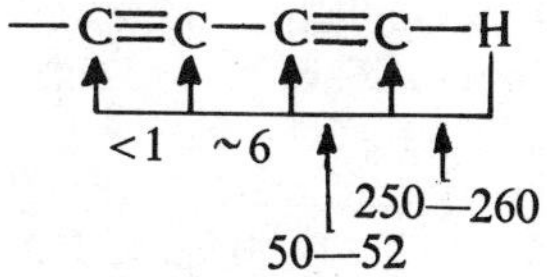

```
        —C≡C—C≡C—H
         ↑  ↑  ↑  ↑
        <1  ~6 |  |
               |  250—260
              50—52
```

Figure 4

11 Through-space Coupling

The theoretical work of Buckingham and Cordle[35] is described in Section 2D, where the related observations of Schaefer and Rowbotham[36] are noted.

Schumann and Cavagna[268] have made careful analyses of the ^{19}F spectra of hexafluoropropene dimers and have discussed their results with particular reference to through-space (F,F) coupling. They find that, in contrast to the work of Hirao et al.,[7,37] ^4J(F,F) has the same sign and, for the same distance, about the same magnitude as ^5J(F,F). It is not possible to conclude from large couplings, i.e. >20 Hz, that the coupled nuclei are close together since the spin information can be transmitted via chemical bonds. This conclusion can, however, be drawn for couplings of ca. 40 Hz. When it is known that there is only one double bond present in the system this limit can be reduced to 20 Hz.

Mallory, Mallory, and Fedarko[303] have reported a study of substituent effects on the (F,F) through-space coupling in 1,8-difluoronaphthalenes with a substituent at position 4 (30). They conclude that the substituent can affect the coupling by chang-

(30)

ing the geometry of the system. The ^5J(F,H) couplings measured in 2-fluorobenzamides (24) by Fritz and Winkler[275] showed a solvent dependence. The authors suggest that this indicates coupling through hydrogen bonds. This would appear to fall within the present rather vague definition of through-space coupling.

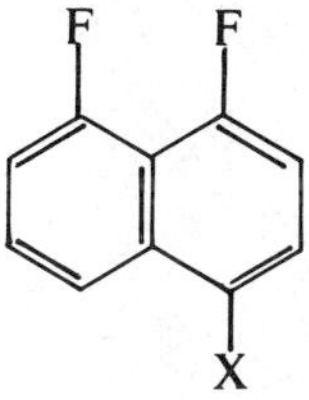

12 Recent Experimental Advances

Following the practice established in Volume 4, the chapter concludes with a brief list of references to experimental work, and associated theory, of proven or potential relevance to the measurement and/or interpretation of nuclear spin–spin coupling constants.

[303] F. B. Mallory, C. W. Mallory, and M.-C. Fedarko, *J. Amer. Chem. Soc.*, 1974, **96**, 3536.
[304] H. Yamada, *Rev. Sci. Instr.*, 1974, **45**, 640.
[305] J. D. Cargioli and E. A. Williams, *Rev. Sci. Instr.*, 1974, **45**, 1263.
[306] J. J. Grimaldi and B. D. Sykes, *J. Amer. Chem. Soc.*, 1975, **97**, 273.
[307] T. Merrick, S. Smiriga, and A. Pines, *J. Magn. Resonance*, 1974, **14**, 270.

(i) General experimental (essentially non-electronic): refs. 304—306.
(ii) Electronic, data acquisition *etc.*: refs. 164, 274, and 307.
(iii) Spectral analysis: refs. 181 and 308—311.
(iv) Double resonance methods: refs. 60, 187, and 312—317.
(v) Bandshape analysis: refs. 137, 274, 318 and 319.
(vi) Use of liquid crystalline solvents: ref. 118.
(vii) Selective population inversion: refs. 299, 317, 320, and 321.
(viii) Use of paramagnetic additives: ref. 322.
(ix) Theory of techniques: refs. 187, 316, and 323.

[308] G. Hägele, W. Peters, and M. Weidenbruch, *Ber. Bunsengesellschaft Phys. Chem.*, 1974, **78**, 781.
[309] L. Ernst and D. N. Lincoln, *J. Magn. Resonance*, 1974, **16**, 190.
[310] D. W. Aksnes, *Acta Chem. Scand (A)*, 1974, **28**, 375.
[311] V. S. Tumanov, *Russ. J. Struct. Chem.*, 1974, **15**, 480.
[312] V. Barboiu, *Mol. Phys.*, 1974, **28**, 707.
[313] V. S. Tumanov, *Zhur. strukt. Khim.*, 1974, **15**, 561.
[314] D. Canet, J.-P. Marchal, and J.-P. Sarteaux, *Compt. rend.*, 1974, **279**, B, 71.
[315] K. Kushida, K. Aoki, and S. Satoh, *J. Amer. Chem. Soc.*, 1975, **97**, 443.
[316] J. M. Anderson, F. B. Mallory, and C. W. Mallory, *J. Magn. Resonance*, 1975, **17**, 340.
[317] K. G. R. Pachler and P. L. Wessels, *J.C.S. Chem. Comm.*, 1974, 1038.
[318] D. G. Hughes, M. R. Smith, and D. A. M. Switzer, *J. Chem. Phys.*, 1974, **60**, 4890.
[319] J. Heidberg and H. Scheve, *Z. phys. Chem.*, 1974, **92**, 295.
[320] A. A. Chalmers, K. G. R. Pachler, and P. L. Wessels, *Org. Magn. Resonance*, 1974, **6**, 445.
[321] H. J. Jakobsen, S. A. Linde, and S. Sørensen, *J. Magn. Resonance*, 1974, **15**, 385.
[322] U. Séquin and A. I. Scott, *J.C.S. Chem. Comm.*, 1974, 1041.
[323] T. P. Sheahen, *J. Magn. Resonance*, 1974, **15**, 69.

3
Nuclear Spin Relaxation in Fluids

BY M. HOLZ AND M. D. ZEIDLER

1 Introduction

The total number of papers concerned with nuclear spin relaxation in fluids which have appeared during this review period is about the same as in previous periods. It is remarkable that within the topics for liquids, papers on ^{13}C relaxation are now dominant. On the other hand the trend to relaxation studies of heavy nuclei, mentioned in last year's Report, has not continued. A few studies of ^{39}K, ^{41}K, and ^{113}Cd relaxation have been reported, as well as the conventional resonances. In ^{13}C work more authors have employed PRFT and thus have given relaxation data for individual carbon atoms. These results have led to a growing interest in the determination of segmental motions in large molecules. Relaxation studies of ionic nuclei in non-aqueous electrolyte solutions have become more popular.

In contrast to last year this Report contains relaxation studies of gases, and in this respect covers the literature of the last two years. Progress in relaxation theory for non-linear polyatomic gases should be noted especially. On the experimental side, more accurate T_1 measurements in gases are now obtainable, allowing a better test of existing theories, and for the first time reliable T_1 data of ^{3}He have been reported.

2 Theoretical Work

During the review period a few theoretical contributions have appeared which are relevant for relaxation studies in liquids.

The effect of multiple internal rotations on the DD relaxation rate has been considered by several authors (see Vol. 4, p. 123). Among these were Levine et al.[1] who treated multiple internal rotations superimposed on the isotropic rotation of the molecule as a whole. In a new paper by Levine et al.[2] their previous rotational model is extended to include anisotropic rotations of the molecule as a whole. The relevant expression for the angular autocorrelation function as derived in a basic paper by Huntress[3] is used. Numerical results are presented by the authors only for the special case of an axially symmetric prolate ellipsoid for which expressions had been derived earlier.[4]

[1] Y. K. Levine, P. Partington, and G. C. K. Roberts, *Mol. Phys.*, 1973, **25**, 497.
[2] Y. K. Levine, N. J. M. Birdsall, A. G. Lee, J. C. Metcalfe, P. Partington, and G. C. K. Roberts, *J. Chem. Phys.*, 1974, **60**, 2890.
[3] W. T. Huntress, jun., *J. Chem. Phys.*, 1968, **48**, 3524.
[4] H. Versmold, *Z. Naturforsch.*, 1970, **25a**, 367.

Two papers concerned with SR relaxation have appeared, one on the limit of diffusional motion,[5] the other on the limit of inertial motion.[6] It should be remembered that the diffusion model is applicable if the angular momentum correlation time τ_J is much shorter than the orientational correlation time τ_2. In this limit the times show an inverse relation whereas in the other extreme of inertial motion the relationship becomes linear.

SR relaxation in the diffusional models due to isotropic rotation was considered long ago.[7] Extension to anisotropic rotation was worked out later by Huntress,[3] and his erroneous result was finally corrected for symmetric-top molecules.[8,9] Wang's new paper[5] is concerned with the extension to asymmetric-top molecules. The result is rather complicated for the general case where the principal axes of the diffusion tensor and the moment of inertia tensor do not coincide, and an explicit formula is only given for the special case of coincidence of all axes.

Rao and Mishra[6] considered the temperature dependence of the SR relaxation rate when the molecular orientation is inertial. In this case the relaxation rate is governed by the orientational correlation time rather than by the angular momentum correlation time. Since, for the former, theoretical calculations in the inertial limit are available from Steele[10] the authors concluded that the SR relaxation rate increases with temperature approximately as $T^{1/2}$. The authors point out that experimental verification is difficult since SR and DD relaxation show opposite temperature dependence, and thus little variation may be observed in the relevant temperature range.

The effect of cross-correlation terms on the nuclear magnetic relaxation rate was discussed in two publications by Werbelow and Marshall.[11,12] The first paper (see Vol. 4, p. 208) is concerned with cross-correlation terms arising from a single relaxation mechanism (Hubbard terms). A methyl group (three-spin system) performing internal rotations and attached to a symmetric-top molecule was considered. Their result was that non-exponential (*i.e.* multi-exponential) relaxation due to intramolecular DD interaction occurs if the methyl group rotates fast and is attached to a molecule performing markedly anisotropic rotation. Practical applications therefore should be observable, especially for macromolecules in solution, unless the SR mechanism masks any non-exponential effect. In the later report[12] cross-correlation terms arising from different relaxation mechanisms (Blicharski terms) are treated. Here a two-spin system is considered. In general it is found that any type of cross-correlation term tends to slow down relaxation. Anisotropic reorientations magnify the effect of Hubbard terms but have a much smaller effect on Blicharski terms. Further peculiarities arise if the extreme-narrowing condition no longer holds and specific geometries occur; here Blicharski terms are much more affected.

A research note by Lynden-Bell is of general interest.[13] From textbooks on n.m.r.

[5] C. H. Wang, *Mol. Phys.*, 1974, **28**, 801.
[6] B. D. N. Rao and P. K. Mishra, *Chem. Phys. Letters*, 1974, **27**, 592.
[7] P. S. Hubbard, *Phys. Rev.*, 1963, **131**, 1155.
[8] H. J. Bender and M. D. Zeidler, *Ber. Bunsengesellschaft phys. Chem.*, 1971, **75**, 236.
[9] C. H. Wang, D. M. Grant, and J. R. Lyerla, jun., *J. Chem. Phys.*, 1971, **55**, 4674.
[10] W. A. Steele, *J. Chem. Phys.*, 1963, **38**, 2404 and 2411.
[11] L. G. Werbelow and A. G. Marshall, *J. Magn. Resonance*, 1973, **11**, 299.
[12] L. G. Werbelow and A. G. Marshall, *Mol. Phys.*, 1974, **28**, 113.
[13] R. M. Lynden-Bell, *Mol. Phys.*, 1975, **29**, 301.

one learns that the transverse relaxation time T_2 cannot be larger than the longitudinal relaxation time T_1. However, Vold and Chan[14] some years ago showed theoretically that T_2 can indeed be larger than T_1, by including cross-relaxation rates into the rate equations for transverse relaxation as with the case of longitudinal relaxation; they also showed this experimentally by measuring proton T_1 and T_2 (the latter by using the Carr–Purcell sequence and varying the pulse spacings) in a mixture of chloroform and benzene. Now Lynden-Bell[13] has demonstrated theoretically that $T_1 \geqslant \frac{1}{2} T_2$ is the true condition if CSA is the relevant relaxation mechanism. The nuclear shielding tensor may be resolved into an isotropic part (which governs the observed chemical shift of the nucleus in an isotropic fluid), a first rank antisymmetric tensor σ_1 and a symmetric second rank tensor σ_2. Under conditions of extreme narrowing and assuming isotropic diffusion, it is found that for ratios $(\sigma_1/\sigma_2)^2$ exceeding 1/15 (with σ_1 and σ_2 defined as in ref. 13), the ratio T_1/T_2 is less than 1. Non-zero values of σ_1 can occur if the local symmetry at the nucleus contains not more than one plane of symmetry, which must be perpendicular to any axis of symmetry. Such a condition may be found in a ligand of an unsymmetrical transition-metal complex.

One of the important applications of n.m.r. relaxation studies is the determination of chemical rate constants, e.g. in electrolyte solutions. In two theoretical treatments[15, 16] the influence of the chemical exchange on the n.m.r. relaxation times has been further investigated. Jen[15] has analysed the decay of the amplitude for a Carr–Purcell train of echoes in a system with two exchanging sites. Earlier, Luz and Meiboom[17] and, in an improved version, Allerhand and Gutowsky[18] had shown that lifetimes τ_A, τ_B of exchanging nuclei may be determined by measuring the dependence of T_2 on the pulse rate in a Carr–Purcell experiment. In the treatments of both groups, however, it was assumed that, in the absence of exchange, $T_{2A} = T_{2B}$, where T_{2A}, T_{2B} are the relaxation times in the states A and B respectively. But there are reactions, e.g. transfer of the nucleus to a paramagnetic particle or exchange of quadrupolar relaxing nuclei between positions of different symmetry of the electrical field,[19] where T_{2A} and T_{2B} may be very different. Jen has now derived a general solution for $T_{2A} \neq T_{2B}$, and he gives closed expressions which are valid for all values of the chemical shift, the two transverse relaxation times, and the two average lifetimes. The previously published solutions to this problem turn out to be special cases of the general solution. It could be shown that high pulse rates eliminate only the effect of the chemical shift on the relaxation, whereas the influence of the chemical exchange disappears only in special cases.

In the second treatment Strehlow and Frahm[16] have investigated theoretically and experimentally the influence of the chemical exchange on the n.m.r. spin-lattice relaxation time T_1. Simple reactions like $B + HA \rightleftharpoons BH + A$ with $T_{1A} \neq T_{1B}$ are treated, as well as more complex reactions; also some important limiting cases are summarized. It turns out that the average lifetimes can be determined from T_1 measurements if the inequality $T_{1B} \leqslant \tau_A, \tau_B \leqslant 5\, T_{1A}$ holds. The exchange

[14] R. L. Vold and S. O. Chan, *J. Chem. Phys.*, 1972, **56**, 28.
[15] J. Jen, *Adv. Mol. Relaxation Processes*, 1974, **6**, 171.
[16] H. Strehlow and J. Frahm, *Ber. Bunsengesellschaft phys. Chem.*, 1975, **79**, 57
[17] Z. Luz and S. Meiboom, *J. Chem. Phys.*, 1963, **39**, 366.
[18] A. Allerhand and H. S. Gutowsky, *J. Chem. Phys.*, 1964, **41**, 2115; *ibid.*, 1965, **42**, 1587.
[19] H. G. Hertz, *Z. Elektrochem.*, 1960, **64**, 53.

rate of H_2O molecules in concentrated solutions of $Ni(NO_3)_2$ was experimentally determined in this way. Finally, Strehlow and Frahm point out that a combination of T_2 (as described by Jen) and T_1 measurements may considerably improve the precision of kinetic data derived from n.m.r. relaxation studies.

3 Relaxation Studies in Pure Liquids

A. Proton and Deuteron Relaxation.—Proton relaxation in liquid acetonitrile has been studied by Harrell.[20] His purpose was to detect non-exponential relaxation due to cross-correlations within the three-spin system (Hubbard terms, in the terminology of the previous section). It is known from Hubbard's theoretical work[21] that cross-correlation effects become increasingly important if the three-spin system is fixed, not to a molecule which undergoes isotropic rotation, but to a molecule which undergoes strongly anisotropic orientation. Bopp's[22] and Woessner's[23] experimental studies showed that acetonitrile is such an anisotropic rotator, and this suggested a search for cross-correlation effects in this liquid. Since the cross-correlations arise from DD interactions between the three spins, it is clear that mechanisms other than intramolecular DD relaxation tend to obscure the effect one is looking for. These competing mechanisms are intermolecular DD and SR relaxation. Logically, Harrell therefore measured T_1 at low temperatures and in mixtures of CH_3CN–CD_3CN, and thus was able to prove non-exponential relaxation below 0 °C.

Another origin of non-exponential relaxation is cross-relaxation, which is not to be confused with cross-correlation effects just discussed. This cross-relaxation effect was briefly mentioned in the preceding section in connection with Vold's work.[14] It arises if unlike spins or spin groups relax by DD interaction. Apart from double resonance techniques, careful analysis of the non-exponential decay[24] can be employed to study cross-relaxation. In a new paper by Cutnell et al.[25] a third method is pointed out, the use of T_1 measurements on partially deuteriated analogues in studies of cross-relaxation in proton systems. The method is illustrated with data on anisole and its partially deuteriated analogues $C_6H_5OCD_3$ and $C_6D_5OCH_3$. Since the DD interaction between hydrogen and deuterium nuclei is much less than that between two hydrogen nuclei, cross-relaxation is absent in the partially deuteriated compounds, and the relaxation rates of the separate spin groups can be obtained. Comparison with non-deuteriated anisole PhOMe then allows the effect of cross-relaxation to be deduced. Inclusion of dilution experiments with perdeuteriated anisole permits the statement that cross-relaxation proceeds primarily by intramolecular DD interaction. It was found that the cross-relaxation rate is positive (interacting spins flip in the same sense) and increases with decreasing temperature. Correlation times for overall molecular rotation were calculated and compared to dielectric relaxation times; the existence of large-step rotational motion was concluded from this comparison.

[20] J. W. Harrell, jun., *J. Magn. Resonance*, 1974, **15**, 157.
[21] P. S. Hubbard, *J. Chem. Phys.*, 1969, **51**, 1647.
[22] T. T. Bopp, *J. Chem. Phys.*, 1968, **47**, 3621.
[23] D. E. Woessner, B. S. Snowden, jun., and E. T. Strom, *Mol. Phys.*, 1968, **14**, 265.
[24] A. A. Brooks, J. D. Cutnell, E. O. Stejskal, and V. W. Weiss, *J. Chem. Phys.*, 1968, **49**, 1571.
[25] J. D. Cutnell, S. B. W. Roeder, S. L. Tignor, and R. S. Smith, *J. Chem. Phys.*, 1975, **62**, 879.

A deuteron relaxation study has been carried out by Versmold[26] in ethanol. A wide temperature range including the supercooled liquid was covered. The purpose of this paper was to differentiate between overall and internal rotation of the ethanol molecule. Maxima of the relaxation rates were found at temperatures around 125—140 K: one maximum corresponds to the hydroxy-group while the maxima of both the methyl and methylene groups occur at the same, lower temperature. This led to the conclusion that the hydroxy-group shows the slowest motion and that the ethyl group, which behaves as a rather rigid entity, reorients faster. Quantitative data of rotational diffusion coefficients, activation energies, and quadrupole coupling constants were obtained from the fit to expressions derived earlier for internal-rotational-diffusion models.[4] Even more interesting is the behaviour of ethanol in mixtures with glycerol, where the motions of the methyl and methylene group become decoupled (see Section 4A).

The pressure and temperature dependence of T_1 for protons in water has been studied by Russian workers,[27] slightly extending the experimental range covered by previous experiments.[28]

O'Reilly[29] has published a paper concerned with the pressure dependence of the viscosity and the spin–lattice relaxation time in H_2O and D_2O. In both liquids at low temperatures and pressures, the coefficient of viscosity (η) decreases with increase in pressure. Likewise the proton and deuteron T_1 values increase with increase in pressure. On the basis of his water model, described in earlier papers,[30, 31] O'Reilly offers a ready explanation for the pressure dependence, using a relationship between the pressure coefficients of T_1 and η and the experimental pressure-dependent T_1 values between 0° and 90 °C obtained by Hertz and Rädle[28] and Lee and Jonas.[32] Water and deuterium oxide exhibit differences, particularly with regard to $(\partial \ln T_1/\partial p)_T$. This fact is explained by taking into account the variation of χ (χ = effective quadrupolar coupling constant of ^2H in liquid D_2O) with pressure and intermolecular contributions to the proton relaxation in H_2O. In O'Reilly's model the effective value of χ is 237 kHz at 0 °C and 245 kHz at 100 °C.

In the following, three papers are mentioned in which proton-relaxation time measurements have been presented in connection with a description of complications which may arise in the determination of the transverse relaxation time when the system under investigation is homonuclear coupled. It is well known that measurements of the T_2 values of individual lines in a complex spin system may be performed by spin-echo techniques using either weak selective pulses or strong non-selective pulses (e.g. see ref. 33). The Meiboom–Gill modified Carr–Purcell pulse trains (CPMG) are normally used, with Fourier transformation of the last half echo. In most cases large pulse spacings are difficult to use because of stringent requirements for field stability and homogeneity. In addition, if non-selective pulses are applied, complications arise in coupled spin systems due to the modulation of

[26] H. Versmold, *Ber. Bunsengesellschaft phys. Chem.*, 1974, **78**, 1319.
[27] V. V. Kisel'nik, N. G. Malyuk, and A. I. Toryanik, *Zhur strukt. Khim.*, 1974, **15**, 697.
[28] H. G. Hertz and C. Rädle, *Z. phys. Chem. (Frankfurt)*, 1969, **68**, 324.
[29] D. E. O'Reilly, *J. Phys. Chem.*, 1974, **78**, 1674.
[30] D. E. O'Reilly, *Phys. Rev. (A)*, 1973, **7**, 1659.
[31] D. E. O'Reilly, *J. Chem. Phys.*, 1974, **60**, 1607.
[32] Y. Lee and J. Jonas, *J. Chem. Phys.*, 1972, **57**, 4233.
[33] R. Freeman and H. D. W. Hill, in 'Dynamic N.M.R. Spectroscopy' ed. L. M. Jackman and F. A. Cotton, Academic Press, New York, 1975.

the echo amplitudes.[34] In order to avoid such modulations and errors due to field instability and inhomogeneity, closely-spaced non-selective pulses are often applied. If the pulse rate $1/2\tau$ in a CPMG experiment is much larger than the frequency spread of the spectrum, spin locking occurs, and the observed transverse decay times may also be designated as $T_{1\rho}$.[35] The behaviour of the transverse relaxation for this experimental case in a system with two coupled nuclei of spin 1/2 has been studied and analysed by Vold and Vold.[36, 37] They show that under the spin-lock conditions in the CPMG experiment, the existence of scalar coupling renders the spins effectively equivalent; consequently all lines in the coupled system will relax with the *same* transverse relaxation time T_2. The measured relaxation time is the average of relaxation times expected for each spin in the absence of scalar coupling. This was demonstrated in the first paper[36] on the proton relaxation rates of isopropyl iodide and isopropyl acetate. The measured T_2 values were considerably shorter than the corresponding T_1's. The T_2 values could be explained by the averaging effect mentioned above, and it was in this way demonstrated that scalar relaxation by coupling to iodine, a possible source of T_1–T_2 differences, is unimportant for isopropyl iodide. In the second paper[37] the relaxation of 2,3-dibromothiophen was shown to be due to a combination of intramolecular dipole–dipole interactions and random fluctuating fields. Different T_1 values, but the same transverse relaxation times, were found. Consequently the non-selective CPMG spin-echo method cannot be used in the short pulse-spacing limit to determine distinct contributions from sources of transverse relaxation outside the spin system being considered. The authors emphasize that in more complex spin systems the averaging extends only over transitions of the same symmetry, which would allow relaxation mechanisms with different symmetry properties to be distinguished.

In the third paper under discussion, Alexandre and Rigny[38] have studied the decrease of the envelope of the first spin-echo after a $\pi/2,\pi$-pulse pair in a system with coupled spins. In contrast to the situation described above, they studied the case where the different lines of the spectrum are not resolved, which means that the frequency differences of the various lines are of the same order of magnitude as the linewidths. This situation occurs with liquid alkanes, for example. Two cases are distinguished: (i) slow relaxation ($1/T_1 \ll v_i - v_j$, where $v_{i,j}$ are the frequencies of the i and j components), in which case, as demonstrated with n-hexane, the envelope of the echo is non-exponential, meaning that it is not dependent upon relaxation but only on the static Hamiltonian; (ii) fast relaxation ($1/T_1 \gg v_i - v_j$): here, as was shown by a detailed calculation,[39] the presence of scalar coupling influences transverse relaxation. For strong coupling one obtains:

$$1/T_2 = 1/T_1 + Bv_0^2 T_1$$

(v_0 is the Larmor frequency). The validity of this equation was demonstrated by frequency-dependent measurements (1.5—10 MHz) and by variation of T_1 in n-dodecane. The expression $1/T_2 - 1/T_1$ vs. v_0^2 is linear, and the slope at the origin is proportional to T_1.

[34] R. L. Vold and S. O. Chan, *J. Magn. Resonance*, 1971, **4**, 208.
[35] R. Freeman and S. Wittekoek, *J. Magn. Resonance*, 1969, **1**, 238.
[36] R. L. Vold and R. R. Vold, *J. Amer. Chem. Soc.*, 1974, **96**, 4043.
[37] R. L. Vold and R. R. Vold, *J. Chem. Phys.*, 1974, **61**, 2525.
[38] M. Alexandre and P. Rigny, *Compt. rend.*, 1974, **278**, B, 949.
[39] M. Alexandre, Thesis, Orsay, France, 1974.

Briguet, Duplan, and Delmau[40] have measured proton-relaxation rates in $CHBr_3$, $SiHCl_3$, and PhSeH. Their work forms an indirect investigation of the spin–lattice relaxation of ^{13}C, ^{29}Si, and ^{77}Se and the reader is referred to Sections 3B and 3C for a report of this paper.

B. ^{13}C Relaxation.—Burnett and Roeder[41] have conducted a very interesting study of glycerol. The ^{13}C spin–lattice relaxation rates of the two outer and the inner carbon atoms were measured separately over the temperature range 267—339 K at a fixed frequency of 25 MHz. The results were interpreted by taking account of intramolecular DD interaction with the directly attached protons, and the anisotropic rotational diffusion of the molecule,[42] with the modification that conformational changes were included by averaging Woessner's expressions over all angles. In order to analyse the temperature dependence of the relaxation rates, the functional dependence of the three diffusion constants of the anisotropic rotor on temperature was taken to obey a modified Arrhenius behaviour, with a temperature close to the glass transition as a parameter. Satisfactory fits to the experimental data were obtained with three different rotational diffusion constants (in the ratio 1:5:200) but with equal activation energies of 10.0 kJ mol^{-1}. It is remarkable that with this diffusion model the distribution of correlation times (Cole–Davidson distribution) commonly invoked for the interpretation of dielectric and n.m.r. relaxation data of glycerol is no longer needed. A conclusive test for the reliability of the proposed model is whether the intramolecular proton-relaxation rate, which is available in the literature[43] as a function of temperature and frequency, can also be fitted with the same parameters. The authors claim, not convincingly, that such is the case. They also point out that the activation energy for translational diffusion[44] equals their adjusted value of the activation energy, and thus conclude that rotational and translational motions are strongly coupled.

In an extension to their study on three alkanes (see Vol. 4, p. 129) Levine et al.[2] have performed separate ^{13}C relaxation studies of all resolved carbon resonances in n-alkanes from C_6 to C_{18} and in n-alkyl bromides from C_4 to C_{15}. The data were interpreted on the basis of their multiple internal-anisotropic-rotation model presented in the same paper and mentioned above. The anisotropic rotation of the molecule as a whole was described by two diffusion coefficients assumed to be in the ratio of the corresponding moments of inertia. Except for the terminal methyl group, which shows the fastest rotation, all other groups down the chain can be described by the same internal diffusion constant.

The work by von Goldammer et al.[45] on ^{13}C spin–lattice relaxation of acetonitrile was discussed in Volume 4 (p. 128). It should be pointed out here that a slight change in the SR interaction constants of the methyl carbon, due to computational error, must be made.[46] In the same paper ^{13}C relaxation in acetone was investigated. Again both types of carbons were treated separately. SR and DD interactions are

---

[40] A. Briguet, J. C. Duplan, and J. Delmau, Mol. Phys., 1975, **29**, 837.
[41] L. J. Burnett and S. B. W. Roeder, J. Chem. Phys., 1974, **60**, 2420.
[42] D. E. Woessner, J. Chem. Phys., 1962, **37**, 647.
[43] J. P. Kintzinger and M. D. Zeidler, Ber. Bunsengesellschaft phys. Chem., 1973, **77**, 98.
[44] D. J. Tomlinson, Mol. Phys., 1972, **25**, 735.
[45] E. von Goldammer, H. D. Lüdemann, and A. Müller, J. Chem. Phys., 1974, **61**, 4590.
[46] E. von Goldammer, H. D. Lüdemann, and A. Müller, J. Chem. Phys., 1974, **61**, 3493.

the dominant relaxation mechanisms; for the methyl carbon only intramolecular DD interaction is important whereas for the carbonyl carbon intermolecular interactions with protons also occur.

In another paper by these authors[47] values of T_1 of ^{13}C nuclei in pyridine and THF were measured at various temperatures. Again SR and DD interactions are the relevant relaxation mechanisms. They were separated by using previous relaxation data from ^2H and ^{14}N.[48-50] Results for the SR interaction constants are given in the paper.

Moreland and Carroll[51] have given an example of the importance of (^{13}C, ^{13}C) intramolecular DD interaction. These authors compared ^{13}C relaxation in normal

$$\text{Et O}-^{13}\text{C}-^{13}\text{CH}_2-\text{C}-\text{OEt}$$
$$\underset{\text{(1)}}{\overset{\overset{\displaystyle O}{\|}\overset{\displaystyle O}{\|}}{}}$$

diethyl malonate with that in the 90% 1,2-^{13}C-enriched compound (1). The relaxation rate of the carbonyl carbon increased noticeably in the enriched compound, but no change was observed for the methylene carbon. The conclusion, which is supported by the determination of NOE factors, is that intramolecular ^{13}C interaction is important for the carbonyl carbon whereas the methylene carbon, as expected, relaxes by DD interaction with its directly-attached protons.

Berezhnoi and Sergeev[52] measured ^{13}C spin–lattice relaxation times in benzene and naphthalene. In the case of benzene the NOE factor showed that DD interaction is the dominating relaxation mechanism. The temperature dependence of the $C_{10}H_8$ relaxation time deviates only a little from Arrhenius behaviour.

In connection with a study on n-alkylammonium ions, Levy and co-workers[53] used ^{13}C T_1 values of individual carbons to monitor segmental motions in several n-alkylamines. A number of trends were noted and discussed qualitatively. The value of T_1 generally decreases all along the chain as the size of the chain increases. This behaviour reflects the slower overall reorientation due to increased size and molecular weight. For a given amine there is very little change in T_1 along the chain, indicating that the effective correlation time is governed by the correlation time for the overall reorientation. For the larger aliphatic chains a slight decrease in correlation time is observed as one proceeds from the centre to the ends of the chains. Obviously, for these large aliphatic chains, the central bulk of the chain restricts motion of the CH_2 units; thus the observed effect reflects the fact that segmental motion increases towards the ends.

The isotope effect in the n.m.r. relaxation (*e.g.* see ref. 54) was utilized by Briguet *et al.*[40] to measure indirectly the spin–lattice relaxation times of ^{13}C (and ^{29}Si and ^{77}Se, see Section 3C). They compared the proton relaxation in the isotopomeric pairs of molecules $^{13}C^1HBr_3$ and $^{12}C^1HBr_3$, and from these measurements they

[47] E. von Goldammer, H. D. Lüdemann, and O. Röder, *Chem. Phys. Letters*, 1974, **26**, 387.
[48] J. P. Kintzinger and J. M. Lehn, *Mol. Phys.*, 1971, **22**, 273.
[49] E. von Goldammer and M. D. Zeidler, *Ber. Bunsengesellschaft phys. Chem.*, 1969, **73**, 4.
[50] E. von Goldammer and H. G. Hertz, *J. Phys. Chem.*, 1970, **74**, 3734.
[51] C. G. Moreland and F. I. Carroll, *J. Magn. Resonance*, 1974, **15**, 596.
[52] V. G. Berezhnoi and N. M. Sergeev, *Zhur. strukt. Khim.*, 1975, **16**, 136.
[53] G. C. Levy, K. A. Komoroski, and J. A. Halstead, *J. Amer. Chem. Soc.*, 1974, **96**, 5456.
[54] H. Ozawa, Y. Arata, and S. Fujawara, *J. Chem. Phys.*, 1972, **57**, 1613.

derived the ^{13}C relaxation times. The effective relaxation mechanisms were specified through an AX ($I = \frac{1}{2}$) system analysis. The advantage of the method lies in the possibility of studying the relaxation of weak nuclei with the sensitivity of the proton resonance. However, this advantage is mitigated by the rather complicated time dependence of the evolution of lines. Nevertheless, the authors showed that with this method it is possible to investigate the different contributions to the relaxation of nuclei with low natural abundance using a conventional continuous-wave HR-spectrometer. This was done by comparing the indirectly measured relaxation times of ^{13}C of bromoform at two different fields with values[40, 55] measured directly by FT. The comparison gave satisfactory agreement. Analysis of the relaxation contribution to ^{13}C in CHBr$_3$ showed that besides the DD relaxation, relaxation due to SC is dominant. From the DD term a reorientational correlation time around the molecular axis of 3 ps was obtained. The SC term allows calculation of a spin–lattice relaxation time of ^{79}Br, $T_1 = 1.5$ μs, and a coupling constant $^1J(^{79}\text{Br},^1\text{H}) \approx 41$ Hz.

Finally in this section we report on a paper also concerned with ^{13}C relaxation, but in a different way. Faller and co-workers[56] have proposed the application of lanthanide relaxation reagents to improve the assignment of ^{13}C resonances. (The complementary use of relaxation reagents for the determination of structure from proton spectra has already been suggested.[57, 58]) The utility of shift reagents in ^{13}C n.m.r. is often mitigated by large contact contributions from delocalization of the electronic spin. Additionally, large quantities of expensive shift reagent often need to be added to produce suitable shifts. The authors believe that the use of relaxation reagents will have advantages over shift reagents. Addition of a small amount of Gd(dpm)$_3$ (dpm is the anion of dipivaloylmethane) drastically shortens the T_1 of those nuclei near the binding site of the metal. This was demonstrated on a n-butanol spectrum, where the T_1 ordering in the presence of the reagent is C-1 < C-2 < C-3 < C-4. The relaxation rate due to Gd(dpm)$_3$ should be proportional to r^{-6}. This distance dependence was tested with a rigid molecule (4-picoline), for which relative distances can be predicted accurately. The agreement was reasonable. Finally the authors discuss possible errors due to relaxation mechanisms other than DD interaction. They emphasize that extensive studies are required to test the working of the method in a large number of cases.

C. **Relaxation of Other Nuclei:** ^{14}N, ^{15}N, ^{17}O, ^{19}F, ^{29}Si, ^{77}Se.—Liquid nitrogen was studied by ^{14}N and ^{15}N relaxation.[59] The relevant interactions are QF and SR, and thus the correlation times τ_2 and τ_J, using appropriate interaction constants (χ = 5.55 MHz, $C_\perp$ = 22 kHz), were calculated from the respective relaxation rates. The data cover a large temperature range (boiling point to critical point along the liquid–vapour coexistence line) and in most of the range τ_J is greater than τ_2, by a factor of 9 at the critical point. This led to the conclusion that the nitrogen molecule reorients more or less freely by large angles. Neither the Hubbard relation[7] nor the extended J- or M-diffusion theories[60] satisfactorily describe the experimental data; conse-

---

[55] T. C. Farrar, S. J. Druck, R. R. Shoup, and E. D. Becker, *J. Amer. Chem. Soc.*, 1972, **94**, 699.
[56] J. W. Faller, M. A. Adams, and G. N. La Mar, *Tetrahedron Letters*, 1974, 699.
[57] G. N. La Mar and J. W. Faller, *J. Amer. Chem. Soc.*, 1973, **95**, 3817.
[58] J. W. Faller and G. N. La Mar, *Tetrahedron Letters*, 1973, 1381.
[59] K. Krynicki, E. J. Rahkamaa, and J. G. Powles, *Mol. Phys.*, 1975, **29**, 539.
[60] R. E. D. McClung, *J. Chem. Phys.*, 1969, **51**, 3842.

quently the authors point out that the reorientation process may be quantum mechanical to some extent. In the lower temperature range, computer simulations[61] (based, however, on classical mechanics) are available, and they show good agreement with experimental τ_2 and τ_y values.

The possibility of obtaining relaxation data for water at low temperature using emulsions with n-heptane was reported in Volume 4 (p. 125). Hindman[62] has now applied this technique to the ^{17}O relaxation of water, and thereby extended ^{17}O data[63] down to -31 °C. The experimental T_1 results were fitted to a double exponential with activation energies of 49.6 and 14.8 kJ mol^{-1} respectively. These energies are in fair agreement with activation data obtained from viscosity and self-diffusion measurements. The low-temperature data, especially, are comparable to those observed in ice, indicating that the molecules are in well-defined hydrogen-bonded structures. Also, activation entropies based on transition-state rate theory were extracted from the relaxation rates, and the large entropy term for the low-temperature process was interpreted[62] as indicating that the relaxation involves the co-operative dissolution of a small cluster of hydrogen-bonded molecules. In contrast, the high-temperature process relaxation involves breaking of a single hydrogen bond only.

DeZwaan and Jonas[64] have used ^{19}F relaxation data to determine the hard-sphere diameter of CCl_3F, which in turn was used to calculate self-diffusion constants of this liquid. Since SR interaction provides the dominant contribution to relaxation, the data were analysed to yield τ_J. Assuming uncorrelated binary collisions in the liquid, τ_J can be related to the Enskog relaxation time,[65] which depends on the hard-sphere diameter. With the extensive experimental data (temperature range 341—460 K, pressure range 100—1600 bar) it was possible to eliminate reliably the other parameters occurring in the theoretical expressions, and it was shown that the hard-sphere diameter decreases slightly with increasing temperature (0.503 nm at 341 K, 0.490 nm at 460 K). On the other hand no density dependence of the hard-sphere diameter was found.

Another ^{19}F relaxation study by Alexandre and Rigny[66] had the purpose of completing data on ClF_3 and $ClOF_3$. These substances are difficult to handle due to their reactivity, and therefore few structural studies had been performed previously. Additionally, if there is fast chemical exchange between the different fluorides, only one line is observed in the HR spectra. Under the condition of fast chemical exchange and at a sufficiently high Larmor frequency $1/T_2 - 1/T_1$ is proportional to $(\sigma_1 - \sigma_2)^2\tau$, where $\sigma_1 - \sigma_2$ is the shielding difference between the non-equivalent fluorine atoms and τ is the exchange time. If one measures, under the same conditions, $T_{1\rho}$ as a function of B_1^2 (B_1 = r.f. field strength) the slope is given by $\tau/(\sigma_1 - \sigma_2)^2$. In this way for $ClOF_3$ the shielding difference was found to be 50 p.p.m. and the exchange time to be 25 μs at 22.5 °C. By measuring T_2 at low fields and determining the ^{35}Cl linewidth the following (^{35}Cl, ^{19}F) mean coupling constants were obtained:

---

[61] J. Barojas, D. Levesque, and B. Quentrec, *Phys. Rev. (A)*, 1973, **7**, 1092.
[62] J. C. Hindman, *J. Chem. Phys.*, 1974, **60**, 4488.
[63] J. C. Hindman, A. J. Zielen, A. Svirmickas, and M. Wood, *J. Chem. Phys.*, 1971, **54**, 621.
[64] J. DeZwaan and J. Jonas, *J. Chem. Phys.*, 1975, **62**, 4036.
[65] D. Chandler, *J. Chem. Phys.*, 1974, **60**, 3508.
[66] M. Alexandre and P. Rigny, *Canad. J. Chem.*, 1974, **52**, 3676.

$$\text{ClF}_3 \quad (\langle J_{\text{ClF}}^2 \rangle)^{1/2} = 260 \text{ Hz}$$

$$\text{ClOF}_3 \quad (\langle J_{\text{ClF}}^2 \rangle)^{1/2} = 195 \text{ Hz}$$

By interpretation of ^{19}F band shapes, the coupling constants $J(^{35}\text{Cl},\text{F})$ of 192 Hz for the axial fluorine and $|J_{\text{ClF}}| \leq 20$ Hz for the equatorial fluorines were derived.

In a communication by Kimber and Harris[67] in which a method is proposed for measuring spin–lattice relaxation in the null-signal case (NOE factor = -1), the ^{29}Si–{^{1}H} relaxation time of methyltriethoxysilane is reported to be 92.6 s. This value was obtained by utilizing the time dependence of the NOE following the switching on of the decoupler.

Briguet, Duplan, and Delmau[40] have measured indirectly, in the manner described in Section 3B, the spin–lattice relaxation time of ^{29}Si in trichlorosilane. Taking 0.1465 nm as the ^{29}Si–^{1}H distance they obtained a reorientation time around the molecular symmetry axis of about 2 ps. From the measured temperature dependence of the homonuclear Overhauser effect, the authors draw the conclusion that above 10 °C SR interaction is playing a role, a phenomenon which has been observed previously in TMS[68] (see Vol. 4, p. 129).

Briguet and co-workers[40] have applied their isotope-effect measurements not only to ^{13}C and ^{29}Si relaxation studies, but also to ^{77}Se in selenophenol (PhSeH). Assuming a Se—H distance of 0.147 nm, they derived from the measured DD relaxation rate a correlation time for the reorientation of the Se–H vector of 2.7 ps. Other relaxation mechanisms besides DD are present. Measurements of the NOE factor led to the conclusion that, at least partly, relaxation by CSA is effective.

D. **Studies of Several Resonant Nuclei: ^{2}H, ^{13}C, ^{14}N, ^{15}N, ^{19}F.**—By studying ^{2}H and ^{13}C relaxation in deuteriated and undeuteriated compounds respectively, the deuteron quadrupole coupling constant in liquids can be determined. This type of approach was tested by Saitŏ et al.[69] and was reported in Vol. 4 (p. 126). Jackman et al.[70] used this method to determine χ in phenyl[^{2}H$_1$]acetylene and in [^{2}H$_6$]-benzene; the results were 227 and 186 kHz, respectively. In contrast to Saitŏ's work, where much too large a coupling constant (254 kHz) was calculated for phenylacetylene [and consequently another competing relaxation mechanism (CSA) was postulated], no discrepancy was found by the present work. The different results between the authors arise from slight differences in measured relaxation times for ^{13}C and ^{2}H and different values for the C—H bond length. Jackman et al., moreover, determined the NOE factor, from which the pure dipolar relaxation rate for ^{13}C relaxation could be derived.

Bull[71] has measured ^{2}H and ^{19}F relaxation rates in 1,3,5-trifluorobenzene, and ^{2}H and ^{14}N relaxation rates in acetonitrile over a large temperature range. In the former case the quadrupolar relaxing ^{2}H nuclei yielded the orientational correlation time τ_2, while the fluorine nuclei relax by SR and consequently yielded the correlation time τ_J. These times can be related to each other, e.g. through the extended

[67] B. J. Kimber and R. K. Harris, *J. Magn. Resonance*, 1974, **16**, 354.
[68] G. C. Levy, J. D. Cargioli, P. C. Juliano, and T. D. Mitchell, 1973, **95**, 3445.
[69] H. Saitŏ, H. H. Mantch, and I. C. P. Smith, *J. Amer. Chem. Soc.*, 1973, **95**, 8453.
[70] L. M. Jackman, E. S. Greenberg, N. M. Szeverenyi, and G. K. Schnorr, *J.C.S. Chem. Comm.*, 1974, 141.
[71] T. E. Bull, *J. Chem. Phys.*, 1975, **62**, 222.

diffusion theory of McClung,[60, 72] which is applicable to symmetric-top molecules. Bull showed that McClung's relation is consistent with the experimental data, and he discovered the amazing fact that even for symmetric-top molecules a single spin-rotational correlation time is sufficient to describe the motion. This remarkable observation means that the moments of inertia and not anisotropic intermolecular forces are responsible for anisotropic rotations in liquids, a result to be regretted from the point of view of learning about the structure of liquids from relaxation measurements. Using this finding, he checked the motion of acetonitrile, where the reorientational correlation time was found from both ^{2}H and ^{14}N (quadrupolar) relaxation and thus is overdetermined. Again it turned out that anisotropic intermolecular forces, which should be pronounced in this liquid due to the existence of a large electric dipole moment, are ineffective. More work along this line is necessary to clarify the situation.

Liquid pyridine has been studied by ^{15}N and ^{13}C relaxation. Schweitzer and Spiess[73] measured T_1 over the range − 60 to 55 °C at two frequencies. The observed frequency dependence of ^{15}N relaxation is due to CSA and was eliminated by extrapolation to zero frequency. The remaining contributions, SR and intra- and intermolecular DD interactions, were separated in the following way: (i) ^{13}C relaxation, which is due to intramolecular DD interaction with the directly attached protons, was used to obtain τ_2, which was then used to calculate the intramolecular DD interaction of ^{15}N with the protons; since all relevant dipole–dipole vectors lie in the same plane, no complication due to anisotropic rotation could occur. (ii) From the powder pattern of ^{15}N in solid pyridine, the chemical-shift tensor was obtained; the numerical data were in agreement with the CSA relaxation contribution as determined above using the same τ_2. (iii) The SR interaction constants were calculated from the chemical-shift tensor components.[74, 75] (iv) The SR contribution was calculated from these interaction constants and a SR correlation time was obtained from the above τ_2 using the Hubbard relation, valid in the diffusion limit.[7] (v) After subtraction of the intramolecular DD and SR contributions from the extrapolated ^{15}N rate, the intermolecular DD contribution remains. The interesting result is that this contribution accounts for 90% of the dipolar rate, which is easily understandable since protons from other molecules interact with the non-bonding orbital at the hydrogen.

4 Relaxation Studies in Liquid Mixtures of Non-electrolytes

A. Proton and Deuteron Relaxation.—In a 1:1 mixture of trichloroacetic acid and DMSO a strong hydrogen-bonded complex forms, which has been investigated using proton and deuteron relaxation.[76] Proton relaxation in the system CCl_3CO_2H–[$^2H_6$]DMSO could be measured over the frequency range 4—100 MHz, and the results were fitted to a combination of intramolecular and intermolecular DD contributions. Below 50 MHz the intramolecular term dominates, and a rotational correlation time of 10^{-7} to 10^{-8} s is deduced for the complex. This time is in fair

[72] R. E. D. McClung, *J. Chem. Phys.*, 1972, **57**, 5478.
[73] D. Schweitzer and H. W. Spiess, *J. Magn. Resonance*, 1974, **15**, 529.
[74] W. H. Flygare, *J. Chem. Phys.*, 1964, **41**, 793.
[75] W. H. Flygare and J. Goodisman, *J. Chem. Phys.*, 1968, **49**, 3122.
[76] D. Pumpernik, G. Lahajnar, D. Hadži, and A. Ažman, *Chem. Phys. Letters*, 1974, **26**, 53.

agreement with the deuteron relaxation in the system CCl_3CO_2D–DMSO measured between 4 and 10 MHz, using for the calculation a quadrupole coupling constant of 100—150 kHz. From the intermolecular proton relaxation rate a translational diffusion constant of 1.3×10^{-7} cm^2 s^{-1} was calculated. Contributions to T_1 from the association–dissociation process are negligible.

Another investigation with DMSO was carried out in admixture with water.[77] The authors determined proton relaxation rates in binary mixtures H_2O–DMSO, H_2O–[$^2H_6$]DMSO and D_2O–DMSO at temperatures between 10 and 41 °C. They also studied ternary mixtures D_2O–DMSO–[$^2H_6$]DMSO in order to separate intra- and inter-molecular contributions to the proton relaxation rate of the organic compound.[78] Their results are in agreement with previous information for this system at 26—31 °C, [79] i.e. the correlation time of DMSO in the aqueous mixture is up to 40% greater than that of the pure liquid, and this maximum is reached at 60 mole % water, which means that the aqueous mixture shows more structure than neat DMSO. This effect is more pronounced at lower temperature, as expected. The intermolecular rate behaves as predicted from the macroscopic viscosity. In addition to these previously known data, the intermolecular DD interaction between water and DMSO protons could be extracted from separate T_1 measurements on the non-deuteriated mixture. A noticeable contribution of this term was discovered only at high water content and low temperature, which means that at low water content and higher temperature the water protons are on the average far removed from the DMSO protons.

Spin–lattice relaxation of protons in the mixtures of formamide, N-methylformamide, and NN-dimethylformamide with water has been studied.[80]

Suchanski and Pajak[81] have investigated the spin–internal rotation relaxation contribution in toluene, and m- and p-xylene dissolved in CCl_4 and CS_2. Since they performed selective T_1 measurements for the methyl and ring protons, a separation of the desired relaxation contribution was possible. From the intramolecular relaxation rate of the ring protons obtained by extrapolation in the respective solvents, a value for τ_2 was obtained which could be used to estimate the DD contribution to the methyl protons, assuming fast internal rotation,[82] as indicated by the low barriers reported in the literature. After subtraction of this estimated rate from the intramolecular methyl proton rate, the SR contribution is left. It was found to be equal for all three compounds. Evaluation in terms of the Burke–Chan model[83] gave a SR interaction constant and a correlation time in good agreement with literature values.

The influence of the kind of substitution of the aromatic ring on the overall molecular motion and on the energy barriers to the internal methyl- and methoxy-reorientation of substituted benzenes was investigated by Bovée and Smidt.[84] They

---

[77] T. Tokuhiro, L. Menafra, and H. H. Szmant, *J. Chem. Phys.*, 1974, **61**, 2275.
[78] See ref. 49.
[79] F. Franks, 'Water—A Comprehensive Treatise', Plenum Press, New York, 1973, Vol. 2, p. 563.
[80] Yu. M. Kessler, A. I. Mishustin, P. S. Yastremskii, E. S. Verstakov, and V. P. Emelin, *Zhur. strukt. Khim.*, 1975, **16**, 130.
[81] W. Suchanski and Z. Pajak, *Chem. Phys. Letters*, 1974, **26**, 182.
[82] D. E. Woessner, *J. Chem. Phys.*, 1962, **36**, 1.
[83] T. E. Burke and S. I. Chan, *J. Magn. Resonance*, 1970, **2**, 120.
[84] W. M. M. J. Bovée and J. Smidt, *Mol. Phys.*, 1974, **28**, 1617.

extended their previous study (see Vol. 4, p. 132) to the solutions in deuteriated chloroform of the substituted benzenes (2)—(8).

(2) (3) (4) (5)

(6) (7) (8)

Selective T_1 measurements of the substituent and ring protons were made between $-20°$ and 90 °C in the solutions at three or four different concentrations. With the exception of the phenolic protons and the protons of compound (2), the temperature dependence of the T_1 values could be described very well by an Arrhenius equation. The intramolecular contribution was separated by extrapolation to infinite dilution in chloroform. Since no contribution of SR interaction was to be seen, the authors concluded that the DD interaction is the only effective relaxation mechanism, except for the phenolic protons of compounds (2), (6), and (8). The relaxation of these protons is additionally influenced by other effects, e.g. proton exchange with traces of water. The separate contributions of different substituents to the dipolar relaxation of the ring proton were determined by NOE.[85] The results were evaluated by application of Woessner's formula,[86] which means that the molecules are regarded as symmetric-top rotors with superimposed internal rotation. By using the χ-test, in which the experimental and free-rotation correlation times are compared,[87] Bovée and Smidt showed that the reorientation of the molecules is diffusional. Anisotropic reorientation is important if the molecules contain two polar substituents *para* to each other. The correlation times and the energy barriers for methyl rotation of methyl and methoxy-groups were determined, and the remarkable fact was found that the reorientation of methoxy-groups around the aryl–oxygen bond is of the same order of magnitude as the molecular motion. The

[85] J. H. Noggle and R. E. Schirmer, 'The Nuclear Overhauser Effect', Academic Press, New York, 1971.
[86] D. E. Woessner, B. S. Snowden, and G. H. Meyer, *J. Chem. Phys.*, 1969, **50**, 719.
[87] W. T. Huntress, *Adv. Magn. Resonance*, 1970, **4**, 10.

anomalies in the temperature dependence of the T_1 values in a 2-methyl-substituted phenol are explained by intermolecular hydrogen bonds. In an appendix the authors give expressions for intramolecular DD relaxation of a ring proton by another methyl or methoxy-group in the case of axially symmetric behaviour of the molecule.

Versmold's deuteron relaxation studies of ethanol[26] (Section 3A) were also done in mixtures with glycerol. Here the behaviour of the relaxation rates of the methyl and methylene groups becomes more distinct with increasing glycerol content: whereas the CD_2 rate still exhibits a maximum, now at higher temperature, the rate of the CD_3 group only shows a shoulder at the same temperature but increases further with decreasing temperature. This result is explained in the following way: due to the admixture of glycerol, the rotational motion of the hydroxy-group is expected to slow down, and this effect is also transferred to the methylene group, whereas the methyl group remains unaffected. Thus the slower tumbling motion of the C_3 axis and the faster spinning motion around this axis become so different that separate maxima for the relaxation rate appear. The maximum at lower temperature, which corresponds to the faster motion, falls below the lowest accessible experimental temperature. Using previously derived theoretical expressions[4] quantitative fits to the experimental curves were obtained.

Voelkel and Sillescu[88] investigated hydrogen bonding in the system pyrrole–camphor. They measured deuteron relaxation rates of [N-^{2}H]pyrrole and of [1,2,3,4-^{2}H$_4$]pyrrole in solutions of camphor and camphene. Comparison between the results for the two solvents (in camphene the polar carbonyl group is absent and thus no hydrogen bonding is expected to occur) shows that reorientation of pyrrole in camphor is slower and more anisotropic than in the pure liquid. The latter conclusion was drawn from the different rates for the two non-equivalent types of deuterons in pyrrole.

B. ^{13}C Relaxation.—Rotational motion of the symmetric-top molecules benzene and mesitylene and of the asymmetric molecules toluene and nitrobenzene was studied in various solvents (isopentane, cyclo-octane, cyclohexanol, and CCl$_4$) and also in the pure liquids by a combination of ^{13}C relaxation and depolarized Rayleigh scattering.[89] Similar work on pure benzene was reported in Volume 4 (p. 131). From the ^{13}C relaxation, where DD interaction with the directly attached protons dominates (as confirmed by determination of NOE factors), the correlation time τ_2 is obtained. For the asymmetric molecules toluene and nitrobenzene separate correlation times can be derived from ^{13}C in *ortho–meta* (unresolved) and *para* positions. Together with the correlation time from light scattering, sufficient data exist in order to calculate $\tau_\parallel$ and $\tau_\perp$ or τ_x, τ_y, and τ_z (correlation times for molecular rotations around the different molecular axes for symmetric-top and asymmetric molecules) by application of Huntress's theoretical relation.[3] The results are discussed in relation to the viscosity of the respective mixtures. Linear relationships are found and $\tau_\perp$ (rotational correlation time perpendicular to the symmetry axis) is always more dependent on viscosity than $\tau_\parallel$ in benzene and mesitylene; similarly, τ_y (rotation axis in the plane but not passing through the substituent) is always more dependent on viscosity than τ_z (rotation axis perpendicular to the plane) in toluene

[88] R. Voelkel and H. Sillescu, *Z. phys. Chem. (Frankfurt)*, 1975, **95**, 73.
[89] D. R. Bauer, G. R. Alms, J. I. Brauman, and R. Pecora, *J. Chem. Phys.*, 1974, **61**, 2255.

and nitrobenzene. Thus the rotation that needs the smallest volume shows the least viscosity dependence and is the fastest. It is noted that the viscosity dependence of τ_y in nitrobenzene is twice that of τ_y in toluene; this might indicate formation of a weak dimer which hinders rotation about the y-axis.

In two papers[90, 91] Berger et al. report measurements of ^{13}C relaxation rates and NOE factors of all resolved carbons in the cycloalkanes C_nH_{2n} ($n = 3$—10) and their methyl derivatives ($n = 3$—8) for solutions in CDCl$_3$. Following the work of Grant et al. (see Volume 4, p. 129), the data were analysed in terms of an asymmetric diffusive rotor,[3, 42] and the three diffusion constants were calculated. As the authors note, the more spherical molecules methylcyclo-octane and methylcycloheptane undergo more or less isotropic reorientation whereas the smaller molecules prefer to reorient more rapidly along their long axis. For the larger methylcycloalkanes ($n > 4$) the methyl group relaxes via the overall molecular motion; consequently the barriers to internal rotation of the methyl group were estimated[92] to be larger than 9.6 kJ mol^{-1}.

The assumption that ^{13}C relaxation time measurements may provide a powerful tool for testing association models in liquids, thus giving insights into hydrogen bonding, was reinforced by a paper of Tucker et al.[93] They measured ^{13}C spin–lattice relaxation in t-butyl alcohol, enriched at the carbinol carbon, in [3,4-^{2}H$_2$]-n-hexadecane and in CCl$_4$. In addition, T_1's of phenol in cyclohexane were measured. Intermolecular associations, such as hydrogen bonds, should increase τ_c and hence T_1^{-1}. If the hydrogen-bonded liquid is diluted, e.g. by CCl$_4$, the hydrogen-bond aggregates are broken up and T_1 is increased, as thoroughly studied by Grüner and Hertz[93b] and also demonstrated by Levy et al.,[94] e.g. for phenol. In this connection it is important to be able to perform the measurements in down to very low concentrations. By enriching the carbon under study t-butyl alcohol Tucker and co-workers measured the T_1 at concentrations as low as 0.06 mol l^{-1}. They found a large increase of the carbinol carbon relaxation time on dilution in the solvents mentioned. The data in $C_{16}D_{34}$ were analysed in detail on the basis of a previously proposed hydrogen-bonding association model, the so-called 1-3-6 model.[95] In this model it is assumed that only mono-, tri-, and hexa-meric species are possible, and that internal motions are unimportant, and that only a single rotational correlation time, proportional to the molecular weight of n-mer species, i.e. with $\tau_c^{(n)} = n \times \tau_c^{(1)}$, is present ($\tau_c^{(1)}$ refers to the monomer). Under these conditions an average τ_c^r is calculated in which the contributions of the different species at different concentrations are considered. The proportionality between the calculated τ_c and the observed T_1^{-1} is fairly good. By a fitting procedure, $(1/T_1)^{(1)}$ for the monomer (0.0112 s^{-1}) was found; calculated values of $1/T_1$ agreed quite well with the measured values. The authors[93] emphasize that the model used is in some respects oversimplified and that it represents only one of several possibilities. The main

---

[90] S. Berger, F. R. Kreissl, and J. D. Roberts, J. Amer. Chem. Soc., 1974, 96, 4348.
[91] S. Berger, F. R. Kreissl, D. M. Grant, and J. D. Roberts, J. Amer. Chem. Soc., 1975, 97, 1805.
[92] G. A. Gray and S. E. Cremer, J. Magn. Resonance, 1973, 12, 5.
[93] (a) E. E. Tucker, T. R. Clem, J. I. Seeman, and E. D. Becker, J. Phys. Chem., 1975, 79 1005; (b) M. Grüner and H. G. Hertz, Adv. Mol. Relaxation Processes, 1972, 3, 75.
[94] G. C. Levy, J. D. Cargioli, and F. A. Anet, J. Amer. Chem. Soc., 1973, 95, 1527.
[95] E. E. Tucker and E. D. Becker, J. Phys. Chem., 1973, 77, 1783.

purpose of the analysis was to show that relaxation rates are quite sensitive to molecular size in hydrogen-bonded systems and may prove valuable in elucidating details of multiple equilibria.

In a preliminary paper by Howarth,[96] ^{13}C spin–lattice relaxation data for n-propanol in cyclohexane, H_2O, D_2O, $CDCl_3$, C_6D_6, and $(CD_3)_2SO$ have been presented, together with data for 2-(n-butoxy)ethanol in H_2O, D_2O, and $(CD_3)_2SO$ (values for the two neat liquids are also given). The paper contributes to a study of the very interesting problem of hydrophobic bonding in liquids. The study of solvent effects on ^{13}C spin–lattice relaxation has the advantage that one has the ability to distinguish the motions of different parts of a molecule. The general variation of relaxation times along the carbon chains in both alcohols is as expected, especially the enhanced mobility of the methyl group. However, the comparatively short T_1's in aqueous solution are surprising. Compared to C_6D_6 and $(CD_3)_2SO$, H_2O and D_2O reduce the molecular motion by up to one-third, even after corrections for viscosity. The effect of D_2O is greater than that of H_2O. Thus, Howarth interprets his results as a suggestion that at least some of the entropy decrease connected with hydrophobic bonding comes from a restriction of the solute molecular motions, a suggestion advanced by Aranow and Witten[96b] a long time ago, but never verified experimentally up to now. A private communication from the author[96] suggests that extended study in about eight solvents lends confirmation and reinforcement to the described effects.

In connection with their study of solvation and segmental motions of n-alkyl-ammonium ions (also see Sections 3B and 5B), Levy, Komoroski, and Halstead[53] have measured the ^{13}C spin–lattice relaxation times of the individual carbons of the n-butyl-, n-hexyl-, n-octyl-, n-decyl-, and tri-n-octyl-amines in a variety of solvents, including some where they exist as the corresponding ammonium ions. The relaxation behaviour in the amines is similar to that observed previously for n-alkanes,[97] where a moderate increase in motion is observed near the chain ends.

Some other mixtures of organic liquids with water have been studied. In addition to their ^{13}C relaxation studies in the pure liquids, von.Goldammer et al.[45–47] also investigated the binary mixtures acetone–water, THF–water, and pyridine–water. As reported in Section 3B for the carbonyl carbon of acetone, intermolecular DD interaction with protons contributes significantly to relaxation. Therefore measurement of ^{13}C relaxation for this carbon atom in the mixtures $(CH_3)_2CO–D_2O$ and $(CD_3)_2CO–D_2O$ yielded the total DD rate, from which the intramolecular contribution, calculated from proton, deuteron, and ^{17}O relaxation data,[49,50] was subtracted to obtain the intermolecular one. The intermolecular contribution was found to be significant in acetone-rich mixtures, supporting the idea of microheterogeneities in this concentration range. For the other two mixtures the authors showed that rotational diffusion of the organic component becomes more anisotropic upon addition of water. They analysed their data in terms of the symmetric-top diffusion model.[3,42]

Armitage et al.[98] have measured $^{13}C\ T_1$ of dioxan in mixtures with D_2O. In this

[96] (a) O. W. Howarth, J.C.S. Chem. Comm., 1974, 286; (b) R. H. Aranow and L. Witten, J. Phys. Chem., 1960, 64, 1643.
[97] J. R. Lyerla, jun., H. M. McIntyre, and D. A. Torchia, Macromolecules, 1974, 7, 11.
[98] I. M. Armitage, H. Huber, D. H. Live, H. Pearson, and J. D. Roberts, J. Magn. Resonance, 1974, 15, 142.

more technically oriented paper the rates were reported for the whole composition range at 30 °C.

C. Studies of Several Resonant Nuclei: ^{1}H, ^{2}H, ^{13}C, ^{14}N.—The molecular structures of *cis*- and *trans*-crotonaldehyde [(9) and (10)] were studied in dilute solutions with acetone by a combination of proton relaxation, including ^{1}H–{^{1}H} NOE work, and

(9) *cis*-crotonaldehyde   (10) *trans*-crotonaldehyde

^{13}C relaxation.[99] The reason for working in dilute solutions was to exclude intermolecular interaction. Upon neglect of relaxation mechanisms other than intramolecular DD and of cross-correlation effects one can obtain,[85] from separate relaxation rates of all proton groups and from NOE factors between all of them, the quantity

$$\rho_{ij} = \gamma_i^2 \gamma_j^2 \hbar^2 \tau_c(ij) r_{ij}^{-6}$$

where r_{ij} is the distance between protons i and j and $\tau_c(ij)$ is the correlation time of this internuclear vector. This expression becomes more complicated if one of the protons belongs to the methyl group since it performs internal motion and therefore r_{ij} is no longer constant. The authors derived appropriate expressions for this case, starting from the general approach of Woessner.[82, 100] The proton–proton distance r_{ij} now can be determined from ρ_{ij} if the correlation time is known. To obtain the latter, ^{13}C relaxation rates were measured. Thus all proton–proton distances in the two molecules could be determined relative to the known C—H bond distance (r_{CH} being needed to get τ_c from the ^{13}C rates). In general the results are in very good agreement with the reported microwave structure for *trans*-crotonaldehyde and with a theoretically optimized structure, starting from the X-ray structure, for *cis*-crotonaldehyde. But of course longer distances ($\gtrsim$ 0.35 nm) are determined less accurately than smaller ones, since they correspond to ρ_{ij}'s which are not well determined.

In order to study the dynamics of motion and the nitrogen quadrupole interaction of glycine in aqueous solution, Tzalmona and Loewenthal[101] have performed ^{2}H and ^{14}N relaxation measurements. They measured T_2 of ^{14}N in $^+$NH$_3$CH$_2$CO$_2^-$ (the zwitterion) and T_1 of ^{2}H in $^+$NH$_3$CD$_2$CO$_2^-$ as a function of the pH of the aqueous solution. In the acid region of the aqueous glycine solution the T_2 of ^{14}N is affected by exchange of the protons of the NH$_3$ group. Thus, T_2 is pH dependent and smaller than T_1. However, at pH of 5.6 it was found $T_1 = T_2$ for ^{14}N. Using a NQCC of 1.25 MHz and an asymmetry factor $\eta = 0.51$ measured in the solid

---
[99] R. Rowan, tert., J. A. McCammon, and B. D. Sykes, *J. Amer. Chem. Soc.*, 1974, **96**, 4773.
[100] D. E. Woessner, *J. Chem. Phys.*, 1965, **42**, 1855.
[101] A. Tzalmona and E. Loewenthal, *J. Chem. Phys.*, 1974, **61**, 2637.

compound,[102] the correlation time τ_2 was calculated. The T_1 of deuterium in $^+NH_3CD_2CO_2^-$ is independent of pH. With the value for the NQCC of solid $CD_2Cl_2 (\chi = 169.6$ kHz[103]) a correlation time $\tau_2 = 6.18$ ps was obtained. The results of both independent determinations of τ_2 agreed. Only the NQCC of the zwitterion was known previously. From their measurements the authors were able to determine the ^{14}N NQCC of $NH_2CH_2CO_2^-$ (anionic form) to be 3.48 MHz. Sheinblatt and Gutowsky[104] calculated the average time τ that a proton spends on the NH_2 group (in the anionic form) and on the NH_3 group (in the zwitterion). Using these values the authors[101] calculated the pH dependence of the ^{14}N spin–spin relaxation time. Calculated T_2 values were in good agreement with experiment. Finally, the activation energy for the ^{14}N relaxation (at pH = 5.6) was determined to be 15.0 kJ mol^{-1}. Since the water fluidity is determined by the same activation energy, the authors conclude that the reorientation time of the glycine molecule is dependent on the viscosity of the solution.

5 Relaxation Studies in Electrolyte Solutions

A. Proton and Deuteron Relaxation.—Relaxation rates of water in electrolyte solutions have been studied extensively[105] but recently Jonas and co-workers[106] have extended this work into the high-pressure region. The authors measured T_1 of water deuterons between 1 and 4500 bar in 4.5 mol kg^{-1} electrolyte solutions of CsBr, RbBr, KBr, LiCl, $CaCl_2$, and $LaCl_3$ at 30 °C. For the structure-breaking electrolytes, minima in the relaxation rate were found around 2000 bar, whereas for LiCl a continuous decrease, and for the other structure-forming electrolytes a continuous increase, with increasing pressure were noted. The results were compared with viscosity data[106] and interpreted in terms of changes of the NQCC and/or changes in the effective volume occurring in the Debye equation [which is equivalent to variations in κ of the modified Debye equation (see Volume 4, p. 127)]. Although a reliable distinction between the two effects cannot be made, the authors tended towards ascribing the decrease of the product $(T_1 \cdot \eta)^{-1}$, which is observed for all electrolytes in case of the structure-breaking electrolytes, to a decrease of the NQCC. No such interpretation was given in the case of the structure-forming electrolytes.

Proton relaxation studies at normal pressure in electrolytes containing Na^+, K^+, Cs^+, Zn^{2+}, Cd^{2+}, Hg^{2+}, and Pb^{2+} have been published recently.[107, 108]

The orientation of the water molecules in the hydration sphere of the tetraethylammonium ion in aqueous solution has been studied by Hertz and Wen.[109] They applied a technique described by Hertz and Rädle[110] (see Vol. 4, p. 135) which uses the comparison of intermolecular DD interactions between a nucleus residing on a solute molecule, on the one hand, and the protons of H_2O or the ^{17}O of $D_2^{17}O$, respectively, on the other. As solute species they used 'one proton'-tetraethyl-

---

[102] D. T. Edmonds and P. A. Speight, *Phys. Letters (A)*, 1971, **34**, 325.
[103] F. S. Millet and B. P. Dailey, *J. Chem. Phys.*, 1972, **56**, 3249.
[104] M. Sheinblatt and H. S. Gutowsky, *J. Amer. Chem. Soc.*, 1964, **86**, 4814.
[105] L. Endom, H. G. Hertz, B. Thül, and M. D. Zeidler, *Ber. Bunsengesellschaft phys. Chem.*, 1967, **71**, 1008.
[106] Y. K. Lee, J. H. Campbell, and J. Jonas, *J. Chem. Phys.*, 1974, **60**, 3537.
[107] A. N. Voronovich, L. S. Lilich, and M. K. Khripun, *Yad. Magn. Rezon.*, 1974, **5**, 112.
[108] V. B. Kolokol'tsev and V. A. Shcherbakov, *Zhur. fiz. Khim.*, 1974, **48**, 1566.
[109] H. G. Hertz and W. Y. Wen, *Z. phys. Chem. (Frankfurt)*, 1974, **93**, 313.
[110] H. G. Hertz and C. Rädle, *Ber. Bunsengesellschaft phys. Chem.*, 1973, **77**, 521.

ammonium ions (CHD_2CD_2) $(C_2D_5)_3N^+$ in order to reduce the intramolecular relaxation rate as far as possible. These ions were first dissolved in $D_2^{16}O + D_2^{17}O$ and the relaxation rate of the single proton was measured as a function of the mole percentage of $D_2^{17}O$. In the second experiment the relaxation rate in $D_2O + H_2O$ was determined as a function of the H_2O content. Comparison of the relaxation rates allowed calculation of the H—O distance relative to the H—H distance. In this manner the orientation of the water molecule relative to the methyl–water oxygen vector was derived. Angle β was found to be 116°, which means that on average the water protons preferentially point into the bulk water. A similar result has been found by Hertz and Rädle[110] for the hydrophobic hydration of propionic acid. The systems investigated up to now were tabulated by Hertz and Wen;[109] this list shows that angle β increases with an increase in the number of hydrophobic groups in a molecule. Such a trend supports the assumption of Hertz and Holz[111] that in a hydrophobic hydration sphere the water molecules are preferentially oriented, so causing the relatively high electrical field gradients which strongly influence the relaxation rates of nearby quadrupolar nuclei. In a discussion of their results Hertz and Wen emphasize that, owing to the experimental need to use relatively high ion concentrations, they could not make a final decision whether the clathrate hydrate model or a preferential orientation by an interface effect more appropriately describes the hydrophobic hydration.

Related to the problem of hydrophobic hydration is that of 'hydrophobic interaction'. In past years many efforts have been undertaken to elucidate this effect, which is involved with the attractive forces operating between hydrophobic solutes in aqueous solutions. Hertz[112] has described the use of intermolecular proton spin-lattice relaxation rates to investigate hydrophobic interactions. In a preliminary paper Jolicoeur et al.[113] present experimental results obtained by a related method. They measured the ^1H relaxation times of the methyl protons of t-butyl alcohol as influenced by several paramagnetic hydrophobic species in water and various solvents. Their results indicate that the relaxation induced by these paramagnetic species is approximately twice as large in water as in other solvents of similar viscosity. They interpreted their data as a manifestation of the *pairwise* hydrophobic interaction between t-butyl alcohol and the hydrophobic radical molecule in the aqueous solution.

B. Relaxation of the Nuclei: ^{13}C, ^{17}O, ^{69}Ga, ^{71}Ga, ^{113}Cd.—^{13}C Relaxation in a 1 mol l^{-1} aqueous solution of [*carboxy*-^{13}C]acetic acid was measured[114] as a function of pH without and with addition of variable amounts (3.8 × 10^{-4} to 7.6 × 10^{-4} mol l^{-1}) of Cu^{2+} ions. The rate shows a strong pH dependence in solutions containing Cu^{2+}, with a maximum at pH 4—5, whereas no effect is observable in Cu^{2+}-free solutions. Relaxation of the carboxyl carbon is dominated by DD interaction with the paramagnetic ion. At intermediate pH the rate is raised by complexation of Cu^{2+} with the carboxylate anions, whereas at low pH protons compete effectively with Cu^{2+} for the acetate, and at high pH hydroxy-ions compete with the acetate for the metal ions.

[111] H. G. Hertz and M. Holz, *J. Phys. Chem.*, 1974, **78**, 1002.
[112] H. G. Hertz, *J. Solution Chem.*, 1973, **2**, 239.
[113] J. Jolicoeur, E. Firkins, and P. Bernier, *J. Phys. Chem.*, 1974, **78**, 851.
[114] J. S. Cohen, R. B. Bradley, and T. R. Clem, *J. Amer. Chem. Soc.*, 1975, **97**, 908.

In order to obtain information about segmental motions in n-alkylammonium ions in D_2O and various organic solvents, Levy et al.[53] (also see Section 3B) performed ^{13}C relaxation-time measurements of the individual carbons as a function of concentration. It was found that the T_1's are greatly reduced relative to the parent amines, indicating much slower overall reorientation of the ions in solution. The authors draw the conclusion that the ionic site is fixed relative to the free amine, allowing increasing segmental motion to be observed for methylene units as a function of distance from the ionic site. In D_2O, increasing the concentration of ammonium ions results in an increased restriction of the polar end, with a reduced effect as one moves towards the terminal methyl. Restrictions on the movement of the solvent-ion are markedly smaller for the shorter chain ammonium ions like butyl and hexyl at relatively high dilution in polar solvents. For the longer chains the motion of the aliphatic chain terminus is essentially 'decoupled' from the polar head group and also from macroscopic viscosity. Comparison of the motions of n-butyl- and n-hexyl-ammonium ions in polar (CD_3OD, D_2O, acetone) and in non-polar (e.g. $CDCl_3$, benzene, dioxan) media shows that in the latter solvents the overall motion is restricted relative to the same concentrations in D_2O and that segmental motion is present to a considerable extent. Segmental motion is less pronounced in polar solvents than in non-polar solvents. It seems that the ion is more easily stabilized in polar media and ion–counter-ion interactions are reduced. Finally, the authors[53] have calculated the rate of terminal methyl spinning, applying the same treatment as Lyerla et al.[97] In the butylammonium case they found a decrease of this rate as ion concentration increases.

Applying FT techniques, Vold and Vold[115] were able to study the ^{17}O resonance in a 2 mol l^{-1} Na_2MoO_4 (0.8% ^{17}O) aqueous solution. They found an intense singlet and six satellite lines arising from scalar coupling to ^{95}Mo. Measurements of T_1 and T_2 yielded the result $T_1 = T_2 = 220$ ms, corresponding to a natural linewidth of 1.45 Hz for the centre line. The relaxation time in molybdate is considerably longer than that observed earlier in permanganate and perrhenate, evidently due to a much smaller ^{17}O NQCC. By first estimating the reorientational correlation time to be 9 ps, using the Debye relation and the Gierer–Wirtz correction,[116] the authors[115] derived an approximate NQCC of 720 kHz for ^{17}O. This exceptionally small value indicates almost spherical symmetry of the electron distribution around oxygen in molybdate. The authors gave also an estimate of the ^{95}Mo NQCC, ca. 400 kHz.

Buslaev and co-workers[117] have investigated non-aqueous solutions of $GaCl_3$, $GaBr_3$, and GaI_3 in MeCN, C_6H_6, $MeNO_2$, MeCN–C_6H_6, and MeCN–$MeNO_3$ mixtures by ^{69}Ga, ^{71}Ga, and ^1H n.m.r. They found that the average solvation number of Ga^{3+} increases on replacing Cl^- by I^-. The authors concluded that this is caused by the easier replacement of I^- by the solvent. In another section of their paper the authors discussed in more detail the peculiar spin relaxation of ^{69}Ga and ^{71}Ga in $Ga(MeCN)_n^{3+}$ complexes.

In connection with an extensive FT investigation of ^{113}Cd Cardin et al.[118]

---

[115] R. R. Vold and R. L. Vold, J. Chem. Phys., 1974, 61, 4360.
[116] A. Gierer and K. Wirtz, Z. Naturforsch., 1953, 8a, 532.
[117] Ya. A. Buslaev, V. P. Tarasov, S. P. Petrosyants, and N. N. Melnikow, Zhur. strukt. Khim., 1974, 15, 617.
[118] A. D. Cardin, P. D. Ellis, J. D. Odom, and J. W. Howard, jun., J. Amer. Chem. Soc., 1975, 97, 1672.

determined some T_1 data for organocadmium compounds and aqueous cadmium halide solutions. They make some effort to clarify the relevant relaxation mechanisms but on the whole the work remains unfinished. They show that in the aqueous solutions intermolecular DD interactions between ^{113}Cd and the water protons are effective and that SC interactions between the halogen nuclei and the cadmium nucleus exist. On the other hand, relaxation in dimethylcadmium is dominated by SR at higher temperatures and probably by CSA at lower temperatures.

C. **Relaxation of Alkali-metal and Halogen Nuclei:** ^{7}Li, ^{23}Na, ^{39}K, ^{41}K, ^{87}Rb, ^{133}Cs, ^{35}Cl, ^{79}Br, ^{81}Br.—All these nuclei possess a nuclear quadrupole moment. Due to this fact, their relaxation behaviour is dominated by QF interaction (only in the case of Li, DD interaction may play a noticeable role). As described in Volume 4 (p. 139), Hertz[119] has derived a new electrostatic theory of QF relaxation for nuclei of ions with noble-gas structure in electrolyte solutions. In succeeding experimental studies (see Vol. 4, p. 140) it was shown that this theory is capable of interpreting the relaxation rates of the above-mentioned nuclei in aqueous solution. The next step was to test the applicability of the electrostatic theory to non-aqueous electrolyte solutions. Melendres and Hertz[120] did this for ^{7}Li, ^{23}Na, ^{87}Rb, and ^{133}Cs in methanol and for ^{87}Rb in ethanol, ethylene glycol, and formamide. The measured relaxation rates, extrapolated to zero ion concentration (in this case only ion–solvent interactions must be considered), were compared with relaxation data calculated for three different models underlying the theory for the state of solvation. It turned out that the calculated values were of the correct order of magnitude if, for the nuclei in question, the appropriate solvation model was used. Generally, it could be seen that for the large ions such as Rb$^+$, Cs$^+$, (and Br$^-$), which are 'structure breaking' in water, a tighter packing of the solvation sphere in the organic solvents (especially in MeOH and EtOH) compared to water has to be assumed. This result is quite reasonable if the entropies of solvation of the ions in the organic solvents are considered, as these are more negative than the entropies of hydration in water. In a separate section of their paper the authors have calculated, for finite electrolyte concentrations, the ion–ion contributions to the relaxation rate as a function of concentration. Summarizing, the authors point out that in principle all observed effects could be interpreted in terms of the electrostatic theory.

Mazitov and Itkina[121] applied ^{7}Li spin–lattice relaxation-time measurements in order to investigate the system LiOH–NaOH–H$_2$O. They obtained evidence of complex formation. For each complex they found a corresponding branch in the relaxation–concentration curve.

In a paper by Sahm and Schwenk[122] which is mainly concerned with the determination of the absolute shielding of the potassium ions in water as well as of the hyperfine structure anomalies, relaxation data of ^{39}K and ^{41}K are also reported. It should be mentioned that the ^{40}K resonance signal in a 31 mol kg^{-1} solution of KNO$_2$ in D$_2$O was detected during this work. According to the authors this was the first time that the ^{40}K resonance had been found. Transverse relaxation times of ^{39}K for infinite dilution in H$_2$O, D$_2$O, MeOH, and ethylenediamine are given.

[119] H. G. Hertz, *Ber. Bunsengesellschaft phys. Chem.*, 1973, **77**, 531, 688.
[120] C. A. Melendres and H. G. Hertz, *J. Chem. Phys.*, 1974, **61**, 4156.
[121] R. K. Mazitov and L. S. Itkina, *Zhur. neorg. Khim.*, 1974, **19**, 3203.
[122] W. Sahm and A. Schwenk, *Z. Naturforsch.*, 1974, **29a**, 1754.

The difference between the values in D_2O and H_2O is only about 8%. In the Reporters' opinion this finding is in contradiction to experience with the relaxation of other alkali-metal-ion nuclei and halide nuclei, where a difference of about 20%, corresponding to the viscosity difference of H_2O and D_2O, has been found (see e.g. refs. 123, 124). This discrepancy may be caused by the relatively high experimental uncertainties of about 10% estimated for the ^{39}K relaxation data. Sahm and Schwenk also measured the T_2's of ^{41}K and ^{39}K in some highly concentrated solutions.[122] They found $T_2(^{39}K)/T_2(^{41}K) = 1.36$, a value not too far from 1.48, the ratio of the squared quadrupolar moments of the two isotopes. This finding indicates that only the QF interaction contributes to the relaxation. The ^{39}K transverse relaxation times of nine potassium salts in H_2O, for KOH in MeOH, and for KI in ethylenediamine were also determined as a function of concentration, but the authors give no interpretation of the data. The T_2 value of ^{39}K in MeOH at low concentration is in fairly good agreement with the T_1 value of ^{39}K in a 0.48 mol l^{-1} KI solution in MeOH at 298 K, reported in a paper by Shporer and Luz.[125] These authors[125] have studied the kinetics of complexation of K^+ ions with dibenzo-18-crown-6 (DBC), a macrocyclic ionophore, in methanol by means of ^{39}K relaxation-time measurements. Those ionophores are capable of forming stable complexes with alkali-metal ions. Shporer and Luz have measured the temperature dependence of T_1 between + 40 and − 90 °C for a 0.48 mol l^{-1} KI sample and a sample of 0.48 mol l^{-1} KI + 0.14 mol l^{-1} DBC. The occurrence of a kinetic process in which K^+ exchanges between the solvated and complexed form is indicated. Below − 55 °C the exchange is slow, and only the signal of the solvated K^+ can be seen, whereas above − 10 °C the exchange is fast and the measured rate is the average of the rates in the two environments. The exchange rate was derived by a quantitative analysis of the n.m.r. data. A similar experiment performed on Rb^+ ions in methanol showed that even at −50 °C the exchange rate in this case was sufficiently fast for the measured ^{87}Rb transverse relaxation times to be averaged values of the two environments.

Berman and Stengle[126] have made a study of the tendency of ClO_4^- to form contact ion pairs as a function of cation and solvent by measuring the ^{35}Cl resonance linewidth $\Delta v_{\frac{1}{2}}$. The shape of the resonance line was found to be Lorentzian; thus the linewidth is directly proportional to the transverse relaxation rate. By comparing the $\Delta v_{\frac{1}{2}}/\eta$ (η = viscosity) values of 0.5 mol l^{-1} solutions of magnesium, lithium, and sodium perchlorates in different solvents, they distinguished in a very qualitative way three classes of solvents according to their promotion of contact-ion association. The basic assumption of their treatment, that large field gradients and therefore large linewidths are only due to contact ion pairing, is based on a simple calculation in which the only source of field gradients at the ^{35}Cl nucleus is assumed to be the point charge of the counter-ion. However, it can be calculated that asymmetries in the solvation sphere may cause large field gradients (e.g. see refs. 111, 119, and 120); in the Reporters' opinion such asymmetries may occur in the case of solvent-separated and solvent-shared ion pairs. Additionally, it is possible that the solvation sphere in distinct solvents has no complete cubic symmetry, even without the influ-

[123] M. Eisenstadt and H. L. Friedman, *J. Chem. Phys.*, 1966, **44**, 1407.
[124] D. E. O'Reilly and E. M. Peterson, *J. Chem. Phys.*, 1969, **51**, 4906.
[125] M. Shporer and Z. Luz, *J. Amer. Chem. Soc.*, 1975, **97**, 665.
[126] H. A. Berman and T. R. Stengle, *J. Phys. Chem.*, 1975, **79**, 1001.

ence of a counter-ion. In this connection it should be mentioned that Berman and Stengle also measured the salt concentration dependence of the $\Delta v_{\frac{1}{2}}/\eta$ ratio and, in all cases investigated, there was no strong decrease (often even an increase is found) of this ratio at low concentrations. If ion–ion interactions were alone responsible for the linewidth, one would expect a very much lower value of $\Delta v_{\frac{1}{2}}/\eta$ at infinite dilution, where ion–ion contributions are absent. It may be argued that such a decrease of the linewidth occurs in *all* systems investigated, but only at concentrations too low to be accessible for n.m.r. investigations; as long as this point is unclarified, such conclusions as the authors' should only be drawn with care. Another point where one has to be careful, in the Reporters' opinion, is in the comparison of the relaxation in different solvents having the same concentration (0.5 mol l⁻¹) on the molarity scale. In the case of water there are many more solvent molecules per ion at this concentration than, *e.g.*, ethyl acetate, where almost all solvent molecules are needed to build up solvation spheres for $Mg(ClO_4)_2$.

Applying the so-called 'halide probe technique', Garnett and Halstead[127] used bromine-relaxation measurements to study the reorientation of a bromomercury compound (bromomercurio-cinnamic acid) in aqueous solutions to which KBr was added. The bromide ion exchanges rapidly between the bromomercury compound (bound state) and the 'free' state in the bulk water. The measured relaxation time is an average of the values in the two states.[128] By temperature-dependent measurements they showed that the relaxation time in the bound state is much longer than the exchange time, and they were able to calculate the relaxation times in the two states. Using the value of T_1 in the bound state, and applying the formula for intramolecular quadrupole relaxation, they give an estimate of τ_2, the reorientation correlation time of the bound state, as *ca*. 10^{-10} s at 298 K.[127] In a pure KBr solution the activation energy for the ^{79}Br and ^{81}Br relaxation process was determined: two values were obtained; below 310 K, 19 kJ mol⁻¹, and above 310 K, < 3.5 kJ mol⁻¹. The last value differs considerably from earlier measured values *e.g.* for NaBr or CsBr solutions[105] and for KBr solutions.[111, 129]

6 Relaxation Studies in Gas–Liquid Solutions

Three papers concerned with proton and deuteron relaxation of liquids in which gases are dissolved will be discussed under this heading.

Assink[130] has studied the deuteron relaxation rate of n-octane. He applied a rather insoluble gas, helium, and a relatively soluble gas, argon. These two gases with different solubilities were chosen to contrast the effect of pressure (up to 700 bar) with the effect of solution. Whereas the rate increases upon compression by helium, there is a decrease if pressure is applied in the case of argon. These results have been compared with spin-echo self-diffusion constants of octane (measured by the same author) and with viscosity data. Viscosity increases with increasing pressure, and the diffusion constant shows the expected behaviour (*i.e.* decrease of diffusion constant) for helium, but again opposite behaviour is shown for argon. Thus the primary effect of helium is to compress the liquid and to restrict rotational

---

[127] M. W. Garnett and T. K. Halstead, *J.C.S. Faraday II*, 1974, **70**, 1920.
[128] H. G. Hertz, *Z. Elektrochem.*, 1961, **65**, 20.
[129] D. E. O'Reilly, G. E. Schacher, and K. Schug, *J. Chem. Phys.*, 1963, **39**, 1756.
[130] R. A. Assink, *J. Chem. Phys.*, 1975, **62**, 790.

and translational motions of the molecules in the liquid. On the other hand argon forms a solution with octane in which the octane molecules are able to rotate and translate more freely. Quantitative conclusions reached in this paper should be considered with caution, in the Reporters' opinion. The internal motions are treated rather crudely, following a proposal by Lyerla et al.,[97] and therefore the derived over-all molecular rotational correlation times are dubious. These were related to viscosity by a modified Debye equation based on the approach of Gierer and Wirtz;[116] however, viscosity was not measured in the argon solutions but estimated from the self-diffusion data. The author's conclusion that the Gierer–Wirtz microviscosity theory explains the experimental relaxation rates satisfactorily is therefore not well established.

The problem of proton relaxation in liquid water still containing dissolved oxygen gas has been reconsidered by Parker and Harmon.[131] Temperature-dependent spin–lattice relaxation data at 4 MHz and the data at 28 MHz of Hausser and Noack[132] have been interpreted in terms of Torrey's modified theory of relaxation by translational diffusion.[133, 134] Using independently measured O_2–H_2O diffusion data, Parker and Harmon showed that electron–proton dipolar coupling modulated by translational diffusion is alone required to explain the relaxation rates between 10 and 120 °C. Scalar interaction, considered as a relaxation mechanism by other authors, need not be taken into account.

In this connection a short paper by Lewa[135] may be mentioned, where an almost technical application of T_2 measurements is described. Lewa has measured T_2 in two water samples containing dissolved oxygen. To one sample he added a layer of oil, and the other sample served as standard. The difference between the relaxation times of the two samples, ΔT_2, as a function of time was used to study the decrease of the oxygen concentration in the water under the oil layer.

7 Relaxation Studies of Gases

A. Theoretical Work.—In a series of three papers, the first of which was reported in Volume 3 (p. 107), Shizgal[136-138] derived theoretical relations between T_1 and parameters of the potential functions describing interactions between atoms in helium (^{3}He) and xenon (^{129}Xe) gas. His calculations are based on the kinetic theory approach of Chen and Snider[139] but employ a distorted-wave Born approximation for the transition operator of collision theory. Whereas in ^{3}He only DD interactions are responsible for relaxation, SR is the dominating mechanism in ^{129}Xe. In the second paper[137] the Lennard-Jones potential is assumed to describe interatomic interactions, and by numerical evaluation of the theoretical expressions derived in the first publication[136] two linear relations between the minimum value of T_1 and the corresponding temperature on the one hand (from a plot of T_1 versus

---

[131] D. S. Parker and J. F. Harmon, *Chem. Phys. Letters*, 1974, **25**, 505.
[132] R. Hausser and F. Noack, *Z. Naturforsch.*, 1965, **20a**, 1668.
[133] H. C. Torrey, *Phys. Rev.*, 1953, **92**, 962.
[134] J. F. Harmon and B. H. Muller, *Phys. Rev.*, 1969, **182**, 400.
[135] C. J. Lewa, *Acta Phys. Polon.* (*A*), 1974, **46**, 643.
[136] B. Shizgal, *J. Chem. Phys.*, 1973, **58**, 3424.
[137] B. Shizgal, *Chem. Phys. Letters*, 1973, **20**, 265.
[138] B. Shizgal, *Chem. Phys.*, 1974, **5**, 464.
[139] F. M. Chen and R. F. Snider, *J. Chem. Phys.*, 1967, **46**, 3937.

temperature), and the parameters ε and σ in the Lennard-Jones potential on the other, were discovered. Thus from measurements of T_1 min and T_{min} both ε and σ can be determined. In the third paper[138] Shizgal again assumed a Lennard-Jones potential, but now in addition the radial dependence of the SR interaction between nuclei comes into play. The form of this dependence is uncertain, and only a semi-empirical approach was followed, including experimental results from temperature-dependent chemical shifts. The calculated temperature dependence of T_1 between 200 and 450 K shows a monotonic decrease with increasing temperature. For this calculation best known Lennard-Jones parameters were used, but curves for different parameters in the spin-dependent interaction term were presented.

Difficulties have always been encountered in rationalizations of the experimental relaxation data of polyatomic symmetric-top molecules; e.g., as reported in Volume 1 (p. 119), polyatomic gases at pressures of a few atmospheres would be expected to give a density dependence of the relaxation time of the form $T_1 \propto \rho$ if SR interactions dominate, but instead a stepped T_1/ρ versus ρ plot was observed experimentally for the proton relaxation in CHF_3 and CH_3F[140] [although in more recent re-examinations these steps are absent or greatly reduced[141, 142] (see Section 7C)]. Also, as reported in Volume 3 (p. 110), if the experimental data are used to calculate SR interaction constants by employing conventional SR theory,[143] the results differ greatly from the values derived from molecular-beam studies. Therefore, re-examination of the relaxation theory in symmetric-top molecules was needed, and this has recently been done by Sanctuary and Snider. Their theory is laid down in a series of three papers.[144-146] Their work has resulted finally[146] in a satisfactory theoretical description of recent experimental proton data for CHF_3[141, 142] with a non-linear density dependence. They even demonstrate that steps can arise theoretically, although not as sharp as those reported by Dong and Bloom.[140] As already discussed by Bloom et al.,[143] such non-linear density dependence arises from high-frequency contributions which NOE experiments have confirmed to be mainly intramolecular DD interactions, and just these are considered in the new theory by Sanctuary and Snider. However, in contrast to the earlier theory, it was shown that spin-symmetry should be included[145] and that collisional shifts of spectral components are very important.[146] If one remembers that the relaxation rate is proportional to the spectral density at the Zeeman frequency, a description in the frequency domain is best suited for an understanding of the relaxation behaviour. At low density, sharp peaks occur at the different transition frequencies for a free symmetric-top rotor, but with increasing pressure they broaden, overlap each other, and shift in frequency. Thus the spectral density in the range of interest increases, resulting in an increase of the relaxation rate. Three adjustable parameters for width, shift, and overlap of the spectral components remain to account for a fit to the experimental data, and the authors show that reasonable figures can be obtained for them.

[140] R. Y. Dong and M. Bloom, *Phys. Rev. Letters*, 1968, **20**, 981.
[141] R. L. Armstrong and J. A. Courtney, *Canad. J. Phys.*, 1972, **50**, 1262.
[142] S. Pausak and J. S. Waugh, *J. Chem. Phys.*, 1974, **61**, 2165.
[143] M. Bloom, F. Bridges, and W. N. Hardy, *Canad. J. Phys.*, 1967, **45**, 3533.
[144] B. C. Sanctuary and R. F. Snider, *Canad. J. Phys.*, 1975, **53**, 707.
[145] B. C. Sanctuary and R. F. Snider, *Canad. J. Phys.*, 1975, **53**, 723.
[146] B. C. Sanctuary and R. F. Snider, *Canad. J. Phys.*, 1975, **53**, 739.

Two other theoretical papers by Sanctuary are concerned with QF relaxation[147] and SR relaxation [148] in the gas phase. In the first paper homonuclear diatomic molecules, which exist as *ortho* and *para* species, are discussed. In the search for deviations from single relaxation processes, nuclear spins of 1/2 need not be considered; thus spins greater than 1/2, for which QF relaxation dominates, are the subject of this study. Expressions were obtained for the relaxation rates of both *ortho* and *para* X_2 molecules, and it was concluded that a dual relaxation process should be observed for diatomic gases with spin greater than 1. The second of Sanctuary's papers was written to clarify the role of nuclear spin symmetry in polyatomic symmetric-top molecules of the types ZX_3Y and ZX_3, which possess C_{3v} symmetry, and in this way to resolve the aforementioned discrepancy between SR interaction constants found from spin relaxation and molecular beam studies. He came to the conclusion that spin-symmetry aspects cannot resolve this disagreement.

B. Mono- and Di-atomic Gases.—For the first time in this series of Reports spin–lattice relaxation-time measurements of a monatomic gas, namely ^{3}He, are discussed. The interactions in this gas causing T_1-relaxation are relatively weak, since they exist only for the duration of the collision between two atoms. According to Bloembergen's[149] suggestion, the dilute ^{3}He gas relaxes *via* the collisional modulation of the DD interaction between pairs of atoms. On this assumption (at 1 atm pressure and room temperature) T_1 may be calculated to be about 10^7 s. In order to shorten the relaxation time Karra and Longenecker[150] added oxygen gas and measured the dependence of ^{3}He relaxation on the density of oxygen at room temperature at a pressure of 25 atm. They found a linear relationship, consistent with the theory of Oppenheim and Bloom.[151] Unexpectedly, however, the linear dependence showed a distinct break at about 9 atom % oxygen; a second straight line, shifted to higher $1/T_1$ values, is observed above 9 atom %. The intercept of these two lines at zero percentage concentration of oxygen yielded ^{3}He relaxation times of 10 and 2.5 s respectively. Traces of oxygen still present account for the short times (1 ˙p.p.m. concentration of oxygen significantly affects $1/T_1$). The authors did not explain the difference in the two extrapolated values nor the occurrence of the break in the oxygen concentration dependence.

Chapman and Richards[152] had more success in the determination of the ^{3}He spin–lattice relaxation time in ^{3}He gas. They performed their measurements at 4.2 K, where all impurities other than ^{4}He will be condensed out on the walls. The cleaning procedure of the sample chamber was similar to that described by Horvitz.[153] Measurements were made over a density range varying from the dilute-gas limit to a dense-gas region ($\rho \approx 3 \times 10^{-3} - 7 \times 10^{-2}$ g cm^{-3}). For the intermediate region (0.01 < ρ < 0.05 g cm^{-3}), where the authors found the correct density dependence for bulk relaxation, the data were fitted by $\rho T_1 = 61 \pm 2$

---

[147] B. C. Sanctuary, *Canad. J. Phys.*, 1973, **51**, 2488.
[148] B. C. Sanctuary, *Canad. J. Phys.*, 1974, **52**, 387.
[149] N. Bloembergen, 'Nuclear Magnetic Relaxation', Benjamin, New York, 1961.
[150] J. S. Karra and D. U. Longenecker, *Phys. Letters*, 1973, **44A**, 301.
[151] I. Oppenheim and M. Bloom, *Canad. J. Phys.*, 1961, **39**, 845.
[152] R. Chapman and M. G. Richards, *Phys. Rev. Letters*, 1974, **33**, 18.
[153] E. P. Horvitz, *Phys. Rev.*, 1970, **A1**, 1708.

s g cm^{-3}. This value may be compared with a theoretical value of 76 s g cm^{-3} of Shizgal,[137] derived using the best interatomic potentials currently available. The satisfactory agreement indicates that the DD interaction plays the major role in the ^{3}He longitudinal relaxation in the bulk gas. At lower densities wall relaxation predominates and at higher densities the binary collision theory is inadequate. Finally, Chapman and Richards have investigated in more detail the wall relaxation present at low densities, and they describe a model in which $(T_1)_{wall}$ is related to $(T_1)_{ad}$, the relaxation time of ^{3}He in the adsorbed phase.

The only diatomic gas to be studied during the period of this Report was molecular hydrogen. As already pointed out in Volume 2 (p. 114) and Volume 3 (p. 107) molecular hydrogen is an excellent system to test the theory of nuclear spin relaxation in molecular gases. There are two theoretical approaches to a description of the nuclear spin relaxation in molecular hydrogen at low densities, namely the Bloom–Oppenheim[154] and the Chen–Snider[155] theories, which lead to equivalent results. At low temperatures only the $J = 1$ rotational state of H_2 needs to be considered (collisions cause only transitions between states of different M) and theory predicts that the ratio τ_J/τ_2 of the correlation times associated with SR and DD interactions should be numerically equal to 0.6. To test the basic approximations of the theory it is therefore important to measure this ratio experimentally. In the room-temperature region H_2 may be treated as a two-J-level system ($J = 1$ and $J = 3$). Transitions between states of different J are possible. Starting from the Bloom–Oppenheim theory one may introduce a parameter R which is a measure of the relative importance to relaxation of collisions involving transitions between the $J = 1$ and the $J = 3$ states and collisions involving transitions within the $J = 1$ state. By measuring the nuclear spin–lattice relaxation times of H_2 at low density at 49 MHz through the region of the characteristic minimum in the T_1 vs. ρ plot at both 296 and 385 K, Kisman and Armstrong[156] have shown that their experimental data are consistent with R values in the ranges $1 < R < 10$ at 296 K and $0.6 < R < 20$ at 385 K, i.e. the Bloom–Oppenheim theory, including inelastic collisions, is able to explain the relaxation in H_2 for temperatures up to about 400 K. In this work an improved n.m.r. pulsed spectrometer was used and consequently the scatter in the data is considerably reduced from that of earlier studies; this has allowed the determination of τ_J/τ_2 with higher accuracy, and thus a closer test of the theory. Hardy[157] had concluded in 1966 from his data that $0.6 \leq \tau_J/\tau_2 \leq 1.0$. By measuring T_1 of H_2 at 77.5 K in the density range from 0.1 to 10.0 amagat Kisman and Armstrong now arrived at $0.58 \leq \tau_J/\tau_2 \leq 0.83$. In the authors' opinion this result means that the basic approximations of the relaxation theory for the one J-level system have now been verified by experiment.

In a later paper Armstrong et al.[158] have reported the next step in testing the Bloom–Oppenheim theory. They investigated the longitudinal relaxation of H_2 in hydrogen gas mixtures at low densities; experiments were performed at 298 K, under experimental conditions similar to those described above, on the systems H_2—He, H_2–Ar, and H_2–N_2. Using the same fitting procedure as before,[156] the authors

[154] M. Bloom and I. Oppenheim, 'Intermolecular Forces', Interscience, New York, 1967, p. 549.
[155] R. M. Chen and R. F. Snider, *J. Chem. Phys.*, 1968, **48**, 3185.
[156] K. E. Kisman and R. L. Armstrong, *Canad. J. Phys.*, 1974, **52**, 1555.
[157] W. N. Hardy, *Canad. J. Phys.*, 1966, **44**, 265.
[158] R. L. Armstrong, K. E. Kisman, and W. Kalechstein, *Canad. J. Phys.*, 1975, **53**, 1.

showed that the theoretical curves according to the Bloom–Oppenheim theory also provide a good representation of the data for mixtures. Since a lack of a comprehensive treatment of inelastic collisions is a major theoretical weakness at this time; any quantitative information from experiments concerning the importance of such collisions in the relaxation process is very helpful. Therefore the authors calculated the probable ranges of values for R in all the systems investigated, and it turned out that R for pure H_2 is greater than those for H_2–He and H_2–Ar, which are greater than that for the H_2–N_2 mixture. This means that the role of inelastic collisions is less important in mixtures of H_2 with other gases than in pure H_2. This result is to be expected if certain types of inelastic collisions, like the so-called 'resonant collisions' $(1,3) \rightleftharpoons (3,1)$ and 'quasi-resonant collisions', do play an important role.

C. Polyatomic Gases.—Methane is the most studied polyatomic gas. The CH_4-proton relaxation is dominated by SR interaction. The SR interaction in spherical-top molecules is characterized by a scalar term and a tensor term. As described in Volume 2 (p. 116), Beckmann et al.[159] showed that in gaseous CH_4 the assumption of a simple exponential correlation function is not consistent with the measurement of T_1 as a function of pressure. The reason for this is that the tensor part of the SR interaction couples the rotational states split by the centrifugal distortion. Lalita[160] has investigated the validity of the correlation functions used by Beckmann et al. for pure methane, this time using CH_4–He mixtures. He measured T_1 as a function of ρ for 85% CH_4–15% He and 29.5% CH_4–70.5% He mixtures at 298 K, and has discussed the results using the formula given in Volume 2 (p. 116). It turns out that it is necessary also to take the centrifugal distortion effects into account in CH_4–He mixtures. Lalita's results yielded an average centrifugal distortion frequency ω_k in the range 200 MHz $< \omega_k <$ 400 MHz, as compared to 160 MHz $< \omega_k <$ 250 MHz obtained by Beckman et al. The cross-section for molecular reorientation due to CH_4–He collisions was also calculated. Comparison with the kinetic cross-section indicates that on average it takes about five CH_4–He collisions to randomize the rotational angular momentum of the CH_4 molecule.

It is well known that nuclear spin–lattice relaxation-time measurements in gases may be used to obtain a deeper insight into the intermolecular potential, especially the anisotropic part. An interesting paper by Rajan et al.[161] on T_1 relaxation in CH_4–inert gas mixtures is concerned with this problem. They determined proton spin–lattice relaxation times in CH_4–He, CH_4–Ne, and CH_4–Ar mixtures as a function of temperature T, density ρ, and the composition in the region where $T_1 \propto \rho$. The T_1/ρ values were then extrapolated to 100% inert gases, yielding values for $(T_1/\rho)_{CH_4-X}$ (X is the inert gas), the contribution due to CH_4–inert gas collisions alone. In the case of SR interaction one expects $T_1/\rho \propto T^{-3/2}$. This temperature dependence has been found in many earlier studies on pure gases for which the SR interaction is dominant and is in accord with the Bloom–Oppenheim theory, which is based on the hard-sphere intermolecular potential and the 'weak-collision approximation'. In 1971 Lalita and Bloom[162] found the first exception to

---

[159] P. A. Beckmann, M. Bloom, and E. E. Burnell, *Canad. J. Phys.*, 1972, **50**, 251.
[160] K. Lalita, *Canad. J. Phys.*, 1974, **52**, 876.
[161] S. Rajan, K. Lalita, and S. V. Babu, *J. Magn. Resonance*, 1974, **16**, 115.
[162] K. Lalita and M. Bloom, *Chem. Phys. Letters*, 1971, **8**, 285.

this 3/2 power law, CH_4 infinitely diluted with He. The later experiments of Rajan and co-workers have yielded $(T_1/\rho)_{CH_4-X} \propto T^{-n}$, with $n = 0.28$ for He, $n = 1.1$ for Ne, and $n = 1.25$ for Ar. This significant departure from the $n = 3/2$ value implies, in the opinion of the authors, that the hard-sphere potential is not a good approximation for these mixtures, since the 'weak collision approximation' is still expected to be valid for collisions between CH_4 and inert-gas molecules. On the basis of their data, the authors have proposed a model for V_1, the anisotropic part of the intermolecular potential for these mixtures. They assume the 6-12 Lennard-Jones potential for the isotropic part, and for the attractive part of V_1, a potential proportional to r^{-7} derived by Buckingham,[163] which depends on the hyperpolarizability A of the CH_4 molecule. The repulsive part of V_1 is taken to have the same angular dependence as that of the attractive part, plus an r^{-12} dependence on the distance. Additionally they assume that the correlation time of the SR interaction can be approximated by the average lifetime of the molecule in the given J-state. The value of A can then be obtained by interpreting the temperature dependence of T_1 in terms of this potential.[143] From their experimental data Rajan et al. obtained values for A (10^{32} cm^4) of 2.71, 0.88, and 0.89 for CH_4–He, CH_4–Ne, and CH_4–Ar mixtures, respectively. The last two values are in good agreement with a theoretical estimate[163] of $A = 0.97 \times 10^{32}$ cm^4, whereas the value for CH_4–He differs by a factor 3. In the authors' opinion this difference may be due to the much lighter He atom probing more of the repulsive part of V_1, which might have a different angle dependence from the attractive part.

Previously reported proton relaxation rates in methyl fluoride and fluoroform[140] have been remeasured by Pausak and Waugh.[142] They covered a pressure range up to 80 amagat at 26—28 °C. The new T_1 data are much larger than the old ones, and no evidence of the step-like behaviour in the T_1/ρ versus ρ plot noted earlier was found. However, as reported in Section 7A, a new theory by Sanctuary and Snider[146] is able to account for such steps.

---

[163] A. D. Buckingham, Adv. Chem. Phys., 1967, **12**, 131.

4
Experimental Techniques

BY D. I. HOULT

1 Introduction

This chapter covers the development of the experimental techniques associated with nuclear magnetic resonance over the period from mid-1973 to mid-1975. Its scope is necessarily narrowed by the limitations of space and those of the author's linguistic abilities. It is therefore confined largely to reports in the English language with occasional reference to publications in German and French; regrettably, material in the Slavonic languages and in Japanese has been ignored.

By far the most dominant influence on n.m.r. practice in every sphere has been the advance of digitization and the consequent move away from analogue techniques. The necessity of having a computer to perform Fourier transforms has of course forced the pace somewhat, not always to advantage, but the trend is in line with that of the electronics industry as a whole and it is now standard, for example, to see a frequency synthesizer advertised as a rack-mounting box with BCD (binary coded decimal) control lines. The front panel knobs are an extra, and are decidedly unfashionable! Correspondingly, the computer is no longer just a Fourier analyser; rather it is the control centre of the spectrometer, and the 'interface' between the spectrometer and the human in control (?) is a typewriter. Whether this line of development represents the most satisfactory march of progress remains to be seen, but before examining the techniques pertinent to the resonance phenomenon itself, it is perhaps useful to observe briefly some ways in which that phenomenon has been employed, if only to demonstrate that the computer is not, as some publications would have us believe, a substitute for original thought.

2 Diverse Applications of N.M.R.

A. General.—By using a complexed lanthanide shift reagent, Schneider *et al.*[1] have made a thermometer for use in ^{13}C experiments. The complex, which involves hexadeuterioacetone, gives an almost linear 15 p.p.m. shift over the range 200—300 K and should prove useful in experiments involving large heat dissipation. At the other end of the temperature scale, Aalto *et al.*[2] have used a pulsed spectrometer to measure the magnetization and spin–lattice relaxation time of a powdered platinum sample. Using a frequency of 413 kHz ($B_0 = 0.045$ T), they found that below 2 K, the amplitude of the free induction decay (FID) was proportional to T^{-1}, and also

[1] H. J. Schneider, W. Freitag, and M. Schommer, *J. Magn. Resonance*, 1975, **18**, 393.
[2] M. I. Aalto, H. K. Collan, R. G. Gylling, and K. O. Nores, *Rev. Sci. Instr.*, 1973, **44**, 1075.

that the Korringa relation $T_1T = \kappa$ was valid, giving two ways of measuring low temperatures with an accuracy of 2%. In two other applications of n.m.r. at these temperatures, Schmidt[3] has used the technique for the continuous measurement of the concentration of orthohydrogen in the solid phase, whilst Wampler et al.[4] have devised a tunable marginal oscillator for magnetic field measurements. The oscillator is of interest, as, surprisingly, they used a silicon n-channel dual gate MOSFET (3N187) at 4.2 K. Whilst carrier freeze-out did not totally inhibit the MOSFET, it rendered the use, for tuning purposes, of a Si varactor diode impractical, and a satisfactory alternative was found to be a GaAsP light-emitting diode.

At room temperature, Cook et al.[5] have also used a marginal oscillator for magnetic field measurements, but in addition have employed a computer in a feed-back loop in order to sample the field at suitable times in a swept e.s.r. experiment, and then to correct for any drift. In this way, control to better than 0.3 p.p.m. was obtained. On the other hand, Hawk et al.[6] have opted for field stabilization with the aid of a crude analogue system which uses the 'time-sharing' technique. The method of probe tuning used throws away signal-to-noise ratio, and the reason why non-orthogonal transmitter and receiving coils are used is not explained, nor is the fact that an emitter follower precedes the pre-amplifier. In addition, their broad-band r.f. buffer amplifier is thermally unstable, having no emitter resistor, although this may well be a printing error. The feed-back loop itself does not employ integration of the error signal from the spectrometer, and thus field drift is only reduced, not cancelled. Finally, their analysis of the mode of operation relies on an analogy with field modulation, and only in very limited circumstances, which unfortunately do not apply, can amplitude and frequency modulation be equated.

N.m.r. has been used for chemical analysis in various ways, and a typical application is the determination of the solid fat content in partially crystallized fats[7] by inspection of the FID, a method which relies on a difference of spin–spin interactions. Le Dang Khoi and Veillet[8] have studied impurities in ferromagnetic compounds, and their novel spin-echo spectrometer, which operates up to 1700 MHz, is capable of detecting very small concentrations. Grimaldi and Sykes[9] have described a stopped-flow Fourier-transform (FT) spectrometer, whilst Lawler and Halfon[10] have published details of a flow reactor for use with an unmodified high-resolution machine in which there is free access from above the sample. Their system allows normal sample spinning, and when applied to CIDNP experiments has a transfer time, from a reaction cell in an external magnetic field to the probe, of less than a second. Hoult et al.[11] have used ^{31}P n.m.r. to study the course of reactions in intact biological tissues, and have obtained directly, without destruction of the sample, the concentrations of phosphorus metabolites in rat leg muscle at various times after excision. The experiments at 129 MHz ($B_0 = 7.5$ T) have also

[3] F. Schmidt, *Rev. Sci. Instr.*, 1974, **45**, 1430.
[4] W. R. Wampler, S. Matula, B. Lengeler, and G. Durcansky, *Rev. Sci Instr.*, 1975, **46**, 58.
[5] T. J. Cook, B. R. Zegarski, and T. A. Miller, *Rev. Sci. Instr.*, 1975, **46**, 425.
[6] R. M. Hawk, R. R. Sharp, and J. W. Tolan, *Rev. Sci. Instr.*, 1974, **45**, 96.
[7] K. P. A. M. van Putte and J. van den Enden, *J. Phys. (E)*, 1973, **6**, 910.
[8] Le Dang Khoi and P. Veillet, *Rev. Sci. Instr.*, 1974, **45**, 759.
[9] J. J. Grimaldi and B. D. Sykes, *J. Amer. Chem. Soc.*, 1975, **97**, 273.
[10] R. G. Lawler and M. Halfon, *Rev. Sci. Instr.*, 1974, **45**, 84.
[11] D. I. Hoult, S. J. W. Busby, D. G. Gadian, G. K. Radda, R. E. Richards, and P. J. Seeley, *Nature*, 1974, **252**, 285.

shown the sensitivity of the chemical shift of inorganic phosphate to pH, and have enabled the spectrometer to be used as a pH meter when conventional techniques were not applicable. Kaplan and Laczynski[12] believe in the use of n.m.r. as a new instrumental tool for the analysis of cosmetic ingredients, whilst Richards and Evans[13] have produced a flow cell for electrolysis within a spectrometer probe.

B. Neutron Resonance and Diffraction.—As representatives of a growing area of application of magnetic resonance which is probably unfamiliar to many, two experiments in the literature merit particular attention. The first involves sub-nuclear resonance and necessitates a word of explanation concerning neutron diffraction by a polarized target. Abragam et al.[14, 15] explain the change of polarization of a neutron beam diffracted by a polarized sample (e.g., protons at low temperatures in a steady magnetic field B_0) in terms of a pseudo magnetic field B^* created in the

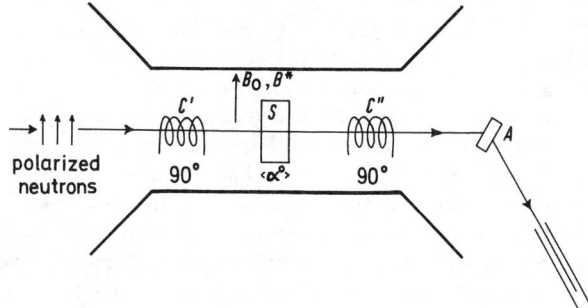

Figure 1 *Apparatus used by Abragam et al. for the observation of neutron diffraction by neutron magnetic resonance*

sample by a pseudo magnetic moment μ^*. The pseudo-field changes the effective Larmor frequency of a neutron within the sample by an amount γB^* at low temperatures, and B^*, which is dependent upon the polarization of the sample, can be quite large, e.g. 1 T. The measurement of B^* gives direct information about the scattering process, and the method demonstrated by Abragam and co-workers is best understood with reference to Figure 1. The incident neutron beam, which is polarized parallel to B_0, passes through two coils C' and C'', and the sample S. Radiofrequency (r.f.) power is applied to C' and C'' at the Larmor frequency of the neutron so that, during the time of flight through a coil, the neutron magnetic moment is flipped through 90° in the rotating frame. In this frame, the neutrons, once within the sample, experience, for their time of flight through it, the pseudo-field B^*, which, being parallel with B_0, causes precession through an angle a, i.e. the phase of the neutron in the rotating xy plane is changed. Only if $B^* = 0$ does the subsequent 90° flip in C'' align the neutrons antiparallel to B_0; precession through

---

[12] R. Kaplan and S. F. Laczynski, *J. Soc. Cosmetic Chemists*, 1974, **25**, 507.
[13] A. J. Richards and D. H. Evans, *Analyt. Chem.*, 1975, **47**, 964.
[14] A. Abragam, G. L. Bacchella, H. Glättli, P. Meriel, M. Pinot, and J. Piesvaux, *Phys. Rev. Letters*, 1973, **31**, 776.
[15] A. Abragom, G. L. Bacchella, H. Glättli, P. Meriel, M. Pinot, and J. Piesvaux, *Phys. Rev. Letters*, 1974, **33**, 102.

angle a reduces their final polarization, which is detected by analyser A. Clearly, B^* can be found, and as added proof, the temperature of the sample S may be varied. The correct dependence of B^* (as T^{-1}) was found, and the values of μ^* obtained for ^{1}H and ^{51}V were in agreement with previous results.

Of course, the usual role of n.m.r. in such experiments is to monitor the nuclear polarization of protons, deuterons, *etc.*, and the majority of instruments in the field use some sort of Q-meter system.[16] However, in a novel inversion of the usual roles, Hayter, Jenkin, and White[17] show a very convincing solid-state proton spectrum of crystalline lanthanum magnesium nitrate; whilst the abscissa is naturally frequency, the ordinate is a neutron count! This is possible because the neutron diffraction pattern is very sensitive to selective depolarization of nuclei, *e.g.* by removal of the microwave power which maintains polarization by the Overhauser effect. Thus irradiation of the sample at the Larmor frequency of the protons naturally changes the count from a Bragg diffraction peak.

C. Image Formation by Zeugmatography.

—Possibly the most exciting development in the period covered by this Report is the demonstration of image formation by various workers. Though at the moment crude, the potential for an n.m.r. equivalent of X-rays is clearly enormous, for, unlike X-rays, magnetic fields are (it is hoped) harmless. In a letter to *Nature*, Lauterbur,[18] the inventor of the technique, points out that image formation usually requires that the object under consideration interact with a field characterized by a wavelength less than, or of the order of, the smallest feature to be distinguished. However, this limitation on the field may be removed, and a new class of image generated, by taking advantage of induced local interactions. In the case of n.m.r., the interaction between the applied r.f. irradiation and the sample can be localized by application of a field gradient. Considering for the moment CW spectroscopy, the field gradient ensures that signal is received only from a slice of sample whose thickness corresponds to a frequency range which is the natural linewidth of the sample. Thus there is a sense in which the sample couples the radiation field B_1 and the main field plus gradient, B_0. From the Greek '$\zeta\varepsilon\nu\gamma\mu\alpha$' ('that which is used for joining') Lauterbur thus coins his title of 'zeugmatography'.

The main problem with the technique is that it is one-dimensional. By applying a field gradient in the z direction and performing a Fourier-transform experiment, one may obtain, say, N bits of information about the distribution of magnetization in the z direction, but each of those bits is an integral of the magnetization in the relevant xy plane. To form an image, N^3 bits of information are required. (Lauterbur has obtained a two-dimensional image by rotating the sample, recording the spectra at various orientations, and then applying an image-reconstruction routine,[19] but clearly this is very time-consuming.) On the other hand, however, the contrast of the image can be T_1-dependent, with its implications for cancer research, and there is also the possibility of the technique being applied to other regions of the spectrum and to solids.

As has been implied above, the resolution obtainable is dependent upon the strength of the field gradient and the natural linewidth of the resonances. Thus for

[16] V. Petriceli, *Nuclear Instr. Methods*, 1968, **58**, 111.
[17] J. B. Hayter, G. T. Jenkin, and J. W. White, *Phys. Rev. Letters*, 1974, **33**, 696.
[18] P. C. Lauterbur, *Nature*, 1973, **242**, 190.
[19] R. Gordon and G. T. Herman, *Comm. Assoc. Comput. Mach.*, 1971, **14**, 759.

solids, with present technology, resolution is poor. For example, to obtain a resolution of 3×10^{-10} m with a field gradient of 10 T m^{-1}, a proton linewidth of *ca.* 0.1 Hz would be required. Even with the use of multiple-pulse sequences[20, 21] to remove dipolar broadening, the linewidth of *ca.* 10 Hz obtained by Rhim *et al.*[22] would not be good enough and so, before zeugmatography can be used as a crystallographic technique, considerable experimental refinements will be required. However, it is worth noting that as one approaches resolutions comparable with the size of a unit cell, the distribution of magnetization can no longer be considered continuous, and in mathematical terms this means that a transition from a Fourier transform to Fourier analysis is required in order to give an accurate description of the signal received by the spectrometer. With this transition to a quantized frequency distribution, one would expect diffraction effects in the time domain, and Mansfield and Grannell[23] have demonstrated that this is so by using a sample which comprised layers of synthetic camphor. The FID obtained with the aid of a multiple-pulse sequence showed the diffraction effects quite clearly, and a subsequent Fourier transform gave the spatial distribution, showing up the layers. The authors also point out, however, that the presence of a field gradient degrades the efficiency of the line-narrowing pulse sequence, and so only small samples may be used.

From the point of view of diffraction, the one-dimensional aspect of zeugmatography is useful, but for image formation the problem must be circumvented. Garroway and co-workers[24] have done so by preparing the spin system prior to observation. They outline a number of methods, but the principle used is to define a plane in the sample by its frequency in a field gradient, and then to distinguish the spins in that plane by, for example, a frequency-selective 90° pulse which operates on all spins *outside* the region of interest. The flipped spins dephase rapidly, and only the spins in the defined plane are left aligned with the main field. By repeating the process with the direction of the gradient changed by 90°, only a *line* of aligned spins is left, and after the gradient direction has been changed yet again by 90° onto the third axis, a free induction decay may be initiated and recorded in order to obtain the spin distribution along that line. Clearly, by a scanning procedure, a multidimensional spin map may also be obtained.

Hinshaw,[25] in a rather different approach, considers the use of three orthogonal field gradients, $\dfrac{dB_z}{dx}$, $\dfrac{dB_z}{dy}$, and $\dfrac{dB_z}{dz}$. He points out that if the three gradients are applied simultaneously, the result is merely a single gradient in an intermediate direction, but that if the three gradients have time dependences which do not correlate, then the field at each point in the sample will have a unique time dependence. Thus the FID after a 90° pulse *can* contain all the information needed to construct the image of a sample. Unfortunately, he does not discuss at length the requirements necessary to ensure that 'can contain' becomes 'does contain', and the importance of the time-dependence principle is overshadowed by one applica-

---

[20] J. S. Waugh, L. M. Huber, and U. Haeberlen, *Phys. Rev. Letters*, 1968, **20**, 180.
[21] P. Mansfield, M. J. Orchard, D. C. Stalker, and K. H. B. Richards, *Phys. Rev. (B)*, 1973, **7**, 90.
[22] W.-K. Rhim, D. D. Elleman, and R. W. Vaughan, *J. Chem. Phys.*, 1973, **59**, 3740.
[23] P. Mansfield and P. K. Grannell, *J. Phys. (C)*, 1973, **6**, CL 422.
[24] A. N. Garroway, P. K. Grannell, and P. Mansfield, *J. Phys. (C)*, 1974, **7**, CL457.
[25] W. S. Hinshaw, *Phys. Letters (A)*, 1974, **48**, 87.

tion — the 'sensitive point' method. By varying the alternating currents flowing in the two halves of a gradient coil, the plane in which there is no field modulation can be moved about at will. Thus, by using the three orthogonal gradients, an elementary volume free from field modulation can be positioned anywhere in the sample, and this 'sensitive point' can be scanned. Hinshaw has demonstrated that this essentially CW method works satisfactorily, but Kumar et al.[26] have also shown quite clearly that the time-dependence principle is viable. Using a pulse method, the three orthogonal field gradients are applied successively for differing times (t_x, t_y, t_z) during the FID, and the latter is recorded during time t_z. The resulting signal $s(t)$, where time t is now a three-dimensional vector, is recorded for a full set of equally spaced components (t_x, t_y, t_z) and the three-dimensional Fourier transform of the complete set $S(t)$ is $S(\omega)$, where the components of ω are ω_x, ω_y, and ω_z. This is of course the required 'zeugmatogram'. Kumar and co-workers, because of the enormous computer store required to perform a three-dimensional transform, worked only in two dimensions and hence obtained not a cross-section, but a projection, albeit very convincingly. However, a possible criticism of their method is that by not using the FID during times t_x and t_y, they are wasting information. Thus, one must ask whether the use of pulsed gradients represents the best employment of the time-dependence technique or whether a stochastic variation of gradients might not yield more information in a given time—an important criterion if the subject under observation is alive and kicking.

In summary therefore, zeugmatography is in its infancy, but it is to be hoped that the many technical problems associated with its development will be surmounted in order to produce a method of image formation which is safe, rapid, and three-dimensional.

3 Temperature and Pressure Studies

Jonas[27] has reviewed the studies of n.m.r. in liquids at high pressures, and his article contains a good introduction to the subject as well as an experimental section detailing the equipment needed and the considerations involved. The references are particularly comprehensive. Several publications give details of vessels suitable for use at high pressures and temperatures. Sawyer and Gale[28] have described a cell of all-glass construction which has a moving mercury seal, through which pressure is transmitted hydrostatically up to a maximum of 2 kbar (2×10^8 N m^{-2}) over the temperature range from -15 to $+250$ °C. Yamaha[29] has also described a glass cell, for use in an ordinary high-resolution spectrometer, which will withstand similar pressures. An obvious problem associated with the use of a metallic cell is that the probe coils have to be within the cell, and hence some sort of electrical feedthrough which will withstand the operating conditions is required. Van der Hart[30] has given details of a 7 pF capacitive feedthrough and a simple re-usable closure seal which can withstand 7 kbar at room temperature and 5 kbar at 200 °C. The design of a probe to work at 30 MHz and up to 0.2 kbar

[26] A. Kumar, D. Welti, and R. R. Ernst, *J. Magn. Resonance*, 1975, **18**, 69.
[27] J. Jonas, *Adv. Magn. Resonance*, 1973, **6**, 73.
[28] D. W. Sawyer and B. N. Gale, *J. Phys. (E)*, 1973, **6**, 1205.
[29] H. Yamaha, *Rev. Sci. Instr.*, 1974, **45**, 640.
[30] D. L. van der Hart, *Rev. Sci. Instr.*, 1974, **45**, 111.

with CH_4-inert gas mixtures has been described by Rajan et al.,[31] and illustrates well the difficulties of working with a metallic cell. The probe coil is close to the conducting walls and thus its Q is probably low. On the other hand, a reduction of size reduces the *effective* sample volume. Presumably an optimum was found.

With regard to temperature control, Dahlstrom and co-workers[32] have modified the Varian HA-100 probe by re-siting the Varian heater/sensor and using a thermocouple to measure the temperature in its place. Apart from giving a more accurate temperature reading, this modification seems to have little to commend it.

4 Measurement of Relaxation Times

A. Systematic Errors.—Much has been published concerning the errors involved in the measurement of relaxation times by the various pulse procedures available, and it is perhaps a sign of the times that very little has been published concerning CW techniques. Errors arise from a variety of sources, principally inhomogeneities in B_0 and B_1, incorrect pulse powers, incorrect timing, and inaccuracies in phasing, leading to a variety of manifestations and a profusion of correctives. Occasionally the method of measurement may be suspect, or chemical exchange may prevent effective compensation of errors. Dealing first with spin–lattice relaxation-time (T_1) measurements, Demco and co-workers[33] have performed a density matrix analysis of the effects of phase modification, in a 'non-ideal' inversion-recovery pulse sequence, upon the apparent recovery of the magnetization. The aim is of course to obtain the true recovery curve, be it exponential or otherwise. They tabulate an omnipotent correction factor of dubious practical value, but come to the useful conclusions that for each line in a spectrum recovering truly exponentially, (i) an apparent exponential recovery occurs when $T_2^* \ll T_1$ (a well-known result), (ii) if $T_1 \approx T_2^*$, the interference effect produced by transverse magnetization from the imperfect 180° pulse can, in the on-resonance case, be avoided by using a quadrature phase for the reading pulse, and (iii) in a complex spectrum, errors can be eliminated by taking the average of two inversion-recovery sequences. In the first, the 180° and 90° pulses are in phase, but in the second they are out of phase.

Progressing from the work of Kumar and Johnson,[34] Assink[35] has shown that a reliable value of T_1 may be obtained by the null-method if the phase of the 90° pulse is inverted. To illustrate his point, the phase inversion corrected a null value of 0.4 ms to 1.0 ms – the correct answer. Similarly, Demco et al.[36] have shown that, with a suitable correction, a quadrature phase shift can also give good answers with the null method.

As a less time-consuming method of measuring T_1, the single-scan method has also received attention. Demco et al.[37] have proposed a 'flip-flop' pulse sequence: $180°\text{-}(90°_x\text{-}\tau'\text{-}90°_x\text{-}\tau)_n$, which periodically 'flips' the magnetization into the transverse plane in order to examine it and then 'flops' it back again. However, in a major article analysing the use of pulse multiplets for the single-scan determination of T_1,

---

[31] S. Rajan, K. Lalita, and S. V. Babu, *J. Magn. Resonance*, 1974, **16**, 115.
[32] P. L. Dahlstrom, R. L. Harris, and J. L. Loo, *J. Magn. Resonance*, 1975, **17**, 350.
[33] D. E. Demco, P. van Hecke, and J. S. Waugh, *J. Magn. Resonance*, 1974, **16**, 467.
[34] A. Kumar and C. S. Johnson, *J. Magn. Resonance*, 1972, **7**, 55.
[35] R. A. Assink, *J. Magn. Resonance*, 1973, **12**, 88.
[36] D. E. Demco, V. Sîmplăceanu, and I. Ursu, *J. Magn. Resonance*, 1974, **13**, 310.
[37] D. E. Demco, V. Sîmplăceanu, and I. Ursu, *J. Magn. Resonance*, 1974, **15**, 166.

Edzes[38] heavily criticizes the method on the grounds that it is degraded by B_0 inhomogeneity, resonance offset, and incorrect phasing. Further, he attributes the method to Gerritsma and Trappeniers.[39] Edzes analyses theoretically the influences of transverse relaxation (T_2) and pulse error, shows that the magnetization decays with a hybrid time-contant T_{12}, and that errors arising from inhomogeneities in the static and r.f. magnetic fields, from improper r.f. phases or from resonance offset, can largely be compensated by the choice of a proper pulse multiplet. However, he does point out that diffusion and chemical exchange can prevent effective compensation of errors. Edzes' calculations have been verified by Kurland and Parrish[40] with a 'half-wave' triplet pulse sequence:

$$[(90°_{-y}\text{-}\tau\text{-}180°_{-y}\text{-}\tau\text{-}90°_{-y})\text{-}\Delta t\text{-}(90°_{-y}\text{-}\tau\text{-}180°_{+y}\text{-}\tau\text{-}90°_{-y})\text{-}\Delta t]_n$$

which was off-resonance by an amount $\Delta v = 1/2\ \tau$. Good results were obtained.

In a purely empirical paper dealing with the method of progressive saturation, Wilkins and co-workers[41] have found that deviations of up to 7° are tolerable in the 90° flip-angle required, but that, in some experiments, up to ten pre-acquisition pulses are necessary in order to attain equilibrium. They make the good point, which is easily overlooked, that in an experiment which uses the amplitude of a spectral line as the basis of calculations, the main field homogeneity must remain constant. This is of course an excellent argument for using the *integral* of a line as the basis of calculation. Another good point, also of general applicability, is made by Armitage *et al.*,[42] who remind us that a pulse may well be longer than supposed, due to ringing in, for example, a tuned transmitter, and thus a 90° pulse is not necessarily half the supposed 180° pulse.

Turning now to the measurement of T_2 by the Carr–Purcell sequence with Meiboom–Gill modification, Vold *et al.*[43] find that, in general, the measured value of T_2 is too large, and that to obtain a good result the following conditions must be fulfilled: (i) all pulse rotation parameters must be within 5% of the requisite values, and phase-glitch greater than 5% must be avoided; (ii) the fractional inhomogeneity in B_1 must be better than 0.2; (iii) the resonance frequency must be stable to within a few Hertz; (iv) the baseline must be found, as there is a steady-state component which can lead to a 5—15% error; (v) a full semi-log plot must be made as imperfections are most obvious in the initial slope; (vi) all pulse widths and delays must be free of jitter, and r.f. phase coherence is necessary at low frequencies; (vii) the sample should not be spun, as field modulation, caused by the rotation, influences the result.

Not all these conditions have been analysed theoretically, but Bull[44] has made a very thorough analysis, from first principles, of the effect of B_1 inhomogeneity. He demonstrates that the errors are much more significant when the modified Bloch equations apply, but that normally the errors can be decreased considerably by the

---

[38] H. T. Edzes, *J. Magn. Resonance*, 1975, **17**, 301.
[39] C. J. Gerritsma and N. J. Trappeniers, *Physica*, 1970, **51**, 365.
[40] R. J. Kurland and R. G. Parrish, *J. Magn. Resonance*, 1975, **17**, 295.
[41] C. L. Wilkins, T. R. Brunner, and D. J. Thoennes, *J. Magn. Resonance*, 1975, **17**, 373.
[42] M. Armitage, H. Huber, D. H. Live, H. Pearson, and J. D. Roberts, *J. Magn. Resonance*, 1974, **15**, 142.
[43] R. L. Vold, R. R. Vold, and H. E. Simon, *J. Magn. Resonance*, 1973, **11**, 283.
[44] T. E. Bull, *Rev. Sci. Instr.*, 1974, **45**, 232.

use of a homogeneous, on-resonance B_0 field. Not surprisingly, he also shows that errors vanish when $T_1 = T_2$. Hughes and Lindblom[45] have analysed the baseline error and have shown that it is due to imperfections in the 180° pulses (principally phase-glitch). They recommend that the phase of the 90° pulse should be inverted on alternate runs, and that the decays should be alternately added and subtracted into the computer store to remove the effect. If this is done, they believe that the initial change of baseline causes no error in the measurement of T_2, but Bull[46] disputes this result, claiming it is only true for $T_1 = T_2$. They also discuss the use of a phase-alternating version of the Carr–Purcell sequence, with compensation for errors in the phases.

Attention has also been paid to the problem of analysing the collected data, and the solutions increase in sophistication chronologically. Moore and Yalcin,[47] in a neat paper, have shown that a good baseline is a valuable asset: for example, a 1% error in the initial signal height leads to a 3% error in the measured time constant. However, if no baseline is available, it can be restored from the data at the expense of a threefold decrease in time-constant accuracy. Smith and Buckmaster[48] have used somewhat different computational techniques, and in a comparison with reference 47 claim to get better results, but the most novel method of calculation is due to de Fontaine et al.[49] By equating derivatives (in T_1, and the infinity value of the signal) of the mean-square error in a decay, they form an equation in T_1 which is independent of the baseline. This gives a very fast method of calculating T_1 from data covering a time of the order of T_1 rather than $5T_1$, but the accuracy of the method is not examined by comparison, and thus no real idea of its worth can be obtained. At this point, work by Chan[50] should be mentioned. He has developed a computer program which performs detailed curve-fitting to a spin echo. The program computes the echo shape and iterates to optimize the fit, working from the variables of resonant frequency, the natural T_2, the instrumental time-constant, chemical exchange rates, and diffusion coefficients. It is not made clear what the advantages are of this type of analysis over, say, Fourier analysis followed by line-shape analysis.

B. Quantization and Other Errors.—The temptation to treat a magnetic resonance system as a linear system to which Fourier analysis can be strictly applied is all too often overpowering, and, despite the warnings which abound, many still fall into the trap. The discussion of this topic by Schäublin et al.[51] is described adequately in Chapter 7, but it is worth noting here that the discussion includes non-equilibrium states created by the large pulses used in relaxation experiments. It is well illustrated in an adjacent paper by Vold and co-workers,[52] where they show the non-exponential recovery of a coupled system (ethyl acetate). Campbell and Freeman[53] have also highlighted the effect and have shown that it is particularly bad in the case where I

---

[45] D. C. Hughes and C. Lindblom, *J. Magn. Resonance*, 1974, **13**, 142.
[46] T. E. Bull, *J. Magn. Resonance*, 1974, **16**, 192.
[47] W. S. Moore and T. Yalcin, *J. Magn. Resonance*, 1973, **11**, 50.
[48] M. R. Smith and H. A. Buckmaster, *J. Magn. Resonance*, 1975, **17**, 29.
[49] D. L. de Fontaine, D. K. Ross, and B. Ternai, *J. Magn. Resonance*, 1975, **18**, 276.
[50] S. O. Chan, *J. Chem. Phys.*, 1975, **62**, 2031.
[51] S. Schäublin, A. Höhener, and R. R. Ernst, *J. Magn. Resonance*, 1974, **13**, 196.
[52] R. L. Vold, R. R. Vold, and H. E. Simon, *J. Magn. Resonance*, 1974, **13**, 226.
[53] I. D. Campbell and R. Freeman, *J. Magn. Resonance*, 1973, **11**, 143.

and S spins are coupled and both are given 180° pulses. Mehring and Raber[54] have reported an orientation dependence in the non-exponentiality of a T_1 recovery for a three-spin system, due to cross-correlation of the intramolecular dipole intraction. For rather different reasons, Waelder et al.[55] have also observed non-exponential behaviour whilst measuring the apparent T_1 of the tryptophan indole nitrogen proton. The experiment was performed in aqueous solution, with the aid of selective long pulses, by using the saturation-recovery method. The water magnetization was thus unperturbed and acted as a reservoir for chemical exchange of the N protons. By performing a pH titration, the approximate chemical exchange rate was found.

C. Technique.—A novel method of measuring T_1 has been reported by Kimber and Harris.[56] Silicon-29 has a negative magnetogyric ratio, so that whilst proton decoupling may be necessary in order to collapse a multiplet, the nuclear Overhauser effect reduces the signal intensity, possibly even giving a null value. However, whilst the decoupling commences upon application of irradiation, the reduction of the ^{29}Si signal has a characteristic time of T_1, which can therefore be measured by varying the length of time the decoupler is on prior to observation. Cornell and Pope[57] have described a pulse sequence which measures T_1 in the presence of a large rotating field distant from resonance by 10^5—10^7 radian s^{-1}. Assuming a weak collision approximation (i.e. $\tau_c \ll T_2$), these workers have interpreted the relaxation time found in such a way as to provide a direct measure of correlation times τ_c in the range 10^{-5}—10^{-7} s – results which are difficult to obtain by conventional measurements of T_1, or $T_{1\rho}$. Their data for solid SF$_6$ are in good agreement with theory.

Canet et al.[58] have pointed out that in an inversion-recovery measurement of T_1, better sensitivity can be obtained by not waiting $5T_1$ after the 90° pulse before repeating. However, no attempt is made to quantify the improvement, and they fail to point out that in a Fourier-transform experiment where several T_1 values are present, information concerning the relative intensities of spectral lines is lost. Gupta and co-workers[59] have shown that with the aid of the saturation-recovery method, T_1 measurements can be made using rapid-scan spectroscopy,[60] whilst Bangerter[61] has described a simple pulsed field-gradient circuit for use after the 180° pulse in an inversion-recovery experiment in order to destroy the residual transverse signal. Briguet et al.[62,63] have shown that a CW high-resolution spectrometer is well fitted for selective relaxation-time measurements where long, low-power selective pulses are required, for the modulation of the field can be used effectively to control the pulse power, length, and phase if the irradiating frequency is at one of the Larmor sideband frequencies. The modifications needed to use this method with an HA 100 spectrometer are outlined in reference 63. Rogers and Wright[64] have

[54] M. Mehring and H. Raber, J. Chem. Phys., 1973, **59**, 1116.
[55] S. Waelder, L. Lee, and A. G. Redfield, J. Amer. Chem., Soc., 1975, **97**, 2927.
[56] B. J. Kimber and R. K. Harris, J. Magn. Resonance, 1974, **16**, 354.
[57] B. A. Cornell and J. M. Pope, J. Magn. Resonance, 1974, **16**, 172.
[58] D. Canet, G. C. Levy, and I. R. Peat, J. Magn. Resonance, 1975, **18**, 199.
[59] R. K. Gupta, J. A. Ferretti, and E. D. Becker, J. Magn. Resonance, 1974, **16**, 505.
[60] J. Dadok and R. F. Sprecher, J. Magn. Resonance, 1974, **13**, 243.
[61] B. W. Bangerter, J. Magn. Resonance, 1974, **13**, 87.
[62] J. C. Duplan, A. Briguet, and J. Delmau, J. Magn. Resonance, 1973, **12**, 270.
[63] A. Briguet, J.-L. Culty, J.-C. Duplan, and J. Delmau, J. Phys. (E), 1974, **7**, 791.
[64] M. T. Rogers and D. A. Wright, Rev. Sci. Instr., 1973, **44**, 1189.

Experimental Techniques

shown how two-pulse experiments may largely be automated by interfacing a mini-computer to a pulse spectrometer, but such is the march of progress that this is now a common feature of commercial machines.

To complete this section, a report by Gerig and co-workers[65] on an automated data-collection system for T_1 and T_2 determinations must be included. To measure T_1, the sequence $(90°\text{-}\tau\text{-}90°\text{-}5T_1)_n$ is used and the two free induction decays are stored. The integral of the first decay is proportional to the equilibrium magnetization M_0 plus a constant due to d.c. offset. The integral of the second decay is proportional to M_τ plus a constant. The logical way of eliminating the constant term is to record the decays down to their baselines, and then perform 'baseline corrections' which subtract off the d.c. terms prior to integration. Gerig *et al.*, however, choose to use a 'null sample' (which is solvent but no solute) to determine the offset. Thus the experimental conditions have altered, *e.g.* the addition of solute may change the Q and tuning of the probe, and hence alter the leakage. Further, this method offers no guarantee against a change of conditions during the course of an experiment. The next mistake they make is to accumulate for a length of time the decays for a given value of τ and then to progress to a new value and repeat the procedure. With each progression the quantity $(M_0 - M_\tau)/M_0$ is calculated, in case conditions change during the experiment, but of course this quantity is highly susceptible to any change in the d.c. offset. Surely a preferable procedure would be to vary τ with each repeat, and to accumulate the results of each *integration* without attempting to compensate for the offset. As a very accurate baseline would be available, the offset would be of no consequence. With regard to the integration itself, the integral is truncated, but no information is given as to when this occurs. Further, the aim of truncation is to improve the signal-to-noise ratio of the integral; a superior method of doing this is by convolution with an exponential of time constant $-T_2$. The T_1 is found by a least-squares fit, but it is not stated that any sort of weighting function is used to improve the accuracy. As to their measurement of T_2, when it is very long, the multichannel analyser employed does not have a sufficient number of points to sample every echo. At the cost of a few flip-flops and some minutes soldering, division of the 180° pulse frequency by a suitable number solves the problem. However, the solution employed is to use an analogue delay, which is adjusted so that the desired number of echoes is skipped. Presumably it is then immediately retriggered – a classic recipe for instability. No wonder 'the analogue devices used to produce the delays are not reliable enough in situations involving long delays and very fast pulsing rates, and procedures to digitally count the echoes to be neglected are now under development'.

5 Broad-line and Solid-state Spectroscopy

The general problems associated with the removal of the effects of dipolar broadening from solid-state spectra have been analysed in two papers by Rhim and co-workers.[22, 66] The effects of finite pulse-width and B_1 inhomogeneity are shown to have limited the resolution of previous pulse cycles, and a new eight-pulse cycle designed to minimize these problems is discussed. Spectra for ^{19}F in CaF_2, taken

---

[65] J. T. Gerig, G. B. Matson, and A. D. Stock, *J. Magn. Resonance*, 1973, **12**, 48.
[66] W.-K. Rhim, D. D. Elleman, R. W. Vaughan, and L. B. Schreiber, *J. Chem. Phys.*, 1974, **60**, 4595.

with the aid of the cycle, show a residual linewidth near 10 Hz to prove the point, and the papers also contain good practical information on how to set up the spectrometer so as to optimize the various parameters, such as pulse width, phase glitch, *etc.* In a cross-polarization experiment employing proton-enhanced n.m.r. spectroscopy, Pines *et al.*[67] have brought two sets of spins into thermal contact. By satisfying the Hartmann–Hahn condition ($\gamma' B_{1S} = \gamma_n B_{1S}$) whilst spin-locking, mutual I and S spin-flips, occasioned by the I–S dipole interaction, become energy-conserving, and thus the two sets of spins come rapidly to thermal equilibrium. If the S spins are rare, the effect on abundant I spins is small, and the experiment can be repeated many times, the S spins being disturbed periodically by the observation pulse. Clearly, under such conditions, the polarization of the rare S spins is greatly enhanced, and Pines and co-workers have obtained about a factor of a thousand gain in signal-to-noise ratio for ^{13}C in organic solids whilst irradiating the protons. However, as the power required at the two frequencies was *ca.* 200 W each, there were problems, and in particular the ^{13}C receiver had to be carefully isolated from the proton irradiation (which, during accumulation, decoupled) by a crossed-coil system and a multistage filter.

Janzen[68] has reported linewidth narrowing in solids, whilst working in the dispersion mode, caused by saturation due to the r.f. field, or to the combination of modulation and r.f. fields. In a later paper[69] he describes the use of the dispersion signal to obtain T_1 values by the method of progressive saturation. Dziergwa *et al.*[70] have devised a method for obtaining absorptive or dispersive signals from *bulk* metallic samples, where the skin effect usually precludes such measurements. By measuring the signal, which is proportional to $\chi' + \chi''$, and the inductance of the r.f. coil, which is proportional to $\chi' - \chi''$, both quantities are clearly obtainable. Modulation broadening of narrow lines present in broad-line spectra can obscure details in a derivative spectrum. On the other hand, decreasing the modulation power means that sensitivity is lost for the broad lines. To use conventional modulation techniques to obtain the absorption-mode spectrum would require an exceedingly large modulation frequency (*e.g.* 100 kHz). However, if this is not available, Pearson and Walter[71] have shown that, provided all the sidebands (down to negligible intensity) are recorded, and the modulation index is known, the absorption mode spectrum can be calculated by deconvolution. The major drawback of the method is that it assumes a linear system. Once saturation commences, the sidebands reappear; further, getting the phase correct can be a problem too. Probably the best solution to the dilemma is a Fourier-transform spectrometer equipped with quadrature detection, and Hatch and collaborators[72] have taken a step towards this solution by constructing their own analogue interface to a PDP 12 computer. Unfortunately, they work in the single phase mode, but their admirable paper describes how, with the aid of an external hardware adder and direct memory access, an inter-point time of 6μs may be achieved. The ingenious manner in which they eliminate systematic noise and obtain a derivative spectrum is discussed in Chapter 7.

[67] A. Pines, M. G. Gibby, and J. S. Waugh, *J. Chem. Phys.*, 1973, **59**, 569.
[68] W. R. Janzen, *J. Magn. Resonance*, 1973, **10**, 263.
[69] W. R. Janzen, *J. Magn. Resonance*, 1973, **12**, 276.
[70] U. Dziergwa, D. Hechtfischer, R. Karcher, and K. Luders, *J. Phys.* (*E*), 1974, **7**, 255.
[71] G. A. Pearson and R. I. Walter, *J. Magn. Resonance*, 1974, **16**, 348.
[72] G. F. Hatch, J. W. Neely, and R. W. Kreilick, *J. Magn. Resonance*, 1974, **16**, 408.

A paper by Schoenberger et al.[73] indicates the gulf which sometimes exists between 'broad-line' and 'high-resolution' spectroscopists. They report that asymmetric lineshapes may be caused by non-random B_0 inhomogeneities, and that confusion with chemical shift anisotropy or Knight shift may result. A high-resolution man would immediately reach for the shims!

6 Instrumentation

In this section of the Report, the division into sub-headings traces a course through the 'average' spectrometer from the audiofrequency end (as typified by a computer) *via* the probe and sample to the transmitter and decoupler, and thence to the pulse programmer, which may be, going full circle, the computer. Published details of complete systems are then reviewed, and finally modifications to commercial instruments are considered.

A. Signal Processing.—In a paper which commences by considering quadrature Fourier transformation (*q.v.*), Stejskal and Schaefer[74] go on to consider the baseline errors produced in a Fourier-transformed spectrum by the recovery time of the spectrometer and by foldover of the spectrum about plus and minus the Nyquist frequency in the quadrature case, and zero frequency and the Nyquist frequency in the single-phase case. They rightly point out that distortion is more likely in the single-phase case as the absorption spectrum can have a dispersive foldover about the origin of frequency (*N.B.* the absorption wing varies as ω^{-2}, the dispersion wing as ω^{-1}), and that this may be significant for moderately broad lines. However, it is vitally important to realise that with *quadrature* detection, in conjunction with good filters at the Nyquist frequency, foldover is *not* a major source of baseline error, contrary to Stejskal and Schaefer's findings. In computing the foldover, they use an artificially generated exponential decay. The use of such a decay, with its infinitely fast rising edge, implies that the bandwidth used is also infinite, which is certainly not true if filters at the Nyquist frequency are applied. Hence the computed foldovers and the subsequent conclusions as to the best phase corrections to use are wrong. In practice, baseline errors are due to two sources, (i) the use of a linear phase-correction to compensate for a delay prior to accumulation, and (ii) the transient response of the filters to the signal itself. The first factor gives a $(-\sin x)/x$ term under each spectral line, whilst the second gives the filter characteristic (with foldover at the frequency extrema). A combination of linear phase-shift with the transient response may well give a sloping baseline. In a later paper[75] the authors, having realised the significance of the transient response to the signal, compare the relative merits of Butterworth and Bessel filters and find that Bessel filters ring for a much shorter time. The price paid for this gain is, of course, that the frequency cutoff is not so sharp.

Ramadam[76] has described a fast recovery d.c. amplifier with diode-limited feedback, and Torgeson[77] has discussed an interface amplifier for use in sweep experi-

---

[73] R. J. Schoenberger, R. B. Creel, B. K. Lunde, and R. G. Barnes, *J. Magn. Resonance*, 1973, **11**, 20.
[74] E. O. Stejskal and J. Schaefer, *J. Magn. Resonance*, 1974, **14**, 160.
[75] E. O. Stejskal and J. Schaefer, *J. Magn. Resonance*, 1974, **15**, 173.
[76] B. Ramadan, T. C. Ng, and E. Tward, *Rev. Sci. Instr.*, 1974, **45**, 1174.
[77] D. R. Torgeson, *Rev. Sci. Instr.*, 1973, **44**, 982.

ments. It acts merely as a buffer between the X digital-to-analogue converter of an instrument and the swept function. Sandu,[78] in an article marred by discrepancies between the text and the diagrams, the omission of a vital capacitor from one of the schematics, and some poor logic circuitry, has described what in principle is a very useful addition to a spin-echo spectrometer – an inexpensive integrating digital voltmeter with a linearity of 0.1%. Complementing this instrument, Burnett and Winther[79] have made a linear envelope-detector which functions at the standard i.f. of 450 kHz. Full details are given and there is a particularly simple integrated-circuit construction. However, for the fast recording of transients, Merrick et al.[80] have described the interfacing of a Biomation 802 transient recorder to a PDP/e computer.

Quadrature accumulation of data has several advantages in a Fourier-transform experiment, in addition to the absence of folding about the frequency origin already mentioned, and the improvement in signal-to-noise ratio by $\sqrt{2}$. These extra advantages[82] stem from the fact that the transmitter frequency may be in the midst, rather than to one side, of the spectrum. However, errors in orthonormality of the two signals collected by the computer result in 'ghost' signals in the spectrum – a line at $+100$ Hz, for example, has a ghost at -100 Hz, whose amplitude is dependent upon the errors. Hoult[81,82] has shown that the errors may be cancelled to a high degree ($<0.1\%$) and the two signals orthonormalized, by the use of a Cyclically Ordered Phase Sequence (CYCLOPS) with the observation pulses, coupled with appropriate data routing of the two signals into the two halves of the computer store. The sequence, the principle of which is shown in Figure 2, also cancels systematic noise, including d.c. offsets. Stejskal and Schaefer[74] have used the basic principle successfully, and have also described a computational method of orthonormalization of the two signals.

B. *The Probe Area.*—A major problem in probe design is the conflict between fast recovery, for the observation of rapidly decaying signals, and high Q, for minimization of Johnson noise. Lowe and Engelberg[83] have proposed an elegant solution, which goes some way to alleviating the problem, in the form of a lumped parameter delay line. The line, which is formed by placing capacitors to ground on each turn of the probe solenoid, has a characteristic impedance of 50 Ω at the requisite frequency, and when terminated correctly has rise and fall times of less than 100 ns at 32 MHz. This allows the signal to be observed within a microsecond after the end of the pulse. From a signal-to-noise point of view, the equivalent Q of the system is *ca.* 13, which is not high, and a major drawback is that the r.f. field produced is inhomogeneous due to the wavelength along the line being less than the coil radius. However, the decay time is certainly good, and the bandwidth is also large.

A rather complex single-coil system for use in pulse work has been described by Clark and McNeil.[84] This paper gives a description of, and design optimization

---

[78] H. S. Sandu, *J. Magn. Resonance*, 1974, **13**, 379.
[79] L. J. Burnett and C. R. Winther, *J. Magn. Resonance*, 1975, **17**, 344.
[80] T. Merrick, S. Smiriga, and A. Pines, *J. Magn. Resonance*, 1974, **14**, 270.
[81] D. I. Hoult, D. Phil. Thesis, Oxford, 1973.
[82] D. I. Hoult and R. E. Richards, *Proc. Roy. Soc.*, 1975, **A344**, 311.
[83] I. J. Lowe and M. Engleberg, *Rev. Sci. Instr.*, 1974, **45**, 631.
[84] W. G. Clark and J. A. McNeil, *Rev. Sci. Instr.*, 1973, **44**, 844.

Experimental Techniques

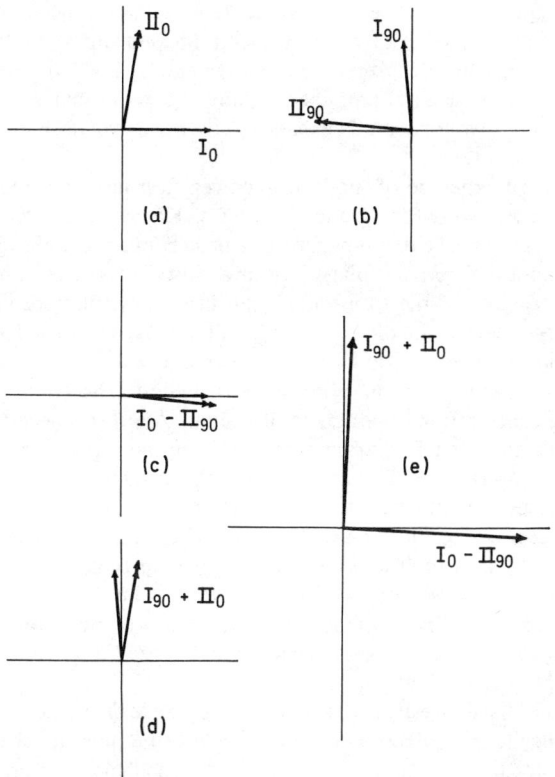

Figure 2 *The orthonormalization of two signals* (a) *nominally in quadrature and nominally equal in amplitude. The use of a transmitter phase-shift of nominally* $90°$ (b) *plus appropriate data routing into the computer store* (c *and* d) *gives resultants* (e) *which are accurately orthonormal. The roman numerals refer to the signal channel, whilst the subscripts indicate the transmitter phase. Further phase-shifts of* $180°$ *and* $270°$ *generate a Cyclically Ordered Phase Sequence* (CYCLOPS), *which also cancels systematic noise and d.c. offsets*

criteria for, a series resonant system with protection for the receiver and isolation of the transmitter by the usual device of crossed diodes. However, the coaxial lines used are not terminated in any way by their characteristic impedance, but rather, are incorporated in the resonant system. This must lead, by virtue of the extra resistance introduced, to a deterioration in signal-to-noise ratio, and their use of extra inductances is open to the same objection. However, at 10 MHz, the loss in signal-to-noise ratio caused by the extra resistance was only 14%, and thus the system should prove useful where, for reasons of space or temperature, it is not possible to tune at the probe coil. A more serious criticism concerns their analysis of noise performance, for they assume that the noise contributed by the preamplifier is independent of the source impedance. If this were so, it would be logical to transform the impedance of the probe coil, with a parallel capacitor, to the highest resistance value possible, and this indeed is the 'ideal case' against which they judge the performance of their system. Unfortunately, at high frequency most semiconductors give their best noise performance when working from source resistances of 2 kΩ or less, a condition which, of necessity in a good amplifier, entails a gross mismatch of impedances. Their arrangement therefore could be made to give better results, particularly at high frequency.

An alternative to the use of diodes for power steering is an electronic switch, which can be made with PIN diodes or FET's. Kisman and Armstrong[85] have used PIN diodes for probe damping, receiver protection, and transmitter isolation. The scheme employed is particularly useful at low frequencies, where transients (including the ringing of the transmitter) can last many microseconds. However, the use of active devices in the receiver circuitry must be treated with a certain amount of suspicion for PIN diodes have a low r.f. resistance when direct current is flowing through them and hence shot noise is present. The authors do not state what the noise contribution is – only that it is small. Another objection to the use of active devices is that switching transients can cause the probe and preamplifier to ring. By using a π-section FET switch, however, Grannel et al.[86] have succeeded in reducing transient breakthrough by two orders of magnitude. As complementary switching pulses are used, the transients cancel. The isolation, at 10 kΩ impedance and 15 MHz, was measured as 84 dB, and, not surprisingly at these impedance levels, there was negligible noise and insertion loss. This would not, however, be true for a 50 Ω system. The amplifier paralysis caused by pulse breakthrough was completely avoided, and at 9 MHz a decrease in the 'dead time' (due to probe ringing) from 12 μs to 7μs was accomplished.

Speight et al.[87] have investigated a spurious resonance phenomenon which occurs at low frequency (e.g. 4 MHz) without a sample being present. The effect, which causes ringing for up to milliseconds, was found to be dependent on the proximity of conductor round the probe coil, and the authors have no real explanation. However, they have shown that the effect can be removed by winding a shield of 26-gauge enamelled copper wire, grounded at each turn, coaxially about the probe coil. Clark[88] had previously suggested that an acoustic resonance in the probe coil could be the culprit, but, clearly, the success of the shielding shows this hypothesis to be incorrect. However, Müller,[89] with what is essentially a n.m.r. spectrometer, has demonstrated the existence of *nuclear* acoustic resonances in this frequency range, and so this may be a possible explanation, the resonance phenomenon taking place in the walls round the probe.

Pratt and Smith[90] have described a sample transfer device, for use principally in double-resonance experiments, which utilizes pneumatic logic. The whole device is easily assembled from commercially available units, and it flips a sample quickly, and without bounce, from one position to another. With suitable computer interfacing, the day of complete automation is at hand.

[85] K. E. Kisman and R. L. Armstrong, *Rev. Sci. Instr.*, 1974, **45**, 1159.
[86] P. K. Grannell, M. J. Orchard, P. Mansfield, A. N. Garroway, and D. C. Stalker, *J. Phys. (E)*, 1973, **6**, 1202.
[87] P. A. Speight, K. R. Jeffrey, and J. A. Courtney, *J. Phys. (E)*, 1974, **7**, 801.
[88] W. G. Clark, *Rev. Sci. Instr.*, 1964, **35**, 316.
[89] V. Müller, *J. Phys. (E)*, 1975, **8**, 127.
[90] J. C. Pratt and J. A. S. Smith, *J. Phys. (E)*, 1973, **6**, 525.

C. **The Transmitter.**—Narath and Barham,[91] in a most professional communication, have described the design and construction of a high-power transmitter for the frequency range 2—200 MHz. Frequencies below 40 MHz are provided by a multi-stage 15 kW (peak r.m.s. power) tuned amplifier, for which full constructional details are given, and the power output can be increased to 100 kW by means of a single-stage booster. Frequencies between 40 and 200 MHz are generated by a three-stage 20 kW frequency multiplier ($\times 3$ or $\times 5$), and extension of the high-power multiplication scheme to higher frequencies is discussed.

El-Hanany[92] has concentrated on the design of a transformer-coupled matching stage for use in a high-power pulsed amplifier. He calculates the optimum primary and secondary inductances and the coupling factor required for low Q, and also makes the very important point that, in a class C circuit, the output impedance of the valve goes high when conduction stops. The damping on the tuned circuit is thus removed to a large extent, causing prolonged ringing. However, by inserting in the tuned circuit a resistance bypassed with crossed diodes, the ringing is rapidly quenched once it has died down to about one volt. This is a very elegant use of crossed diodes, and can almost certainly be used to advantage in the earlier stages of a transmitter as well as in the final output.

D. **Modulation Control.**—The use of a computer to control two-pulse experiments has already been mentioned,[64] but in general the interfacing of a computer with a spectrometer in order to provide a totally flexible system of timing, phase, and frequency control presents a formidable problem about which little has been published. Taylor and co-workers,[93] however, have given details of a modular pulse programmer which, although manually controlled, demonstrates well the principles involved. The pulse-widths are under analogue control, as, for pulses < 1 µs wide, a 100 MHz clock would be needed to obtain the required resolution digitally. The programmer as a whole is made from three basic modules. The first is a 10 MHz clock plus 11 cascaded dividers, which generate periods of up to 10^3 seconds for the pulse-delay circuits. This module also contains the reset circuitry and the first pulse channel, for which no delay is required. The second type of module (for a single-pulse channel) contains the period input gate, a programmable counter to multiply the chosen period by 0—999, and the pulse-width monostable. Any number of pulse-channel modules may be used. The third type of module controls the number of sequences and the sequence repetition period, and also includes facilities for setting up multiple-pulse sequences by recycling any number of pulse channels a predetermined number of times.

Levy and collaborators[94] have used digital techniques to produce a high-efficiency broad-band decoupling scheme. Noise is converted into a random binary bit stream, which is then strobed to create a $(\sin x)/x$ distribution. A variety of gating schemes may be used, but, more importantly when digitized noise is used for modulation purposes, an oscillatory component may be mixed in, which, by creation of a sideband system, broadens out the sphere of influence of the decoupling. The useful point is made that the maximum decoupling power is only needed during accumula-

[91] A. Narath and D. C. Barham, *Rev. Sci. Instr.*, 1974, **45**, 100.
[92] U. El-Hanany, *Rev. Sci. Instr.*, 1973, **44**, 1069.
[93] D. G. Taylor, S. Booth, and P. S. Allen, *J. Phys. (E)*, 1974, **7**, 105.
[94] G. C. Levy, I. R. Peat, R. Rosanske, and S. Parks, *J. Magn. Resonance*, 1975, **18**, 205.

tion, and that the Overhauser effect, for example, needs much less; therefore the sample need not be grossly overheated.

E. Complete Spectrometers.—Vold et al.[43] have given brief details of a home-made spectrometer which works at 55 MHz and employs a commercial oscillator, amplifier, and phase detector, 1 kW power amplifier, and wideband receiver. The probe is of the crossed-coil variety. The transmitter coil is a one-turn Helmholtz pair, whilst the receiving coil is a four-turn solenoid, 5.5 mm in diameter, which can be rotated to minimize cross-talk. A 180° pulse is 20 μs long, and, 200 Hz off-resonance, the null from a sample 1 cm long is 3—5% of the 90° pulse FID height. The spectrometer employs an external deuterium field-frequency lock, which maintains a stability of 1—2 Hz (^{1}H) over several hours. The receiver is interfaced to a Nicolet 1085 computer with disc, and the period used for the generation of Carr–Purcell sequences is derived from the computer's 10 MHz clock to ensure phase coherence.

In an interesting communication dealing with the electronics required for a high-resolution spectrometer, Kan and co-workers[95] have proposed several novel techniques. In proton n.m.r. several similar frequencies are often required – for example for observation, for decoupling, for homonuclear locking, and for solvent-peak saturation (see Chapter 7). Instead of using several synthesizers, Kan et al. commence with a basic frequency, and then, using a digital single-sideband technique, generate the requisite frequencies with the aid of audiofrequency synthesizers – an obvious economy. The sideband system used effectively advances the phase of the r.f. by 90° every $\tau/4$ seconds with the aid of two ring modulators, and thus, if the basic frequency is v_0, the modulated signal contains frequencies $(v_0 + v_1) \pm 4nv_1$, where n is an integer and $v_1 = 1/\tau$. The frequency v_1 is generated not from a commercial unit but rather by the use of rate multipliers followed by division by 100 ($25 \times 2 \times 2$), so as to smooth out phase noise (which is an inherent feature of rate multipliers) and to obtain two signals in quadrature for the sideband process. The various frequencies, one from each modulator, are summed in a power combiner and fed to the probe by a circulator which isolates the receiver from the transmitter, but neither from the probe. (A criticism here is that unless the circulator is cooled in liquid nitrogen, it will reduce the signal-to-noise ratio by 2 to 3 dB. Circulators are also very sensitive to magnetic field, and so care has to be taken with positioning when the magnet is superconducting.)

The probe, designed by Lafond,[96] was a coaxial cavity for which, at 240 MHz, full constructional details are given. A Q of only 70 is obtained, but without a knowledge of the filling factor, which for an ordinary Helmholtz coil is poor ($\lesssim 0.16$), this is no criterion by which to judge. A better criterion is the signal-to-noise ratio, which for the ^{1}H resonance of 1% ethylbenzene, under standard CW conditions of 1 Hz s^{-1} sweep and 1 s filter, is 160:1. Radiofrequency single-phase-sensitive detection occurs at frequency v_0, giving an intermediate frequency of v_1. As the spectrometer works in the pulsed mode (time-sharing for CW measurements), and the modulation in the transmitter is gated, leakage occurs at frequency v_0 and is thus rejected at the intermediate-frequency amplifier. Audio-phase detection is accom-

[95] S. Kan, P. Gonord, C. Duret, J. Salset, and C. Vibet, *Rev. Sci. Instr.*, 1973, **44**, 1725.
[96] C. Lafond, 3rd Cycle Thesis, Orsay, 1972.

plished with an aperiodic synchronous detector, which the authors, in a slightly confused manner, claim has superior linearity.

In a paper which examines some of the more fundamental aspects of high-resolution spectrometer design, Hoult and Richards[82] have described a multinuclear instrument working with a field of 7.5 T (^{1}H = 320 MHz). The analysis of magnetic field in terms of spherical harmonics in order to form a correcting coil system is considered, and an alternative method of correcting axial inhomogeneity in a superconducting magnet, using a matrix of coils, is described. It is shown that tesseral inhomogeneity gives rise to spinning sidebands, and that the order of the spinning sidebands present is a reflection of the degree of the spherical harmonic which causes them (*e.g.* a magnetic field which varies round the axis as sin 2ϕ causes second spinning sidebands). However, it is also pointed out that B_1 inhomogeneity can, with a Helmholtz coil, also cause sidebands, and that to minimize these sidebands, whose amplitude is *independent* of spinning speed, the angular width of the coil should be 120°.

The transmitter gating is accomplished with the aid of an FET circuit, which in the off state has an isolation of 140 dB between 5 and 200 MHz. The switching time of the gate is, however, slow ($\sim$300 ns) but adequate for high-resolution work. The pulse widths, times, and phases may be controlled manually or directly from a Bruker-Nicolet BNC 12 computer which has been modified so that a variety of pulse sequences may be programmed without resort to an external programmer.

The receiver does not employ an intermediate frequency, and the design of low-noise preamplifiers using FET's, and broad-band post-amplifiers is discussed. The preamplifiers, which have noise figures, for a 50 Ω source, in the range 0.3—2 dB, incorporate diode protection against the transmitter pulses, and no $\lambda/4$ lines are required. The broad-band amplifier used has a bandwidth of 200 MHz and a recovery from overload, due to the absence of capacitive coupling, of < 1 µs. Phase-sensitive detection in quadrature at one-third of the Larmor frequency is employed. As the preamplifier is tuned, this device reduces leakage greatly, and the ring modulators used for the detection function well at the third harmonic of the reference frequency. A further protection against leakage (or rather the change of leakage when pulses are applied and the receiver is gated off) is the use of a sample-and-hold circuit to carry the signal across the gated period. This is particularly useful for reducing the effects of gating at the time-sharing frequency of the CW deuterium lock. The computer has been further modified to accept two simultaneous signals, and data routing for use with the CYCLOPS method of signal orthonormalization is accomplished by hardware as the signals arrive. A useful aid to setting up the receiver is a computer program which calculates the r.m.s. values of the two signals and the phase difference between them. No sample or pulse is required – the program works on the noise. The paper ends with an assessment of the spectrometer's capabilities, and, in particular, the excellent signal-to-noise performance with ^{31}P (129 MHz) is demonstrated by a spectrum (of 1 mmol l^{-1} inorganic phosphate) obtained in 15 min. The signal-to-noise ratio is *ca.* 20:1.

Finally, the spectrometer of Le Dang Khoi and Veillet[8] mentioned at the beginning of this Report is considered. It is a spin-echo machine, and, because their work involves ferromagnetics, it has to cover a frequency range from 170 MHz to 1.7 GHz. The transmitter is a valve oscillator which uses a stub tuner or a cavity as the resonant element. It is gated on the high-tension line, and r.f. pulses are obtained by the application of 1 kV pulses of duration 1—6 μs. The sample sits in a cavity, and a coupling loop connects the signal to the preamplifier, which employs PIN diode protection. At s.h.f. the preamplifier is replaced with a grounded-base mixer – a u.h.f. transistor is used. No details of the noise performance are given, though the authors do claim high sensitivity. The instrument is used for the determination of magnetic structure and for hyperfine interaction studies as well as for the impurity measurements referred to already.

F. Modification of Commercial Instruments.—A number of publications have appeared concerned with multinuclear operation of spectrometers. All of them employ a frequency synthesizer of some form, using it either in the lock or the signal channel. Where the synthesizer is used on the signal channel, various frequency-mixing schemes are used so that the spectrometer i.f. can still be obtained. Typical is a report by Peters et al.,[97] who use the Varian 'Gyrocode' spin decoupler as a frequency synthesizer. Normally the spin decoupling frequency is derived by phase-locking an oscillator to the master oscillator used for field-frequency locking with deuterium (~ 15.4 MHz). However, the phase-sensitive detection which provides the error signal does not take place at the deuterium frequency; instead, both frequencies are first divided. Thus, if the division factors are p and q:

$$\frac{v}{Nq} = \frac{v_D}{p} \text{ and therefore } v = \frac{Nq}{p} v_D$$

where v is the desired frequency, v_D is the deuterium frequency, and N is an additional factor which is dependent upon the r.f. band required. Thus, by suitable choice of the integers, a frequency close to that required can be obtained. (Wilkins and Thoennes[98] have described in detail the calculations and the computer interfacing required.) The phase-noise inherent in such a system of synthesis is likely to be bad, but it does not appear to have been measured. Peters and collaborators use the frequency v for the main transmitter, whilst the local oscillator frequency is obtained by mixing v with 10.7 MHz in a balanced mixer and selecting the high-frequency sideband.

Gupta[99] has also used the 'Gyrocode' decoupler, but in order to generate the lock frequency in the region of 15.4 ± 2 MHz. Several other frequencies are of course standard on the Varian XL 100 machine (namely those of the more commonly observed nuclei), and Gupta has shown that by varying the magnetic field a very large number of pairs of nuclei can be made to have frequencies within the range of the lock and within the range of one of the standard frequencies.

Traficante et al.[100] have approached the multinuclear problem from a slightly different angle by aiming for wide-band operation with a Bruker HFX90 spectrometer. The latter was set for ^{31}P at 36.5 MHz, and a frequency synthesizer was used to generate two sidebands by mixing. The transmitter was a broad-band class A amplifier, and thus 50% of its power capability was thrown away by not using a

[97] C. S. Peters, R. Codrington, H. C. Walsh, and P.D. Ellis, *J. Magn. Resonance*, 1973, **11**, 431.
[98] C. L. Wilkins and D. J. Thoennes, *Rev. Sci. Instr.*, 1973, **44**, 1311.
[99] R. K. Gupta, *J. Magn. Resonance*, 1974, **16**, 185.
[100] D. D. Traficante, J. A. Simms, and M. Mulcay, *J. Magn. Resonance*, 1974, **15**, 484.

single-sideband technique. It is not made clear whether the wide-band probe was single- or cross-coil, but as only one coil is described, it was presumably the former. The question then arises as to what steps were taken to stop noise from the class A amplifier reaching the receiver. Crossed diodes are usually insufficient at higher frequencies, but this problem was not considered. Their analysis of the variation of Q of their coil with frequency is a little vague, particularly as they imply that skin effect varies the inductance. It is well known that, considerably below the self-resonant frequency, the inductance of a solenoid is essentially independent of the conductor used, being a function only of the length, diameter, and number of turns. They make much of the fact that the Q passes through a maximum, and in comparing it with standard Bruker probes at various frequencies note that these have the same behaviour, e.g. $Q = 111.5$ at 36 MHz and $Q = 129$ at 22.6 MHz. There are too many factors involved to draw this sort of conclusion, as is borne out by the fact that it is easy to make coils with a Q of the order of 200 at 130 MHz. A maximum in Q at a particular frequency is normally caused by the onset of self-resonance, or, equally likely, the fact that many a good tuning capacitor has a poor Q when fully open.

Throughout the paper, the distinction between impedance and resistance is blurred. The authors measured the noise figure of the system for various source impedances, but unless these impedances were purely real (i.e. resistive) the measurements do not mean much. They also refer to the matching of the probe impedance to that of the preamplifier: an error, as has already been mentioned, which is prevalent. However, in their case it is probably an error of terminology rather than of fact. What is in error, however, is the assumption that the probe tuning is correct when the maximum signal issues from the spectrometer. The signal may well be a maximum, but it is unlikely that the ratio of signal-to-noise also is. It is preferable to check the tuning with a bridge or a vector impedance meter. After preamplification, the signal frequency is returned to 36.5 MHz by mixing, and by this technique there is a capability for 71 nuclei with but little loss in signal-to-noise ratio. The main advantage of the method is that there is no need to re-shim the field, as the probe is not changed, and also, of course, only one decoupling and one lock frequency are needed. It is of interest to note that the authors have used similar techniques on a Perkin–Elmer R-20B spectrometer.[101] Because of the type of circuitry employed, CW spectrometers often do not take kindly to pulses, and exhibit long paralysis times. Unfortunately, this aspect of the conversion to pulsed operation is not discussed.

Marshall et al.[102] have modified a Varian XL-100 machine in a similar manner, as have Dorn et al.[103] a Brüker instrument. However, in the latter case, the unwanted sideband in the transmitter was filtered out. Wilson and co-workers[104] have described a microcell for use with the XL-100 for sub-milligram natural-abundance ^{13}C analyses. The use of the cell, with the aid of quadrature detection, has increased the sensitivity of the instrument by a factor of 4—5. Unfortunately, the two quadrature signals cannot be sampled simultaneously, and so they have to be transformed separately and different linear phase corrections applied, prior to addition and sub-

[101] D. D. Traficante and J. A. Simms, *Rev. Sci. Instr.*, 1974, **45**, 1063.
[102] A. G. Marshall, L. D. Hall, M. Hatton, and J. Sallos, *J. Magn. Resonance*, 1974, **13**, 392.
[103] H. C. Dorn, L. Simeral, J. Natterstad, and G. E. Maciel, *J. Magn. Resonance*, 1975, **18**, 1.
[104] D. M. Wilson, R. W. Olsen, and A. L. Burlingame, *Rev. Sci. Instr.*, 1974, **45**, 1095.

traction to obtain the two halves of the spectrum. This is tedious, and the use of a sample-and-hold circuit in one of the signal channels would improve the operation enormously, allowing effectively simultaneous data acquisition and the use of a complex Fourier-transform routine. Davies and Manning[105] have replaced the valve preamplifier, in a Varian HA series spectrometer, by a cascode MOSFET amplifier. Circuit values are given, and it is shown that at 60 MHz the sensitivity is doubled, but the noise figure of the amplifier is not given, and so no absolute assessment is feasible.

7 Miscellaneous

Correlation spectroscopy, of which, strictly speaking, Fourier spectroscopy is a part, is considered in detail in Chapter 7. Ziessow[106] has published what promises to be a standard text on the subject (in German), and Kaiser[107] has analysed the application of the Hadamard transform to n.m.r. For a linear system, an extra Fellgett (multichannel) advantage in sensitivity over Fourier-transform methods is obtained, but as an n.m.r. system is non-linear (it saturates) the advantage is manifest not in signal-to-noise improvement, but rather in a reduction of applied power. Kaiser demonstrates the use of a pseudo-random binary sequence using a shift register with modulo-2 addition to provide feed-back, and shows that, if the register is part of the computer, the binary contents can be used to address the memory for the incoming signal. This effects a considerable reduction in the number of computations required.

Delayre[108] has described a 'CHIRP' pulse (an analogy with radar techniques) which, by the use of frequency modulation, has an approximately square frequency spectrum. The pulse is generated with digital techniques, and is essentially for use with a CW spectrometer to effect a conversion into the Fourier mode. As it lasts much longer than a conventional 90° pulse, its use is restricted, with the power available, to high-resolution studies. Cargioli and Williams[109] have given details of a new microcell design for use with large n.m.r. tubes. A glass insert squeezes the sample already in the bottom of the tube into an annulus cut out of the insert. Jameson and Jameson,[110] by assuming that the resonance frequency of ^{129}Xe, extrapolated to zero density, is temperature-independent, have monitored the temperature dependence of the chemical shift of various lock solvents in the range 240—440 K. Takagi[111] has observed the n.m.r. properties of the A phase of liquid ^{3}He. Remarkable properties abound, and the article contains a good list of references for those interested. Al-Rawi et al.[112] have continued their investigations of ^{3}H. Pachler and Wessels[113] have investigated selective population transference in the ^{13}C Fourier experiment by use of a selective 180° pulse applied to one of the

[105] T. K. Davies and B. C. Manning, *Rev. Sci. Instr.*, 1973, **44**, 1513.
[106] D. Ziessow, 'On-line Rechner in der Chemie, Grundlagen und Anwendungen in der Fourier-Spektroskopie', De Gruyter, Berlin, 1973.
[107] R. Kaiser, *J. Magn. Resonance*, 1974, **15**, 44.
[108] J. Delayre, Third International Meeting on Nuclear Magnetic Resonance Spectroscopy, St. Andrews, Scotland, 1975.
[109] J. D. Cargioli and E. A. Williams, *Rev. Sci. Instr.*, 1974, **45**, 1263.
[110] A. K. Jameson and C. J. Jameson, *J. Amer. Chem. Soc.*, 1973, **95**, 8559.
[111] S. Takagi, *J. Phys. (C)*, 1975, **8**, 1507.
[112] J. M. A. Al-Rawi, J. A. Elvidge, J. R. Jones, and E. A. Evans, *J.C.S. Perkin*, 1975, **2**, 449.
[113] R. G. R. Pachler and P. L. Wessels, *J. Magn. Resonance*, 1973, **12**, 337.

Experimental Techniques

coupled ^{1}H lines. This gives information as to the relative sign of the coupling constant, and also can effect a considerable improvement in sensitivity. Linde and co-workers[114] have applied the same technique to studies of ^{29}Si, where, because of the negative Overhauser effect, it is particularly useful. Other double-resonance techniques, including Campbell's method of solvent peak saturation, [115, 116] are further considered in Chapter 7.

[114] S. A. Linde, H. J. Jakobsen, and B. J. Kimber, *J. Amer. Chem. Soc.*, 1975, **97**, 3219.
[115] I. D. Campbell, C. M. Dobson, and R. J. P. Williams, *Proc. Roy. Soc.*, 1975, **B189**, 485.
[116] H. E. Bleich and J. A. Glasel, *J. Magn. Resonance*, 1975, **18**, 401.

5
Spectral Analysis

BY R. G. JONES

1 Introduction

Spectral analysis has followed closely the technological development of n.m.r. The advent of Fourier transform methods has led to improved precision in measuring n.m.r. parameters[1] thanks to the inherent peak definition created by using 32 K store for a sweep width of 250 Hz! Similarly, improved sensitivity has led authors to employ the established ^{13}C satellite technique in a greater number of cases where deceptive simplicity defies comprehensive analysis of spectra, or where the interest prompts this approach.

This review covers a two-year period; the format of the report in Volume 3 has now been revised and emphasis is now placed on the spin system. A list of the compounds studied has been retained, however.

2 New Approaches to Known Spin Systems

A. $[A]_2X$ and ABX Systems.—The determination of relative signs of J_{AX} and $J_{AX'}$ coupling constants has been illustrated[2] for $[A]_2X$ systems using off-resonance decoupling. Changes in the X pattern are considered when continuous irradiation of power is applied at a known frequency offset from v_A. The spacing of the lines separated by $|J_{AX} + J_{A'X}|$ goes to zero for increasing decoupling efficiency. If the signs of J_{AX} and $J_{A'X}$ are the same this pair form the outermost lines of the X part. Since the combination lines always maintain a finite spacing (though with diminishing intensity) there is a cross-over between the two sets of lines. No such cross-over is expected for different signs, where the order of the two sets is reversed. Proper choice of decoupling power and offset is not critical. The technique is illustrated for $[A]_2X$ using the ^{13}C resonance lines of the olefinic carbons in maleic anhydride and diethyl fumarate.

The ABX double-resonance spectrum has been described in some detail,[3] and the determination of relative signs illustrated with reference to *trans*-3-chloroacrylic acid, ethyl fumarate, hexylmaleate, and dimethyl maleate. An advantage over the single-resonance experiment is only gained for values of $|v_A - v_B|$ either much smaller or much larger than

$$\{[(J_{AX} - J_{BX})/2]^2 - J_{AB}^2\}^{\frac{1}{2}}$$

The influence of unequal effective coupling constants in the AB part on the X

---

[1] L. Ernst and D. N. Lincoln, *J. Magn. Resonance*, 1974, **16**, 190.
[2] H. Fritz and H. Sauter, *J. Magn. Resonance*, 1974, **15**, 177.
[3] H. Fritz and H. Sauter, *J. Magn. Resonance*, 1975, **18**, 527.

Spectral Analysis

part is discussed as a means of determining the sign of J_{AB} relative to that of J_{AX}–J_{BX}. Asymmetry in the intensity distribution in the X part can be used to determine on which side of the proton resonance (AB part) the decoupling field is applied. The method is limited to simple systems.

B. [AB]$_2$ System.—A method of analysis has been described[3] which is not new but the method of recording the spectrum took advantage of the resolution inherently possible in the Fourier transform technique. The spectrum of a 30 mol % solution of o-phenylenephosphorochloridite was recorded using 100 Hz spectral width and 32 K data points (0.006 Hz per point). The spectrum was run three times (ten 90° pulses each), so that six frequency values were obtained for each transition (the six worst measurements of the total number were disregarded!). The probable errors in line positions were given as 0.002—0.006 Hz. The line positions were obtained using a parabolic centroid function

$$v = v_0 + \frac{I_+ + I_-}{2(2I_0 - I_+ - I_-)}$$

where v_-, v_0, v_+ represent frequencies of the channels below, at, and above the peak maximum, and I_-, I_0, and I_+ the respective intensities; v is the true central frequency. Twenty out of 24 lines were assigned and LAOCOON 3 was used to give r.m.s. errors of 0.002 Hz and a calculated probable error of 0.001 Hz in the spectral parameters.

C. The X-{A} INDOR Spectrum of the [AX$_n$]$_2$ Spin System.[4]—The X-{A} INDOR spectrum of the [AX$_n$]$_2$ spin system has been treated as the part of the A spectrum related to a particular X transition.[4] This assumption was partially justified by the low power ('tickling' level) of the irradiating field used, and also by the agreement between observed spectra and those predicted on this basis. Previous work has not related the A transitions to the X transitions in the way required.

The interest centred particularly around A transitions related to the strong doublet in the X spectrum, i.e. X transitions of the $\alpha\alpha(m,p)$ and $\beta\beta(m,p)$ states.

$$\alpha\alpha(m,p): v_A + \tfrac{1}{2}\{J + (m+p)N \pm [(m-p)^2L^2 + J^2]^{\frac{1}{2}}\}$$

$$\beta\beta(m,p): v_A + \tfrac{1}{2}\{-J + (m+p)N \pm [(m-p)^2L^2 + J^2]^{\frac{1}{2}}\}$$

with $J \equiv J_{AA'}$; $N = J_{AX} + J_{AX'}$ and $L = J_{AX} - J_{AX'}$; m and p are sums of spin eigenvalues for the X_n and X'_n groups, respectively.

For the transitions to $\alpha\alpha(m,p)$, there is a set of 'inner' lines where the J and square-root terms have opposite signs, and a set of outer lines, J away from the inner lines, on one side only. Thus if J is sufficiently large the sign can be determined immediately. The signs of $J_{AA'}$ and $(J_{AX} + J_{AX'})$ are taken to be the same if the outer A line observed lies to the same side of v_A as the X line monitored does of v_X (provided γ_A and γ_X have the same sign).

Inner and outer lines overlap for lower values of J, and in this case the best way of obtaining the sign of J is by comparing observed spectra with those calculated from the formulae given.

[4] R. J. Goodfellow and B. F. Taylor, *J. C. S. Dalton*, 1974, 1676.

The method has been applied to bis(phosphine) complexes cis-[MX$_2$(PMe$_3$)$_2$] (M = Pd or Pt; X = Cl, Br, or I).

Analytical expressions have been derived for transition energies where $m_T(A) = \pm 2, \pm 1$ in the [AX$_9$]$_4$ system arising within tetrakis(phosphine) complexes trans-[MX$_2$(PMe$_3$)$_4$]$^+$ (M = Rh or Ir) and [Pt(PMe$_3$)$_4$]$^{2+}$.

D. [[A]$_2$X$_n$]$_2$ and [ABX$_n$]$_2$ Spin Systems with n = 1, 2, and 3.[5]—The theory of spectral analysis for these spin systems has been conveniently summarized[5] using the combined symmetry–composite particle–good quantum number approach. Some of the past work on n = 1, 2, and 3 systems is noted in this paper. The spin states for the three systems are defined in composite particle notation, and the total relative intensities (normalized to $n \times 2^{n-1}$ for an n-spin system) attributable to these four generalized spin states are given in each case.

States with maximum and minimum total spin possess $\tilde{C}_{2v}$ and $\tilde{C}_2$ symmetry for the respective systems. The intermediate total spin states of the [[A]$_2$X$_n$]$_2$ systems possess $\tilde{C}_2$ symmetry. The basic molecular wavefunctions have been constructed from the basic group symmetry wavefunctions by taking all possible products of the group wavefunctions and reclassifying these product functions under the total molecular symmetry. The breakdown of the [[A]$_2$X$_n$]$_2$ system into the possible subsystems is illustrated using good quantum numbers and C_{2v} symmetry. The further reduction of 2:4:8:4:2 and 4:4:4 sub-systems when $J_{XX'} = 0$ is emphasized but not referenced.

The schematic diagram representing the factorization of the secular matrices for [ABX$_n$]$_2$ systems is generated from that for the [[A]$_2$X$_n$]$_2$ systems. The calculation of matrix elements is shown in some detail. Subspectral analysis is highlighted as a useful approach, and the different parts of the spectra discussed from this standpoint. The spectrum of 1,4-dibromobutane is analysed as an example of the [[A]$_2$X$_2$]$_2$ system and trans,trans-hexa-2,4-diene is presented as an [ABX$_3$]$_2$ system.

3 More Examples of known Spin Systems

A. Three-spin Systems.
ABC: 7-Methyl-2-naphthol and its sodium salt;[6] methylnaphthalenes-{Me};[7] 2-pyrazolines;[8] unique determination of n.m.r. spectral parameters;[9] 1,2,4-substituted, 3-phenylcycl[3,2,2]azines;[10] 2,4-dimethylpenta-2,4-diene-{Me};[11] 8-hydroxyquinoline;[12] cyclohepta [c or b] furan-6-ones;[13] methylnitronaphthalenes;[14] 2,6-dialkyl-4-oxo-1,3-dioxans;[15] 2,3-diphenylpropanoic acid;[16] 3-(methoxycarbonyl)thio-

---

[5] D. W. Aksnes, *Acta Chem. Scand. (A)*, 1974, **28**, 375.
[6] T. Bisanz, M. Bukowska, and A. Wojcicki, *Roczniki Chem.*, 1973, **47**, 1659.
[7] N. K. Wilson and J. B. Stothers, *J. Magn. Resonance*, 1974, **15**, 31.
[8] J. Elguero and A. Fruchier, *Anales de Quim.*, 1974, **70**, 141.
[9] V. S. Tumanov, *Isvest. V. U. Z. Fiz.*, 1974, **17**, 17.
[10] C. M. Gupta, B. B. P. Srivastava, R. K. Rizvi, and N. Anand, *Indian J. Chem.*, 1974, **12**, 674.
[11] A. V. Cunliffe and R. K. Harris, *Org. Magn. Resonance*, 1974, **6**, 121.
[12] S. Katayama, Y. Akahori, H. Mori, *Chem. and Pharm. Bull. (Japan)*, 1973, **22**, 568.
[13] M. El Borai, R. Guilard, and P. Fournari, *Bull. Soc. chim. France*, 1974, 1383.
[14] V. Lucchini and P. R. Wells, *J. C. S. Perkin II*, 1974, 1434.
[15] P. Äyräs and K. Pihlaja, *Tetrahedron*, 1973, **29**, 1311.
[16] S. L. Spassov, A. S. Orahavato, S. M. Mishov, and J. Schraml, *Tetrahedron*, 1974, **30**, 365.

phen;[17] pleiadiene;[18] phenylsubstituted azole derivatives;[19] 4,6,6-trimethyltrimethylene sulphite;[20] 1,2-dichloro-2-iodoethane;[21] N-vinylpyrroles;[22] chloro-, bromo-, and methyl-cyclobutadiene iron tricarbonyl.[23]

ABX: Trisodium monofluorodiphosphate;[24] 1,4:3,6-bis-thioanhydro-D-iditol derivatives;[25] 4-azacarbazoles;[26] calculation of transition frequencies using Green's functions;[27] 2,2,6-trialkyl-4-oxo-1,3-dioxans;[15] [5,6,6-^{2}H$_3$]-D-glucose;[28] metal carbonyl complexes of di(tertiary arsines).[29]

AB$_2$: Na$_5$P$_3$O$_{10}$;[24] proton relaxation and symmetry of spin functions in the AB$_2$ system.[30]

B. Four-spin Systems.

ABCD: 1,3-Thiazolidines;[31] [^{2}H$_6$]-camphor;[32] neostrychnine and 18-oxostrychnine;[33] flavone and deuteriated analogues;[34] [4,4-^{2}H$_2$]-α-tetralone and [2,2-^{2}H$_2$]-α-tetralone;[35] ethylene sulphites;[36] 1-methyl[1,2,2,3,4-^{2}H$_5$]cyclopentanes;[37] methylnaphthalenes;[7] dithieno[3,2-e:3′,2′-e']benzothieno[2,3-b] benzothiophen;[38] [11a,11b,16-^{2}H$_3$]strychnine;[39] 3-methylindano[2,1-c]piperidine;[40] substituted oxetans [41] benzo[b]fluoroanthrene and dibenzothiophen;[42] 1,3-oxazolidines.[43]

A$_2$BC: CF$_3$OOSF$_4$OCF$_3$.[44]

ABCX: Cyclopropane phosphonate;[45] [1-^{15}N]pyrazole;[46] vinylstyrene (X ≡

[17] S. Nagata, T. Yamake, and K. Fukui, *Tetrahedron*, 1974, **30**, 1315.
[18] D. Wendisch, W. Hartman, and H. G. Heine, *Tetrahedron*, 1974, **30**, 295.
[19] M. Begtrup, *Acta Chem. Scand.* (*B*), 1974, **28**, 61.
[20] P. Albriktsen, *Acta Chem. Scand.*, 1972, **26**, 3678.
[21] Y. P. Wong and H. F. White, *J. Magn. Resonance*, 1973, **12**, 1.
[22] B. Trofimov, G. A. Kalakin, A. S. Atavin, A. I. Mikhaleva, and E. G. Chebotareva, *Khim. geterotsikl. Soedinenii*, 1975, 360.
[23] H. A. Brure, H. Hanebeck, G. Horlbeck, and H. Huether, *Z. Naturforsch.*, 1974, **29b**, 219.
[24] H. Falius and M. Murray, *J. Magn. Resonance*, 1973, **10**, 127.
[25] P. Sohar and J. Kuszman, *Org. Magn. Resonance*, 1974, **6**, 407.
[26] F. Balkan and M. L. Heffernan, *Austral. J. Chem.*, 1973, **26**, 1501.
[27] M. W. Grant, *Diss. Abs. Internat.* (*B*), 1975, **35**, 3852.
[28] N. Cyr, G. S. Ritchie, T. M. Spotswood, and A. S. Perlin, *Canad. J. Spectroscopy*, 1974, **19**, 190.
[29] W. R. Cullen, L. D. Hall, J. T. Price, and G. Spendijan, *J. Amer. Chem. Soc.*, 1974, **96**, 412.
[30] S. Fujiwara, Y. Umezawa, Y. Arata, and H. Furuyama, *Chem. Letters*, 1974, 555.
[31] G. E. Wilson, jun. and T. J. Bazzone, *J. Amer. Chem. Soc.*, 1974, **96**, 1465.
[32] J. L. Marshel and S. R. Walter, *J. Amer. Chem. Soc.*, 1975, **96**, 6358.
[33] G. W. Luther, J. Valentini, and J. C. Carter, *J. Magn. Resonance*, 1974, **15**, 132.
[34] P. Joseph-Nathan, J. Mares, M. C. Hernandez, and J. N. Shoolery, *J. Magn. Resonance*, 1974, **16**, 447.
[35] K. Galts, J. D. Reinheimer, K. Shafer, and J. T. Gerig, *Org. Magn. Resonance*, 1974, **6**, 577.
[36] C. H. Green and D. G. Hellier, *J. C. S. Perkin II*, 1973, 1966.
[37] R. L. Lipwick, *J. Amer. Chem. Soc.*, 1974, **96**, 2941.
[38] J. H. Dopper, D. Oudman, and H. Wynberg, *J. Amer. Chem. Soc.*, 1973, **95**, 3692.
[39] P. E. Hansen, J. Feeney, and G. C. K. Roberts, *J. Magn. Resonance*, 1975, **17**, 249.
[40] V. F. Zakharov, V. P. Zvolinskii, D. A. Fesenko, and N. S. Prostakov, *Zhur. strukt. Khim.*, 1974, **15**, 774.
[41] J. Jokisaari, *Z. Naturforsch.*, 1974, **29a**, 1902.
[42] D. W. Jones, R. S. Mathews, and K. Bartle, *Spectrochim. Acta*, 1974, **A30**, 489.
[43] J. Devillers, D. H. Giao, and J. Navech, *Compt. rend*, 1973, **277**, *C*, 1067.
[44] F. A. Hohorst, D. D. Desmartean, C. R. Anderson, D. E. Gould, and W. B. Fox, *J. Amer. Chem. Soc.*, 1973, **95**, 3866.
[45] E. Hellmuth, J. A. Kaczynski, and L. L. McCoy, *J. Org. Chem.*, 1974, **39**, 3125.
[46] J. P. Jacobsen, O. Snerling, E. J. Pedersen, J. T. Nielsen, and K. Schaumberg, *J. Mag. Res.*, 1973, **10**, 130.

^{13}C);[28] thiophen-3-carboxaldehyde;[17] flavone;[34] 1,2-dibromo-1-fluoroethane and 1-chloro-1-fluoro-2-iodoethane;[21] aspartic acid and homoserine;[39] chelate complexes $\overline{CR^1R^2As(Me)_2M(CO)_2As(Me)_2CR^3R^4}$.[47]

ABXY: 2 - Phenyl - 1,3,2 - oxathiaphospholan;[48]   4 - germa - 1,3 - dioxans;[49] $\overline{CH_2As(Me)_2Mn(CO)_3(X)As(Me)_2CF_2}$;[50] phenyl - 3 - oxo - 2 - oxathiazolidine;[51] [1]benzothieno[4,5-d]thieno[3,2-d']benzo[1,2-b:4,3-b']dithiophen.[38]

[AB]$_2$: 2-Pyrazolines;[8] anthracene-related compounds;[52] 1,1,2,3,4,4-hexachlorobutane;[53] substituted 2-furyl-1,3-dioxans;[54] 1,2-disubstituted ethanes;[55] 1,3,2-dioxastibinane derivatives;[56] 1,3,2-dioxa-arsolans, 1,3,2-oxathioarsolans, and 1,3,2-dithia-arsolanes;[57] 3,3,6,6-tetramethyloctanedioic anhydride;[58] 2-phenyl-1,3,2-dithiaphospholane;[48] choline tryptamine;[59] 2-methoxyethanol;[60] NN'-di-imidyl systems;[61] p-methylbenzylbromide;[62] α-tetralones;[35] substituted ferrocenes;[63] D-(tertiary arsines) metal carbonyl complexes;[29] pleiadiene.[18]

[AX]$_2$: Diazadiphosphetidines;[64] dipotassium difluorodiphosphate;[24] cyclohepta[c or b]furan-6-ones;[13] cyclopentadienyl-silicon and -germanium complexes.[65]

ABTX: 1-Bromo-2-chloro-1,1,2-trifluoroethane.[66]

AMTX: Anthracene-related compounds;[52] 1-bromo-3,4,4-trifluorobut-4-ene.[67]

C. Five-spin System.
ABCDE: Substituted oxetans;[68] 2,3,4-trisubstituted thiophens;[69] [2(exo),3,6,7,7-^{2}H$_5$]camphor;[32] thienothiapyrylium perchlorates;[70] 2-dichlorophenyloxetans;[71] 2-chlorophenyloxetans;[41] substituted 2-furyl-1,3-dioxans;[54] polyfluorinated 1,1-

---

[47] W. R. Cullen, L. D. Hall, and J. E. H. Ward, *J. Amer. Chem. Soc.*, 1975, **96**, 3422.
[48] K. Bergesen, M. Bjoroy, and T. Gramstad, *Acta Chem. Scand.*, 1972, **26**, 3037.
[49] L. Casaux, J. Barreau, and M. Massol, *Org. Magn. Resonance*, 1975, **7**, 31.
[50] W. R. Cullen, L. D. Hall, and J. E. H. Ward, *J. Amer. Chem. Soc.*, 1975, **96**, 3431.
[51] J. Devillers, M. Cornus, J. Navech, and L. Casaux, *Org. Magn. Resonance*, 1975, **7**, 296.
[52] N. E. Sharpless, R. B. Bradley, and J. A. Ferretti, *Org. Magn. Resonance*, 1974, **6**, 115.
[53] R. J. Abraham and J. R. Monasterios, *J. C. S. Perkin II*, 1975, 699.
[54] Y. Y. Samitov, Z. I. Zelikman, A. I. Shkrebets, and V. G. Kulnevich, *Khim. geterotsikl. Soedinenii*, 1974, 319.
[55] Li-Shu Chen and Wei Chewanhin, *Hue Hsueh*, 1972, 141.
[56] B. A. Arburov, Yu. Yu. Samitov, and Yu. M. Mareev, *Doklady Akad. Nauk S.S.S.R.*, 1974, **219**, 863.
[57] Yu. Yu. Samitov, N. K. Tazeeva, and N. A. Chadaeva, *Zhur. strukt. Khim.*, 1975, **16**, 34.
[58] G. Borgen, *Acta Chem. Scand.* (*B*), 1974, **28**, 13.
[59] C. C. J. Culvenor and N. S. Ham, *Austral. J. Chem.*, 1974, **27**, 2191.
[60] V. Viti, P. L. Indovina, F. Podo, and L. Radics, *Mol. Phys.*, 1974, **27**, 541.
[61] S. M. Verma, K. O. P. Sinher, and K. C. Rao, *Canad. J. Chem.*, 1974, **52**, 2399.
[62] K. Chum, J. B. Rowbotham, and T. Schaefer, *Canad. J. Chem.*, 1974, **52**, 3489.
[63] N. Kamezawa, *J. Magn. Resonance*, 1973, **11**, 88.
[64] R. K. Harris, M. I. M. Wazeer, O. Schlak, and R. Schmutzler, *J. C. S. Dalton*, 1974, 1912.
[65] H. Köpf, W. Kahl, *J. Organometallic Chem.*, 1974, **64**, C37.
[66] Soon Ng, *J. Magn. Resonance*, 1975, **17**, 244.
[67] J. F. Hinton and L. W. Jaques, *J. Magn. Resonance*, 1973, **11**, 229.
[68] K. Pihlaja, J. Jokisaari, P. Olari, I. Virtanen, H. Ruotselainen, and M. Anteunis, *Org. Magn. Resonance*, 1975, **7**, 286.
[69] T. M. Filippova, B. D. Lavrukhin, and I. K. Shmyrev, *Org. Magn. Resonance*, 1974, **6**, 92.
[70] F. C. Boccuzzi and R. Fochi, *Gazzetta*, 1974, **104**, 671.
[71] J. Jokisaari and E. Rahkamaa, *Z. Naturforsch.*, 1973, **28a**, 30.

difluoro-4-hydroxyarenium ions;[72] 1,2,4-substituted-3-phenylcycl[3,2,2]azines;[73] substituted 4-piperidinones;[74] [11a, 11b,16-^{2}H$_3$]strychnine.[39]

ABCDX: Phenoxaphosphinic acids;[75] flavone (X ≡ ^{13}C);[34] pyridine-2- and -3-aldehydes;[76] 2-fluorobenzamides;[77] 2-bromopyridine (X ≡ ^{13}C);[78] [1 – ^{15}N]pyrid[^{14}N]azine;[46] furan (X ≡ ^{13}C).[79]

ABCXY: Thieno[2,3-b]pyridine;[80] 2,5-difluorophenyl methyl sulphide;[81] diazadiphosphetidines.[64]

ABCX$_2$: Vinyl ferrocene;[63] thiiranes;[82] ethylene sulphites.[36]

ABCMX: Polyfluorinated cyclohexadienones.[83]

ABMTX: Methyl- and methoxy-fluoro-1,3,2,4-diazadiphosphetidines.[64]

ABCD$_2$: Trialkenyl-aluminium and -gallium derivatives.[84]

A$_2$MTX: Thiete 1,1-dioxide.[85]

AB$_2$MX: Platinum hydrides.[86]

[[A]$_2$]$_2$X: 1,2-Dimethoxyethane (X ≡ ^{13}C).[60]

[AB]$_2$C: Ethyleneimine;[87] 2-alkyl and 2-phenyl-5-t-butyl-1,3,2-dioxaphosphorinans;[88] 1,3-dioxa-2-silacyclohexanes;[89] benzonitrile;[90] thietan 1-oxide-3-carboxylic acid;[91] phenylcarbonylfluorides;[92] flavone and deuteriated analogues;[34] 1-t-butyl-4-methoxycarbonyl[3,3,6,6-^{2}H$_4$]cyclohexane;[93] phenylferrocenes.[63]

[AB]$_2$X: Pyridine-4-aldehyde;[76] 4-alkylnitrobenzenes;[94] p-substituted triphenylphosphines;[95] sodium (p-methoxyphenyl)-methylphosphinodithionate dihydrate;[96] 2,5,5-trimethyl-1,3,2-dioxaphosphorinans;[97] naphthalene (X ≡ ^{13}C);[98] substi-

---

[72] N. G. Kostina and V. D. Shteingarts, *Zhur. org. Khim.*, 1974, **10**, 1705.
[73] C. M. Gupta, B. B. P. Srivastava, R. K. Rizvi, and N. Anard, *Indian J. Chem.*, 1974, **12**, 674.
[74] E. Luedtke and R. Heller, *Chem.-Ztg.*, 1974, **98**, 371.
[75] J. B. Levy, *Israel J. Chem.*, 1974, **12**, 779.
[76] W. Dauchna, T. Schaefer, J. B. Rowbotham, and D. J. Wood, *Canad. J. Chem.*, 1974, **52**, 3986.
[77] H. Fritz and T. Winkler, *Helv. Chim. Acta*, 1974, **57**, 836.
[78] M. Hansen and H. J. Jakobsen, *J. Magn. Resonance*, 1973, **10**, 74.
[79] M. Hansen, R. S. Hansen, and H. J. Jakobsen, *J. Magn. Resonance*, 1974, **13**, 386.
[80] L. H. Klemm, R. E. Merrill, and F. H. W. Lee, *J. Heterocyclic Chem.*, 1974, **11**, 535.
[81] G. Hägele, J. Richter, and M. Peach, *Z. Naturforsch.*, 1974, **29b**, 619.
[82] K. J. Ivin, E. D. Lillie, and I. H. Petersen, *Internat. J. Sulfur Chem.*, 1973, **8**, 411.
[83] N. E. Akhmetova, N. G. Kostina, V. I. Mametyuk, A. A. Shtark, and V. D. Shteingarts, *Khim. Nauk*, 1973, 86.
[84] T. W. Dolzine and J. P. Oliver, *J. Amer. Chem. Soc.*, 1974, **96**, 1737.
[85] G. Fronza, A. Gamba, R. Mondelli, and G. Pagani, *J. Magn. Resonance*, 1973, **12**, 231.
[86] T. W. Dingle and K. R. Dixon, *J. Inorg. Chem.*, 1974, **13**, 846.
[87] H. Nakanishi and O. Yamamoto, *Chem. Letters*, 1973, **5**, 465.
[88] W. G. Bentrude, H. W. Tan, and K. C. Yu, *J. Amer. Chem. Soc.*, 1975, **97**, 573.
[89] P. Albriktsen and S. Heggelund, *Acta Chem. Scand. (B)*, 1974, **28**, 573.
[90] J. P. Jacobsen and K. Schaumburg, *Mol. Phys.*, 1974, **28**, 1505.
[91] C. Cistano, G. Fronza, R. Mondelli, S. Bradamonte, and G. A. Pagani, *J. Magn. Resonance*, 1974, **15**, 367.
[92] H. Bildsoe and K. Schaumburg, *J. Magn. Resonance*, 1974, **14**, 223.
[93] J. P. Aycard and H. Bodot, *Org. Magn. Resonance*, 1975, **7**, 35.
[94] D. W. Beistel and H. E. Chen, *J. Chem. Phys.*, 1973, **59**, 4642.
[95] R. Benessi, M. L. Schenetti, F. Taddei, P. Vivarelli, and P. Bembech, *J. C. S. Perkin II*, 1974, 1338.
[96] G. Hägele, K. Diemart, and W. Kuchen, *Z. Naturforsch.*, 1973, **28a**, 30.
[97] J. P. Albrand, J. P. Dutasta, and J. B. Robert, *J. Amer. Chem. Soc.*, 1975, **96**, 4584.
[98] H. Gunther, H. Schmickler, and G. Jikeli, *J. Magn. Resonance*, 1973, **11**, 344.

tuted-1,3,2-dioxaphosphorinan;[99] phenylcarbonyl fluorides;[92] [1-^{2}H]cyclopropanecarbonyl fluoride;[100] [^{15}N]diazacyclopentadiene;[101] substituted-4-piperidones;[74] methyl-substituted histamines;[102] 2-thioxo-1,3,2-dioxaphosphorinanes.[103]

A[BX]$_2$: Cyclotriphosphazatrienes.[104]

[AM]$_2$X: 3-Substituted thietan 1,1-dioxides;[105] polyfluorinated cyclohexadienones;[83] polyfluorinated-1,1-difluoro-4-hydroxyarenonium ions.[72]

AB$_2$CX: [1-^{15}N]-3,5-Dimethylborazine.[106]

D. Six-spin Systems.

ABCDEF: 1,3-Dithian 1-oxide;[107] 1,7-diacetoxynapthalenes;[108] N-acetyl-4-hydroxyproline.[109]

ABCDEX: 4-Substituted 1,3,2-dithiaphospholans;[110] [2-^{13}C]pyridine.[78]

ABCDXY: [1,3-^{15}N$_2$]-4-Amino-5-methylborazine.[106]

ABCMXY: 4-Bromo-3-chloro-3,4,4-trifluorobut-1-ene.[111]

ABCD$_2$X: Pyrimidine nucleoside.[112]

ABCD$_3$: 2-Chloro- and 2-phenyl-4-methyl-1,3,2-dioxarsolans and -1,3,2-dithiarsolans;[113] 1-methyl-[2,2,3,3,4,4-^{2}H$_6$]cyclopentane.[37]

ABCX$_3$: 4-Methylethylenesulphite;[114] ethylene sulphites.[36]

[AB]$_2$CD: Thietan 1-oxide and 1,1-dioxide;[91] 1,3-dithian;[115] 2-substituted 1,3-thiolane-1,1,3,3-tetraoxides.[116]

[AB]$_2$CX: 2-Substituted 5-methyl-1,3,2-oxathiaphospholan;[117] triphenylphosphine;[118,119] [3-^{13}C]- and [4-^{13}C]-pyridine;[78] [^{15}N]pyrrole;[120] cyclopropane-

---

[99] H. Cogne, A. G. Guimaraes, J. Martin, R. Nardin, J. B. Robert, and W. J. Stec, *Org. Magn. Resonance*, 1974, **6**, 629.
[100] H. Bildsoe and K. Schaumburg, *J. Magn. Resonance*, 1973, **12**, 255.
[101] J. P. Jacobsen, K. Schaumburg, and J. T. Nielson, *J. Magn. Resonance*, 1973, **12**, 372.
[102] C. R. Ganellin, E. S. Pepper, G. N. J. Port, and W. G. Richards, *J. Medicin. Chem.*, 1973, **16**, 610.
[103] J. P. Dutasta, A. Grand, T. B. Roberts, and M. Taieb, *Tetrahedron Letters*, 1974, **31**, 2659.
[104] P. Clare, D. B. Sowerby, R. K. Harris, and M. I. M. Wazeer, *J. C. S. Dalton*, 1975, 625.
[105] C. Cistano, G. Fronza, R. Mondelli, S. Bradamante, and G. A. Pagani, *J. Magn. Resonance*, 1975, **17**, 219.
[106] L. J. Turbini and R. F. Porter, *Org. Magn. Resonance*, 1974, **6**, 456.
[107] M. J. Cook and A. P. Tonge, *J. C. S. Perkin II*, 1974, 762.
[108] P. Granger and M. Mangras, *Chem. Phys. Letters*, 1974, **24**, 331.
[109] T. Prange, C. Garbay-Jaureguiberry, B. Roques, and M. Anteunis, *Biochem. Biophys. Res. Comm.*, 1974, **61**, 104.
[110] M. Revel, J. Roussel, H. Bordjabel, and J. Navech, *Compt. rend.*, 1973, **277**, C, 57.
[111] J. F. Hinton and L. W. Jaques, *J. Magn. Resonance*, 1974, **15**, 564.
[112] M. P. Schweizer, E. B. Bauta, J. T. Wikowski, and R. K. Robins, *J. Amer. Chem. Soc.*, 1973, **95**, 3770.
[113] D. W. Aksnes and O. Vikane, *Acta Chem. Scand.*, 1973, **27**, 2135.
[114] P. Albriktsen, *Acta Chem. Scand.*, 1972, **26**, 3671.
[115] J. Martin and J. B. Robert, *Org. Magn. Resonance*, 1975, **7**, 76.
[116] L. A. Sternson, L. C. Martinelli, and R. S. Egan, *J. Heterocyclic Chem.*, 1974, **11**, 117.
[117] K. Bergesen and M. Bjoroy, *Acta Chem. Scand.*, 1973, **27**, 3477.
[118] S. Sorenson and H. J. Jakobsen, *Acta Chem. Scand. (A)*, 1974, **28**, 249.
[119] L. Radics, E. Baitz-Gaes, and A. Neszmelyi, *Org. Magn. Resonance*, 1974, **6**, 60.
[120] J. M. Briggs, E. Rahkamaa, and E. W. Randall, *J. Magn. Resonance*, 1973, **12**, 40.

carbonylfluoride;[100] benzaldehyde and phenylcarbonylfluoride;[121] pentafluorophenylsulphur fluorides.[122]

ABMX$_3$: Hexafluoropropene dimers;[123] [1-^{13}C]crotonic and isocrotonic acids.[124]

[AB]$_2$XY: 2-Amino-5-t-butyl-1,3,2-dioxaphosphorinans.[125]

[ABC]$_3$: 3,4-Disubstituted thiophens;[126] 1,2,3,4-tetrachlorobutanes;[53] reaction of o-styryldiphenylphosphine with Ru$_3$(CO)$_{12}$;[127] 2,2-dimethyldithiane;[107] 2,7-diacetoxynaphthalenes.[108]

[ABX]$_2$: Me$_2$M(C$_5$H$_4$)$_2$TiS$_5$ (M = Si or Ge).[128]

[AX]$_2$BY: Cyclotriphosphazatrienes.[104]

[A[B]$_2$]$_2$: Cyclobutanone;[129] 2,5-dihydrofuran, 2,5-dihydrothiophen, and butadiene sulphone.[130]

[A[X]$_2$]$_2$: Diazadiphosphetidines;[64] 1,3-difluoroacetone and 1,1,3,3-tetrafluoroacetone;[131] chlorinated dimethyltetrafluorodiazadiphosphetidines;[132] 1,2-di-iodo- and 1,2-dichloro-hexafluorocyclopent-1-ene.[133]

[AMX]$_2$: 4,8-Dioxabicyclo[5,1,0]octa-2,5-diene.[134]

A$_2$B$_3$X: Low-temperature (slow-exchange limit) spectrum of Rh$^+$[P(OMe)$_3$]$_5$-BO$_4$$^-$.[135]

AB$_4$X: PF$_5$L (L = NMe$_3$, pyridine, or 4-methyl pyridine).[136]

[AX]$_3$: Cyclotriphosphazatrienes.[104]

E. Seven-spin Systems.

ABCDEFG: [2,2,5-exo or 5-endo-^{2}H$_3$]Camphor;[32] 2-, 3-, and 4-vinylpyridine.[137]

ABCDEFX: cis- and trans-4-fluoro-L-proline;[138] 4-substituted 1,8-difluoronaphthalene.[139]

ABCDEXY: [^{15}N, ^{13}C]Pyrrole.[120]

[121] H. Bildsoe and K. Schaumburg, J. Magn. Resonance, 1974, 14, 223.
[122] P. Meakin, D. W. Overall, W. A. Sheppard, and J. P. Jesson, J. Amer. Chem. Soc., 1975, 97, 522.
[123] C. Schumann and F. Cavagna, J. Magn. Resonance, 1975, 18, 172.
[124] J. L. Marshall and R. Seiwell, J. Magn. Resonance, 1974, 15, 150.
[125] W. G. Bentrude and H.-W. Tan, J. Amer. Chem. Soc., 1973, 95, 4666.
[126] T. M. Filippova, B. D. Lavrukhin, and I. K. Shmyner, Org. Magn. Resonance, 1974, 6, 92.
[127] M. A. Bennett, R. N. Johnson, and I. B. Tomkins, J. Amer. Chem. Soc., 1974, 96, 61.
[128] H. Köpf and W. Kahl, J. Organometallic Chem., 1974, 64, C37.
[129] R. G. Green, P. Shaw, L. H. Sutcliffe, B. Taylor, and S. M. Walker, Spectrochim. Acta (A), 1974, 30, 529.
[130] R. Lozac'l and B. Braillon, J. Magn. Resonance, 1973, 12, 244.
[131] B. L. Shapiro, W. A. Thomas, J. L. McClanahan, and M. O. Johnston, jun., J. Magn. Resonance, 1973, 11, 355.
[132] R. K. Harris, M. Lewellyn, M. I. M. Wazeer, J. R. Woplin, R. E. Dunmur, M. J. C. Hewson, and R. Schmutzler, J. C. S. Dalton, 1975, 61.
[133] L. Cavalli and R. K. Harris, J. Magn. Resonance, 1974, 15, 150.
[134] H. Klein and W. Grimme, Angew. Chem., 1974, 13, 672.
[135] P. Meakin and J. P. Jesson, J. Amer. Chem. Soc., 1973, 95, 7272.
[136] K. P. John and R. Schmutzler, Z. Naturforsch., 1974, 29b, 730.
[137] J. B. Rowbotham and T. Schaefer, Canad. J. Chem., 1974, 52, 136.
[138] J. T. Gerig and R. S. McLeod, J. Amer. Chem. Soc., 1973, 95, 5725.
[139] J. M. Anderson, F. B. Mallory, and C. W. Mallory, J. Magn. Resonance, 1975, 17, 340.

ABCDMXY: 1,4-Dibromo-2-chloro-1,2,2-trifluorobutane.[140]
ABCDRTX: L-Proline.[141]
ABCDX$_3$: m-Toluidine.[142]
ABCPX$_3$: 2-Substituted 4-methyl-1,3,2-dithiaphospholanes.[143]
ABX$_3$Y$_2$: Diazadiphosphetidines.[64]
[AB]$_2$CDX: 2-Phenyl-1,3,2-dithiaphosphorinane.[115]
[AB]$_2$MTX: 1-Bromo-3,4,4-trifluorobut-4-ene.[67]
AM]$_2$BNX: sym-Trifluorobenzene (X ≡ ^{13}C).[144]
[AM]$_2$RTX: 3-Methyl-1,1,2,4,5,6-hexafluorobenzenonium ion.[145]
[A[B]$_2$]$_2$X: Platinum–cyclopropane compounds.[146]
[ABX]$_2$C: [1,3-^{15}N$_2$]-5-Methylborazine.[106]
[AMT]$_2$X: 1-Iodoheptafluorocyclopent-1-ene.[147]
AB$_6$: (CH$_3$)$_2$SiCH.[148]

F. Eight-spin Systems.
ABCDEFGH: Camphor [4-$endo$,5-$endo$-^{2}H$_2$]camphor;[32] 2-[3,4(and 2,4)-dichlorophenyl]oxetans.[71]
ABX$_3$Y$_3$: Ethylene sulphites;[36] $trans$-2-isopropyl-1,3-dithian 1-oxide;[149] 5,6-dimethyl-2-oxo-1,4-dioxan;[150] 4,5-dimethylethylene sulphite.[114]
[AB]$_2$CX$_3$: Toluene and tolylcarbonylfluoride.[121]
[AB]$_2$KMTX: 1-Bromo-3,4,4-trifluorobut-4-ene.[67]
[AM]$_2$RX$_3$: Perfluorotoluene.[151]
A$_2$[KP]$_2$TX: F$_4$P^{15}NH$_2$.[152]
[ABCD]$_2$: 1,4:3,6-Bis(thioanhydro)-D-iditol derivatives.[153]
[A[X]$_2$M]$_2$: [ClCHF$_2$PNMe]$_2$.[132]
[AB]$_4$: Tetrahydrofuran;[154, 155] 1:4-dibromobutane.[5]
[AX]$_4$: Polyfluoro-oxathian.[156]

[140] J. F. Hinton and L. W. Jaques, *J. Mag. Resonance*, 1975, **17**, 95.
[141] L. Pogliami, M. Ellenburger, and J. Valet, *Org. Magn. Resonance*, 1975, **7**, 61.
[142] C. Pascual, H. Huber, and E. Wullschleger, *Afinidad*, 1974, **31**, 751.
[143] K. Bergesen and M. Bjoroy, *Acta Chem. Scand.*, 1973, **27**, 1103.
[144] V. Wray and D. N. Lincoln, *J. Magn. Resonance*, 1975, **18**, 374.
[145] T. V. Chuikova, A. A. Shtark, and V. D. Shteingarts, *Zhur. org. Khim.*, 1974, **10**, 1712.
[146] D. W. Hall, R. J. Puddephatt, and C. F. H. Tipper, *J. Organometallic Chem.*, 1974, **71**, 145.
[147] L. Cavalli, and R. K. Harris, *J. Mag. Resonance*, 1973, **10**, 355.
[148] G. Hägele, W. Peters, and M. Weidenbruch, *Ber. Bunsengesellschaft phys. Chem.*, 1974, **78**, 781.
[149] M. Anteunis, L. Van Acker, and D. Danneels, *Bull. Soc. chim. belges*, 1974, **83**, 301.
[150] P. Äyräs, *Org. Magn. Resonance*, 1975, **7**, 177.
[151] C. W. Haigh, J. Hietan, L. H. Sutcliffe, and G. J. T. Tiddy, *J. Magn. Resonance*, 1975, **18**, 241.
[152] A. H. Cowley and J. R. Schweiger, *J. Amer. Chem. Soc.*, 1973, **95**, 4719.
[153] P. Sohar and J. Kuszmann, *Org. Magn. Resonance*, 1974, **6**, 407.
[154] E. Diez, A. L. Esteban and M. Rice, *J. Magn. Resonance*, 1974, **16**, 136.
[155] R. G. Jones and S. M. Walker, *Mol. Phys.*, 1966, **10**, 363.
[156] L. Philips and V. Wray, *J. C. S. Perkin II*, 1974, 928.

G. Spin Systems of nine nuclei or more.

[A$_3$]$_2$B$_2$X: Propane (X ≡ ^{13}C).[157]

A$_6$B$_2$X: Propane (X ≡ ^{13}C).[157]

ABC... (ten spins): cis-Cyclononatetraene iron tricarbonyl.[158]

[A$_3$[B]$_2$]$_2$: Butane.[159]

[A$_3$[X]$_2$]$_2$: 2,3,5,6-Tetrafluoro-p-xylene;[160] 1,1,1,4,4,4-hexafluorobutane.[161]

[AMX$_3$]$_2$: 1,2-Dithiomethoxy-4,5(and 3,6)-difluorobenzenes.[162]

[AB]$_2$R$_3$T$_3$X: p-Substituted ββ-dimethylstyrenes.[163]

ABR$_3$X$_6$: Bisphosphinoyl amines.[164]

AKR$_3$X$_6$: Bisphosphinoyl amines.[164]

[AC]$_2$BX$_9$: C$_6$F$_5$SiMe$_3$.[165]

ABK$_3$P$_3$T$_3$X$_3$: Bisphosphinoyl amines.[164]

[AX$_3$Y$_3$]$_2$: Bisphosphinoyl amines.[164]

[[AX]$_2$M]$_2$: [(ClCH$_2$)F$_2$PNMe]$_2$.[132]

AKR$_2$X$_{12}$: Bisphosphinoyl amines.[164]

[AX$_6$]$_2$R$_3$: Bisphosphinoyl amines.[164]

ABX$_6$Y$_6$R$_3$: Bisphosphinoyl amines.[164]

A$_2$MRX$_{12}$: Perfluoropinacolyltrifluorophosphorans.[166]

AKR$_3$X$_{12}$: Bisphosphinoyl amines.[164]

[AX$_6$]$_2$R$_3$: Bisphosphinoyl amines.[164]

ABX$_6$Y$_6$R$_3$: Bisphosphinoyl amines.[164]

[ACM$_3$]$_2$BX$_9$: C$_6$F$_5$SiMe$_2$SiMe$_3$.[165]

[[AC]$_2$B[X$_3$]$_2$]$_2$: [C$_6$F$_5$SiMe$_2$]$_2$.[165]

A$_3$FKN$_2$RX$_{12}$: Perfluoropinacolyltrifluorophosphorans.[166]

[AX$_9$]$_2$, [AX$_9$]$_3$, [AX$_9$]$_4$, and Related Systems: Complexes of Pd, Pt, Rh, Ir, Hg and Au containing two or more PMe$_3$ ligands (or related ligands).[4]

[AX$_9$R$_3$]$_2$: cis[PtMe$_2$(PMe$_3$)$_2$].[4]

[[A]$_2$X$_2$]$_2$: 1,4-Dibromobutane.[5]

[ABX$_3$]$_2$: trans,trans-2,4-Hexadiene.[5]

[A$_n$X]$_2$: Phosphorus-containing compounds.[167]

[157] R. E. Wasylishen and T. Schaefer, *Canad. J. Chem.*, 1974, **52**, 3247.
[158] E. J. Reardon, jun. and M. Brookhart, *J. Amer. Chem. Soc.* 1973, **95**, 4311
[159] D. W. Aksnes and P. Abriktsen, *Acta Chem. Scand.*, 1972, **26**, 3021.
[160] F. A. M. Ayanbadejo, *Spectrochim. Acta*, 1974, **30**, 271.
[161] R. J. Abraham and P. Loftus, *J. C. S. Perkin II*, 1975, **6**, 535.
[162] G. Hägele, J. Richter, and M. Peach, *Org. Magn. Resonance*, 1974, **6**, 374.
[163] E. R. Curry and D. J. Sardella, *J. Amer. Chem. Soc.*, 1974, **96**, 1823.
[164] G. Hägele, R. K. Harris, M. I. M. Wazeer, and R. Keat, *J.C.S. Dalton*, 1974, 1985.
[165] G. Hägele and M. Weidenbruch, *Org. Magn. Resonance*, 1974, **6**, 66.
[166] J. A. Gibson, G. V. Roschenthaler and R. Schmutzler, *J. C. S. Dalton*, 1975, 918.
[167] W. MacFarlane and D. S. Rycroft, *J. C. S. Faraday II*, 1974, **70**, 377.

$[A\Sigma X_{n_i}^i]_2$: Theory for the A region only.[168]

$[[AC]_2B]_n$ and $[[AX_2]M]_n$: Silicon compounds containing C_6F_5 groups.[165]

A computer program, NMR MRG, has been designed particularly for simulating polymer spectra involving many spins. Its use has been illustrated for analysing proton spectra of polypropylenes.[169]

4 Conclusions

Some trends have emerged in the past two years. One is the re-examination of the simpler molecules incorporating spin-$\frac{1}{2}$ isotopes hitherto undetected. Another manifests itself in the complexity of systems which are now being tackled using increasingly efficient computer programmes. Spectral analysis provides the research worker with a wealth of molecular information which is of prime importance in studies of molecular structure.

[168] G. Hägele, *Ber. Bunsengesellschaft Phys. Chem.*, 1974, **78**, 43.
[169] R. C. Ferguson, *J. Magn. Resonance*, 1973, **12**, 296.

6
Bandshape Phenomena in Liquids

BY T. DRAKENBERG AND H. WENNERSTRÖM

1 Introduction

The general outline of this Report is the same as in Vol. 4. In section 2 we have chosen to discuss bandshapes at slow molecular tumbling rates since this might be an area for future experimental work. Section 3 on chemical exchange effects is intended to cover all total bandshape analysis. The development in this area is still towards applications using nuclei other than protons. A neat example of this trend is found in an article by Cowley et al.,[1] where the exchange processes in one molecule are investigated by ^{1}H, ^{13}C, ^{19}F, and ^{31}P n.m.r. spectroscopies. The Swift–Connick method is reviewed in Section 4, but except for the work of Fiat and co-workers no real progress has been made in this area. We have not been able to find a single article describing bandshape analysis in the gas phase, and consequently this area is not reviewed.

We should also mention that an extensive review[2] dealing with chemical exchange effects has recently appeared.

The Report is intended to cover those articles published between June 1974 and the end of May 1975 in English, German, or French. For translated Russian journals the Report essentially covers the 1974 issues.

2 General Theory

In general, a non-Lorentzian shape of the n.m.r. signal in fluids is only obtained in the presence of some sort of slow motion in the system. The slow motions usually studied by n.m.r. techniques in solution are of the chemical exchange type. However, with the current interest in biological applications of n.m.r., it can be anticipated that cases when, for example, diffusional motion is slow will be encountered in the future. These phenomena are well known in e.s.r. spectroscopy, and the basic theoretical problems have been solved during recent years. Freed et al.[3] have presented a treatment of diffusional rotation. Although in principle exact, their method leads to rather tedious calculations, and it is based on a series expansion that converges slowly towards the rigid-lattice limit. In a series of papers, Alexander et al.[4-6]

---

[1] A. H. Cowley, R. W. Braun, and J. W. Gilje, *J. Amer. Chem. Soc.*, 1975, **97**, 434.
[2] 'Dynamic Nuclear Magnetic Resonance Spectroscopy', ed. L. M. Jackman and F. A. Cotton, Academic Press, New York, 1975.
[3] J. H. Freed, G. V. Bruno, and C. F. Polnazek, *J. Phys. Chem.*, 1971, **75**, 3386.
[4] A. Baram, Z. Luz, and S. Alexander, *J. Chem. Phys.*, 1973, **58**, 4558.
[5] S. Alexander, A. Baram, and Z. Luz, *Mol. Phys.*, 1974, **27**, 441.
[6] S. Alexander, A. Baram, and Z. Luz, *J. Chem. Phys.*, 1974, **61**, 992.

have refined the method. In addition to the diffusion motion they consider[5] simultaneous discrete jumps. Properly generalized, these calculations should be applicable to the description of the n.m.r. spectrum of a rotating macromolecule that changes conformation on the same timescale. In the most recent paper in the series Alexander et al.[6] show how one can modify the series expansion of Freed et al.,[3] calculating only the first few terms exactly and then using an analytical approximation for the remaining ones. This approach reduces the computational effort considerably. Another way of solving the problem of long computational time has been attempted by Atkins and Hills.[7] They calculated the magnetic resonance spectrum starting from a correlation function and using projection operator techniques. This leads to a perturbation expansion which converges rapidly at moderately slow motions, but it becomes difficult to handle when the correlation times grows and one approaches the rigid-lattice region. However, for $2\pi\tau_c\nu_{int} \leqslant 2$ (ν_{int} = interaction that is modulated by the motion) Atkins and Hills give analytical expressions for the bandshape that could prove very useful in simple applications. The reader is warned that their paper unfortunately contains several printing errors.

In all these papers[3-7] the authors have extensively used symmetry arguments to simplify the calculations. A clear example of these techniques is contained in a paper by Harris and Pyper[8] dealing with calculations of n.m.r. bandshapes of spin-$\frac{1}{2}$ nuclei (A,B) that have a scalar coupling to nuclei X which possess a quadrupole moment. The slow motion in this case is the relaxation of the nuclei X. As the number of degrees of freedom is finite and small, a more exact solution of the slow-motion problem can be obtained. The paper consists of a detailed description of the computer program that performs the calculation, combined with two illustrative examples. It is concluded that non-trivial effects, which cannot be described by the simple chemical-exchange analogy,[9] occur in the bandshape when there is a coupling between the A and B nuclei and/or between two X-type nuclei.

An experimental demonstration of the measurement of ^{2}H relaxation times through a scalar coupling to ^{1}H has been given by Drakenberg.[10] He studied isopropyl alcohol that was deuterated in the 2-position in a solution containing the alcohol-binding enzyme alcohol dehydrogenase. In the presence of the enzyme, extreme narrowing conditions do not apply. As a consequence, the simple Pople theory[9] has to be modified to account for the $\Delta m=1$ and $\Delta m=2$ T_1 processes separately. A noticeable consequence of this is that the spectral densities at ω_0 and $2\omega_0$ are measured independently, and not as a linear combination, as they are in ordinary T_1 measurements.

Relaxation effects in spectra of spin-$\frac{1}{2}$ nuclei with spin couplings to other spin-$\frac{1}{2}$ nuclei have been reported. Rao and Randall[11] have given a theoretical explanation of the observation by Fronza et al.[12] of differential broadening in ^{13}C multiplets. These observations were made in solutions containing a paramagnetic relaxation agent, and Rao and Randall showed that the results were in accordance with a random field relaxation mechanism, in spite of the fact that triplets and quartets gave

[7] P. W. Atkins and B. P. Hills, *Mol. Phys.*, 1975, 29, 761.
[8] R. K. Harris and N. C. Pyper, *Mol. Phys.*, 1975, 29, 205.
[9] J. A. Pople, *Mol. Phys.*, 1958, 7, 168.
[10] T. Drakenberg, *J. Magn. Resonance*, 1974, 15, 354.
[11] B. D. Nageswara Rao and E. W. Randall, *Chem. Phys. Letters*, 1975, 33, 191.
[12] G. Frónza, R. Modelli, and E. W. Randall, *J.C.S. Chem. Comm.*, 1974, 196.

qualitatively different differential broadening effects. Polak and Navon[13] have studied the interaction between organic molecules and oxygen through the ^{1}H n.m.r. signals. They found that the singlets gave the narrowest peaks, then the doublets, and so on. This could be nicely explained in a model with only statistical encounters between the oxygen molecules and the organic molecules. The greater broadening effects in the multiplets are due to the T_1 relaxation of the nucleus responsible for the multiplet structure. Hughes et al.[14] have calculated the bandshape for the n.m.r. signal of a spin-$\frac{1}{2}$ nucleus in the AX and ABX (with $\nu_A = \nu_B$) systems, with full regard being taken to relaxation effects. The calculations are very similar to those described by Pyper and Harris,[8] the only difference being in the value of I for the X nucleus. The program of Harris and Pyper has, however, an option where I_X is set equal to $\frac{1}{2}$, and hence this treatment is probably able to deal with the problem considered by Hughes et al. in a more general way.

3 Exchange of Magnetic Sites

A. General Theory.—The most important theoretical development in the calculation of n.m.r. bandshapes in the presence of chemical exchange is found in a paper by Kaplan et al.[15] dealing with effects of intense r.f. fields. The authors show how their previously presented formalism[16] is well adapted to deal with problems beyond the linear-response approximation. It is shown in an illustrative example that the n.m.r. bandshape is more sensitive to exchange effects in the slow-exchange region at higher r.f. fields. This is not unexpected, and this high-r.f.-power technique bridges, in a way, the gap between the conventional bandshape method and the Forsén–Hoffman technique of measuring exchange rates. There are, however, a couple of difficulties associated with the method suggested by Kaplan et al. Firstly, a rather accurate value for the magnitude of the r.f. field is needed. Secondly, it is known[17] that relaxation effects become non-trivial at high r.f. powers, and this must also influence the bandshape in the presence of exchange. As yet, a practical application of the method has not appeared, which makes it difficult to assess how great these problems are.

Several papers have appeared that deal with the practical problem of evaluating a rate constant from a spectrum showing chemical exchange effects. Szymanski and Gryff-Keller[18] have suggested a transformation of the basis functions normally used[19] so that the resulting matrices are always symmetrical. Since it is normally easier to invert or diagonalize such matrices, this transformation could lead to less computation. With a specific example, the authors find that they can calculate the intensity of a spectrum at 400 points using the same computer time as it takes to diagonalize the complex matrix, which is the most time-consuming step in the calculation of a bandshape in the conventional procedure. A possible advantage with the transformation suggested by Szymanski and Gryff-Keller is that the condition of

---

[13] M. Polak and G. Navon, *J. Phys. Chem.*, 1974, **78**, 1747.
[14] D. G. Hughes, M. R. Smith, and D. A. M. Switzer, *J. Chem. Phys.*, 1974, **60**, 4890.
[15] J. Kaplan, P. P. Yang, and G. Fraenkel, *J. Chem. Phys.*, 1974, **60**, 4840.
[16] G. Fraenkel and J. Kaplan, *J. Amer. Chem. Soc.*, 1972, **94**, 2907.
[17] R. A. Hoffman, *Adv. Magn. Resonance*, 1970, **4**, 87.
[18] S. Szymanski and A. Gryff-Keller, *J. Magn. Resonance*, 1974, **16**, 182.
[19] G. Binsch, *J. Amer. Chem. Soc.*, 1969, **91**, 1304.

detailed balance is automatically satisfied and should remain so when approximations are invoked. This is not always the case in the conventional procedure.[20]

In an investigation of the internal rotation around the amide bond in N-acetyl-N-methyl-2-chloro-4,6-dinitroaniline Heidberg and Scheve[21] found that their computational efforts were made a lot easier by evaluating analytical expressions for the bandshape. They studied the AB system of the aromatic protons and give explicit expressions for the bandshape for the exchange $(AB)_k \rightleftarrows (AB)_l$ with different populations in sites k and l. Al'ber et al.[22] have presented a density-matrix treatment of the intermolecular exchange case, which, the authors claim, has not been treated previously. This seems not to be true, however.[16] Lozach et al.[23] have presented an approximate procedure, based on peak heights, for determining rate constants. In the Reporters' opinion such approximate methods are only useful in a preliminary analysis of the experimental data.

B. Intramolecular Exchange Processes in Organic Molecules.—(i) *Rotation about Sterically Hindered Single Bonds.* Bushweller and co-workers[24,25] have continued their work on t-butyl rotation, using ^{1}H n.m.r. total bandshape analysis. They found a trend in the barrier height of 2-t-butyl-1,3-diheteroatomic rings (1) and (2) as a function of ring structure and heteroatoms but with one exception. This could be rationalized on the basis of ring geometry obtained from approximate MO calculations. Greenberg and Laszlo[26] obtained similar results from a total bandshape analysis of the t-butyl proton signal in (3).

(1) a; X = Y = O   (2)   (3)
     b; X = Y = S
     c; X = O, Y = S

The Japanese group working on restricted rotation about bonds between tetrahedral atoms have continued their work with a study of the torsional barrier in systems like (4).[27] The barrier height was found to decrease drastically on lengthening the C—M bond. Oki and co-workers[28,29] have also studied the barrier to rota-

---

[20] P. Meakin and J. P. Jesson, *J. Amer. Chem. Soc.*, 1973, **95**, 7272.
[21] J. Heidberg and H. Scheve, *Z. phys. Chem. (Frankfurt)*, 1974, **92**, 295.
[22] S. I. Al'ber, G. V. Lagodzinskaya, G. B. Manelis, and E. B. Fel'dman, *Izvest. Akad. Nauk S.S.S.R., Ser. khim.*, 1974, 1719.
[23] R. Lozach, L. Legrand, and J. Sandström, *Org. Magn. Resonance*, 1975, **7**, 54.
[24] P. E. Stevenson, G. Bhat, C. H. Bushweller, and W. G. Anderson, *J. Amer. Chem. Soc.*, 1974, **96**, 1067.
[25] C. H. Bushweller, S. Hoogasian, W. G. Anderson, and L. J. Letendre, *J.C.S. Chem. Comm.*, 1975, 152.
[26] A. Greenberg and P. Laszlo, *Tetrahedron Letters*, 1974, 2355.
[27] F. Susuki, M. Oki, and H. Nakanishi, *Bull. Chem. Soc. Japan*, 1973, **46**, 2858.
[28] M. Nakamura, M. Oki, H. Nakanishi, and O. Yamamoto, *Bull. Chem. Soc. Japan*, 1974, **47**, 2415.
[29] F. Suzuki, M. Oki, and H. Nakanishi, *Bull. Chem. Soc. Japan*, 1974, **47**, 3114.

tion of the methyl and isopropyl groups in 9-methyl- and 9-isopropyl-triptycene derivatives. As far as $\Delta G^{\ddagger}$ is concerned, the data for substituted 9-isopropyltriptycenes (5) are in good agreement with data on neopentyl rotation in trineopentylbenzenes[30] and $CHCl_2$ rotation in $\alpha\alpha$-dichlorobenzenes.[31] [There is an error in the data given for the barrier in (5a); probably the activation entropy should be negative.]

(4) a; M = C
b; M = Si
c; M = Ge
d; M = Sn

(5) a; X = Z = H, Y = W = Me
b; X = W = OMe, Y = Z = H
c; X = W = Me, Y = Z = H
d; X = Y = Z = W = Cl

Proshutinskii et al.[32] have estimated the rate of internal rotation for some trimesitylgermanium compounds by means of proton bandshape analysis. Only the Arrhenius energy of activation was reported, but to the Reporters' eyes the activation entropy is unexpectedly low (-40 J mol^{-1} K^{-1}).

In the continued study of the isopropyl rotation in 3-isopropylthiazole derivatives Lidén et al.[33] found that for 3,4-di-isopropylthiazoline-2-thione (6) three different rotamers exist, and they can be frozen out at low temperature. The analogous three rotamers were also found in two other similar compounds. The different torsional barriers were evaluated by means of ^{1}H n.m.r. total bandshape analysis. Three different isopropyl rotamers were also found for Se-methyl-NN-di-isopropyldiselenocarbamate and the sulphur analogue.[34] The assignments of the rotamers are, however, probably in error according to force-field calculations.[35] In tetraisopropylethylene the four isopropyl groups seem to rotate together like a set of cogwheels, giving rise to two sets of signals at room temperature that coalesce to one set at higher temperatures.[36]

The torsional barrier in three chloropropanes has been studied by means of ^{1}H n.m.r. total bandshape analysis.[37] The rate constants were obtained from visual

---

[30] B. Nilsson, P. Martinsson, K. Olsson, and R. E. Carter, *J. Amer. Chem. Soc.*, 1974, **96**, 3190.
[31] J. Peeling, J. B. Rowbotham, L. Ernst, and T. Schaefer, *Canad. J. Chem.*, 1974, **52**, 2414.
[32] V. I. Proshutinskii, I. I. Lapin, and V. A. Dumler, *J. Struct. Chem.*, 1974, **15**, 206.
[33] A. Lidén, C. Roussel, M. Chanon, J. Metzger, and J. Sandström, *Tetrahedron Letters*, 1974, 3629.
[34] Y. Takeda and T. Tanaka, *Org. Magn. Resonance*, 1975, **7**, 107.
[35] T. Liljefors, personal communication.
[36] D. S. Bomse and T. H. Morton, *Tetrahedron Letters*, 1975, 781.
[37] A. B. Dempster, K. Price, and N. Sheppard, *J. Magn. Resonance*, 1974, **16**, 235.

(6a)

(6b)

(6c)

fitting of calculated to experimental spectra. The results were compared with earlier data on analogous chloroethanes. The enthalpy of activation was used in this comparison, which can be a little misleading owing to the difficulty in obtaining accurate values for this parameter. In this case, for two of the compounds no significant difference is found in $\Delta H^{\ddagger}$, whereas in $\Delta G^{\ddagger}$ there is a difference of 4.2 kJ mol^{-1}, which is normally considered as a significant difference. This fact is not mentioned in the paper. Schaefer and co-workers[31, 38] have continued their work on barriers to internal rotation in 2,6-dihalogenobenzal halides (7) and in some $\alpha\alpha'$-dihalogeno-2,4,5,6-tetrachloro-m-xylenes (8) by means of ^{1}H n.m.r. bandshape analysis. From a comparison of the barrier heights in the two classes of molecules they concluded that the transition state for internal rotation in dihalogenobenzal halides has the side-chain proton perpendicular to the aromatic ring, which is in agreement with CNDO/2 and INDO calculations.[38] Barriers of the same type were studied by Baas et al.[39] by means of measurements of the proton coalescence temperature. Dahlberg et al.[40] have calculated the barrier to internal rotation of the 2-acyl and 2-alkyl substituents in 1,3,5-trineopentylbenzenes (9), using proton total bandshape analysis. For one compound only was this analysis performed over a temperature range large enough to give reliable values of the entropy and enthalpy of activation. This analysis gave a slightly negative $\Delta S^{\ddagger}$.

It has once more been shown that the steric size of a group is dependent on the system studied. In a ^{1}H n.m.r. bandshape study, Sakamoto and Oki[41] thus showed that the methoxy-group acts as a far larger group than methyl does, as far as the hindrance of benzene rotation in hetera-p-carbophanes (10) is concerned. They also found,[42] using ^{1}H n.m.r. coalescence measurements, that this rotation is dependent on the state of sulphur oxygenation in (11), and they explained this as being due

[38] J. B. Rowbotham, A. F. Janzen, J. Peeling, and T. Schaefer, *Canad. J. Chem.*, 1974, **52**, 481.
[39] J. M. A. Baas, J. M. van der Toorn, and B. M. Wepster, *Rec. Trav. chim.*, 1974, **93**, 133.
[40] E. Dahlberg, B. Nilsson, K. Olsson, and P. Martinsson, *Acta Chem. Scand.* (*B*), 1975, **29**, 300.
[41] K. Sakamoto and M. Oki, *Bull. Chem. Soc. Japan*, 1975, **48**, 497.
[42] K. Sakamoto and M. Oki, *Chem. Letters*, 1974, 1173.

(7) X = Cl, Br, or I

(8) X = Cl, Br, or I

(9)

to various degrees of attraction between the X group and the benzene ring in the ground state.

von Voithenberg et al.[43] have used both ^{1}H and ^{13}C n.m.r. total bandshape analysis to study the rotation about the glycosidic (C—N) bond in some nucleosides.

(10) R = Me or OMe

(11) X = S, SO, or SO$_2$

They found a pronounced decrease in the height of the torsional barrier in (12) compared to (13), which indicates that the barrier is mainly determined by interactions between the aglycon carbonyls and the nearest oxygens in the sugar moiety.

(12) R = COPh

(13) R = COMe

[43] H. von Voithenberg, A. Skrzelewski, J. Jochims, and W. Pfleidere, *Tetrahedron Letters*, 1974, 4063.

Rotation about Single Bonds with some Double-bond Character. Much attention is still paid to the study of amide bonds. Spassov *et al.*[44] have thus continued to use the method of computerized fitting of calculated to experimental bandshapes. They have reported activation parameters for three acetylenic amides (14). The free energies of activation for these compounds appeared to be almost identical, whereas there was a larger variation in values of $\Delta S^{\ddagger}$ and $\Delta H^{\ddagger}$. For (14b), studies were performed at both 60 and 90 MHz, resulting in differences in $\Delta S^{\ddagger}$ and $\Delta H^{\ddagger}$ that were much larger than the given statistical errors, indicating that some systematic errors also exist, at least for one of the measurements. This is not discussed in the paper.

$$\begin{array}{c} \text{Me} \\ \diagdown \\ \text{N}\!-\!\text{C} \\ \diagup \quad \diagdown \\ \text{Me} \quad \text{C} \\ \quad \diagdown \\ \quad \text{C} \\ \quad \diagdown \\ \quad \text{R} \end{array} \quad \begin{array}{c} \text{O} \end{array}$$

(14) a; R = H
b; R = Me
c; R = Ph

Nakanishi and Yamamoto[45] have studied the torsional barrier in formanilide by ^{13}C bandshape analysis and found no solvent effect on the barrier when changing from CDCl$_3$ to dioxan. Elguero *et al.*[46] obtained the rates of internal rotation in [(15)—(17)] from visual comparison of calculated with experimental spectra. In the latter work activation entropies close to zero were obtained, in accordance with expectations.

The amide torsional barrier in *N*-acetyl-*N*-methyl-L-alanine methyl ester[47] has been studied in various solvents. No pronounced solvent effect on the free energy of activation was found, and positive $\Delta S^{\ddagger}$ values were reported. Montaudo *et al.*[48] have studied the torsional barrier in a series of amides. They used a shift reagent, Eu(fod)$_3$, to enlarge the difference in chemical shift between the exchanging signals, and showed that, within the error limits, no effect from the shift reagent on the barrier could be detected. The rate constants were estimated utilizing the coalescence approximation. The authors discussed the variation in the torsional barrier with

---

[44] S. L. Spassov, V. S. Dimitrov, M. Agova, I. Kantschovska-Dimicoli, and R. Todorova-Mancheva, *Org. Magn. Resonance*, 1974, **6**, 508.
[45] H. Nakanishi and O. Yamamoto, *Chem. Letters*, 1974, 521.
[46] J. Elguero, A. Fruchier, L. Knutsson, R. Lazaro, and J. Sandström, *Canad. J. Chem.*, 1974, **52**, 2744.
[47] M. Goodman, Fu Chen, and Chi-Yu Lee, *J. Amer. Chem. Soc.*, 1974, **96**, 1479.
[48] G. Montaudo, P. Maravigna, S. Caccamese, and L. Librando, *J. Org. Chem.*, 1974, **39**, 2806.

(15) (16) (17)

substituents purely in terms of differences in the conjugative power of the substituent, even though it is obvious, for some cases, that steric effects must also be considered. Hirota and Todokoro[49] have also used measurements of coalescence temperature to study the effect of hydrogen bonding to the amide carbonyl on the C—N rotational barrier in some o-NN-dimethylbenzamides. The hydrogen bond in o-hydroxy-NN-dimethylbenzamide was found to decrease the rotational barrier ($\Delta G^{\ddagger}$) by approximately 20 kJ mol^{-1}. Strelenko et al.[50] have estimated the barrier to internal rotation in NN-bis(dimethylchlorosilylmethyl)carbamyl chloride from rate constants obtained from approximate formulae and have derived a negative entropy of activation. Both C—N and C—aryl rotation barriers in two ortho-substituted NN-dimethylthiobenzamides[83] were obtained from computerized fitting of calculated to experimental spectra. A probable coupling of the two exchange processes was discussed.

Sullivan and Price[51] have studied the thioamide C—N torsional barrier in several thioureas by means of ^{1}H n.m.r. bandshape analysis. In unsymmetrical compounds different torsional barriers were obtained for the two C—N bonds. Total bandshape analysis of the t-butyl signal in NN'-di-t-butylthiourea[52] was performed to obtain the C—N barrier to rotation, resulting in data similar to those for other thioureas.

Azoulay et al.[53] have studied an exchange process in dimeric o-nitrosotoluene (18), using ^{1}H n.m.r. bandshape analysis of the methyl signals. The exchange studied was probably the hindered rotation about the two C—N bonds. With the same bandshape analysis the N=N cis to trans equilibrium could also be studied. It was, moreover, shown that the monomer-to-dimer exchange can be studied by n.m.r. A torsional barrier of the nitroso-group in 5-nitrosotropolone has been studied by means of ^{1}H n.m.r. coalescence.[54]

Wennerbeck[55] has studied the barrier to internal rotation about the exocyclic

---
[49] M. Hirota and K. Todokoro, Chem. Letters, 1974, 777.
[50] Yu. A. Strelenko, A. V. Kisin, V. D. Sheludyakov, E. S. Rodinov, and N. V. Alekseev, J. Struct. Chem., 1974, 15, 826.
[51] R. H. Sullivan and E. Price, Org. Magn. Resonance, 1975, 7, 143.
[52] C. D. Freedman and D. L. Hooper, J. Phys. Chem., 1974, 78, 961.
[53] M. Azoulay, T. Drakenberg, and G. Wettermark, Tetrahedron Letters, 1974, 2243.
[54] Y. Ikegami and T. Asai, Chem. Letters, 1974, 805.
[55] I. Wennerbeck, J. Mol. Structure, 1974, 22, 1.

(18)

C—N bond in 3-dimethylamino-1,2,4-triazine, using computerized fitting of calculated to experimental spectra. The activation entropy turned out to be negative (-36 J mol^{-1} K^{-1}), in agreement with data for other cyclic systems. Fairly good agreement between experimental ΔH values and CNDO/2 calculations was obtained for some NN-dimethylaminoazines. Pitner et al.[56] used ^{1}H n.m.r. bandshape analysis to study the rate of internal rotation about the $C^{(6)}$—$N^{(6)}$ bond in NN-dimethyladenine in aqueous hydrochloride solution. They found it possible to obtain rate constants only for the monoprotonated species in solutions with DCl concentrations from 1.68 to 13.4 mol l^{-1}. The rate constants do not appear to be very accurate, but result in a slightly negative activation entropy, in agreement with data for similar compounds.[55, 57]

Extremely low-temperature ^{1}H n.m.r. has been used to study the conformations and torsional barrier in formic anhydride.[58] The observation of two equally intense signals showed that the dominant conformation was the (EZ) one. The torsional barrier was calculated from the rate constant obtained from complete bandshape analysis of the spectrum recorded at -179 °C. Cooney et al.[59] have studied the barrier to internal rotation in some cyclic N-nitrosamines. The rate of internal rotation was estimated either by the total bandshape method or by the intensity-ratio approximation. The authors performed an error analysis in order to obtain the probable accuracy in E_a. The present Reporters, however, believe there will also be systematic errors in the rate constants, not taken care of, owing to the fact that some approximations were made even in the total bandshape analysis. Otherwise, in the simplest case, a six-spin system would be needed. It has been shown that the inclusion in T_2^* of the effects of broadening due to small splittings will give rise to too high $\Delta S^{\ddagger}$ and $\Delta H^{\ddagger}$ values. This seems to be the case for compounds (solvents) where only a temperature region below the coalescence temperature could be used. The Reporters wish to warn of uncritical use of the total bandshape method. *Just the use of the total bandshape method will not ensure reliable data.*

Owing to the increased use of ^{13}C n.m.r., there has been a renaissance in the studies of rotational isomers and torsional barriers in aromatic aldehydes during the

[56] T. P. Pitner, H. Sternglanz, C. E. Bugg, and J. D. Glickson, *J. Amer. Chem. Soc.*, 1975, **97**, 885.
[57] T. Liljefors, *Org. Magn. Resonance*, 1974, **6**, 144.
[58] E. A. Noe and M. Raban, *J.C.S. Chem. Comm.*, 1974, 479.
[59] J. D. Cooney, S. K. Brownstein, and J. W. ApSimon, *Canad. J. Chem.*, 1974, **52**, 3028.

period covered by this Report. Thus Lunazzi et al.[60] and Drakenberg et al.[61] have studied the torsional barrier in benzaldehydes by means of total bandshape analysis of ^{13}C n.m.r. spectra, extending the study to compounds that are not open to investigation by ^{1}H n.m.r. owing to the low temperature needed. As expected, the barrier height for para-substituted benzaldehydes correlates well with substituent σ-values. Chadwick et al.[62] and Roques et al.[63] have discussed the potential of ^{13}C n.m.r. in the study of rotational isomerism and rotational barriers in heteroaromatic aldehydes. Furthermore, works dealing with torsional barriers in aromatic aldehydes obtained from ^{1}H n.m.r. bandshape analysis have been published.[64, 65]

Bushby and Ferber[66] have studied a dynamic process in the 1,3-diphenylallyl anion (19) by ^{13}C bandshape analysis. From the fact that the barrier height does not vary with either concentration or counter-ion, they concluded that the process is intramolecular in nature (internal rotation). The barrier height was found to be similar to the aldehyde rotational barrier in benzaldehydes. The barrier to internal

(19)

rotation in 3-diazo-2-butanone[67] was estimated from visual fitting of calculated to experimental ^{1}H n.m.r. spectra for two solvents, CDCl$_3$ and CCl$_4$. The unusually large difference between the $\Delta S^{\ddagger}$ values obtained from the two solutions, ca. 80 J mol^{-1} K^{-1}, was rationalized as being due to specific solvation interaction in the CDCl$_3$ solution. The Reporters are, however, not convinced that this difference is not caused by systematic errors in the spectrum obtained for the CCl$_4$ solution, where a small difference between chemical shifts makes the evaluation of accurate activation parameters difficult. Sufficient details regarding the bandshape analysis are, however, not given.

Meakin et al.[68] have used ^{19}F n.m.r. total bandshape analysis to study the exchange process in (pentafluorophenyl)sulphur trifluoride. The rate of exchange was determined independently from the shape of three different sets of signals. The agreement among the calculated activation parameters was satisfactory but not perfect. The nature of the exchange process could not be determined. The simplest one consistent with the observation is a hindered internal rotation about the S—C bond.

[60] L. Lunazzi, D. Macciantelli, and A. C. Boicelli, Tetrahedron Letters, 1975, 1205.
[61] T. Drakenberg, R. Jost, and J. M. Sommer, J.C.S. Chem. Comm., 1974, 1011.
[62] D. J. Chadwick, G. D. Meakins, and E. E. Richards, Tetrahedron Letters, 1974, 3183.
[63] B. P. Roques, S. Combrisson, and F. Wehrli, Tetrahedron Letters, 1975, 1047.
[64] M. Farnier and T. Drakenberg, J.C.S. Perkin II, 1975, 337.
[65] T. B. Grindley, A. R. Katritzky, and R. D. Topsom, J.C.S. Perkin II, 1975, 443.
[66] R. J. Bushby, and G. J. Ferber, Tetrahedron Letters, 1974, 3701.
[67] R. Curci, F. DiFuria, and V. Lucchini, Spectroscopy Letters, 1974, 7, 211.
[68] P. Meakin, D. W. Ovenall, W. A. Sheppard, and J. P. Jesson, J. Amer. Chem. Soc., 1975, 97, 522.

Rotation about Double Bonds. The solvent effects on the barriers to internal rotation about the C=C bond in some polarized ethylenes have been reported.[69] The barrier, $\Delta G^{\ddagger}$, was mostly estimated from the coalescence temperature. For (20) in CD_3CN and C_6D_5Br, total bandshape analysis was employed to obtain $\Delta H^{\ddagger}$ and $\Delta S^{\ddagger}$. Surprisingly, it was found that $\Delta H^{\ddagger}$ and $\Delta G^{\ddagger}$ depend in different ways on the solvent polarity, owing to a large solvent dependence in $\Delta S^{\ddagger}$. One would expect the barrier to decrease with increasing solvent polarity, as found to be the case for $\Delta G^{\ddagger}$ but for $\Delta H^{\ddagger}$ it was the other way around. However, until now, only for one compound has the enthalpy of activation been reported for two different solvents. Until more data are available the Reporters believe that it is best to assume that the observed solvent effect on $\Delta H^{\ddagger}$ may be due to uncertainties in the measurements. Prokof'ev et al.[70] have also studied the barrier to internal rotation about C=C bonds, using the coalescence approximation. For a series of dienic δ-aminocarbonyl compounds (21) they found that $\Delta G^{\ddagger}$ depends strongly on the substituent on the α-carbon, and for two compounds (21b) and (21c) they also note that $\Delta G^{\ddagger}$ decreases with increasing solvent polarity, in agreement with other data.

(20)

(21) a; X = Y = CO_2Me
b; X = MeCO, Y = CO_2Me
c; X = Y = MeCO

Filleux-Blanchard et al.[71] have studied rate processes and conformational equilibria in some enamines. The activation parameters were obtained from the coalescence of several different pairs of signals in both ^1H and ^{13}C n.m.r. spectra. With differences of chemical shift ranging from 5.8 to 140 Hz, the coalescence temperatures were within an interval of 25 °C. Thus reasonable estimates of even the values of $\Delta H^{\ddagger}$ and $\Delta S^{\ddagger}$ could be obtained just from measurements of the coalescence temperature. Probably the rate process studied is not a C—C rotation, as suggested by the authors, but a C=C rotation.

Ring Inversion. Booth and Griffiths[72] have studied an exchange process in N-alkyl-cis-decahydroquinolines by ^{19}F n.m.r. bandshape analysis. They concluded that the rate process studied was ring inversion, partly from the fact that the activation

[69] H.-O. Kalinowski, H. Kessler, and A. Walter, *Tetrahedron*, 1974, **30**, 1137.
[70] E. P. Prokof'ev, Zh.A. Krasnaya, and V. F. Kucherov, *Bull Acad. Sci. U.S.S.R.*, 1973, **22**, 1963.
[71] M. L. Filleux-Blanchard, F. Mabon, and G. J. Martin, *Tetrahedron Letters*, 1974, 3907.
[72] H. Booth and D. V. Griffiths, *J.C.S. Perkin II*, 1975, 111.

energy, E_a, was close to the value for *cis*-decalin. However, as far as the Reporters can judge, the accuracy does not seem to be very good, as can be seen from some duplicate measurements. The inversions of six-membered rings were also studied by Bushweller *et al.*[73] They found that for some tetrathians (22) and (23) both the chair and twist conformations were present in measurable amounts, and that the twist–twist inversion takes place *via* the chair conformation and the chair–chair inversion *via* the twist conformation. For (23) they observed that the barrier to inversion of the cyclohexane rings was slightly higher when the tetrathian adopts the twist conformation. Werner *et al.*[74] have studied the ring inversion in tetra- and hexa-methyl-cyclohexanes. Single-parameter formulae were used to obtain the rate constants from ^{13}C n.m.r. spectra.

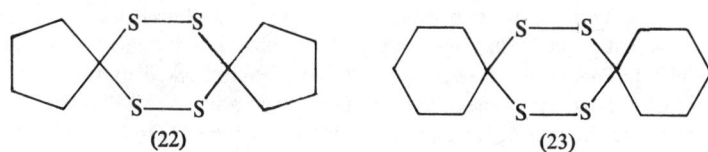

(22)   (23)

Panunzi *et. al.*[75] has studied the ring inversion in hexahydro-1,3,5-trimethoxy-1,3,5-triazine. Computer fitting of calculated to experimental bandshapes was used to obtain the rate constants for ring inversion. The reported entropy of activation is suspiciously high for a normal ring-reversal process (56–88 J mol^{-1} K^{-1}).

Both ^{1}H and ^{13}C n.m.r. bandshape analyses have been utilized in the study of the ring-reversal process in cycloheptene oxide.[76] A difference of 1.6 kJ mol^{-1} in the free energy of activation obtained from the two measurements indicates that there was an error in the temperature measurement (*ca.* 7 °C) in at least one of the experiments. This is not discussed in the paper, but the authors believed that, using the carbon-13 spectrometer, they could measure the temperature within 3 °C; no similar belief was expressed regarding the proton spectrometer. From comparison with data for cycloheptene and cycloheptane they concluded that the stable conformer is of the chair type. Conformational interconversion in seven-membered rings has also been studied by Canuel and St.-Jacques.[77] They apparently performed an accurate bandshape analysis of the proton spectrum from 4,5-benztetrahydro-oxepin. The authors, however, did not describe how T_2 was obtained for the intermediate exchange rates, which often has been shown to be the critical parameter in accurate bandshape analysis, especially when the temperature range studied is extended close to the fast- and slow-exchange limits, which seems to be the case in this work. The error limits given for $\Delta S^{\ddagger}$ (4 J mol^{-1} K^{-1}) and $\Delta H^{\ddagger}$ (0.8 kJ mol^{-1}) should not be taken too seriously. As in the work on cycloheptene oxide, it was assumed that the ground state was the chair conformation, which inverts *via* a boat

---

[73] C. H. Bushweller, G. Bhat, L. J. Letendre, J. A. Brunelle, H. S. Bilofsky, H. Ruben, D. H. Templeton, and A. Zalkin, *J. Amer. Chem. Soc.*, 1975, **97**, 65.
[74] H. Werner, G. Mann, H. Jancke, and G. Engelhardt, *Tetrahedron Letters*, 1975, 1917.
[75] A. Panunzi, L. Paolillo, and P. A. Temussi, *Gazzetta*, 1974, **104**, 1173.
[76] K. L. Servis, E. A. Noe, N. R. Easton, jun., and F. A. L. Anet, *J. Amer. Chem. Soc.*, 1974, **96**, 4185.
[77] L. Canuel and M. St.-Jacques, *Canad. J. Chem.*, 1974, **52**, 3581.

(24) R = COMe, COCH$_2$Cl, or CO$_2$Et

conformation. Abraham et al.[78] have studied the ring inversion and amide rotation in some 5-substituted 10,11-dihydrodibenz[b,f]azepines (24) by means of measurements of the coalescence temperature in both ^{1}H and ^{13}C spectra. Even though the free energies of activation for the two processes differ by at least 8 kJ mol^{-1}, the authors state that they cannot entirely rule out the possibility of this difference being due to an entropy effect. This could, however, easily have been checked by means of a few calculations of the total bandshape. It has previously been shown that the barrier to ring inversion in the benzodiazepam (25) is relatively high.[79] On the other hand, no non-equivalence of chemical shift between the methylene protons could be detected for (26) in CDCl$_3$.[80] Raben et al.[81] have shown that this lack of non-equivalence is due to accidental shift equivalence in chloroform solution and not to fast ring inversion, since in toluene–acetone a small shift non-equivalence could be detected. Complete bandshape analysis was used to obtain rate constants for ring reversal that were in good agreement with the data for (25). Also, Sarrazin et al.[82] have studied the ring inversion in some substituted diazepams.

(25) (26)

Anet and Yavari[83] used ^{13}C n.m.r. to study the conformational behaviour of cis,cis-cyclo-octa-1,3-diene. They found two stable low-temperature conformers, twist-boat and twist-boat-chair, with different barriers to pseudorotation.

Roberts et al.[84] have studied the rate of ring inversion in several cis,cis-cyclo-

---

[78] R. J. Abraham, L. J. Kricka, and A. Ledwith, J.C.S. Perkin II, 1974, 1648.
[79] P. Linschied and J. M. Lehn, Bull. Soc. chim. France, 1967, 992.
[80] W. Sadee, Arch. Pharm., 1969, 302, 49.
[81] M. Raben, E. H. Carlson, J. Szmuszkovics, G. Slomp, C. G. Chidester, and D. J. Duchamp, Tetrahedron Letters, 1975, 139.
[82] M. Sarrazin, M. Bordeaux-Pontier, C. Briant, and E. J. Vincent, Org. Magn. Resonance., 1975, 7, 89.
[83] F. A. L. Anet and I. Yavari, Tetrahedron Letters, 1975, 1567.
[84] B. W. Roberts, J. J. Vollmer, and K. L. Servis, J. Amer. Chem. Soc., 1974, 96, 4578.

deca-1,6-dienes (27). In the slow-exchange region, decoupling of the vinylic protons resulted in an AB system for the allylic protons, which was used for total bandshape analysis. Only the values of free energy of activation were given, even though an Arrhenius plot was shown. From the rate constants given in a figure, the $\Delta S^{\ddagger}$ value can be estimated to be *ca.* 40 J mol^{-1} K^{-1}. This large entropy of activation is not discussed in the paper, and no comment is made as to whether or not similar activation entropies were obtained for all compounds studied. The Reporters believe that this is a serious mistake, since all $\Delta G^{\ddagger}$ values are given at 298 K and are extrapolated from the coalescence region, often more than 100 K away. Differences in the values of $\Delta S^{\ddagger}$ used in such extrapolation will affect the derived free energy of activation, which means that the $\Delta S^{\ddagger}$ values at least should be given.

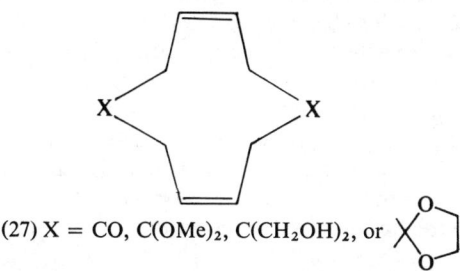

(27) X = CO, C(OMe)$_2$, C(CH$_2$OH)$_2$, or

Anet and Cheng[85] have studied the pseudorotation in cyclohexadecane by ^{13}C n.m.r; from the ^{13}C chemical shifts and force-field calculations they concluded that cyclohexadecane adopts the [4444] conformation.

Nitrogen Inversion. Bushweller and co-workers[86, 87] have studied the rate processes in some *N*-t-butyl-*NN*-dialkylamines, with bandshape analysis of the ^{1}H n.m.r. spectra. For *N*-t-butyl-*N*-methyl-*N*-[*Me*-^{2}H$_3$]ethylamine they were able to analyse both the t-butyl and methylene signals, which yielded the same rate constant. Since the t-butyl methyls are exchanged *via* rotation at the t-butyl group and the methylene protons *via* nitrogen inversion, these results strongly indicate that, at least for this compound, the t-butyl-rotation and nitrogen-inversion processes proceed *via* a common transition state. The shape of the methylene spectrum from *N*-t-butyl-*NN*-di([*Me*-^{2}H$_3$]ethyl)amine indicates that a similar inversion–rotation mechanism operates in this case as well. In the less crowded *N*-isopropyl-*NN*-dialkylamines, on the other hand, the dynamic n.m.r. spectra of the methyl protons of the isopropyl group are consistent with slow nitrogen inversion and fast rotation of the isopropyl group.

Phosphorus Pseudorotation. Eisenhut *et al.*[88] have studied the pseudorotation pro-

---

[85] F. A. L. Anet and A. K. Cheng, *J. Amer. Chem. Soc.*, 1975, **97**, 2420.
[86] C. H. Bushweller, W. G. Anderson, P. E. Stevenson, D. L. Burkey, and J. W. O'Neil, *J. Amer. Chem. Soc.*, 1974, **96**, 3892.
[87] J. Reny, C. Y. Wang, C. H. Bushweller, and W. G. Anderson, *Tetrahedron Letters*, 1975, 503.
[88] M. Eisenhut, H. L. Mitchell, D. D. Traficante, R. J. Kaufman, J. M. Deutch, and G. M. Whitesides, *J. Amer. Chem. Soc.*, 1974, **96**, 5385.

cesses in some XPF$_4$ compounds. From the bandshapes of the ^{31}P n.m.r. spectrum they could discriminate between Berry and non-Berry permutation. For both compounds studied that had sufficiently high barriers, Me$_2$NPF$_4$ and ClPF$_4$, the exchange was found to follow the Berry permutation scheme. The authors also tried to find evidence for a square-pyramidal intermediate, but they concluded that the effect from such an intermediate on the ^{31}P n.m.r. bandshape should be too small to be detected with the resolution they could obtain. However, the possibility of detecting an intermediate might be larger from ^{19}F spectra since this is, with ^{31}P decoupling, an A$_2$B$_2$ spin system for the trigonal-bipyramidal comformation and an A$_4$ system for the square pyramid; the difference in chemical shift between these two conformations might be large. In any case, the effect of a small amount of the square-pyramidal conformation on the ^{19}F bandshape should be different for the various signals in the A$_2$B$_2$ system owing to the large shift difference within this system. The same group also studied the phosphorus pseudorotation in two arylbis-(4,4′-dimethyl-2,2′-biphenylene)phosphoranes (28).[89] For (28a) they found, from ^{1}H n.m.r. bandshape analysis, that the four aromatic methyl groups and the two isopropyl methyls exchange with the same rate constant, which the authors interpreted as an indication of exchange *via* a square-pyramidal intermediate in which the barrier to aryl rotation is relatively low. For (28a) all n.m.r. bandshapes are in agreement with this mechanism, whereas for (28b) the rate constants are not equal for the two processes, which is not in agreement with the proposed mechanism. These latter data are, however, less accurate.

Bond Migration. In two papers the dynamic properties of polymethylbicyclo-[3,1,0]hexenyl cations (29) and 5-acylpentamethylcyclopentadiene aluminium tri-

[89] G. M. Whitesides, M. Eisenhut, and W. M. Bunting, *J. Amer. Chem. Soc.*, 1974, **96**, 5398.

chloride zwitterions (30) have been discussed, largely on the basis of ^{1}H n.m.r. bandshape data.[90,91] It is plausible that (29) isomerizes *via* a transition state similar to (30) and that (30) isomerizes *via* a transition state similar to (29). The bandshape analysis showed clearly that the isomerization proceeds stepwise, and for (29) with complete retention of the configuration of the substituents on the cyclopropyl ring. The relative energies of the two types of cations should depend only on the charge-stabilizing abilities of the substituent when it is on the cyclopropyl ring and on the side-chain, respectively. Curci *et al.*[92] have studied the N_3 migration in (31) with

proton bandshape analysis. They concluded that the mechanism for this exchange was intramolecular in non-polar solvents and intermolecular in polar solvents.

During the period under review two papers dealing with the Cope rearrangement in bullvalene have appeared.[93,94] This illustrates the increased scope for dynamic n.m.r. studies made possible by the use of ^{13}C n.m.r. In both works, carbon-13 total bandshape analysis was used to determine the rate of rearrangement. There is a fairly good agreement among the reported activation parameters,[93-95] especially when it is considered that different solvents and concentrations were used. In both papers, however, there seems to be a tendency to go beyond the range for which it is normally possible to obtain rate constants with high accuracy. The Cope rearrangements in some 9-heterobarbaralanes were studied by measurements of proton coalescence temperatures,[96] and the rearrangements in 2,3-dimethyl-2-norbornyl cations were studied with both ^{1}H and ^{13}C bandshape simulation.[97]

C. **Complex Rearrangements.**—Serpone and Hersh[98] have studied the configurational rearrangements in some tin(IV) acetylacetonato complexes. Both the methine

---

[90] R. F. Childs and S. Winstein, *J. Amer. Chem. Soc.*, 1974, **96**, 6409.
[91] R. F. Childs and M. Zeya, *J. Amer. Chem. Soc.*, 1974, **96**, 6418.
[92] R. Curci, V. Lucchini, G. Modena, P. J. Kocienski, and J. Ciabattoni, *J. Org. Chem.*, 1973, **38**, 3149.
[93] H. Günther and J. Ulmen, *Tetrahedron*, 1974, **30**, 3781.
[94] J. F. M. Oth, K. Müller, J.-M. Gilles, and G. Schröder, *Helv. Chim. Acta*, 1974, **57**, 1415.
[95] H. Nakanishi and O. Yamamoto, *Tetrahedron Letters*, 1974, 1803.
[96] A. G. Anastassiou, E. Reichmais, and J. C. Wetzel, *Tetrahedron Letters*, 1975, 1651.
[97] A. J. Jones, E. Huang, R. Haseltine, and T. S. Sorensen, *J. Amer. Chem. Soc.*, 1975, **97**, 1133.
[98] N. Serpone and K. E. Hersh, *Inorg. Chem.*, 1974, **13**, 2901.

and methyl proton spectra showed non-equivalence, indicating that the *cis* configuration is adopted in solution. The rate of rearrangement was obtained from the combined use of several one-parameter formulae. A total bandshape analysis of the methyl proton spectra from *cis*-MeClSn(acac)$_2$ in CDCl$_3$ solution indicated that the methyl and methine protons are not exchanging at the same rate. Cotton *et al.*[99] observed the coalescence of the acetylacetonato methyl signals in Ni(C$_5$H$_7$O$_2$)(Et)(PPh$_3$). Lindmark and Fay studied the rearrangement processes in some β-diketonato-titanium(IV) complexes. In order to determine the rate constants they compared some bandshape parameters for calculated and experimental spectra instead of using the total bandshape.[100]

Whitesides and Budnik[101] have used ^{31}P n.m.r. to study the ligand-scrambling process in many transition-metal complexes with diene or dienyl and phosphite ligands. Total bandshape analysis of the phosphorus spectra was used to obtain the rate constants. An unexpectedly small variation in the free energy of activation as a function of metal atom or ligand was found. Kreiter and co-workers[102] have continued their studies on rearrangements of metal complexes. From ^{1}H n.m.r. bandshape analysis they deduced both the orientation of the ethylene ligand in (32) and its barrier to rotation about the bond to the metal atom.

(32)

Ashley-Smith *et al.*[103] have used ^{1}H n.m.r. coalescence to estimate the barrier to olefin rotation in several square-planar platinum complexes. From the observation that all ^{13}C n.m.r. signals from cyclo-octatetraene in its complex with Mo(CO)$_3$ broaden by the same amount, Cotton *et al.*[104] concluded that the exchange takes place *via* a transition state with a planar or nearly planar cyclo-octatetraene ring. Kruczynski *et al.*[105] and Wilson *et al.*[106] have both studied the combined process of olefin rotation and ligand scrambling in some five-co-ordinated Fe complexes, and Segal and Johnson[107] have examined the olefin rotation in [Os(CO)NO(C$_2$H$_4$)

[99] F. A. Cotton, B. A. Frenz, and D. L. Hunter, *J. Amer. Chem. Soc.*, 1974, **96**, 4820.
[100] A. F. Lindmark and R. C. Fay, *Inorg. Chem.*, 1975, **14**, 282.
[101] T. H. Whitesides and R. A. Budnik, *Inorg. Chem.*, 1975, **14**, 664.
[102] H. Alt, M. Herberhold, C. G. Kreiter, and H. Strack, *J. Organometallic Chem.*, 1974, **77**, 353.
[103] J. Ashley-Smith, Z. Douek, B. F. G. Johnson, and J. Lewis, *J.C.S. Dalton*, 1974, 128.
[104] F. A. Cotton, D. L. Hunter, and P. Lahuerta, *J. Amer. Chem. Soc.*, 1974, **96**, 4723.
[105] L. Kruczynski, L. K. K. LiShingMar, and J. Takats, *J. Amer. Chem. Soc.*, 1974, **96**, 4006.
[106] S. T. Wilson, N. J. Coville, J. R. Shapely, and J. A. Osborn, *J. Amer. Chem. Soc.*, 1974, **96**, 4038.
[107] J. A. Segal and B. F. G. Johnson, *J.C.S. Dalton*, 1975, 677.

(PPh$_3$)$_2$]PF$_6$. Lallemand et al.[108] have used ^{13}C n.m.r. bandshape analysis to obtain the CO basal–apical exchange barrier in (1-methoxy-1,3-cyclohexadiene)iron tricarbonyl. Meakin and Jesson[109–112] have used ^{31}P n.m.r. to study pentacoordinated species of various metal atoms. In the case of slow intramolecular exchange, the geometry of the complexes could be deduced to be trigonal-bipyramidal for all the metal atoms studied. A careful study of the intermolecular ligand exchange in Rh[P(OMe)$_3$]$_5^+$ showed that the dissociation is the rate-determining step. Bandshape analysis showed best agreement between calculated and experimental spectra when a mechanism involving simultaneous exchange of two axial with two equatorial ligands was considered, as opposed to a single axial–equatorial exchange mechanism. The activation parameters for the exchange were also given, and showed normal small values for the activation entropy. Miller and Caulton[113] have studied the intramolecular rearrangement process in some five- and six-co-ordinate iridium complexes. The rate of exchange was determined from ^{31}P total bandshape analysis. The process of CO scrambling in metal-carbonyl complexes has been studied by several groups. Cotton et al.[114, 115] have studied this process in some dinuclear complexes using ^{31}P n.m.r., whereas Adams and Chodosh[116] used ^{1}H n.m.r. to study the same process in Mn$_2$(CO)$_7$(CNMe). Tri- and tetra-nuclear complexes have also been studied to obtain information on their CO-scrambling process.[117–120]

Cotton and Stanislowski[121] have studied two rearrangement processes in the complex (33). The low-temperature process made protons H$_a$ and H$_b$ equivalent and the high-temperature process made the two ethyl groups equivalent. The low-

(33)   (34)

[108] J. Y. Lallemand, P. Laszlo, C. Muzette, and A. Stockis, *J. Organometallic Chem.*, 1975, **91**, 71.
[109] P. Meakin and J. P. Jesson, *J. Amer. Chem. Soc.*, 1974, **96**, 5751.
[110] J. P. Jesson and P. Meakin, *J. Amer. Chem. Soc.*, 1974, **96**, 5760.
[111] R. R. Schrock and P. Meakin, *J. Amer. Chem. Soc.*, 1974, **96**, 5288.
[112] P. Meakin, A. D. Englosh, S. D. Ittel, and J. P. Jesson, *J. Amer. Chem. Soc.*, 1975, **97**, 1254.
[113] J. S. Miller and K. G. Caulton, *J. Amer. Chem. Soc.*, 1975, **97**, 1067.
[114] F. A. Cotton, D. L. Hunter, and P. Lahuerta, *J. Amer. Chem. Soc.*, 1975, **97**, 1046.
[115] F. A. Cotton, D. L. Hunter, and P. Lahuerta, *J. Organometallic Chem.*, 1975, **87**, C42.
[116] R. D. Adams and D. F. Chodosh, *J. Organometallic Chem.*, 1975, **87**, C48.
[117] F. A. Cotton, L. Kruczynski, and A. J. White, *Inorg. Chem.*, 1974, **13**, 1402.
[118] A. J. Deeming and M. Underhill, *J.C.S. Dalton.*, 1974, 1415.
[119] L. Milone, S. Aime, E. W. Randall, and E. Rosenberg, *J.C.S. Chem. Comm.*, 1975, 452.
[120] J. Evans, B. F. G. Johnson, J. Lewis, and T. W. Matheson, *J. Amer. Chem. Soc.*, 1975, **97**, 1245.
[121] F. A. Cotton and A. G. Stanislowski, *J. Amer. Chem. Soc.*, 1974, **96**, 5074.

temperature process could be followed by the change in the bandshape of several ¹H n.m.r. signals. It is a little surprising that the free energy of activation differs by as much as ca. 4 kJ mol⁻¹ when calculated from various signals, which are supposed to be influenced by the same exchange. The hindered rotation about the M—C bond in some rhodium complexes (34) has been studied by ¹H n.m.r. coalescence measurements.[122] The strongly negative activation entropy obtained might be an artifact that is due to inaccurate calculations. The fluxional behaviour of the {o-C_6H_4(AsMePh)$_2$}Mo(CO)$_3$I$_2$ complex was studied by observing the coalescence of the signals from the two methyl groups,[123] and related studies have been performed for some neutral rhodium–diene complexes.[124]

Yamamoto et al.[125] have studied the bridge–terminal exchange in aluminium trialkyl dimers by ¹³C n.m.r. They performed total bandshape analyses but used T_2 as a variable to obtain the best fit between calculated and experimental spectra which, for the Reporters, is a most questionable procedure for exchange-broadened signals. The authors concluded that the exchange was mostly intramolecular in nature for tetramethylaluminium and for tetraethylaluminium in cyclopropane solution, whereas it was at least partly intermolecular for tetraethylaluminium in toluene. The Reporters do not agree with this last conclusion, as a broadening of the CH$_2$ signal was also observed by the above authors, even at temperatures greater than 70 K above the coalescence of the signals. This was interpreted by the authors as being due to (Al, C) couplings that were incompletely washed out by quadrupolar relaxation.

Day et al.[126] have studied the dynamic behaviour of Ni$_4$(CNCMe$_3$)$_7$. Two exchange processes were found to take place: exchange among the terminal[127] ligands, and bridge–terminal ligand exchange. Hunter and Massey have used ¹H n.m.r. total bandshape analysis to study the fluxional behaviour of the six-membered chelate ring in 1,3-bis(methylseleno)-2,2-dimethylpropane complexes with Cr, Mo, and W, respectively, and Rhee and Zuckerman[128] have studied the fluxional behaviour of 1,1-dicyclopentadienyl-1-bromo-2,3,4,5-tetraphenylstannole.

Fish[129] has studied the keto–enol tautomerization in bis(1,1,1,2,2,3,3-heptafluoro-7,7-dimethyl-4,6-octanedione-5-yl)mercury. ¹H N.m.r. total bandshape analysis was used to obtain the rate of tautomerization, and the process was found to be intramolecular. Lehn and Stubbs[130] have used ¹³C n.m.r. to study the fluxional behaviour of several complexes of cryptates with alkaline-earth cations.

D. **Intermolecular Exchange Processes.**—*Proton Exchange*. Limbach and Seiffert [131, 132] have studied the proton exchange in NN'-dipentadeuteriophenyl-1-amino-3-

---

[122] M. J. Doyle and M. F. Lappert, *J.C.S. Chem. Comm.*, 1974, 679.
[123] K. Hendrick and S. B. Wild, *J.C.S. Dalton*, 1974, 2500.
[124] D. G. van Derveer and R. Eisenberg, *J. Amer. Chem. Soc.*, 1974, **96**, 4994.
[125] O. Yamamoto, K. Hayanizu, and M. Yanagisawa, *J. Organometallic Chem.*, 1974, **73**, 17.
[126] V. W. Day, R. O. Day, J. S. Kristoff, F. J. Hirsekorn, and E. L. Muetterties, *J. Amer. Chem. Soc.*, 1975, **97**, 2571.
[127] G. Hunter and C. Massey, *J.C.S. Dalton*, 1975, 209.
[128] W. Z. M. Rhee and J. J. Zuckerman, *J. Amer. Chem. Soc.*, 1975, **97**, 2291.
[129] R. H. Fish, *J. Amer. Chem. Soc.*, 1974, **96**, 6664.
[130] J. M. Lehn and M. E. Stubbs, *J. Amer. Chem. Soc.*, 1974, **96**, 4011.
[131] H. H. Limbach and W. Seiffert, *Ber. Bunsengesellschaft phys. Chem.*, 1974, **78**, 532.
[132] H. H. Limbach and W. Seiffert, *Ber. Bunsengesellschaft phys. Chem.*, 1974, **78**, 641.

iminopropene. At low temperatures two sets of proton signals were observed, from the *cis*- and *trans*-isomers, respectively. The *trans*-isomer was believed to be a cyclic trimer. The proton exchange rates were obtained from total bandshape analysis, resulting in zero entropy of activation for the *trans*-isomer and *ca.* $-120\,\text{J}\,\text{mol}^{-1}\text{K}^{-1}$ for the *cis*-isomer, indicating that the exchange takes place within the cyclic trimer. The proton spectra, however, show that the *N*-protons are really intermolecularly exchanged since the coupling information is lost. Menger and Lynn[133] have studied the proton exchange in long-chain amines. The authors found that the exchange is much faster in the micelle-forming amine *NN*-dimethyldodecylamine than in the non-aggregating analogue *NN*-dimethylhexylamine. The exchange was also found to increase drastically when the concentration was increased in the neighbourhood of the critical micelle concentration. Perrin[134, 135] has used n.m.r. line-broadening to study the mechanism of proton exchange in primary amides and amidinium ions. He concluded that the exchange proceeds *via* $RCONH_3^+$ and not $RC(OH)\!=\!NH_2^+$. The ^{15}N n.m.r. linewidth was used to estimate the proton exchange in glycine.[136] Computerized fitting of calculated to manually digitized experimental spectra was used to study the rate of proton exchange of the *NN*-dimethylanilinium ion in acidic solutions. The results indicate that the dissociation is a diffusion-controlled process.[137]

The proton exchange in toluene-α-thiol[138] has been studied in DMSO solution under both acidic and basic conditions. The exchange could be studied from the shape of the CH_2 signal, but only in two very limited pH regions, *ca.* pH 1 and *ca.* pH 7.

Electron Transfer. Palazzotto and Pignolet[139] have studied the electron transfer between tris(*NN*-dimethyldithiocarbamato)iron(III) and tris(*NN*-dimethyldithiocarbamato)iron(IV) tetrafluoroborates. The difference in chemical shift between the *N*-methyl signals from the two compounds was 111 p.p.m. at $-91\,°\text{C}$ as obtained from pure solutions of the iron(III) and iron(IV) compounds, respectively. In mixed solutions only a mean signal, very broad at low temperature, was observed, from which the rate of electron transfer was evaluated by total bandshape analysis. Owing to the very large difference in chemical shift, second-order rate constants of the order of $10^8\,\text{l}\,\text{mol}^{-1}\,\text{s}^{-1}$ could be measured.

Ligand Exchange. Fraenkel *et al.*[140] have studied an exchange process in the dineopentylmagnesium–*l*-sparteine complex (35). Total bandshape analysis of the neopentyl ^{1}H n.m.r. spectrum showed that the exchange process was best explained as purely due to the exchange

$(Me_3CCH_2)_2Mg$–*l*-sparteine → $(Me_3CCH_2)_2Mg$ + sparteine

[133] F. M. Menger and J. L. Lynn, *J. Amer. Chem. Soc.*, 1975, **97**, 948.
[134] C. L. Perrin, *J. Amer. Chem. Soc.*, 1974, **96**, 5628.
[135] C. L. Perrin, *J. Amer. Chem. Soc.*, 1974, **96**, 5631.
[136] T. K. Leipert and J. H. Noggle, *J. Amer. Chem. Soc.*, 1975, **97**, 269.
[137] A. J. Kresge and G. L. Capen, *J. Amer. Chem. Soc.*, 1975, **97**, 1795.
[138] J. J. Delpuech and D. Nicole, *J.C.S. Perkin II*, 1974, 1025.
[139] M. C. Palazzotto and L. H. Pignolet, *Inorg. Chem.*, 1974, **13**, 1781.
[140] G. Fraenkel, B. Appleman, and J. G. Ray, *J. Amer. Chem. Soc.*, 1974, **96**, 5113.

with no direct exchange between the neopentyl groups in the complex. The mechanism of the exchange was also discussed. The same group of workers[141] have also studied the ligand exchange between dineopentylmagnesium and diphenylmagnesium.

(35) R = neopentyl

Ham et al.[142] have studied the dissociation of the dimeric acetylacetonatotrimethylplatinum complex (36). Total bandshape analysis of both the Pt-bound methyl proton signals and the methine signals indicated that there is slightly faster exchange for the methyl group than for the methine protons, showing that the methyl groups might be exchanged *via* two different mechanisms.

Chung and Tuck[143] have used ^{19}F n.m.r. coalescence and line broadening to study the ligand exchange in trifluoromethyl-β-diketonato(dimethyl)indium complexes. They found that the exchange was first-order in both complex and free ligand concentration. Abraham et al.[144] have observed a splitting of the ^{1}H n.m.r. signal from Tl(OAc)$_3$ and the ^{19}F n.m.r. signal from Tl(OCOCF$_3$)$_3$ at low temperature. The splitting was identified as being due to spin coupling to Tl. The coalescence of the signals was used to obtain the activation energy for the anion exchange. Raben-

(37)

---

[141] G. Fraenkel and S. H. Yu, *J. Amer. Chem. Soc.*, 1974, **96**, 6658.
[142] N. S. Ham, J. R. Hall, and G. A. Swile, *Austral. J. Chem.*, 1975, **28**, 759.
[143] H. L. Chung and D. G. Tuck, *Canad. J. Chem.*, 1974, **52**, 3944.
[144] R. J. Abraham, G. E. Hawkes, and K. M. Smith, *Tetrahedron Letters*, 1975, 1999.

stein and Fairhurst[145] have used ^{13}C n.m.r. to study the complexes between methylmercury and sulphydryl-containing amino-acids. The ligand-exchange rates were obtained from comparison of calculated and experimental ^{13}C n.m.r. spectra. Fraenkel and Watson[146] have studied the adduct (37) using ^{1}H n.m.r. The N-methyl signals were involved in an exchange that was believed to be nitrogen inversion that is slowed down owing to chelate formation with potassium. The lifetime obtained should thus be an estimate of the lifetime of the ion-pair. Blackborow[147] has studied the exchange in the PhNMe$_2$,BX$_3$ adduct, utilizing the shape of the methyl ^{1}H n.m.r. signal. Both the rate of exchange and the boron relaxation rate were evaluated.

E. **Biochemical Applications.**—Morishima and Iizuka[148, 149] have reported data for restricted rotation of a methyl group on a haem group in myoglobins. From 220 MHz ^{1}H n.m.r. they detected a splitting of one of the methyl-group signals, which collapsed to a singlet upon heating. Total bandshape simulation was used to obtain the rate of internal rotation. The height of the barrier ($\Delta H^{\ddagger} = 81$ kJ mol^{-1}) indicates that the steric interaction on this methyl group is very large. The effects of salt concentration and various inhibitors on the barrier were also studied. Wüthrich and Wagner[150] have shown that it is possible to study the rotation of some of the aromatic residues in basic pancreatic trypsin inhibitor from 360 MHz ^{1}H n.m.r. spectra.

Akasaka and Yamada[151] have studied the pH dependence of the ^{31}P n.m.r. signal from some homopolyribonucleotides. For polyriboguanylic acid a broad signal was observed at neutral pH, which was gradually replaced by a narrow line at higher pH, indicating a slow exchange between two states.

4 Relaxation Effects: Linewidths

A. **Chemical Exchange Rates Determined by the Swift–Connick Approach.**—The Swift–Connick method[152] of determining exchange rates is usually used when the relaxation rate is much faster in one site than in the other(s). Such conditions are most often met in systems containing paramagnetic ions, but they can also occur for nuclei with a quadrupole moment and large differences of quadrupole coupling constant between the sites.

In a purely theoretical paper, Granot and Fiat[153] give an extension of Swift and Connick's original treatment of the chemical exchange contributions to linewidths. Starting from the basic equations of ref. 152, Granot and Fiat determine which conditions give a Lorentzian bandshape. In the second part of the paper the different exchange contributions to T_2 are discussed in some detail. Explicit equations and diagrams are presented for both slow and fast exchange and for small and large differences in chemical shift.

---

[145] D. L. Rabenstein and M. T. Fairhurst, *J. Amer. Chem. Soc.*, 1975, **97**, 2086.
[146] G. Fraenkel and D. Watson, *J. Amer. Chem. Soc.*, 1975, **97**, 231.
[147] J. R. Blackborow, *J. Magn. Resonance*, 1975, **18**, 107.
[148] I. Morishima and T. Iizuka, *J. Amer. Chem. Soc.*, 1974, **96**, 9365.
[149] I. Morishima and T. Iizuka, *Biochim. Biophys. Acta*, 1975, **386**, 542.
[150] K. Wüthrich and G. Wagner, *F.E.B.S. Letters*, 1975, **50**, 265.
[151] K. Akasaka and A. Yamada, *F.E.B.S. Letters*, 1975, **53**, 339.
[152] T. J. Swift and R. E. Connick, *J. Chem. Phys.*, 1962, **37**, 307; 1964, **41**, 2553.
[153] J. Granot and D. Fiat, *J. Magn. Resonance*, 1974, **15**, 540.

In a paper from the same laboratory,[154] an experimental investigation of the solvation of aqueous Ni^{2+} ions is presented. Through a slow cooling of the solutions and a careful recording of the n.m.r. spectra, the authors were able to observe the proton and deuteron resonances of the water molecules of hydration. The solvation water peaks were more clearly seen when the r.f. phase was adjusted to incorporate part of the dispersion signal into the ordinary absorption. In this way the intensity of the large solvent peak was reduced at the frequency of interest. The independent ^1H and ^2H n.m.r. studies gave consistent values for the hyperfine coupling constant and a spin density at the hydrogen atom of 219×10^{-5} electron $a.u.^{-3}$. Reliable rate constants for the exchange between hydration and bulk water were also determined.

Other applications involving the Swift–Connick method using ^1H n.m.r. in paramagnetic systems are given by Rusnak and Jordan,[155] investigating solvent exchange in a nickel–Schiff base complex, by Tomiyasu et al.,[156] studying malonate exchange in an oxovanadium complex, by Chastellain and Merbach,[157] exploring the exchange of a bidentate ligand at a praseodymium ion, and by Karpel et al.,[158] studying purine exchange at a Ni^{2+} ion in aqueous solution. All these applications are fairly straightforward from the n.m.r. point of view. One noteworthy detail is that Rusnak and Jordan[155] find a substantial pseudocontact contribution to the shift of the OH proton in methanol; this is an observation that could be of some importance in the evaluation of hyperfine splitting constants in similar systems.

An example of studies using nuclei other than protons can be found in a paper by Lincoln and West.[159] They have continued their examination of acetonitrile exchange in cobalt complexes using ^{14}N n.m.r. The kinetics of manganese binding to ATP have been studied by Kuntz et al.[160] In a previous[161] similar study the authors used ^{13}C n.m.r., and they have now found that additional and complementary information is obtainable through ^{31}P n.m.r. It was concluded that both exchange with free Mn^{2+} and direct ligand exchange contributed to the observed broadening effects.

Two studies using the Swift–Connick approach and ^{35}Cl n.m.r. have appeared during the year. Lincoln et al.[162] have investigated the Cl^- exchange in aqueous solutions containing Ga^{3+} and Cl^- ions. Since the quadrupole coupling constant for the ^{35}Cl nucleus is vastly different in $GaCl_4^-$ and Cl^-, exchange contributions to the linewidth could be observed.

In proteins containing metal-anion binding sites the relaxation of the bound ^{35}Cl nucleus could be very short indeed. A chemical exchange contribution to the linewidth is then quite probable. This possibility has been tested by Man and Bryant[163] for four different proteins labelled with mercury. In two cases they found large exchange contributions to the widths. It is the Reporters' opinion, however, that in

---

[154] J. Granot, A. M. Achlama, and D. Fiat, J. Chem. Phys., 1974, 61, 3043.
[155] L. L. Rusnak and R. B. Jordan, Inorg. Chem., 1975, 14, 988.
[156] H. Tomiyasu, S. Ito, and S. Tagami, Bull. Chem. Soc. Japan, 1974, 47, 2843.
[157] F. Chastellain and A. Merbach, Helv. Chim. Acta, 1975, 58, 1.
[158] R. Karpel, K. Kustin, and M. A. Wolff, Israel J. Chem., 1973, 11, 735.
[159] S. F. Lincoln and R. J. West, Austral. J. Chem., 1974, 27, 97.
[160] G. P. P. Kuntz, Y.-F. Lam, and G. Kotowycz, Canad. J. Chem., 1975, 53, 926.
[161] Y.-F. Lam, G. P. P. Kuntz, and G. Kotowycz, J. Amer. Chem. Soc., 1974, 96, 1834.
[162] S. F. Lincoln, A. C. Sandercock, and D. R. Stranks, J.C.S. Dalton, 1975, 669.
[163] M. Man and R. G. Bryant, J. Magn. Resonance, 1974, 16, 220.

such studies the intensities should be checked to make sure that transitions other than $m = -\frac{1}{2} \to m = \frac{1}{2}$ are not observed.

In general, it seems that many authors are tempted to determine activation parameters $\Delta S^{\ddagger}$ and $\Delta H^{\ddagger}$ from experimental data that are too inaccurate. In ref. 162, for example, the rate constant at 298 K for four different concentrations differs by at most a factor of three, while the reported $\Delta H^{\ddagger}$ differs by 14 kJ mol^{-1} between two samples.

B. Linewidth Determinations in Paramagnetic Systems.—In a number of systems, paramagnetic contributions to the relaxation can be interpreted to give valuable structural information. This has been systematically used in connection with studies using shift reagents, but that particular aspect will not be discussed here. There have, however, appeared a number of special applications of linewidth determinations to structural problems.

De Boer and co-worker[164] have continued their studies of aromatic hydrocarbon radicals, now using ^{1}H and ^{2}H n.m.r. The measurements of linewidths of corresponding signals of both nuclei make, as usual, more detailed conclusions possible. For protons with large hyperfine coupling constants the relaxation is dominated by the modulation of this coupling through electron spin exchange. For more peripheral protons and for the deuterons the reorientational motion is usually the critical effect determining the correlation times.

The possibility of determining oxygen–oxygen distances in hydrogen-bonded complexes of nitroxide radicals has been discussed by Sysoeva et al.[165] They present a rather extensive theoretical discussion of the problem and also show a couple of practical examples. It is not clear, however, if the accuracy of the method is good enough to give interesting information.

In a short note Cotton et al.[166] point out one danger in the use of paramagnetic relaxation probes in ^{13}C bandshape studies of chemical exchange; it apparently has been neglected previously that the paramagnetic broadening can be strongly temperature-dependent and can give unwanted contributions to the linewidth at low temperatures.

In biological systems paramagnetic broadening effects often give unique structural information. LaMar and Viscio[167] found a strong concentration dependence of the linewidths of some methyl signals in the porphyrin dicyanohaemin. They interpreted this in terms of the formation of a specific dimer where the pyrrole IV of one porphyrin ring was stacked on top of the pyrrole I of another ring. Gelbaum and Engel[168] have investigated the interaction between Ni^{2+} ions and several α-aminoacids as a function of pH. Finally, the apparently very complex binding of transition-metal ions to ATP has been studied by Feldman and Wee[169] in ATP–Cu^{2+} aqueous solutions. They found a 1:1 ATP–Cu^{2+} complex at low ATP concentrations but at higher concentrations 2:1 complexes were formed, probably owing to ATP dimerization.

[164] B. M. P. Hendriks and E. De Boer, *Mol. Phys.*, 1975, **29**, 129.
[165] N. A. Sysoeva, A. Yu. Karmilov, and A. L. Buchachenko, *Chem. Phys.*, 1975, **7**, 123.
[166] F. A. Cotton, D. L. Hunter, and A. J. White, *Inorg. Chem.*, 1975, **14**, 703.
[167] G. N. LaMar and D. B. Viscio, *J. Amer. Chem. Soc.*, 1974, **96**, 7354.
[168] L. Gelbaum and R. Engel, *J. Inorg. Nuclear Chem.*, 1975, **37**, 793.
[169] I. Feldman and V. Wee, *Biochemistry*, 1974, **13**, 1836

7
Fourier Transform N.M.R.

BY D. SHAW

1 Introduction

It is now nearly ten years since Ernst and Anderson[1] first demonstrated that the response of an n.m.r. sample to a pulse could be converted by Fourier transformation into a 'normal' frequency-domain spectrum, and that this approach led to a gain in sensitivity. After a slow start, mainly due to lack of suitable instrumentation, so-called 'Fourier Transform N.M.R.' has developed rapidly, and will soon no doubt become the 'normal' way of recording n.m.r. spectra. The development of Fourier Transform n.m.r. up to mid-1973 has been dealt with previously (ref. 2 and the references therein). This chapter is concerned with the developments since then, and, following the style of these Reports, is concerned mainly with the technique itself and not with its numerous applications.

The interpretation of the title used in this chapter should be defined. The term 'Fourier Transform N.M.R.' has become, for obvious historical reasons, equated with pulsed excitation. *All* n.m.r. experiments contain the following key stages: (i) excitation of the spin system; (ii) detection of the response; (iii) cross-correlation of the response with the excitation (unscrambling the phases); (iv) Fourier transform into the frequency domain. In the case of the slow-sweep experiment, where the phase is constant, the latter two stages are performed by the spectrometer's phase-sensitive detector, effectively one frequency at a time (a discrete Fourier transform). The definition used here for 'Fourier Transform N.M.R.' includes all experiments where the total spectrum is excited on a timescale less than T_1; stage (iv) therefore necessitates a multi-channel Fourier transformation. Using this definition the techniques of Correlation and Hadamard spectroscopy are easily accommodated. In these experiments the phases are quadratic and pseudo-random, respectively; hence the cross-correlation is a non-trivial process. Discussion of T_1 measurements is also included, on the rationale that these experiments are dependent on the short timescale associated with Fourier Transform n.m.r.

During the past two years pulsed FT n.m.r. has developed into a mature technique; nearly all the ^{13}C, ^{15}N *etc.* spectra reported in the literature now come from pulsed spectrometers. The subject of pulsed FT n.m.r. has been reviewed recently by Komoroski and Levy,[3] who give a good survey of the technique and many refer-

---

[1] R. R. Ernst and W. A. Anderson, *Rev. Sci. Instr.*, 1966, **37**, 93.
[2] D. Shaw, in 'Nuclear Magnetic Resonance', ed. R. K. Harris (Specialist Periodical Reports), The Chemical Society, London, 1974, Vol. 3, p. 249.
[3] R. A. Komoroski and G. C. Levy, *Magn. Resonance Rev.*, 1974, **3**, 289.

ences to its various applications. Rummens has discussed the impact of the sensitivity of FT ^{1}H n.m.r. and drawn attention to the problems of solvent purity which occur when working down at the microgram level.[4] Several other general articles have appeared[5-7] which concern themselves with such areas as applications in organic chemistry and biochemistry.[7] The widespread use and acceptance of FT n.m.r. is perhaps illustrated by the publication of an introductory article to the technique in Swedish![8]

Fourier techniques have recently been applied to other magnetic spectroscopies, e.g. e.s.r.,[9,10] n.q.r.,[11] and broad-line n.m.r.,[12] with similar advantages to those achieved in high-resolution n.m.r. In the latter two cases, as well as gaining in sensitivity due to multichannel excitation, even when the linewidths are 10% of the total spectral width,[12] gains in resolution are also achieved due to the absence of modulation side-bands.[11,12] Additional sensitivity is also achieved in broad-line n.m.r. using pulse excitation, since this enables higher mean powers to be applied to the sample. Ion-cyclotron resonance, a technique with many similarities to n.m.r., has also benefitted from the fact that pulsed excitation and Fourier transformation can give increased sensitivity and, since lower pressures can be used with consequently less intermolecular effect, an 'improvement' in resolution.[13]

Consideration of the application of pulsed techniques to obtain high-resolution n.m.r. spectra from solids has not been included within this chapter.

2 New Applications

There are numerous applications of Fourier Transform n.m.r. It is difficult to decide when an application is new and when it is simply an extension of an existing application. The original use of FT n.m.r. was to increase the basic sensitivity of high-resolution experiments. Three interesting recent examples of this facet of the technique are the ability to study easily ^{2}H spectra at natural abundance,[14] gases at low pressure,[15] and tritium resonances.[16,17] In the first case, under proton-noise-decoupling conditions, the spectra produced simply consist of single lines, as is the case with ^{13}C, isotopic dilution removing homonuclear coupling.[14] In the second case Jameson et al.[15] were able, owing to the increased sensitivity of pulse n.m.r., to study gaseous ^{129}Xe chemical shifts at sufficiently low concentration (pressure) to eliminate three-body effects. Effects of this type normally cause problems during studies of the linear (with respect to pressure) chemical shift term. Tritium has a sensitivity even greater than the proton, but is difficult to study owing

---

[4] F. H. A. Rummens, in 'Proceedings of a Symposium on Mass Spectrometry and N.M.R. Spectroscopy in Pesticide Chemistry', 1973, p. 219.
[5] R. Radeglia, Z. Chem., 1974, **14**, 82.
[6] W. Von Philipsborn, Pure Appl. Chem., 1974, **40**, 159.
[7] J. A. Glasel, Fed. Proc. Amer. Soc. Exp. Biol., 1974, 33.
[8] S. Forsén, Kem. Kemi, 1974, **1**, 484.
[9] I. M. Brown and R. W. Kreilick., J. Chem. Phys., 1975, **62**, 1190.
[10] W. E. Blumberg, W. B. Mins, and D. Zuckerman, Rev. Sci. Instr., 1973, **44**, 544.
[11] R. Lenk and E. A. C. Lucken, Pure Appl. Chem., 1974, **40**, 199.
[12] J. Biemond, A. B. Lohman, and C. MacLean, J. Magn. Resonance, 1974, **16**, 408.
[13] M. B. Comisarow and A. G. Marshall, J. Chem. Phys., 1975, **62**, 293.
[14] J. M. Briggs, L. F. Farnell, and E. W. Randall, J. C. S. Chem. Comm., 1973, 70.
[15] C. J. Jameson, K. A. Jameson, and M. S. Cohen, J. Chem. Phys., 1975, **62**, 4224.
[16] J. M. A. Al-Rawi and J. A. Elvidge, J. C. S. Perkin II, 1975, 449.
[17] J. M. A. Al-Rawi and J. P. Bloxidge, J. C. S. Perkin II, 1974, 1635.

to its radioactivity. The increased sensitivity obtainable by Fourier n.m.r. has enabled Al-Rawi et al.[16,17] to expand their previous studies of this interesting nucleus, and to work with samples having much lower radiation levels than those previously used. The lower radiation levels, apart from being safer, induce fewer complications due to radiation damage, etc.

Recently, pulsed n.m.r. has been applied increasingly to the study of quadrupolar nuclei, e.g. ^{27}Al,[18] ^{2}H,[19] and ^{14}N,[20] where, owing to the larger linewidths associated with these nuclei, the achievable increase in sensitivity, though significant, is not as large as for nuclei giving narrow lines.[1]

Kintzinger and Lehn[20] have discussed the merits of CW *versus* FT excitation when applied to ^{14}N, from the sensitivity viewpoint. They also detail the problems of optimizing the experimental parameters in view of the enormous range of T_1's possible, even within one molecule.[20] Another quadrupolar nucleus, ^{11}B, continues to provide the best (only?) examples of resolution-enhancement techniques revealing new splittings[21,22] in spectra, as opposed to merely improving the resolution of already visible multiplicity.

One criticism often levelled at FT n.m.r. is that the precision/resolution is not as high as that obtainable by swept techniques. The origin of this supposed limitation lies in the finite number of data points available; if, as has been shown by Lustig and Lincoln, sufficient data points are utilized, results of unsurpassed precision can be obtained by FT.[23] They digitized the [AB]$_2$ spectrum of *o*-phenylenephosphorochloridite with 32 000 points over a spectral width of 100 Hz (*i.e.* 0.006 Hz per point). The line positions were obtained using a parabolic fitting function, and the spectra were analysed using LAOCOON III to yield J and Δv values with a probable error of 0.001 Hz.

Pulse excitation can yield high-resolution spectra on the timescale of seconds, thus permitting the study of dynamic effects. The effect of rotations on liquid crystals has been studied previously;[24] Fung[25] has used the speed of pulsed n.m.r. to study the return of the liquid crystal order to equilibrium following a sudden rotation. Spectra were taken at various times after a rapid 90° rotation of the sample; these spectra show the dipolar splitting increasing back to its equilibrium value as orientation is re-established within the liquid crystal.

The same advantage of pulsed excitation, namely speed, has been exploited by Grimaldi and Sykes. They have described a stopped-flow system which they used to study reactions with a first-order rate constant of the order of seconds;[26] the example given is the hydrolysis of L-phenylalanine-butyl ester catalysed by α-chymotrypsin at pH 6.9. The reactants were mixed in a rapid mixing cell in the probe. The mixing cell was controlled by standard driving/stopping syringe blocks.

[18] J. W. Akitt and R. M. Duncan, *J. Magn. Resonance*, 1974, **15**, 162.
[19] H. H. Mautsch, H. Saito, L. C. Leitch, and I. C. P. Smith, *J. Amer. Chem. Soc.*, 1974, **19**, 256.
[20] J. P. Kintzinger and J. M. Lehn, *Helv. Chem. Acta*, 1975, **58**, 905.
[21] A. O. Clouse, D. C. Moody, R. R. Reitz, T. Roseberry, and R. Schaeffer, *J. Amer. Chem. Soc.*, 1973, **95**, 2496.
[22] J. W. Akitt and C. G. Savory, *J. Magn. Resonance*, 1975, **17**, 122.
[23] E. Lustig and D. M. Lincoln, *J. Magn. Resonance*, 1974, **16**, 190.
[24] J. W. Emsley, J. C. Lindon, G. R. Luckhurst, and D. Shaw, *Chem. Phys. Letters*, 1973, **19**, 345.
[25] B. M. Fung, *J. Magn. Resonance*, 1974, **15**, 171.
[26] J. J. Grimaldi and B. D. Sykes, *J. Amer. Chem. Soc.*, 1975, **97**, 273.

The detection of the signal uses the technique of difference spectroscopy, described by Ernst,[27] which relies on the FID of the solvent (water) signal as a reference and 'carrier' of the solute information. This operation allows weak signals to be studied in the presence of strong solvent signals, since the latter can be subtracted after the detection stage. Difference spectroscopy also allows successive transients to be averaged, since they are now independent of magnetic field changes, *i.e.* no field/frequency lock is required.[27] The stopped-flow experiment is described by the pulse sequence[26]

$$[[\text{Push} - \text{DP} - [\text{Pulse} - \text{AT} - \text{DT}]_{NT} - \text{DK}]_{NK}]_{NP}$$

where AT is the acquisition time, DT, DK, and DP are respectively the delays between the pulses within a block and between blocks and the time required for equilibrium (chemical and magnetic) after a push. NT is the number of transients in a block, NK the number of blocks from a single push, and NP the number of pushes. Detailed consideration is given to the limits for each of the above parameters, and spectra are shown.

A further elegant example of the speed of FT n.m.r. is given by the direct observation of intramolecular chemical exchange by Dahlqvist *et al.* in [2,2]-2,5-pyrroloparacyclophane.[28] This molecule undergoes an internal rearrangement which interconverts the A and X parts of the $[AX]_2$ system in the pyrrole ring. The A part is subjected to a selective (0.05 s) 180° pulse using the homodecoupler controlled by the gated decoupling software, and then, after a time interval τ later, the spectrum is recorded using a non-selective (25 μs) 90° pulse. At short values of τ a net excess magnetization is seen in the A region which is equilibrated by chemical exchange, as can be followed by increasing τ. A study of magnetization as a function of τ yields the appropriate rate constants. The technique is, of course, analogous to the saturation transfer technique used in CW n.m.r. to study slow exchange. For it to be applicable the rate must be slower than or equal to T_1^{-1}, but slow enough to give resolved lines; also, cross-relaxation effects *etc.* must be negligible.[28]

Saturation transfer has been successfully applied to biological problems like NH exchange.[29-31] For such examples correlation spectroscopy is used (see Section 5). When the H_2O peak is saturated the exchangeable protons decrease in intensity owing to saturation transfer effects. However, the non-exchangeable protons, which are close to the hydration sphere of the molecule, increase in intensity owing to nuclear Overhauser effects,[31] giving a powerful direct way to investigate the difference between these two types of water.

The technique of tailored excitation[32] was discussed in the previous article in this series[2] as a method for homonuclear decoupling and solvent elimination. Freeman *et al.*[33] have utilized the possibility of generating an excitation power spectrum of any required distribution to perform new relaxation studies. Using

[27] R. R. Ernst, *J. Magn. Resonance*, 1971, **5**, 398.
[28] F. W. Dahlqvist, K. J. Longmuir, and R. B. DuVernet, *J. Magn. Resonance*, 1975, **17**, 406.
[29] S. Waelder, L. Lee, and A. G. Redfield, *J. Amer. Chem. Soc.*, 1975, **97**, 2927.
[30] J. D. Glickson, J. Dadok, and G. R. Marshall, *Biochemistry*, 1974, **13**, 11.
[31] T. P. Pinto, J. D. Glickson, J. Dadok, and G. R. Marshall, *Nature*, 1975, **250**, 582.
[32] B. L. Tomlinson and H. D. W. Hill, *Chem. Phys.*, 1973, **59**, 1775.
[33] R. Freeman, H. D. W. Hill, B. L. Tomlinson, and L. D. Hall, *J. Chem. Phys.*, 1974, **61**, 4466.

tailored excitation techniques they studied the relaxation of various protons in glucopyranose as the spectrum was subjected to selective 180° pulses in an inversion-recovery sequence. In this manner they were able to ascertain the contributions of each neighbouring proton to the relaxation rate of an individual proton. They conclude that, for the compounds studied, the relaxation is purely dipolar, and by dynamic NOE measurements[34] they assess the NOE between various protons, which in turn gives internuclear distances. Experiments of this type have great potential in studying stereochemistry in rigid systems.

James and McDonald have used Fourier techniques to measure the self-diffusion coefficient of each component in a complex system.[35] A 90° pulse is applied to the sample, followed after an interval τ by a 180° pulse. A magnetic field gradient pulse along the z axis, of duration δ, is applied in between the pulses and after the 180° pulse. A spin-echo occurs at time 2 τ, which is recorded and Fourier-transformed. In order to avoid phase problems, which could be associated with the inability to define the tip of the echo, power spectra are plotted. The field gradient pulse must be off during the actual pulses and the period of data acquisition. The samples were not spun, in order to avoid modulation artifacts. The technique has the advantages of FT n.m.r. that time-averaging can be used to improve the signal-to-noise ratio and that the individual components give separate lines. The example initially studied was a 1:1 DMSO–water mixture; the field gradient required, 54.2 mT m^{-1}, was obtained on the HR220 spectrometer used simply by increasing the current in the z shim coil. Values obtained on the pure components compare favourably with values obtained by standard means: D for water is $2.35 \pm 0.10 \times 10^{-5}$ cm^2 s^{-1}, and in the mixture is $0.83 \pm 0.04 \times 10^{-5}$ cm^2 s^{-1} at 25 °C. The authors later extended the technique to study phospholipid vesicles.[36]

3 Instrumentation

The aim of this section is not to compete with Chapter 4, but to outline some of the developments in instrumentation which have a specific bearing on the technique of FT n.m.r. itself.

A. Multinuclear Spectrometers.—The increased sensitivity of FT n.m.r. has made feasible the study of many low-sensitivity nuclei. Truly multinuclear spectrometer operation is necessary to make the feasibility a practicality. One rather limited approach is to vary the magnetic field until the desired nucleus resonates at a frequency corresponding to one of the existing observing channels within the spectrometer. The limitation of this technique is that once the field has been changed the frequency of the lock channel is 'wrong', and hence cannot be used. The variable-field method has been reported by Gupta using an XL-100 spectrometer.[37] He minimizes the limitations by locking on a nucleus which at the new field is close (± 2 MHz) to the 15 MHz of the normal deuterium locking system. An example is shown of ^{205}Tl spectra recorded at 25.1 MHz (^{13}C channel) locking on to ^{11}B (boron trifluoride in ether) at 14 MHz. A table is given listing suitable locking

---

[34] R. Freeman, H. D. W. Hill, and R. Kaptein, *J. Magn. Resonance*, 1972, **7**, 327.
[35] T. L. James and G. G. McDonald, *J. Magn. Resonance*, 1973, **11**, 58.
[36] G. G. McDonald and J. M. Vanderkooi, *Biochemistry*, 1975, **14**, 2125.
[37] R. K. Gupta, *J. Magn. Resonance*, 1974, **16**, 185.

nuclei in order to observe 40 nuclei on a spectrometer equipped to detect ^{1}H, ^{19}F, ^{31}P, ^{11}B, ^{13}C, ^{27}Si, ^{17}O, ^{15}N, ^{33}S, and ^{14}N.

A more flexible approach to the same spectrometer is taken by Marshall et al.[38] Their technique uses only one observing channel (e.g. ^{13}C) at a fixed field and hence with the normal locking channel. The output frequency from this channel is mixed with (added to) a frequency obtained from an external synthesizer to give the desired new frequency. The response from the probe (which is tuned to the frequency appropriate to the nucleus being detected), after the preamplifier, is again mixed to produce the original (^{13}C) frequency, which is handled by the spectrometer normally. Very similar schemes have been described for Bruker spectrometers by Dorn et al.[39] and Traficante et al.[40] The work of Traficante does, however, go further and tackles the problem of the insert. In all previous spectrometer systems the receiver coil has only had a narrow frequency range over which it will tune; hence, even if the appropriate frequencies are available 'at the push of a button' time is required to change the probe insert (and quite often also the preamplifier) to that appropriate to the new nucleus if this is more than a few MHz away from the one already being studied. Traficante et al.[40] have described in detail the design of an inductively-coupled insert (as opposed to the normal capacitively-coupled type) which, along with a wideband preamplifier, is usable over the range 6.4—38.2 MHz. They show ^{31}P, ^{13}C, ^{14}N, and ^{2}D spectra taken with their wideband insert, and spectra of the same samples obtained with conventional inserts. The comparison is very impressive

B. Quadrature Phase Detection (QPD).

—The free induction decay is a complex (in the mathematical sense as well!) signal i.e. to define it fully two components, the real and the imaginary, need to be detected. In present commercial spectrometers this is not done, only one component being detected; during the transformation it is assumed to be the real component, the imaginary component being inferred as zero. The consequences of this simplification are that the information in the real and imaginary components following Fourier Transformation are not independent,[41] and that positive and negative frequencies cannot be distinguished. The latter ambiguity can be resolved for the signal, but not the noise, by placing the pulse frequency at one extreme of the spectrum, thus ensuring that all the spectral frequencies have the same sign. Twice the necessary amount of noise is thus present in the spectrum, whose signal-to-noise ratio is therefore only $\sqrt{2}$ that which it might be.

As was pointed out some while ago by Redfield and Gupta,[42] there are two ways round this problem. The first is the use of crystal filters which will only pass frequencies to one side of the basic frequency of the pulse, and the second is the use of two phase detectors 90° out of phase (i.e. in quadrature) to detect the two individual components of the FID. The first method, which is the simplest and cheapest, has been used by Allerhand et al.,[43] who have achieved the expected gain of 40% in

---

[38] A. G. Marshall, L. D. Hall, M. Halton, and J. Sallos, J. Magn. Resonance, 1974, 13, 392.
[39] H. C. Dorn, L. Simiral, J. J. Natterstad, and G. E. Maciel, J. Magn. Resonance, 1975, 18, 1.
[40] D. D. Traficante, J. A. Simms, and M. Mulcay, J. Magn. Resonance, 1974, 15, 484.
[41] E. Bartholdi and R. R. Ernst, J. Magn. Resonance, 1973, 11, 9.
[42] A. G. Redfield and R. K. Gupta, Adv. Magn. Resonance, 1971, 5, 81.
[43] A. A. Allerhand, R. F. Childers, and E. Oldfield, J. Magn. Resonance, 1973, 11, 272.

sensitivity. It is surprising that such a large enhancement in sensitivity has been overlooked for so long in commercial spectrometers.

The second approach, that of quadrature detection, has lain dormant because of instrumental problems. In order to avoid 'ghost' peaks, *i.e.* reflections about zero and spurious signals at the pulse frequency, which in this technique is placed at the middle of the spectrum, it was necessary to have an almost unachievable match between the two detectors. Hoult and Richards[44] have devised a four-pulse sequence which overcomes this problem. Stejskal and Schaefer have successfully used this sequence[45] and produced spectra, using QPD, which also show the 40% gain in sensitivity anticipated. In a later paper, Stejskal and Schaefer[46] compare the two techniques mentioned above. In this paper they also discuss other merits of QPD, *e.g.* that it requires only half the acquisition rate (a 1000 Hz spectrum extends from $+500$ to -500 Hz instead of 0—1000 Hz) and makes more efficient use of the pulse power, since when operating in this mode the pulse may be placed in the middle of the spectrum. Schaefer and Stejskal also investigated the lineshape distortion generated in Fourier Transform spectroscopy by the use of Butterworth Filters prior to the ADC. They suggest that Bessel Filters are more appropriate.[47] All these techniques are more fully discussed in Chapter 4.

C. **Solvent Elimination.**—The topic of solvent 'elimination' in order to avoid dynamic-range problems at the digitization stage, especially in aqueous solutions, still[2] generates some publications. Mooberry and Krugh[48] have reported the rediscovery of WEFT[49] and its modification using homogeneity spoiling pulses[50] and have used them to obtain spectra at the millimolar level from NH protons in H_2O solution. An NH proton would of course not be visible if D_2O had been used as a solvent because of the chemical exchange. WEFT-type techniques have the limitation of requiring a difference in relaxation time between the solvent and the solute,[49,50] and are quite tricky to optimize. The use of a decoupling field to pre-saturate the water resonance does not have this disadvantage, though it does have the problem of obliterating from study certain areas of the spectrum close by the irradiated signal, and is also open to complications due to saturation transfer effects (see above). Campbell *et al.*[51] have reported the successful application of the method to biochemical problems, as latterly have Bleich and Glasel.[52] The relative efficiency of presaturation at removing the residual water depends on the homogeneity of the second r.f. field.[53] The less the homogeneity of this field, the more efficient the dephasing of the magnetization in the yz plane. The logic of this

[44] D. A. Hoult and R. E. Richards, First European Experimental N.M.R. Conference, Kent, 1974; *Proc. Royal Soc.*, 1975, **A344**, 311.
[45] E. O. Stejskal and J. Schaefer, *J. Magn. Resonance*, 1974, **13**, 249.
[46] E. O. Stejskal and J. Schaefer, *J. Magn. Resonance*, 1974, **14**, 173.
[47] J. Schaefer and E. O. Stejskal, *J. Magn. Resonance*, 1974, **14**, 160.
[48] E. S. Mooberry and T. R. Krugh, *J. Magn. Resonance*, 1975, **17**, 128.
[49] S. L. Platt and B. D. Sykes, *J. Chem. Phys.*, 1972, **56**, 3182.
[50] F. W. Benz, J. Feeney, and G. C. K. Roberts, *J. Magn. Resonance*, 1972, **8**, 114.
[51] I. D. Campbell, C. M. Dobson, G. Jeminet and R. J. P. Williams, *F.E.B.S. Letters*, 1974, **49**, 115.
[52] H. E. Bleich and J. A. Glasel, *J. Magn. Resonance*, 1975, **18**, 401.
[53] D. A. Hoult, *J. Magn. Resonance*, 1975, in the press.

can be taken one stage further, as has been done by Kimber *et al.* who use a noise-modulated decoupler in the presaturation experiment.[54]

4 Computer Simulations of Pulsed N.M.R.

The use of computers to simulate pulsed n.m.r. experiments continues, and the models have become more and more sophisticated. The general objective of these simulations is to provide programs which will simulate the response of a nuclear spin system under as many conditions as possible. In this way, it is possible to predict the efficiency of various pulse sequences, to investigate the consequences of various relaxation mechanisms on the spectra produced by pulsed excitation, and to simulate the consequences of instrumental shortcomings.

A. **Multiple Experiments.**—The two main groups involved in work of this type are those of Goodwin and Wallace,[55-57] and of Meakin and Jesson.[58-61] Both groups have followed a similar path, starting with simple models based purely on the Bloch equations,[55, 58] and then, in order to include spin-coupling effects, moving on to density-matrix methods,[56, 59] and latterly including relaxation effects *via* the full Bloch, Wangsness, and Redfield formalism.[57, 60, 61] In all but the simplest cases no analytical solutions to these problems were possible and the reader is faced with a bewildering array of simulations of the effect of this parameter on that spectrum under these approximations, *etc.* Nevertheless, some very useful general conclusions arise from this approach to the study of pulsed n.m.r. The method does not give the deeper understanding which may result from an individual approach to the specific problem,[61] but having a general model permits its easy extension to new examples as and when they become of interest. Both groups are still secretive concerning the computation time required for these simulations, except, that is, to rule out some cases as being too expensive!

Goodwin and Wallace have studied both CW and pulsed n.m.r. experiments and take magnetic field inhomogeneities specifically into account.[55, 57] Part III[57] in their series concentrates on the consequences of including more sophisticated relaxation models than those previously used when studying multispin systems. Following the observation of Harris and Worvill[62] that information on the dominant relaxation mechanisms can be gained from the study of CW saturation behaviour of the spin system, they simulate the behaviour of an AB system during an inversion recovery experiment using various relaxation mechanisms.[57]

Meakin and Jesson have developed a general density-matrix program which specifically includes relaxation from both intramolecular dipole and external random field effects (the latter being used to represent all the non-dipolar mechanisms). They initially apply their treatment to the two-spin systems under single-

---

[54] B. L. J. Kimber, J. Feeney, and G. C. K. Roberts, to be published.
[55] B. W. Goodwin and R. Wallace, *J. Magn. Resonance*, 1972, **8**, 41.
[56] B. W. Goodwin and R. Wallace, *J. Magn. Resonance*, 1973, **9**, 280.
[57] B. W. Goodwin and R. Wallace, *J. Magn. Resonance*, 1973, **12**, 60.
[58] P. Meakin and J. P. Jesson, *J. Magn. Resonance*, 1973, **10**, 290.
[59] P. Meakin and J. P. Jesson, *J. Magn. Resonance*, 1973, **11**, 182.
[60] P. Meakin and J. P. Jesson, *J. Magn. Resonance*, 1974, **13**, 354.
[61] P. Meakin and J. P. Jesson, *J. Magn. Resonance*, 1975, **16**, 411.
[62] R. K. Harris and K. M. Worvill, *J. Magn. Resonance*, 1973, **9**, 394.

pulse conditions.[59] Their program was next extended to include spin systems subjected to a series of repetitive pulses which have established a steady state. The program, however, still maintains the simplifying assumption that all the off-diagonal matrix elements were either always zero or decayed rapidly to zero during the interval between the pulses,[60] a situation which, if not existing in practice, can be enforced by using a homogeneity spoiling pulse, prior to the initiation of an observing pulse. This limitation is lifted in the fourth paper in the series,[61] permitting the demonstration of phase and echo effects, the inclusion of spin coupling into CPMG simulations, and consideration of J spectra. The papers of Meakin and Jesson discuss many aspects of the AB system, including the effect of a simple $180° - \tau - 90°$ sequence,[51] saturation under dynamic conditions,[60] the effects of chemical exchange,[60] finite pulse power[59] (with the reassuring conclusion that with the pulse power available from commercial spectrometers no extra distortions are to be expected from this source), and finally the approach to equilibrium of AB spectra under nearly saturating conditions.[61] Meakin and Jesson have also considered aspects of the gated-decoupling experiment, where the decoupler is on only during a delay period immediately following data acquisition. In this mode spectra are obtained which give a nuclear Overhauser enhancement but without the normally associated spin decoupling. They show that in order to obtain an exponential growth of the NOE during the on period high decoupling powers are needed.[59] They also consider the problem of optimizing the sensitivity of this experiment and show that, whereas the optimum delay depends on the relaxation mechanisms involved, under conditions appropriate to typical ^{13}C work the optimum delay is of the order of the acquisition time.

Kundla has also applied the Bloch, Wangsness, and Redfield theory to pulsed spectra resulting from an AB system.[63]

A more general consideration of the role of the computer in Fourier Transform n.m.r. spectroscopy has been published by Cooper.[64]

B. Diffusion and Field Gradient Effects.—Kaiser et al. have considered the effects of diffusion and magnet inhomogeneity on the echoes which can occur during a steady-state pulse experiment.[65] The formation of echoes in a repetitive pulse experiment was first noted by Hahn,[66] and if they are unchecked they can lead to distortion of the FID and hence the spectrum. Freeman and Hill have previously described these echoes and methods of removing them.[67] Kaiser et al. use two independent approaches to the problem, one based on Fourier expansion techniques, the other partition methods. The results obtained from the two methods are discussed and used to explain measurements made on a one-spin system (undegassed benzene) in an inhomogeneous field. An inhomogeneous field is used since the contribution to the FID from refocusing effects is more marked in this case than in a homogeneous field. It is shown that the echoes depend strongly on the rate of diffusion (D) in the sample; the more rapid the diffusion the quicker the echoes are suppressed. The use of homogeneity-spoiling pulses (HSP) to suppress echoes is also considered.

[63] E. Kundla, *Eesti N.S.V. Tead. Akad. Toim. Fuus. Mat.*, 1974, **23**, 396.
[64] C. Cooper, in 'Topics in ^{13}C n.m.r.', ed. G. C. Levy, Interscience, New York, 1975.
[65] R. Kaiser, E. Bartholdi, and R. R. Ernst, *J. Chem. Phys.*, 1974, **60**, 2966.
[66] E. L. Hahn, *Phys. Rev.*, 1950, **80**, 580.
[67] R. Freeman and H. D. W. Hill, *J. Magn. Resonance*, 1971, **4**, 366.

The application of a regular HSP prior to the observing pulse in a repetitive sequence will not, in the absence of diffusion, suppress echoes; a stochastic process is necessary.[60,67] In the presence of diffusion it is shown that a gradient (G) given by

$$\gamma G = (3/D\,\tau^2 T_{ac})^{\frac{1}{2}}$$

is necessary. Taking a typical value for the diffusion constant of 2.5×10^{-5} cm² s⁻¹ and letting the duration of the HSP (τ) be 1/10th that of the acquisition time (T_{ac}), the modest gradient of 12.6 μT cm⁻¹ is found to be required. In order to maintain a satisfactory lock, HSP's are normally shorter than given above and therefore need proportionally more power. The effect of refocusing on the sensitivity both of repetitively-pulsed FT and DEFT are considered, and the same conclusions reached as those discussed previously (section 6E of ref. 2).

5 The Equivalence of Fourier and Continuous Wave Spectra

Ernst and co-workers have addressed themselves to the important problem of the equivalence of spectra obtained in the frequency domain *via* slow-sweep techniques and those resulting from Fourier transformation of time-domain phenomena.[68-70] Instrumental artifacts associated with each technique *e.g.* the effects of using too rapid a sweep in CW work, and digitization, spin echoes, *etc.* in pulsed excitation are specifically included; then the conditions necessary for equivalence between the two techniques are considered using a density matrix formalism.[68] The authors show that the following three conditions are sufficient and in most cases necessary to guarantee the equivalence between the spectra generated by the two techniques:[70] (i) the density-operator equation must have time-independent coefficients; (ii) the density-operator equation must be linear in the density operator; (iii) the relevant part of the density operator (*i.e.* that responsible for the observed magnetization immediately after a pulse) must be independent of the flip angle except for an arbitrary scaling factor.

The first two conditions are normally met, but the third condition can be frequently violated,[69] especially when non-equilibrium states occur, *e.g.* during CIDNP experiments, chemical exchange, and double resonance.[69,70] Condition (iii) can be factorized into four auxiliary conditions which imply it, without themselves being necessary conditions.[70] These are: (*a*) each strongly coupled subsystem of the spin system which is homogeneously excited by the pulse must be in internal thermodynamic equilibrium prior to the pulse. Additional spins which are not excited by the pulse and which are only weakly coupled to the excited spins can be in an arbitrary non-equilibrium state; (*b*) the high-temperature approximation must fulfilled; (*c*) the high-field approximation must be fulfilled; (*d*) any homonuclear part of the complete system must be exclusively excited by the pulse. These considerations show that for systems without homonuclear coupling, *e.g.* ^{13}C and ^{15}N in natural abundance, equivalence is always fulfilled.[69] For CIDNP studies, however, Schaublin *et al.* show that equivalence between the techniques occurs if small pulse

---
[68] R. R. Ernst, W. P. Ane, E. Bartholdi, A. Hoehener, and S. Schaeublin, *Pure Appl. Chem.*, 1974, **37**, 47.
[69] R. R. Ernst, Proceedings Special College on Pulsed Nuclear Magnetic Resonance Spin in Dynamic Solids, 1973, 40.
[70] S. Schaublin, A. Hohner, and R. R. Ernst, *J. Magn. Resonance*, 1974, **13**, 196.

angles (<20°) are used. For larger angles the net effect remains unchanged, but the multiplet effects are partially (and for a 90° pulse completely) averaged. Similar complications are shown to occur in experiments measuring Overhauser effects in strongly coupled AB and AB_2 systems. Examples are also given of the behaviour of strongly coupled systems during T_1 sequences.[70]

Studies of the type discussed above demonstrate that the saturation behaviour of strongly coupled spin systems excited simultaneously by a pulse and sequentially in a sweep experiment differ. The two excitation techniques are therefore not equivalent under these conditions. With the wealth of information now appearing[56-63] on the behaviour of samples run using pulsed spectrometers under saturating conditions, and on their sensitivity to the various relaxation mechanisms taking place, an interesting area of research seems to be opening up.

The equivalence or otherwise of the spectra obtained by pulsed and swept excitation from systems undergoing chemical exchange has generated some controversy. Kaplan originally questioned the equivalence of the two techniques for protons undergoing intermolecular exchange.[71] Ernst subsequently pointed out that equivalence held provided that the high-temperature approximation held (as it nearly always does) *i.e.* $|E_i - E_k| \ll kT$ where E_i and E_k are any eigenvalues of the Hamiltonian.[68-70, 72] Kaplan maintained that differences may be detected under conditions of very fast exchange.[73] Gupta *et al.*[74] have shown the equivalence of the two methods analytically and experimentally for the case of an uncoupled spin system undergoing exchange between two non-equivalent sites. The experimental example studied is the proton spectrum of trimethylamine in H_2O as a function of pH. Using their computer program discussed above, Meakin and Jesson have simulated a more complicated case of an AB system undergoing chemical exchange.[59] Their simulations show that, even in cases where in the slow-exchange limit line intensities are seriously distorted by steady-state (saturation) effects, spectra equivalent to those obtained by slow-passage techniques result for exchange rates >3 or $4\,T_p^{-1}$. For rates faster than this the exchange process switches the two spin states before they can relax, and effectively equalizes the relaxation times in the two sites, thus restoring their relative intensities to the values observed in a normal single-pulse experiment.[59] As the exchange rate increases, normal coalescence behaviour occurs, resulting, in the fast-exchange limit, in a single sharp line. They show that the programs normally used for bandshape analysis can be used to analyse spectra obtained by pulse techniques.

6 Correlation Spectroscopy

After two or more years in the proceedings of various conferences 'correlation spectroscopy' has officially appeared in the literature,[71] to be followed closely by a slight variant called 'Rapid Scan Fourier Transform Spectroscopy'.[72] Both techniques utilize a fast linear sweep to excite the spectrum, and are thus capable of exciting only the area of interest, thus avoiding exciting unwanted areas, *e.g.*

---

[71] J. I. Kaplan, *J. Chem. Phys.* 1972, **57**, 5615.
[72] R. R. Ernst, *J. Chem. Phys.*, 1973, **59**, 989.
[73] J. I. Kaplan, *J. Chem. Phys.*, 1973, **59**, 990.
[74] J. Dadok and R. F. Sprecher, *J. Magn. Resonance*, 1974, **13**, 243.

solvent lines, *etc.* (see Section 3C). The name 'correlation spectroscopy' is in a way misleading since, as was pointed out in the introduction, all n.m.r. experiments require cross-correlation; it is just that in the case of a rapid sweep it is not a trivial operation, as it is with pulsed and slow-sweep excitation, where the phase is a constant and not a function of frequency.

The spectrum is swept rapidly compared with T_1 (~ 100 Hz s^{-1}), and the resulting signal is phase-detected, filtered, digitized and time-averaged in the normal way. The spectrum at this stage is a function of frequency and contains the basic lines, but it also contains transient effects (ringing) which have to be corrected. The correction is achieved by cross-correlation (deconvolution) with respect to either an experimental or theoretical line.[75] For computational efficiency the cross-correlation is carried out in the co-domain, using the Fourier theorem which states that cross-correlation in one domain is complex multiplication in the co-domain. There are efficient programs for Fourier transformation available for most computers, and complex multiplication is a standard operation. The average response and the excitation are therefore transformed into a co-domain, and the complex conjugate of the former multiplied by the latter. The result is a function, effectively equivalent to the FID following a pulse, *i.e.* an impulse-response function. As such, it can be subjected to weighting functions of the same type as those normally used in FT n.m.r. This function is then Fourier transformed again to give the final spectrum.

Gupta *et al.*[76] have discussed the relative merits of cross-correlating with a theoretical and an experimental response. The former is the more convenient, the latter has the advantage of automatic phase correction and of frequency referencing with respect to the experimental response chosen. To be useful a reference line must be a pure single line; otherwise artifacts are introduced. TMS is therefore not a suitable reference due to the presence of ^{29}Si satellites. The sensitivity per unit time of rapid-sweep excitation compared with that of pulse excitation is shown to be

$$\frac{(S/N)_P}{(S/N)_{RS}} = \left[\frac{(\Delta/a + 3T_2^*)(1 - E_1^P)(1 + E_1^{SR})}{3T_2^*(1 + E_1^P)(1 - E_1^{RS})}\right]^{\frac{1}{2}}$$

where $E_1^P = \exp(-3\,T_2^*/T_1)$ (the pulse experiment is assumed to have a period of $3\,T_2$), $E_1^{RS} = \exp[-(\Delta/a + 3\,T_2^*)/T_1]$, a is the sweep rate, and Δ is the sweep width. Pulsed excitation is always more efficient, but under typical proton conditions $\Delta = 1000$ Hz, $a = 250$ Hz s^{-1}, and $T_1 = T_2^* = 1$ s, so correlation spectroscopy is only 35% less efficient.

It has been shown that rapid-scan techniques can be used to measure relaxation times by a method analogous to saturation recovery.[76] Since the r.f. field is on all the time the relaxation time measured is a hybrid between T_1 and $T_{1\rho}$.

Sheahen has analysed the consequences on the total spectrum of any distortions introduced into a strong dominant signal.[77] He shows that for the case of a rapid linear sweep, the consequences of these distortions can be described analytically, and consequently corrected for. However, no examples are given.

[75] R. K. Gupta, J. A. Ferretti, and E. D. Becker, *J. Magn. Resonance*, 1974, **13**, 275.
[76] R. K. Gupta, J. A. Ferretti, and E. D. Becker, *J. Magn. Resonance*, 1974, **16**, 505.
[77] T. P. Sheahen, *J. Magn. Resonance*, 1974, **15**, 69.

7 Hadamard Transform N.M.R.

Ziessow[78,79] and Kaiser[80] have both considered the application of Hadamard transforms to n.m.r. spectroscopy. The gain in sensitivity achieved by FT n.m.r. is due to multichannel excitation and detection.[1] It would be significant if this gain could be achieved for a second time; one possibility would be by applying a pulse sequence to the spin system, and unscrambling it in such a way that the resulting FID had gained in sensitivity due to multichannel detection of each FID.[80] One possible sequence already used in optical spectroscopy is the Hadamard.[81] Kaiser has considered this possibility and shown that the cascading of the two multichannel processes does not gain any further sensitivity than that obtained by using one only.[80]

Hadamard transforms do, however, have a useful function in n.m.r. The use of stochastic excitation in n.m.r. has been previously discussed.[82,83] The most satisfactory form of this experiment involves using the output of a pseudo-random shift register as the noise source.[78,82] One problem with stochastic n.m.r., due to the random phase of the excitation, is the computation of the cross-correlation between the output of the spin system and the output of the pseudo-random shift register. This cross-correlation can be carried out using double Fourier transformation along the lines outlined above.[74] However, this method does not produce an impulse function in this case, since the original spectrum is in the time domain. As at no stage is an impulse-response function produced, weighting functions of the classical type cannot be used. A triple Fourier transform approach would overcome this limitation, but it is a very time-consuming business and would be prone to cumulative round-off errors in the Fourier transformation process. It has been shown that subjecting the response of a spin system as a result of pseudo-random excitation to a Hadamard transform produces an impulse-response function (*i.e.* effectively acts as a cross-correlator for this particular excitation sequence).[78,80] Weighting functions can now be applied to the signal, which is finally Fourier transformed in the normal manner to produce a frequency-domain spectrum. The advantage of the Hadamard transform over multiple Fourier transforms is that it is faster than even a single Fourier transform, since the algorithms involved do not require any multiplications; they consist only of additions and subtractions. Details of the algorithms and examples are given by both workers.[78-80]

8 Double Resonance in FT N.M.R.

When analysing experiments using Fourier transform techniques, care must be taken not to use a conclusion reached in one domain to infer behaviour in the co-domain, without duly allowing for the consequences of the transformation. One example of experiments where conclusions from one domain can be used in the co-domain is spin-decoupling using time-sharing techniques. Pines and Ellett[84]

[78] D. Ziessow, 'On-line Rechner in der Chemie', De Gruyter-Co, Berlin, 1973.
[79] D. Ziessow and B. Bluenich, *Ber. Bunsengesellschaft phys. Chem.* 1974, **78**, 1168.
[80] R. Kaiser, *J. Magn. Resonance*, 1974, **15**, 44.
[81] J. A. Decker, *Appl. Optics*, 1971, **10**, 510.
[82] R. R. Ernst, *J. Magn. Resonance*, 1970, **3**, 10.
[83] R. Kaiser, *J. Magn. Resonance*, 1970, **3**, 28.
[84] A. Pines and J. D. Ellett, *J. Amer. Chem. Soc.*, 1973, **95**, 4438.

have taken Jesson et al.[85] to task for just such a simplification, but in doing so have generalized the model further than the authors originally intended it to apply. When discussing time-shared decoupling Jesson et al. consider only the centre band of the B_2 field, which is decreased to $B_2\ \tau_w/T_{dwell}$, where τ_w is the duration of the decoupler pulse and T_{dwell} is the interval between these pulses.[85] Pines and Ellett point out that if τ_w were chosen such that it were a 360° pulse for the 'decoupled' nuclei, then no decoupling would in fact occur despite there being a finite B_2 field present at the sample. Such behaviour would not be expected based on the description of the experiment proposed by Jesson et al. If, however, values for B_2, T_{dwell} and τ_w typical of the experiment being considered (i.e. homonuclear decoupling)[85] are taken, then their model does correctly describe the experimental observations.[84]

INDOR can be a useful technique and, since it is essentially a sweep technique, it has no direct equivalent in Fourier spectroscopy. Feeney and Partington[86] have shown that the equivalent information can be obtained by subtracting, in the frequency domain, spectra obtained with and without continuous low-power irradiation of the line which would have been, in the classical INDOR experiment, the monitor line. Only those lines whose intensities are changed by the B_2 field appear in the final spectrum, which therefore looks like a normal INDOR spectrum. Spectra obtained in this way are really more closely related to those obtained by saturation transfer mechanisms than by the generalized NOE.[87, 88]

Equivalent information can also be obtained by selective population transfer (SPT) experiments. In SPT experiments the B_2 field is applied in the form of a selective pulse, i.e. $\gamma B_2 \sim (\pi T_2^*)^{-1}$, prior to the non-selective observing pulse.[89, 90] SPT has been used to assign and determine the sign of (^{13}C, X) coupling constants.[89, 90] If the special case of selective population inversion (SPI), where the B_2 field is a 180° pulse, is combined with subtraction techniques then pseudo-INDOR spectra can again be obtained.[91] Pachler and Wessels compare this technique with the one outlined above due to Feeney and Partington, and show it to be more efficient. In their comparison they use the now-classic molecule 2,3-dibromopropionic acid. Pachler and Wessels have also applied their technique (termed 'difference selective population inversion') to heteronuclear examples, studying the (^{13}C, ^{1}H) coupling constants in 2,3-dibromothiophen.[92] Jakobsen et al. have applied the same basic pulse sequence, with a slight improvement which includes the delay associated with the B_2 pulse in both cycles,[93] to the study of systems with degenerate energy levels, e.g. AX_n systems. They use the ^{13}C spectrum of acetone as an example. The same group has used difference SPT techniques to obtain the ^{13}C lines from $(CH_3)_2CO$ hidden under those of $(CD_3)_2CO$. They had previously used SPT techniques to study the same system,[94] and they report the relative intensities of SPT[94] and difference SPT[93] spectra from the AX_n systems.

[85] J. P. Jesson, P. Meakin, and G. Kneissel, J. Amer. Chem. Soc., 1973, 95, 618.
[86] J. Feeney and P. Partington, J. C. S. Chem. Comm., 1973, 611.
[87] S. Sørensen, R. S. Hausen, and H. J. Jakobsen, J. Magn. Resonance, 1974, 14, 243.
[88] A. A. Chalmers, K. G. R. Pachler, and P. L. Wessels, Org. Magn. Resonance., 1974, 6, 445.
[89] S. Sørensen, M. Hansen, and H. J. Jakobsen, J. Magn. Resonance, 1973, 12, 340.
[90] K. G. R. Pachler and P. L. Wessels, J. Magn. Resonance, 1973, 12, 337.
[91] K. G. R. Pachler and P. L. Wessels, J. C. S. Chem. Comm., 1974, 1038.
[92] A. A. Chalmers, K. G. R. Pachler, and P. L. Wessels, J. Magn. Resonance, 1974, 15, 415.
[93] T. Bundgaard and H. J. Jakobsen, J. Magn. Resonance, 1975, 18, 209.
[94] H. J. Jakobsen, S. A. A. Linde, and S. Sørensen, J. Magn. Resonance, 1974, 15, 385.

9 The Measurement of Relaxation Times

Interest in the measurement of the relaxation times of individual lines in a high-resolution spectrum continues to grow, extending into such areas as high-pressure samples[95] and CIDNP spectra.[96] Not surprisingly, considerable effort has gone into optimizing the efficiency of these very time-consuming experiments.

Levy and Peat have made an extensive experimental study of the factors influencing the accuracy of carbon T_1 measurements, using all the four common sequences.[97] They discuss the factors influencing the choice of a particular pulse sequence for a particular sample, and in general they favour the inversion recovery sequence. They also consider such instrumental effects as coil geometry (single coil spectrometers give T_1 values dependent on the length of the sample used due to r.f. inhomogeneity), pulse power, resolution and gain stability. Also investigated is the value of homogeneity-spoiling pulses to combat phase errors which result from the imperfect nature of the 180° pulse used. Bangerter has given details and examples of circuitry *etc.* applicable to generating suitable pulses for this purpose[98] (see also the work of Ernst[65]).

One of the areas of T_1 measurement discussed by Levy and Peat [97] but often ignored is determination of T_1 direct from the experimental data. This is normally carried out using a semi-log plot of intensity against τ, the slope of which yields T_1. Now, using a semi-log plot, the significance of the error within each determination increases as τ increases. Levy and Peat suggest that for this reason τ values greater than *ca.* 1.5 T_1 should not be used (except of course for the infinity value). De Fontaine *et al.* have proposed a novel and fast alternative approach; this is the use of a non-linear least squares fit program.[99] Since this method does not require an infinity value, T_1 is calculated progressively as τ is increased and the experiment is terminated when no change (within defined limits) is observed in the calculated value of T_1 and/or M_∞. Typically this involves $\tau = 0$ to 0.3 T_1 for the inversion recovery sequence and 0.5 to $2T_1$ for progressive saturation. The method is applied to the measurement of T_1's for the C-1 and C-4 carbons of *cis-* and *trans-*2-methyl-1-phenylcyclopropane.

The progressive saturation method has also been studied experimentally by Wilkins *et al.*[100] with a view to automation. They confirm the sensitivity of the method to the accuracy of the 90° pulse used ($\pm 7°$), and find that over long periods resolution stability can be an important factor; they therefore consider an autoshim system essential. Details are given of the circuitry to 'protect' the autoshim from the consequences of a homogeneity-spoiling pulse. It is found that to obtain reliable results 10 pulses are required before a steady state can be considered to have been established.

Christensen *et al.*[96] have investigated the progressive saturation technique in great detail prior to using it to measure ^{13}C T_1 values during the thermal decomposi-

---

[95] D. J. Wilbur and J. Jonas, *J. Chem. Phys.* 1975, **62**, 2800.
[96] K. A. Christensen, D. M. Grant, E. M. Schalman, and C. Walling, *J. Phys. Chem.* 1974, **78**, 1971.
[97] G. C. Levy and I. R. Peat, *J. Magn. Resonance*, 1975, **18**, 500.
[98] B. W. Bangerter, *J. Magn. Resonance*, 1974, **13**, 87.
[99] D. L. DeFontaine, D. K. Ross, and B. Ternai, *J. Magn. Resonance*, 1975, **18**, 276.
[100] C. L. Wilbins, T. R. Brunner, and D. S. Thoennes, *J. Magn. Resonance*, 1975, **11**, 373.

tion of benzoyl peroxide in various solvents, a CIDNP system. Owing to the low concentration and short lifetimes of the species being measured, considerable attention was given to optimizing the accuracy per unit time of the progressive saturation experiment. The optimization was carried out by considering the sensitivity of the line intensities to changes in $r = T_p/T_1$ and the pulse angle $\Phi°$. In the simple form of the progressive saturation experiment only 90° pulses are used. The method finally adopted for a particular sample was to select a value of r which gave the maximum changes in intensity as a function of Φ for a specific T_1 range. T_1 was then measured from the dependence of the signal intensity on the pulse angle Φ (50° < Φ < 120°). None of the T_1's measured in this way showed a decrease in T_1 due to the presence of free radicals, since under the experimental conditions used the estimated free radical flux is of the order of 10^{-6} mol l^{-1}, much less than the oxygen content of an organic liquid in contact with the air. CIDNP enhancements were also measured.

Canet et al. have developed a sequence they call 'fast inversion recovery' to measure T_1's.[101] This sequence is developed from the classical inversion recovery sequence by decreasing the delay between the sequences to well below the normal 3 to 4 T_1. The fast inversion recovery sequence thus becomes comparable in speed with the progressive saturation and saturation recovery sequences, but it still maintains some of the higher sensitivity associated with inversion recovery. This sequence is the more sensitive since the decay and magnetization is from $-M_0$ to $+M_0$ i.e. the range is $2M_0$. This is to be compared with the range of 0 to M_0 for the other two sequences. In other words the dynamic range, which can be equated with sensitivity, of the inversion recovery sequence is double that of the others. Therefore for a comparable signal-to-noise ratio the inversion recovery sequence takes one quarter of the time of the other two. The equation for the fast inversion recovery sequence is:

$$M = M_0 [1 - (2 - E_1) \exp(-\tau/T_1)]$$

where $E_1 = \exp -(T_p/T_1)$. Normal graphical methods can be used to obtain T_1, the only difference being that the intercept is now reduced from 2 by a factor of E_1 (< 1). The relative sensitivity compared with the inversion recovery technique is therefore reduced by $2/(2-E_1)$, but the sequence takes less time; therefore for a fixed experimental time it may be repeated more times and, by time averaging, picks up some of the lost sensitivity. The effects of systematic errors on the sequence are considered, and an example using 80% (w/v) phenol in D_2O is given. In this example a saving in time by a factor of 7 is claimed (35 min for fast inversion recovery vs. 4.2 h for inversion recovery). Further theoretical consideration will have to be given to this sequence before it can be applied to systems showing homonuclear coupling.

In the spectra of nuclei with a negative magnetogyric ratio, e.g. ^{29}Si, it is possible in the presence of decoupling to get zero signal. This occurs when the NOE = -1. The problem of measuring T_1 under these conditions has been solved by Levy et al.[102] and Kimber and Harris.[103] The first group use a gated decoupling approach:

---

[101] D. Canet, G. C. Levy, and I. R. Peat, *J. Magn. Resonance*, 1975, **18**, 199.
[102] L. C. Levy, J. D. Cargioli, P. C. Julian, and T. D. Mitchell, *J. Amer. Chem. Soc.*, 1973, **95**, 3445.
[103] B. J. Kimber and R. K. Harris, *J. Magn. Resonance*, 1974, **16**, 355.

in the inversion recovery sequence the decoupler is off during the delay prior to the 180° pulse and τ, but is on during data acquisition. The NOE is hence eliminated and signals may be obtained. There are difficulties associated with using the inversion recovery sequence in this way, since the relaxation occurs under spin-coupled conditions and non-exponential recovery due to cross-relaxation effects is possible. Kimber and Harris[103] use the dynamic nuclear Overhauser sequence. The growth of the NOE (which is exponential with time constant T_1) is measured by sampling the spectrum with a 90° pulse at time τ after the decoupler has been switched on. For ^{29}Si this results in a signal decaying with τ which yields T_1. The authors measured the T_1 for MeSi(OEt)$_3$ (92.6 s), a compound which under continuous decoupling conditions gives 'no' signal.

10 Miscellaneous

Fourier transforms are additive, and it is therefore possible to decompose an FID into a series of parts, *i.e.* from 0 to τ_1, from τ_1 to τ_2, *etc.* and transform them separately. The full spectrum can then be created by adding the transformed parts together in the frequency domain. Moniz *et al.*[104, 105] have used this property to study small splittings. For a simple situation consisting of only one chemically shifted resonance, they choose the ^{13}C spectrum of (CD$_3$)$_2$SO.[104] The first node in the FID occurs after J_s^{-1}, where J_s is the smallest coupling present, in this case ^3J(CD). If only the FID prior to the first node is transformed, then no evidence of the smallest splitting will be observed in the spectra. Effectively the acquisition time is not long enough. The resulting spectrum can be thought of as an 'envelope' of the full spectrum. If only the remainder of the FID is transformed, *i.e.* taking the time of the first node as the origin and ignoring negative frequencies, the resulting frequency spectrum consists of the total spectrum minus the 'envelope'. This form of display, which is dispersive in character, exaggerates the small couplings and apparently reduces the linewidth, *i.e.* it is a form of resolution-enhancement. Absorption-like display can be obtained by subtracting a fraction of the envelope spectrum. Moniz *et al.* show as a further example[105] the detection of the long-range proton coupling in ethylene glycol. The coupling, they state, 'is not at all resolved' when the FID is processed in the normal manner. This is hard to believe as evidence of it is clearly visible in the FID, as indeed it must be for a node to be detected and their method to be applied. Unfortunately the authors do not show the effect on their spectrum of a normal resolution enhancement function, which would provide an interesting comparison. The technique is an interesting example of Fourier theory.

Fourier theory and hence Fourier transforms continue to be used in theoretical papers dealing with n.m.r. instrumentation. For example, the statistical analysis of noise in an n.m.r. system by Kaiser,[106] and the work of Buckmaster *et al.*[107] on the choice of sensitivity enhancement filters in CW spectroscopy should be mentioned.

Fourier transform n.m.r. is obviously here to stay. Maybe in the next article in this series the chapter singled out for special treatment as the exception should be swept n.m.r. as opposed to Fourier transform n.m.r.!

[104] W. B. Moniz and S. A. Sojka, *J. Magn. Resonance*, 1973, **12**, 214.
[105] W. B. Moniz, C. F. Pronanski, and S. A. Sojka, *J. Magn. Resonance*, 1974, **13**, 110.
[106] R. Kaiser, *J. Magn. Resonance*, 1973, **11**, 113.
[107] H. A. Buckmaster, S. Cohn-Sfectu, and D. J. I. Fry, *J. Magn. Resonance*, 1974, **14**, 296.

8
Macromolecules

BY I. D. ROBB

1 Introduction

This chapter reviews reports dealing with the n.m.r. spectra of macromolecules published between June 1974 and May 1975. It is divided according to the nature of the material under investigation and then by type of nucleus and technique. The studies under review have in general been concerned with the determination of the structures and conformations of macromolecules or with the interactions between small molecules and macromolecules.

2 Reviews

Reviews have mainly been concerned with the application of various n.m.r. techniques to the elucidation of structure or conformation of biopolymers. Bovey's reviews have dealt with the application of n.m.r. to the structure determination of proteins,[1] as well as recent progress in the study of conformational behaviour and molecular motion in polypeptides.[2] Wüthrich,[3] surveying the application of ^{13}C n.m.r. to haemoproteins, has described how chemical shifts of haem nuclei may provide information on the electronic structure of the haem groups, and how ligand binding may be investigated. Quilley and Webb[4] have reviewed the use of high-resolution techniques with metal-containing enzymes, referring particularly to structural information about the active sites of the enzymes that can be obtained from relaxation data. The contribution of n.m.r. to the study of the various interactions of the imidazole ring of histidine in enzymes and other proteins has been reviewed concisely by Markley,[5] and recent progress in the use of ^{13}C enrichment of biopolymers for the elucidation of structure and biosynthetic pathways has been briefly surveyed by Séquin and Scott.[6] Reuben[7] has illustrated the usefulness of lanthanide ions for protein structure determination. As part of a general review of polymer analysis Mitchell and Chiu[8] have summarized the applications of n.m.r.

---

[1] F. A. Bovey, Proceedings of the Conference on the Critical Evaluation of Chemical and Physical Structural Information, 1973-74, p. 248.
[2] F. A. Bovey, *J. Polymer Sci., Macromol. Rev.*, 1974, **9**, 1.
[3] K. Wüthrich, *Pure Appl. Chem.*, 1974, **40**, 127.
[4] P. J. Quilley and G. A. Webb, *Co-ord. Chem. Rev.*, 1974, **12**, 407.
[5] J. L. Markley, *Accounts Chem. Res.*, 1975, **8**, 70.
[6] U. Séquin and A. I. Scott, *Science*, 1974, **186**, 101.
[7] J. Reuben, *Naturwiss.*, 1975, **62**, 172.
[8] J. Mitchell, jun. and J. Chiu, *Analyt. Chem.*, 1975, **47**, 289R.

to polymer analysis published between November 1972 and November 1974. A useful short review of some recent theoretical and experimental developments in the investigation of solid polymers has been reported by McBrierty,[9] who, with McDonald, has also published a detailed survey[10] of the theoretical and experimental results concerning linear and branched polyethylene.

3 Synthetic Macromolecules

A. ¹H High-resolution N.M.R.—Papers in this section are grouped into studies on the analysis of polymers and other reports.

Analysis. One of the most important and useful applications of high-resolution techniques is in the determination of the structure of polymers and the distribution of various units within the polymer. Table 1 summarizes references to the analyses of various polymers, together with appropriate remarks.

Table 1

Polymer	Comments	Ref.
Poly(vinyl chloride)	Detailed theoretical predictions of dyad, triad, and tetrad structures compared with experiment.	11
Poly(vinyl chloride)	Decoupled α-proton resonance showed all ten possible pentads; differential enthalpy and entropy changes for tactic placements calculated.	12
Poly(vinyl chloride) and polyethylene	Degree of branching determined from spectral analysis, with aid of model compounds.	13
Polypropylene	Dyad and triad tacticities determined from 100 MHz spectra. Assignments made with aid of pure isotactic and syndiotactic polymers.	14
Polybutadiene (Equibinary)	Decoupled 100 MHz spectra interpreted as indicating that distribution of *cis* and *trans* units is almost random.	15
Poly(2,3-dimethyl-1,3-butadiene), prepared using butyl-lithium; polar solvent	220 MHz spectra interpreted in terms of triads; Bernoullian statistics obeyed for low-temperature polymerizations.	16
Styrene oligomers	Spectral assignments made of fractionated materials—end-group effects noted.	17
Poly(*p*-isopropyl-α-methylstyrene)	Methine, α-methyl, and β-methylene resonances of two samples resolved to give tacticities.	18

---

[9] V. J. McBrierty, *Polymer*, 1974, **15**, 503.
[10] V. J. McBrierty and I. R. McDonald, *Polymer*, 1975, **16**, 125.
[11] I. Ando, A. Nishioka, and T. Asakura, *Makromol. Chem.*, 1975, **176**, 411.
[12] A. H. Abdel-Alim, *J. Appl. Polymer. Sci.*, 1975, **19**, 1227.
[13] V. E. Bezdadea, D. Braun, E. Buruianǎ, A. Caraculacu, and G. Istrate-Robilǎ, *Angew. Makromol. Chem.*, 1974, **37**, 35.
[14] K. Mitani, *J. Macromol. Sci., Chem.*, 1974, **A8**, 1033.
[15] K. Hatada, Y. Terawaki, H. Okuda, Y. Tanaka, and H. Sato, *J. Polymer Sci., Polymer Letters*, 1974, **12**, 305.
[16] N. Roy and J. Prud'homme, *Macromolecules*, 1975, **8**, 78.
[17] S. Fujishige and N. Ohguri, *Makromol. Chem.*, 1975, **176**, 233.
[18] J. Leonard and S. L. Malhotra, *J. Polymer. Sci., Polymer Chem.*, 1974, **12**, 2391.

Polymer	Comments	Ref.
Poly(methyl acrylate)	With the aid of deuteriated analogues, the methylene resonance of 300 MHz spectra was interpreted in terms of tetrads.	19
Poly(p-biphenyl acrylate)	With the aid of spectral assignments from stereoregular samples, radical-initiated polymer was shown to be atactic.	20
Polyesters	Spectral assignments of many different polymers made with aid of model compounds.	21
Poly(methyl vinyl ketone), poly(phenyl vinyl ketone)	Isotactic and syndiotactic polymers used to assist assignments of spectra from atactic samples; dyad patterns of β-methylene resonances distinguished.	22
Poly(allyl alcohol)	Methylene resonances assigned to different tacticities, though no spectra shown.	23
Poly(vinyl alcohol)	Degree of short chain branching estimated by comparison with model polymers.	24
Poly(vinyl alcohol)	220 MHz spectral analysis revealed 1,2-glycol content (from head-to-head addition); results in agreement with $NaIO_4$ titration.	25
Poly(4-vinyl-N-phenacyloxime pyridinium bromide)	Claimed that methylene protons of oxime side-group exchange with D_2O solvent.	26
Aromatic polyamides	Simple spectral analyses of eight different polymers; addition of LiCl induced chemical shifts and facilitated assignments.	27
Poly(cyclic formals)	Spectral analysis of polymerized 1,3-dioxan and triethyloxonium tetrafluoroborate indicated propagating species is oxycarbenium ion rather than cyclic trialkyloxonium ion.	28
Poly(propylene sulphide)	With aid of optically active deuteriated compounds, 300 MHz methylene resonance analysed in terms of triads.	29
Polyurethanes	LiCl dissolved in NN-dimethylacetamide used to separate urethane and urea resonances, thus facilitating polymer analysis.	30
Polyazetidine	Analysis of reaction intermediate indicated dimer to be polymerizing unit.	31

[19] T. Suzuki, E. R. Santee, jun., H. J. Harwood, O. Vogl, and T. Tanaka, *J. Polymer Sci., Polymer Letters*, 1974, **12**, 635.
[20] G. Ceccarelli, V. Frosini, P. L. Magagnini, and B. A. Newman, *J. Polymer Sci., Polymer Letters*, 1975, **13**, 101.
[21] H. Tanaka and Y. Mukoyama, *Nippon Kagaku Kaishi*, 1974, **6**, 1083.
[22] L. Merle-Aubry, Y. Merle, and E. Selegny, *Makromol. Chem.*, 1975, **176**, 709.
[23] H. Girard and P. Monjol, *Compt. rend.* 1974, **279**, C, 553.
[24] Y. Morishima, H. Iimuro, Y. Irie, and S. Nozakura, *J. Polymer Sci., Polymer Letters*, 1975, **13**, 157.
[25] R. L. Adelman and R. C. Ferguson, *J. Polymer Chem.*, 1975, **13**, 891.
[26] Yu. E. Kirsh, A. A. Rahnanskaya, G. M. Lukovkin, and V. A. Kabanov, *European Polymer J.*, 1974, **10**, 393.
[27] L. H. Chan, P. Blais, D. J. Carlsson, T. Suprunchuk, and D. M. Wiles, *J. Appl. Polymer Sci.*, 1975, **19**, 69.
[28] Y. Yokoyama, M. Okada, and H. Sumitomo, *Makromol. Chem.*, 1975, **176**, 795.
[29] M. Sepulchre, N. Spassky, D. van Ooteghem, and E. J. Goethals, *J. Polymer Sci., Polymer Chem.*, 1974, **12**, 1683.
[30] Y. Chokki, *Makromol. Chem.*, 175, 3425.
[31] E. H. Schacht and E. J. Goethals, *Makromol. Chem.*, 1974, **175**, 3447.

Polymer	Comments	Ref.
Polyimidazoles	Spectra of polymers prepared at 195 and 273 K showed that lower temperatures produced more syndiotactic polymers.	32
Phenol–formaldehyde resins	Degrees of polymerization calculated from comparisons of intensities of methyl and aromatic protons of acetylated polymers.	33
Polyether–polyol condensates	Alkyl resonances used to determine percentage of primary hydroxyls as end groups.	34
Copolymers of methyl methacrylate (MMA) and methacrylic acid	Extensive analysis of ^{1}H and ^{13}C spectra of syndio-, iso- and a-tactic copolymers in terms of triads; deuteriated compounds used to assist with assignments.	35
Copolymer of MMA and methyl acrylate	α-Methyl resonance analysed in terms of triads.	36
Copolymer of MMA and vinyl chloride (VC)	In VC-rich copolymers spectral analysis indicated isolated MMA units were co-isotactic; confirmation of sequence distribution attempted by examining lactonized samples.	37
Copolymers of chloroprene and MMA	Methoxy resonance split by Eu(tmhd)$_3$ and then interpreted in terms of pentad structures.	38
Copolymer of MMA and butadiene	Methoxy peaks resolved into triad and pentad fractions.	39
Copolymers of MMA and butadiene; MMA and isoprene	Sequence distribution of monomers in both copolymers calculated from α-methyl resonance, showed reasonable agreement with first-order Markov chain model.	40
Copolymer of MMA and acrylonitrile	Olefinic and methine resonances assigned to triad structures; polymers mainly alternating.	41
Copolymer of styrene and acrylonitrile	Attributed change in splitting of aromatic resonance as a function of mole ratios of monomers to different triad structures.	42

Other Studies. In this section, papers dealing mainly with molecular motion or slightly unusual methods of characterizing polymeric systems are considered. Spěváček and Schneider published two papers on complexes formed by poly(methyl methacrylates) of different tacticities in solution. In the first,[43] spectral intensities of isotactic and syndiotactic polymers indicated that the former were in complexes

---

[32] J. B. Lando, M. Litt, N. G. Kumar, and T. M. Shimko, *J. Polymer Sci., Symposia*, 1974, No. 44, 203.
[33] P. R. Steiner, *J. Appl. Polymer Sci.*, 1975, **19**, 215.
[34] Y. Chokki, M. Nakabayashi, K. Kodama, and M. Sumi, *Nippon Kagaku Kaishi*, 1974, 1662.
[35] E. Klesper, A. Johnsen, W. Gronski, and F. W. Wehrli, *Makromol. Chem.*, 1975, **176**, 1071.
[36] Y. Mori, A. Yeda, H. Tanzawa, K. Matsuzaki, and H. Kobayashi, *Makromol. Chem.*, 1975, **176**, 699.
[37] Q.-T. Pham and J. Guillot, *European Polymer J.*, 1974, **10**, 1043.
[38] T. Okada, M. Izuhara, and T. Hashimoto, *Polymer J.*, 1975, **7**, 1.
[39] J. R. Ebdon, *J. Macromol. Sci., Chem.*, 1974, **A8**, 417.
[40] T. Suzuki, K. Mitani, and Y. Takegami, *Polymer J.*, 1974, **6**, 496.
[41] G. A. Lindsay, E. R. Santee, jun., and H. J. Harwood, *Appl. Polymer. Symp.* 1974, **25**, 41.
[42] M. H. Mihailov, S. K. Dirlikov, Z. Georgieva, N. Peeva, and M. Panayotova, *Doklady Bulg. Akad. Nauk*, 1974, **27**, 1227.
[43] J. Spěváček and B. Schneider, *J. Polymer Sci., Polymer Letters*, 1974, **12**, 349.

(aggregates) whose mobility was too slow to contribute to the spectra. For CCl_4, the amount was ca. 20%, showing that care was necessary in stereoregularity determinations. In the second paper[44] it was shown that mixtures of 1% isotactic and syndiotactic polymers formed complexes, and the ratio of isotactic to syndiotactic units which were associated were calculated from intensity measurements. Cohen-Addad and Faure[45] have shown that normal sample rotation can reduce the linewidth of proton resonances of concentrated solutions of cis-1,4-polybutadiene in CS_2 from ca. 60 to ca. 30 Hz. This effect, much larger than field inhomogeneities, was attributed to non-zero average dipolar coupling, which occurs under certain conditions of anisotropic segmental motion. The effects of temperature and solvent on the linewidth were explained by the normal free-volume model. By examining the coupling constants obtained[46] from ^{13}C satellites in ^1H spectra of poly(ethylene oxide) and model compounds in various solvents and at different temperatures, the relative populations of the *trans* and *gauche* forms of the polymer were calculated. The *gauche* form was more stable than the *trans* by ca. 1.50—2.10 kJ mol^{-1}. The cationic polymerization of tetrahydrofuran[47] can produce two isomeric forms (a macro-ion and a macro-ester) of the active chain ends depending on the solvent used. The alkyl resonances, recorded at 300 MHz between 1 and 5 p.p.m. to high frequency of TMS, indicated that in CCl_4 the macro-ester predominates, in CH_2Cl_2 both species exist at comparable concentrations, and for CH_3NO_2 only 8% of the ester is found. Nishioka and Shimazaki have measured[48] the relaxation times of methyl and methylene protons of poly(isobutylene) in various solvents and have calculated their activation energies from their temperature dependence. Both ^1H and ^{13}C relaxation times (T_1 and T_2) were measured[49] for poly(oxymethylene) in the viscous solvent, hexafluoropropan-2-ol. The fast local segmental jumps and slower diffusive co-operative motions were separated and both appeared to be linearly dependent on the solvent viscosity. The temperature dependence of T_1 measured[50] during the polycondensation of furfuryl alcohol allowed some activation energies for molecular motion to be calculated, though it provided little information on the polymerization process.

B. Broadline N.M.R.—Broadline studies have been used largely to estimate the proportions of the various phases present or to identify different molecular processes occurring in solids.

Mathematical expressions have been derived[51] for the resonance linewidth in polymers, after making allowance for the distribution of correlation times described by the Fuoss–Kirkwood, Cole–Cole, Cole–Davidson, and Gavrilyaka–Negami relations. A maximum in the resistance of polyisobutylene and polypropylene to mechanical rupture was noticed[52] at ca. 77 K, a temperature corresponding to a

---

[44] J. Spěváček and B. Schneider, *Makromol. Chem.*, 1974, **175**, 2939.
[45] J. P. Cohen-Addad and J. P. Faure, *J. Chem. Phys.*, 1974, **61**, 1571.
[46] K. Matsuzaki and H. Ito, *J. Polymer Sci., Polymer Phys.*, 1974, **12**, 2507.
[47] K. Matyjaszewski and S. Penczek, *J. Polymer Sci., Polymer Chem.*, 1974, **12**, 1905.
[48] A. Nishioka and K. Shimazaki, *Kobunshi Ronbunshu*, 1974, **31**, 571.
[49] G. Hermann and G. Weill, *Macromolecules*, 1975, **8**, 171.
[50] Z. Dudzik, S. Glowinkowski, Z. Pajak, and Z. Szuba, *Bull. Acad. polon. Sci., Sér. Sci. chim.*, 1974, **22**, 755.
[51] V. P. Grigor'ev and A. I. Maklakov, *Polymer Sci. (U.S.S.R.)*, 1973, **15**, 2922.
[52] G. M. Bartenev and I. S. Lyakhovich, *Polymer Sci. (U.S.S.R.)*, 1974, **16**, 619.

maximum in the second moment of the broadline specrum. A complex relation was described, between the creep or rupture properties of the polymer and the segmental motion of the polymers, as indicated by the n.m.r. data. The degree of crystallinity of linear and branched polyethylene was measured[53] as a function of temperature, and activation energies of ca. 8 kJ mol^{-1} were calculated for molecular motion in the low-temperature region.

The broadline spectra from several differently treated polyethylene samples were interpreted[54] by invoking four possible components, not all of which were present in each sample. The four components, in order of decreasing linewidth, were from crystalline phase or 'tight' tie molecules (α), grain boundaries or relaxed tie molecules (β_1), loose loops (β_2), and chain ends (γ). Isotropic and drawn high-density polyethylene of different molecular weight and thermal treatments have been studied.[55] From the angular dependence of the spectra it was concluded that there were three components present, a broad one arising from crystalline regions of the polymer, an intermediate one corresponding to high molecular weight tie molecules and a narrow isotropic one attributed to low molecular weight fractions or ends of molecules. After γ-irradiation at 77 K methacrylic acid (melting point 289 K) polymerizes[56] if warmed to 273 K. Immediately after bringing to 273 K a single broad line, of width ca. 13 kHz was observed, but after ca. 1.5 h an additional narrow line appeared. Polymerization in the solid phase presumably disrupted the monomeric crystal structure, allowing some rapid motion of monomer molecules below their melting point. The phase separation of PVC and ethylene–vinyl acetate copolymer with increasing temperature was described,[57] though few data are given. Spectra of the same copolymer, recorded[58] between 253 and 113 K showed four peaks, which were attributed to groups undergoing various possible relaxation processes in the polymer. With the aid of spectra recorded using different modulation amplitudes, Wangermann and Zachmann have identified[59] three separate components in solid polyamides. The components were associated with different phases of the solid, and the effects of temperature and drawing on the relative proportions of the phases discussed. The temperature dependence of the broadline spectra of a series of polymers containing poly(vinyl trichloroacetate) crosslinked with poly(methyl methacrylate) showed[60] that the onset of rotation of the α-methyl group of PMMA occurred at different temperatures depending on whether there was significant microphase (of the order of μm size) separation or not. Where this separation occurred, α-methyl rotation started at about the same temperature as for pure PMMA, whereas for those samples where little separation took place it started at slightly lower temperatures. The second moment of mixtures of rubber and fluoropolymers indicated[61] that in solution the polymers remained as separate species, whereas during vulcanization cross-linking took place.

[53] S. R. Ahamad and A. Charlesby, *European Polymer J.*, 1975, **11**, 91.
[54] O. Phaovibul, J. Loboda-Čačković, H. Čačković, and R. Hosemann, *Makromol. Chem.*, 1974, **175**, 2991.
[55] J. B. Smith, A. J. Manuel, and I. M. Ward, *Polymer*, 1976, **16**, 57.
[56] C. Chachaty, M. Latimer, and A. Forchioni, *J. Polymer Sci., Polymer Chem.*, 1975, **13**, 189.
[57] C. Elmqvist and S. E. Svanson, *Colloid and Polymer Sci.*, 1975, **253**, 327.
[58] J. Sobottka, F. Keller, and K. Wunderlich, *Faserforsch. Textiltechnik.*, 1974, **25**, 352.
[59] K. Wangermann and H. G. Zachmann, *Progr. Colloid Polymer Sci.*, 1975, **57**, 236.
[60] C. H. Bamford, G. C. Eastmond, and D. Whittle, *Polymer*, 1975, **16**, 377.
[61] A. S. Shashkov and V. N. Fomin, *Polymer Sci. (U.S.S.R.)*, 1973, **15**, 3143.

C. Spin–echo N.M.R.—Some interesting line-narrowing experiments have been reported, together with studies of the effects of small molecules on glass transition temperatures and molecular motion in polymers.

Dybowski and Vaughan[62] have carried out a multiple-pulse experiment on samples of *cis*-polyisoprene. An eight-pulse cycle was able to reduce the normal linewidth from 340 to 100 Hz, and this was followed by another eight-pulse cycle, containing a phase shift within the cycle. The effect of the phase shift is to average any small off-resonance Hamiltonians and thus to remove chemical shift and magnetic field inhomogeneity effects. Together with the normal T_1 measurements, it was possible to separate the contribution to the proton spectrum of the dipolar Hamiltonian, chemical shift Hamiltonian, and homogeneous broadening due to relaxation processes in the polymer samples. Pulsed, broadline, and high-resolution experiments of concentrated mixtures (10–15% w/v) of isotactic and syndiotactic poly(methyl methacrylates) indicated[63] the presence of three types of segments of different mobility, having T_2 values of 36×10^{-3} s, 0.9×10^{-3} s, and 15×10^{-6} s. By spinning the samples at the magic angle and comparing the intensities of α-Me plus CH_2 resonances with those of the $-OMe$, it was concluded that the mobilities of the ester groups were greatly hindered.

From the temperature dependence of the proton T_1 and T_2 values of poly(dimethyl siloxane), Assink[64] showed that gases such as argon and nitrogen, which under pressure dissolved in the polymer, could change the glass transition temperature, in a similar way to normal solvents. In a similar experiment, from a comparison of the polymer proton T_1 and T_2 values of polystyrene powder and latex, Smith and Robb[65] concluded that water lowered its glass transition temperature by *ca.* 10 K, the effect being slightly greater in the presence of electrolyte.

In a short note[66] the T_1 values of solid poly(vinyl acetate) and poly(methacrylic acid) between 130 and 330 K were reported, though assignment of the minima observed to various torsional stretching motions seems doubtful. The T_1 values for pure poly(ethylene glycols) were measured[67] over the frequency region 3 kHz to 75 MHz. For low molecular weight (≤ 400) samples a single dispersion was observed and attributed to translational motion, whereas for higher molecular weights an additional dispersion was observed, associated with local segmental motion. Both T_1 and T_2 of poly(ethylene oxide) in aqueous urea solutions showed[68] only small gradual changes up to 10M urea, with no indication of any sharp transition in the conformation of the polymer.

D. ^{13}C N.M.R.—Papers dealing with ^{13}C n.m.r. studies are grouped into sections on structural analysis and other studies.

Analysis and Structure Determination. ^{13}C n.m.r. has continued to be used to resolve structures that are mainly inaccessible to proton studies (see Table 2).

---

[62] C. R. Dybowski and R. W. Vaughan, *Macromolecules*, 1975, **8**, 50.
[63] J. Spěváček and B. Schneider, *Makromol. Chem.*, 1975, **176**, 729.
[64] R. A. Assink, *J. Polymer Sci., Polymer Phys.*, 1974, **12**, 2281.
[65] E. G. Smith and I. D. Robb, *Polymer*, 1974, **15**, 713.
[66] B. Fajt, D. Pumpernik, F. Jagodic, M. Penko, and A. Ažman, *Colloid and Polymer Sci.*, 1974, **252**, 997.
[67] G. Preissing and F. Noack, *Progr. Colloid Polymer Sci.*, 1975, **57**, 216.
[68] A. A. Jones and W. H. Stockmayer, *J. Phys. Chem.*, 1974, **78**, 1528.

Table 2

Polymer	Comments	Ref.
Polybutadiene	Improved Grant and Paul parameters obtained by noting the 'corrective terms' can be temperature sensitive.	69
Polybutadiene	Spectral analysis, with aid of hydrogenated analogues, showed random distribution of cis-1,4, trans-1,4, and 1,2 units in polymer.	70
Polybutadiene	Comparison with model compounds allowed assignment of aliphatic resonances in accord with that of Clague et al.[72]	71
Polybutadiene	Simple spectral assignment analysis giving cis and trans contents; results similar to earlier reports.	73
1,2-Polybutadiene	67.88 MHz spectra analysed in terms of pentads for various temperatures of preparation.	74
Polybutadiene (cis–trans equibinary and cis–trans isomerized samples)	Using low molecular weight samples, the authors were able to distinguish cis and trans resonances in olefinic region.	75
cis-1,4-, trans-1,4-, and 1,2-Polybutadienes	Resonances of title compounds in region ca. 130 p.p.m. to high frequency OMTS assigned.	76
cis-1,4-1,2-Polybutadiene	67.88 MHz spectra assigned and the influence of 1,2 units on cis-1,4 resonances estimated.	77
Atactic polystyrene	Methylene resonances interpreted in terms of tetrad structures; ring current effects on methylene resonances limited to adjacent phenyl substituents.	78
Poly(vinyl acetate)	^1H and ^{13}C spectra of samples of known stereoregularity interpreted in terms of tetrads and pentads.	79
Poly(isopropenyl acetate)	Spectral assignments made; use of carboxyl resonance showed the polymer to be largely atactic.	80
Polymethacrylonitriles	In acetone, α-methyl resonances resolved into triads and pentads, methylene into tetrads. In TFA α-methyl resonance resolved into pentads, and cyano resonances into triads.	81
Poly(n-alkyl laurolactams)	Both ^1H and ^{13}C spectra showed syn and anti conformations present at 303 K, with rapid exchange between the two at 353 K.	82

[69] J. C. Randall, *J. Polymer Sci., Polymer Phys.*, 1975, **13**, 901.
[70] A. D. H. Claque, J. A. M. van Broekhoven, and L. P. Blaauw, *Macromolecules*, 1974, **7**, 348.
[71] M. Deneke and H. C. Broecker, *Makromol. Chem.*, 1975, **176**, 1471.
[72] A. D. H. Clague, J. A. M. van Broekhoven, and J. W. de Haan, *J. Polymer Sci., Polymer Letters*, 1973, **11**, 305.
[73] W. Hoffmann, P. Kuzay, and W. Kimmer, *Plaste Kaute.*, 1974, **21**, 423.
[74] K.-F. Elgert, G. Quack, and B. Stützel, *Makromol. Chem.*, 1974, **175**, 1955.
[75] Y. Tanaka, H. Sato, M. Ogawa, K. Hatada, and Y. Terawaki, *J. Polymer Sci., Polymer Letters*, 1974, **12**, 369.
[76] K.-F. Elgert, G. Quack, and B. Stützel, *Polymer*, 1975, **16**, 154.
[77] K.-F. Elgert, G. Quack, and B. Stützel, *Makromol. Chem.*, 1975, **176**, 759.
[78] J. C. Randall, *J. Polymer Sci., Polymer Phys.*, 1975, **13**, 889.
[79] T. K. Wu and D. W. Ovenall, *Macromolecules*, 1974, **7**, 776.
[80] K. Matsuzaki, T. Kawamura, and K. Saito, *J. Polymer Sci., Polymer Chem.*, 1975, **13**, 253.
[81] K. Matsuzaki, T. Kanai, and S. Matsumoto, *J. Polymer Sci., Polymer Chem.*, 1974, **12**, 2377.
[82] S. W. Shalaby, G. E. Babbitt, and R. L. Lapinski, *Spectroscopy Letters*, 1974, **7**, 203.

Macromolecules

Polymer	Comments	Ref.
Copolyamides; repeat units [aminoacylamino-acid]	Carbonyl resonances analysed to give sequence distributions.	83
Copolymer of propylene and butadiene	Fractions of *cis* and *trans* segments calculated from spectra of normal and hydrogenated copolymers.	84
Copolymers of butadiene and styrene	Detailed spectral analysis in terms of triads; both *trans*-1,4-, *cis*-1,4-, and 1,2-butadiene present.	85
Copolymers of α-methyl styrene and acrylonitrile	^{13}C and ^1H spectra indicated alternating structure with random steric placements.	86
Copolymers of ethylene sulphide and propylene sulphide	Spectral analysis in terms of dyad and triad sequences.	87
Copolymers of 1,3,5-trioxan and 1,3-dioxolan	Pentad sequences observed for both methylene oxide and ethylene oxide resonances. ^1H spectra showed triad sequences for methylene oxide, but no sequence dependence for ethylene oxide resonances.	88
Copolymers of ethylene and vinyl acetate	Methine and methylene resonances assigned to triad structures by comparison with model compounds.	89
Copolymer of ethylene and vinyl acetate	^1H and ^{13}C spectra allowed sequence distribution to be determined from methine and methylene resonances.	90

^{13}C: *Other Studies.* In a short note, the high-resolution spectra of solid random and block styrene–butadiene copolymers were presented[91] which indicated that in the block copolymers the polystyrene was confined to domains having low segmental mobility. Spectra of acrylic monomers complexed with Lewis acids were measured,[92] the β-carbon of acrylonitrile and methacrylonitrile showing a high-frequency shift, the α-carbon showing a low-frequency shift, and the nitrile showed no significant shift. These results were discussed in terms of electron distribution on the carbons and of possible alternating polymerization mechanisms.

E. Other Nuclei.—This section is mainly concerned with ^{19}F studies but also includes ^{31}P and ^{29}Si work.

Both ^{19}F and ^1H resonances have been used[93] to analyse *p*-substituted α-methyl styrene polymers. The α-methyl and ^{19}F resonances were interpreted in terms of triad structures, the ^{19}F spectra providing better resolution of the triad patterns. A note[94] reporting ^{19}F chemical shifts of fluorinated acids and esters used to catalyse the polymerization of THF in CCl_4 or CH_3NO_2, discussed whether the macro-ion

[83] H. R. Kricheldorf, E. Leppert, and G. Schilling, *Makromol. Chem.*, 1975, **176**, 81.
[84] C. J. Carman, *Macromolecules*, 1974, **7**, 789.
[85] A. L. Segre, M. Delfini, F. Conti, and A. Boicelli, *Polymer*, 1975, **16**, 338.
[86] I. Kuntz and N. F. Chamberlain, *J. Polymer Sci., Polymer Chem.*, 1974, **12**, 1695.
[87] C. Corno, A. Roggero, and T. Salvatori, *European Polymer J.*, 1974, **10**, 525.
[88] D. Fleischer and R. C. Schulz, *Makromol. Chem.*, 1975, **176**, 677.
[89] W. Hoffmann and F. Keller, *Plaste Kaut.*, 1974, **21**, 359.
[90] F. Keller, *Plaste Kaut.*, 1975, **22**, 8.
[91] F. Conti, M. Delfini, and A. L. Segre, *Polymer*, 1974, **15**, 539.
[92] J. Furukawa, E. Kobayashi, S. Nagata, and T. Moritani, *J. Polymer Sci., Polymer Chem.*, 1974, **12**, 1799.
[93] R. W. Lenz, W. Regel, and L. Westfelt, *Makromol. Chem.*, 1975, **176**, 781.
[94] T. K. Wu and G. Pruckmayr, *Macromolecules*, 1975, **8**, 77.

or ester was the intermediate species. Fluorine attached to the macro-ion produced a resonance ca. 4 p.p.m. to low frequency of the ester, thus facilitating the identification of the active polymeric species. During the polymerization of THF initiated by PF_5, other fluorine-containing species such as PF_6^-, PF_5,THF, and (1) are pro-

$$PF_4[O(CH_2)_4]_{n-1}-\overset{+}{O}\bigcirc$$

(1)

duced. The ^{19}F spectra taken[95] at various times allowed the kinetics of the appearance or disappearance of these species to be measured; rate constants for the polymerization were in reasonable agreement with those determined by other methods. Spectra[96] from copolymers of chloro-trifluoroethylene and propylene showed that the polymer had an alternating structure. The CF_2 resonances were explained in terms of tetrad structures whereas the CFCl resonances corresponded to triads. Multiple pulse experiments[97] have been used for line-narrowing of the ^{19}F resonance from drawn polytetrafluoroethylene. The chemical shift was obtained as a function of orientation of the sample and showed that the most screened shift tensor component lay along the C—F bond whereas the least screened was aligned at ca. 20° to the chain axis. Broadline ^{19}F spectra were used[98] to follow the γ-ray damage induced in polytetrafluoroethylene. Increasing γ-ray dosage caused an unexpected broadening of the absorption line because a rearrangement of chains from an amorphous to a more crystalline state was facilitated. Kinetic relations and activation energies for the processes were calculated. The thermal decomposition of cyclo-di(phosphazanes) was followed[99] by the formation of polymeric phosphazane derivatives. ^{31}P Chemical shifts were used to assist in product identification. The noise-decoupled $^{29}Si-\{H\}$ spectra of a polymeric silicone showed[100] separate resonances for three units at each end and some splitting of resonances, possibly due to tacticity effects.

4 Biological and Related Macromolecules

A. ^1H N.M.R. of Synthetic Polypeptides.—In a brief note,[101] the T_1 of phenyl protons of poly(γ-benzyl-L-glutamates), measured by adiabatic rapid passage in mixtures of $CDCl_3$ and TFA, were presented. The correlation times of ca. 10^{-10} s indicated that side-chain motion rather than macromolecular tumbling was the dominant relaxation process. Imanishi et al. have investigated[102] the conformational changes of poly(N-methyl-L-alanine) in mixtures of CD_2Cl_2 and dichloroacetic acid, using the chemical shift of the NMe and C_α-Me protons to indicate cis or trans

---

[95] F. Andruzzi, A. Pescia, and G. Ceccarelli, *Makromol. Chem.*, 1975, **176**, 977.
[96] K. Ishigure, Y. Tabata, and K. Oshima, *Macromolecules*, 1975, **8**, 177.
[97] A. N. Garroway, D. C. Stalker, and P. Mansfield, *Polymer*, 1975, **16**, 161.
[98] F. D. Trexler and L. N. Mulay, *J. Magn. Resonance* 1974, **15**, 234.
[99] H.-G. Horn, *Z. anorg. Chem.*, 1974, **406**, 199.
[100] R. K. Harris and B. J. Kimber, *J.C.S. Chem. Comm.*, 1974, 559.
[101] E. T. Samulski, *J. Polymer Sci., Symposia*, 1974, No. 46, 335.
[102] Y. Imanishi, K. Kugimiya, and T. Higashimura, *Polymer*, 1975, **16**, 350.

structures. The polymer in CD_2Cl_2 consisted of nearly all *trans* amide bonds, and addition of dichloroacetic acid induced *trans–cis* isomerization. The solvent-induced left handed-right handed helix transition of poly(β-benzyl-L-aspartate) substituted in the *ortho* or *para* position with a nitro-group, was followed[103] by the shift of the α-CH resonance and the splitting of the β-CH$_2$ resonance. Both ^{1}H and ^{13}C n.m.r. spectra were used[104] to analyse the different conformations of poly(L-azetidine-2-carboxylic acid). Separate resonances were observed for *cis*- and *trans*-isomers, the proportion of the former decreasing as the solvent changed from water to formic acid. Goodman *et al.* have published two papers on the conformation of oligopeptides. In the first,[105] studies of alanine oligopeptides containing a methoxyethoxy-ethoxyacetyl blocking group on the *N*-terminal residue and a morpholino-blocking group on the *C*-terminal residue were reported for $CDCl_3$–TFA mixtures. For oligomers with more than five residues, the α-CH resonance showed the well known 'double peak' phenomenon which coalesced on addition of TFA, indicating that the double peak arose from specific folded forms which contained intramolecular hydrogen bonds. In a similar paper[106] an end-blocked alanine tetramer in $CDCl_3$ also showed a 'double peak' for the α-CH resonance. The chemical shifts, as a function of temperature and concentration, suggested the formation of a dimer.

Bradbury *et al.*, in two related papers, have reported the use in two ways of paramagnetic ions such as Gd^{3+} to help elucidate protein structure. Firstly,[107] binding of the Gd^{3+} at the ends of oligopeptides produced progressive broadening down the chain, allowing sequence determination to be made, and secondly,[108] the broadening of *N*-methyl resonances from N-methylated lysozyme with progressive addition of Gd^{3+} allowed methyl-Gd^{3+} distances to be calculated. The temperature and solvent-mixture dependence[109] of the peptide proton chemical shifts of linear and cyclic polymers consisting of Val-Pro-Gly$_1$-Gly$_2$ repeat units suggested a β-turn with Pro-Gly residues at the corners. The possibility was discussed of a similar β-turn occurring in elastin, which contains the tetrapeptide. The use of lanthanide shift reagents for the sequence determination of oligopeptides in aqueous solution has also been discussed.[110] Shifts of both end groups and backbone protons occurred, though in alkaline solutions line-broadening limited the usefulness of the chemical shifts in the structure determination.

B. ^{1}H N.M.R. of Proteins.—Nitroxide paramagnetic labels and difference spectra have been used[111] to identify the amino-acids around the combining site of the FV fragment of immunoglobulin A from the mouse myeloma protein, MOPC315. Similar pH difference spectra were used to obtain pK_a values for some histidine protons and, from water proton relaxation enhancement experiments, Gd^{3+} was

[103] M.-H. Loucheux-Lefebvre, A. Forchioni, and C. Duflot, *Polymer*, 1974, **15**, 474.
[104] R. Boni, R. Di Blasi, and A. S. Verdini, *Macromolecules* 1975, **8**, 140.
[105] M. Goodman, N. Ueyama, F. Naider, and C. Gilon, *Biopolymers*, 1975, **14**, 915.
[106] M. Goodman, N. Ueyama, and F. Naider, *Biopolymers*, 1975, **14**, 901.
[107] J. H. Bradbury, L. R. Brown, M. W. Crompton, and B. Warren, *Pure Appl. Chem.*, 1974, **40**, 83.
[108] J. H. Bradbury, M. W. Crompton, and B. Warren, *Analyt. Biochem.*, 1974, **62**, 310.
[109] D. W. Urry and T. Ohnishi, *Biopolymers*, 1974, **13**, 1223.
[110] M. Anteunis and R. Callens, *J. Magn. Resonance*, 1974, **15**, 317.
[111] R. A. Dwek, J.C.A. Knott, D. Marsh, A. C. McLaughlin, E. M. Press, N. C. Price, and A. I. White, *European J. Biochem.*, 1975, **53**, 25.

shown to bind to the protein. The pH dependence of the chemical shift of three high-frequency histidine resonances from bovine α-lactoglobulin allowed their pK_a values to be obtained.[112] Assignment of the resonances to particular histidine residues was achieved by using their selective reaction to iodoacetate in conjunction with a study of the carboxymethylation of α-N-acetyl-L-histidine. From intensity measurements[113] of the contact-shifted resonances of the haem protons (8—18 p.p.m. to high frequency of HDO), the binding of oxygen and CO to human haemoglobin in the presence of various phosphate complexes was studied. Some preferential binding to the α-chain rather than the β-chain was observed, though the differences often appear to be only just outside experimental error. Comparison of the chemical shifts from the deoxy and ligand-bound forms of haemoglobins A and Chesapeake have shown[114] that the tertiary structures of the α and β haem pockets in the two liganded proteins were equivalent, though there were differences between the α- and β-chains. The marked broadening, with decreasing temperature, of a methyl resonance from horse ferricyanomyoglobin has been attributed[115] to hindered rotation and an activation energy of 61.9 kJ mol^{-1} obtained. In a related paper,[116] the temperature dependence of the linewidth of the haeme methyl resonance of various myoglobins was used to calculate the activation energy for methyl rotation to be 61.9 kJ mol^{-1}. Addition of electrolytes narrowed the methyl resonance, the increased methyl mobility being thought to arise from perturbation of the bond between the haem propionate group and the Lys-45 of the polypeptide. The temperature-dependent chemical shifts[117] of the hyperfine shifted resonances (ca. 15—30 p.p.m. to high frequency of TMS) of sperm whale and horse heart myoglobin and human haemoglobin showed marked deviations from Curie's law predictions. This was shown to be due to the temperature-dependent equilibrium between the high-spin and low-spin states of the iron, causing the haeme resonances to move to either lower or higher frequencies with increasing temperature. In a paper[118] duplicating some of these results, the hyperfine shift and linewidth in the presence of different ligands increased as follows; cyanide < azide < imidazole < deuteroxide.

The spectra of ferro- and ferri-cytochrome C_2 from *Rhodospirillum rubrum* have been analysed,[119] and resonances up to 34 p.p.m. to high frequency of DSS have been assigned. A comparison[120] of the pH dependence of spectra from horse, tuna, and yeast cytochromes c allowed the pKa of the histidine residue 33 of horse cytochrome c to be determined and the assignment of other resonances to particular histidine residues to be made. The exchange rates of slowly exchanging protons on ferredoxin were estimated[121] by examining the ^{1}H spectra after various degrees of D$_2$O exchange with the protein had taken place. Resonances from ca. 27 to 28 protons appeared at 6 to 11 p.p.m. to high frequency of trisilyltetradeuteriopropionate, about 12 of these having half-times for exchange of the order of 10^7 min, suggesting

[112] J. H. Bradbury and R. S. Norton, *European J. Biochem.*, 1975, **53**, 387.
[113] M. E. Johnson and C. Ho, *Biochemistry*, 1974, **13**, 3653.
[114] K. J. Weichelman, S. Charache, and C. Ho, *Biochemistry*, 1974, **13**, 4772.
[115] I. Morishima and T. Iizuka, *J. Amer. Chem. Soc.*, 1974, **96**, 7365.
[116] I. Morishima and T. Iizuka, *Biochim. Biophys. Acta*, 1975, **386**, 542.
[117] I. Morishima and T. Iizuka, *J. Amer. Chem. Soc.*, 1974, **96**, 5279.
[118] T. Iizuka and I. Morishima, *Biochim. Biophys. Acta*, 1974, **371**, 1.
[119] G. M. Smith and M. D. Kamen, *Proc. Nat. Acad. Sci. U.S.A*, 1974, **71**, 4303.
[120] J. S. Cohen and M. B. Hayes, *J. Biol. Chem.*, 1974, **249**, 5472.
[121] H. L. Crespi, A. G. Kostka, and U. H. Smith, *Biochem. Biophys. Res. Comm.*, 1974, **61**, 1407.

the presence of β structure in the proteins. The T_1 and T_2 values of the methyl protons of tetra-p-tolylporphyrin ferric (or chromium) chloride allowed[122] the calculation of correlation times to be made, assuming that relaxation was primarily due to the nuclear–electron dipolar mechanism. For the chromium compound, τ_c was ca. 4×10^{-10} s, suggesting that rotation of the molecule was the main relaxing process whereas for the iron compound, τ_c was ca. 2.5×10^{-11}s, suggesting the electron spin was dominating the relaxation. The non-equivalent sides of the high-spin ferric porphyrins were averaged by a mechanism whereby the iron moves from side to side through the porphyrin 'hole'. Pekary et al. have used[123] high-resolution spectra to obtain information on the solution conformation of histone IV and its fragments. For the intact histone IV, the C-terminal portion had broad resonances, indicating secondary structures, and aggregation of the protein was induced at salt concentrations above about 0.01 mol l^{-1}. In a similar work,[124] two histone IV fragments obtained by cleaving at Met-84 with cyanogen bromide have been examined as a function of temperature, peptide concentration, ionic strength and pD. From the intensities of resonances the N-peptides were concluded to be in extended coils, which aggregated in electrolyte, whereas the C-peptides possessed extensive secondary structure. A detailed analysis[125] of the influence of salts and pH on the chemical shifts of calf thymus histone H1, in the region 0—5 p.p.m. to high frequency of DSS, suggested the existence of a folded structure, stabilized by ionic, hydrogen, and hydrophobic bonds. As the linewidths for the folded form were comparable to the random coil, it was assumed that aggregation did not occur, and no evidence for β structures was noticed. Histone 2A was investigated[126] at different electrolyte concentrations between pH 3 and 7. From simple analysis of the spectra, combined with i.r. and c.d. data, it was concluded that addition of salt induced α-helix formation and intermolecular association in the region of the molecule between residues 25 and 113.

C. ^{1}H N.M.R. of Enzymes.—Patel et al. have published three related papers on the assignment and analysis of resonances from ribonuclease. In the first,[127] chemical derivatives were used to aid assignment of an exchangeable resonance from RNase A occurring at 11—13 p.p.m. to high frequency from DSS. Carboxymethylation was carried out at His-119 and His-19, and the pH dependence of these spectra allowed the assignment of the exchangeable resonance to be made to the imidazole N-1 proton of His-12, which in the crystal is hydrogen-bonded to the carbonyl of Thr-45. In the second[128] the ionization of the His12 N-1 proton was compared with that of the four histidine C-2 protons. The influence of cytidine monophosphates, copper(II), AgNO$_3$, and spin-labels on the resonances allowed a correlation to be made of the exchangeable His-12 N-1 proton with the active-site histidine C-2 proton

[122] G. N. La Mer, Pure Appl. Chem., 1974, 40, 13.
[123] A. E. Pekary, H.-J. Li, S. I. Chan, C.-J. Hsu, and T. E. Wagner, Biochemistry, 1975, 14, 1177.
[124] A. E. Pekary, S. I. Chan, C.-J. Hsu, and T. E. Wagner, Biochemistry, 1975, 14, 1184.
[125] E. M. Bradbury, P. D. Cary, G. E. Chapman, C. Crane-Robinson, S. E. Danby, H. W. E. Rattle, M. Boublik, J. Palau, and F. J. Aviles, European J. Biochem. 1975, 52, 605.
[126] E. M. Bradbury, P. D. Cary, C. Crane-Robinson, H. W. E. Rattle, M. Boublik, and P. Sautière, Biochemistry, 1975, 14, 1876.
[127] D. J. Patel, L. L. Canuel, C. Woodward, and F. A. Bovey, Biopolymers, 1975, 14, 959.
[128] D. J. Patel, C. Woodward, L. L. Canuel, and F. A. Bovey, Biopolymers, 1975, 14, 975.

having the lower pK_a. Further confirmation[129] of the assignment of this exchangeable resonance with the lower pK_a to His-12 rather than His-119 came from selective deuteriation of C-2H of His-12 in RNase S'. This enzyme was regenerated from the S peptide [His-12 C-2(D)], and comparison of the spectra of the deuteriated and undeuteriated samples permitted the above assignment. In two papers[130, 131] Benz and Roberts have reported studies of the unfolding of bovine pancreatic ribonuclease by careful examination of the effects of pH, temperature, and denaturants on the aromatic region of the high-resolution spectrum of the enzyme. The chemical shifts and intensities of the proton resonances from the imidazole groups of the histidine residues provided indications as to whether the enzyme was in a folded or unfolded state, thus allowing the effect of various experimental conditions on the folding process to be estimated. Denaturation by urea or thermally at low pH produced similar behaviour of the histidine groups, and the active site of the protein containing His-12 and His-119 appeared to be disrupted at an early stage in the unfolding. pK Values of methylated amine groups of RNase-A were calculated[132] from the pH dependence of the chemical shift of their resonances. Corrections for the effect of methylation were made, and deviations in pK values from simple titration data suggested interactions between specific residues. In the presence of phosphates, titration curves for the H-2 proton resonances of His-12 and His-119 of methylated RNase-A indicated binding of phosphate at the active site. The addition of $PtCl_4^{2-}$ to ribonuclease at pH 1.5 caused[133] broadening of the proton resonances from the methionine S-methyl group, indicating interaction between the electrolyte and enzyme. At pH 5.5, where the enzyme is more folded, similar but much weaker effects were noted.

A resonance in the spectrum[134] of human carbonic anhydrase occurring at 15 p.p.m. to high frequency of TMS had a pK_a of 7.3 and was chemically shifted by ligands which bound at the active site of the enzyme. The resonance was assigned to a histidine imidazole ring N—H proton near the active site. Nine resonances from human carbonic anhydrase B occurring between 6 and 9.5 p.p.m. to high frequency of DSS were investigated[135] as a function of pH and in the presence of inhibitors. Tentative assignments of these resonances to particular histidine residues were made and linewidth measurements provided information on solvent exchange with these sites. Pesando *et al.* have published three papers discussing the use of high-resolution techniques to investigate the structure and reactions of human carbonic anhydrase. In the first,[136] seven discrete resonances between 7 and 9 p.p.m. to high frequency of DSS were tentatively assigned to protons on histidine residues, and from their pH dependence five were assigned pK_a values between 6.0 and 8.24. In the second paper,[137] addition of ligands or inhibitors and change in pH were shown to shift some resonances to high frequency and to leave unchanged two that were

[129] D. J. Patel, L. L. Canuel, and F. A. Bovey, *Biopolymers*, 1975, **14**, 987.
[130] F. W. Benz and G. C. K. Roberts, *J. Mol. Biol.*, 1975, **91**, 345.
[131] F. W. Benz and G. C. K. Roberts, *J. Mol. Biol.*, 1975, **91**, 367.
[132] L. R. Brown and J. H. Bradbury, *European J. Biochem.*, 1975, **54**, 219.
[133] P. J. Sadler, F. W. Benz, and G. C. K. Roberts, *Biochim. Biophys. Acta*, 1974, **359**, 13.
[134] R. K. Gupta and J. M. Pesando, *J. Biol. Chem.*, 1975, **250**, 2630.
[135] I. D. Campbell, S. Lindskog, and A. I. White, *J. Mol. Biol.*, 1974, **90**, 469.
[136] J. M. Pesando, *Biochemistry*, 1975, **14**, 675.
[137] J. M. Pesando, *Biochemistry*, 1975, **14**, 681.

located at the active site. A model for the catalytic mechanism of the enzyme was proposed. Finally, the binding of sulphonamide inhibitors has been studied[138] by difference spectra. The enzyme spectra with and without the inhibitors suggested a conformational change, though the large number of unaltered resonances indicated the involvement of only a few residues. The 220 MHz spectra of dihydrofolate reductase, a low molecular weight enzyme, consisted[139] of several distinct resonances, particularly in the aromatic region, and these were considerably narrowed by denaturation in 6M urea. Addition of ligands such as dihydrofolate to the apoenzyme converted its relatively featureless aromatic region into one indistinguishable from the original enzyme-dihydrofolate binary complex, and different forms of the enzyme, separated by affinity chromatography were distinguished by their resonances in the aromatic region. The 220 MHz spectrum[140] of staphylococcal protease was found to resemble the sum of the spectra of its component aminoacids and a large number of slowly exchanging protons was noted, whose resonance occurred in the aromatic region. The pK values of three histidine residues were found to be 7.19, 6.85, and 6.69, and the pH titrations were best explained if it was assumed that two histidines interacted together.

Campbell et al.[141] have shown how the addition of varying amounts of Gd^{3+} to lysozyme can be used to increase the enzyme's spectral resolution. Resonances from groups close to the bound Gd^{3+} were broadened through interaction with the free spins of Gd^{3+}, and subtraction of the broadened spectrum from the normal spectrum provided better resolved spectra.

D. **^{1}H N.M.R. of Natural and Related Polypeptides and Polysaccharides.**—The resonances in the spectra from the cyclic decapeptide mono-N-methyl-leucine gramicidin S have been assigned[142] to specific protons by making use of their temperature- and solvent-dependent chemical shifts. These assignments, together with (HNC$^\alpha$H) coupling constants, enabled a conformation of the molecule to be derived. Assignments and partial conformational analysis of the 220 MHz spectrum of arginine vasopressin and arginine vasotocin in deuteriated DMSO have been reported.[143] From the chemical shifts, coupling constants, and proton–deuterium exchange rates, it was concluded that their backbone structures were similar and also resembled the related hormones lysine vasopressin and oxytocin. In a short paper,[144] an unequivocal assignment of NH resonances was made of spectra of oxytocin, using selectively enriched ^{2}H and ^{15}N derivatives. The high-resolution spectra of fragments produced by digesting native collagen with cyanogen bromide showed[145]

[138] J. M. Pesando and A. P. Grollman, *Biochemistry*, 1975, **14**, 689.
[139] E. J. Pastore, R. L. Kisliuk, L. T. Plaute, J. M. Wright, and N. O. Kaplan, *Proc. Nat. Acad. Sci. U.S.A.*, 1974, **71**, 3849.
[140] J. L. Markley, W. R. Finkenstadt, H. Dugas, P. Leduc, and G. R. Drapeau, *Biochemistry*, 1975, **14**, 998.
[141] I. D. Campbell, C. M. Dobson, R. J. P. Williams, and A. V. Xavier, Proceedings of the Lysozyme Conference, 1974, p. 219.
[142] N. G. Kumar, N. Izumiya, M. Miyoshi, H. Suganok, and D. W. Urry, *Biochemistry*, 1975, **14**, 2197.
[143] R. Walter, A. Ballardin, I. L. Schwartz, W. A. Gibbons, and H. R. Wyssbrod, *Proc. Nat. Acad. Sci. U.S.A.*, 1974, **71**, 4528.
[144] A. F. Bradbury, A. S. V. Burgen, J. Feeney, G. C. K. Roberts, and D. G. Smyth, *F.E.B.S. Letters*, 1974, **42**, 179.
[145] B. D. Batts, R. A. Brown, P. Mason, and A. R. Ward, *Austral. J. Chem.*, 1975, **28**, 41.

both aromatic resonances due to phenylalanine and several peaks from aliphatic protons. The temperature dependence of the areas and chemical shifts of these peaks over the melting region for the fractions were correlated roughly with optical rotation data. The reduced wool fraction, S-carboxymethyl keratin, was investigated[146] in aqueous and TFA solutions and the spectra assigned, knowing the amino-acid composition of the fraction. A simple spectral analysis has been used[147] to determine relative phenyl, methoxy, and acetyl contents of different lignins.

E. ^{1}H N.M.R. of Polynucleotides.—Shulman et al.[148] showed that the intensity of the high-frequency resonances of hydrogen-bonded protons of tRNA^{Glu} or tRNA^{Phe} when complexed with elongation factor Tu, were almost unchanged, indicating that the helical arms of the proteins did not open up on complex formation. In a similar paper[149] opening of the helical arms of tRNA^{Glu} was not observed when it complexed with glutamyl tRNA synthetase. The chemical shifts of resonances from 13 methyl groups and four methylene groups belonging to 12 modified bases of tRNA$^{Phe}_{yeast}$ were measured[150] between 294 and 353 K. The results indicated that the anticodon loop was not associated with other components of the molecule, and environments of methyl groups on the remainder of the molecule were suggested. The shifts induced by Eu^{3+} on the resonances from protons on specific Watson–Crick base pairs of tRNA^{Phe} occurring 11—15 p.p.m. to high frequency of DSS were studied[151] at various Eu^{3+} concentrations. Assignment of all the resonances that shifted was not possible, though a tentative correlation between the probable binding sites of Eu^{3+} and the structure determined by X-ray diffraction was made. The 300 MHz spectra of hydrogen-bonded protons (resonances again between 11 and 15 p.p.m. to high frequency of DSS) in tRNA^{Glu} were measured[152] between 298 and 347 K. The disappearances of various resonances at different temperatures were assigned to progressive melting of the molecule, and assisted in the assignment of the spectra. Crothers et al. have made an extensive investigation[153] of the thermal unfolding of E. Coli tRNA^{fMet} by observing the temperature dependence of the chemical shifts and intensities of the hydrogen-bonded ring NH proton resonances. Temperature-jump measurements indicated the presence of five co-operative melting transitions, which were correlated with melting behaviour observed by n.m.r. spectroscopy. A mechanism proposed for the reversible thermal unfolding was an initial transient opening of the dihydrouridine helix followed by simultaneous melting of the dihydrouridine helix and an unspecified 'tertiary' interaction. The 300 MHz spectra of E. coli tRNA^{fMet} with and without Mg^{2+} has been studied[154] in the high-frequency region, 11—15 p.p.m. from DSS. Some resonances, previously assigned to secondary structure base pairs, were reassigned to a tertiary structure base pair and a protected uridine residue in the anticodon loop.

[146] B. J. Dale and D. W. Jones, *Textile Res. J.*, 1974, **44**, 778.
[147] O. Faix and W. Schweers, *Inst. Holzchem. Chem. Technol. Holzes.*, 1974, **28**, 179.
[148] R. G. Shulman, C. W. Hilbers, D. Söll, and S. K. Yang, *J. Mol. Biol.*, 1974, **90**, 609.
[149] R. G. Shulman, C. W. Hilbers, and D. L. Miller, *J. Mol. Biol.*, 1974, **90**, 601.
[150] L. S. Kan, P. O. P. Ts'O, F. van der Haar, M. Sprinzl, and F. Cramer, *Biochem. Biophys. Res. Comm.*, 1974, **59**, 22.
[151] C. R. Jones and D. R. Kearns, *Proc. Nat. Acad. Sci. U.S.A.*, 1974, **71**, 4237.
[152] C. W. Hilbers and R. G. Shulman, *Proc. Nat. Acad. Sci. U.S.A.*, 1974, **71**, 3239.
[153] D. M. Crothers, P. E. Cole, C. W. Hilbers, and R. G. Shulman, *J. Mol. Biol.*, 1974, **87**, 63.
[154] K. L. Wong, Y. P. Wong, and D. R. Kearns, *Biopolymers*, 1975, **14**, 749.

Analysis of the temperature dependence of the spectra indicated that the tertiary structure remained intact up to 313 K in the absence of Mg^{2+} but up to 343 K in the presence of Mg^{2+}. The high-frequency region of the spectrum from yeast $tRNA_3^{Leu}$, containing resonances from the base pairs, was studied[155] as a function of temperature. It was shown that the DHU stem was the first to melt, with the amino-acid acceptor stem the last, and various models for the denatured conformer were examined. In dealing with a similar problem, Kearns and Wong[156] measured the temperature dependence of the spectra from hydrogen-bonded NH protons of 5S RNA from *E. coli*, and estimated the change in the number of base pairs with temperature from the intensity of the resonances. From a comparison with computer simulations, the base pairs involved in helices at different temperatures were identified, and a model for the complete secondary structure of 5S RNA was proposed. Several methods, such as extended Hückel theory, CNDO, and classical potential functions, have been used[157] to calculate probability distributions of the torsional angles of the nucleotide backbone and hence several n.m.r. coupling constants. The results suggested considerable flexibility around the O—P bond and were compared with relevant experimental data. The 300 MHz spectra, between 11 and 14 p.p.m. to high frequency of DSS, of several double-stranded deoxynucleotides in H_2O were recorded[158] and these resonances were assigned to $G-N_1H$ and $T-N_3H$ protons from specific base pairs. Ring-current calculations were used to evaluate low-frequency shifts of the $G-N_1H$ and $T-N_3H$ of Watson–Crick base pairs occurring in regular DNA-B and RNA-A type helices.

F. ^{15}C N.M.R.—This section includes reports dealing with proteins, polypeptides, and polysaccharides.

T_1^C and T_2^C values of cyanoferrimyoglobin were measured[159] at 1.41 and 2.35 T. A correlation time of 22×10^{-9} s was calculated for the α-carbon resonances, corresponding to overall tumbling of the molecule, and a similar time was calculated for a major resonance in the aromatic region. The interaction between CO_2 and aminoacids was investigated by measuring[160] the chemical shifts of carbon nuclei adjacent to the amine group as a function of CO_2 content. The results were used to calculate the fraction of amine groups existing as carbamate in sperm whale myoglobin. In a short communication,[161] the effect on the ^{13}C spectrum of removing Ca^{2+} from the parvalbumin protein of carp was noted, and deductions made about possible conformational changes. The imidazole C-2 atom of His-12 in bovine pancreatic ribonuclease-S was enriched[162] with ^{13}C, and the pH dependence of its chemical shift measured. A pK of 6.77 was obtained, and, when complexed with ribonuclease-S-(21—124), broadening of the enriched ^{13}C resonance was observed. The aromatic regions of *Clostridium acidi-urici* [Phe]² ferredoxin (a chemically modified ferrodoxin in which a phenylalanine replaces a tyrosyl residue) and *Clostridium pas-*

---

[155] D. R. Kearns, Y. P. Wong, S. H. Chang, and E. Hawkins, *Biochemistry*, 1974, **13**, 4736.
[156] D. R. Kearns and Y. P. Wong, *J. Mol. Biol.*, 1974, **87**, 755.
[157] R. Tewari, R. K. Nanda, and G. Govil, *Biopolymers*, 1974, **13**, 2015.
[158] D. J. Patel and A. E. Tonelli, *Biopolymers*, 1974, **13**, 1943.
[159] R. B. Visscher and F. R. N. Gurd, *J. Biol. Chem.*, 1975, **250**, 2238.
[160] J. S. Morrow, P. Keim, and F. R. N. Gurd, *J. Biol. Chem.*, 1974, **249**, 7484.
[161] S. J. Opella, D. J. Nelson, and O. Jardetzky, *J. Amer. Chem. Soc.*, 1974, **96**, 7157.
[162] I. M. Chaiken, J. S. Cohen, and E. A. Sokolski, *J. Amer. Chem. Soc.*, 1974, **96**, 4703.

teurianum ferredoxin were sufficiently different to allow assignment[163] of most of the resonances appearing. The ^{13}C chemical shifts, T_1^C, T_2^C, and ^{13}C–{^{1}H} NOE have been determined[164] for the random coil and triple helical states of the α1-CB2 fragment of rat skin collagen, a 36-residue polypeptide. Correlation times for the C_α nuclei, linewidths, and large NOE values (2.8 ± 0.3) indicated greater mobility in the coil than the helix form, and for the latter it was concluded that the Ala, Leu, and Phe residues located in the interiors of the sequence undergo rapid motion. Complete assignments of the spectra[165] of synthetic oligopeptides related to elastin have been reported including the peptide carbonyls. Selective ^{13}C enrichment together with solvent titration using trifluoroethanol and DMSO were used in the assignment. The high-resolution spectrum of the repeat tetrapeptide of elastin, BOC-Val$_1$-Pro$_2$-Gly$_3$-Gly$_4$-OMe was recorded[166] in the solvent pairs $(CD_3)_2SO$–CD_3OD and $(CD_3)_2SO$–D_2O. The carbonyl chemical shift of Val$_1$ was almost independent of solvent composition, indicating it was internally hydrogen bonded and that Val$_1$ was involved in a β turn of the molecule.

Deslauriers *et al.*[167] measured the T_1 values of the octapeptide angiotensin and interpreted them in two possible ways; isotropic overall motion with internal motion of some residues and, in less detail, anisotropic overall motion. The α-carbons of the peptide backbone showed equally restricted motion whereas side-chain carbons, except for the Tyr residue, were more mobile than the backbone. Chemical shifts detected no conformational constraints in the molecule and showed that the proline residue was in the *trans* conformation about this His-Pro bond. With the aid of model compounds, many resonances from actinomycin D were assigned,[168] and from the relative shielding and deshielding effects some intramolecular interactions such as hydrogen bonding between C(2)NH$_2$ and C(3)=O were indicated. The non-equivalent T_1 values of carbon nuclei in proline were interpreted[169] as indicating rapid interconversion between various ring-puckered conformations. Steric constraints imposed on non-terminal proline residues in hormones such as oxytocin resulted in equal T_1 values for the α and β carbons. Urry *et al.*[170] showed that solvent titrations, using appropriate solvent pairs, produce less chemical shift in intramolecularly bonded peptide carbonyl resonances than in non-bonded carbonyl groups. This provided structural information on some short-chain natural and synthetic polypeptides. Resonances from both backbone carbon nuclei of polydisperse poly-N^5-(3-hydroxypropyl)-L-glutamine showed[171] two separate peaks from residues in the different conformations occurring during a helix–coil transformation. In a more homogeneous low molecular weight sample the fact that the lower frequency C_α resonance is much less intense during the transition is quoted as showing the dominant influence of molecular weight polydispersity on the double

---

[163] E. L. Packer, H. Sternlicht, E. T. Lode, and J. C. Rabinowitz, *J. Biol. Chem.*, 1975, **250**, 2062.
[164] D. A. Torchia, J. R. Lyerla, jun., and A. J. Quattrone, *Biochemistry*, 1975, **14**, 887.
[165] D. W. Urry, L. W. Mitchell, and T. Ohnishi, *Biochemistry*, 1974, **13**, 4083.
[166] D. W. Urry, L. W. Mitchell, and T. Ohnishi, *Biochem. Biophys. Res. Comm.*, 1974, **59**, 62.
[167] R. Deslauriers, A. C. M. Paiva, K. Schaumburg, and I. C. P. Smith, *Biochemistry*, 1975, **14**, 878.
[168] U. Hollstein, E. Breitmaier, and G. Jung, *J. Amer. Chem. Soc.*, 1974, **96**, 8036.
[169] R. Deslauriers, I. C. P. Smith, and R. Walter, *J. Biol. Chem.*, 1974, **249**, 7006.
[170] D. W. Urry, L. W. Mitchell, and T. Ohnishi, *Proc. Nat. Acad. Sci. U.S.A.*, 1974, **71**, 3265.
[171] R. Di Blasi and A. S. Verdini, *Biopolymers*, 1974, **13**, 2209.

peak phenomenon. The ^{13}C relaxation times of oligo-lysines[172] (up to 12 residues) showed that the T_1 of the side-chain carbons (C_β to C_ε) became independent of chain length after about five residues whereas the carbonyl T_1's, being more sensitive to backbone motions, required about 12 residues to become independent of chain length. Chemical shifts and T_1 values for pentapeptides containing L-methionine, L-proline, L-arginine, or L-lysine as the central residue in the pattern Gly-Gly-X-Gly-Gly were used[173] to estimate the degree of sensitivity of pentapeptide resonances to the state of protonation of terminal and side-chain groups. The pyrrolidone ring flexibility appears to be sterically hindered by the adjacent glycine residue on the NH$_2$-terminal side. The spectra of polymers[174] prepared by dehydrogenative polymerization of p-hydroxycinnamic alcohols were compared to that of spruce lignins. A similarity between the spectra, when polymerizing conditions resemble those *in vivo*, led to the conclusion that the lignins may have been prepared by polymerization of the alcohols. A complete assignment[175] of the ^{13}C spectra of some sialic acid-containing meningococcal polysaccharide antigens has been made from comparison with the chemical shifts of related monomers. The data indicated that the serogroup B polysaccharide was a 2→8-α linked homopolymer of sialic acid, whereas the de-O-acetylated serogroup C polysaccharide is a 2→9-α linked homopolymer. The spectra[176] of the mannan from *Rhodotorula glutinis*, a polysaccharide containing (1→3) and (1→4) linked β-D-mannopyranose residues, consisted of 12 resonances. Assignment of these was made by use of specifically deuteriated D-glucoses and comparison between the effects of O-methylation and O-mannosylation. The high-resolution spectra[177] of dextrans isolated from cane sugar showed separate resonances for 1→6 and non-1→6 linkages, allowing the degree of branching to be estimated. With the aid of spectra of glucose and some of its O-methylated derivatives, the ^{13}C spectra of two cyclodextrins and several linear glucans have been completely assigned.[178] A comparison of spectra of lignins[179] extracted from mistletoes that were grown on different types of tree showed that the lignin structure and composition depended slightly on the nature of the tree.

G. Other Nuclei.—This section includes studies made using ^{19}F and ^{31}P resonances. Rabbit muscle glyceraldehyde-3-phosphate dehydrogenase was labelled at Cys-149 with 3,3,3-trifluorobromoacetone and the ^{19}F intensities and chemical shifts used[180] to monitor enzyme interactions. The effects of addition of coenzyme and NADH were studied and a dynamic model for the enzyme structure was proposed. Alkaline phosphatase was labelled[181] with m-fluorotyrosine and a separate resonance was observed for each of the 11 known tyrosines per subunit. Relaxation times were

[172] H. Saito and I. C. P. Smith, *Arch. Biochem. Biophys.*, 1974, **163**, 699.
[173] P. Keim, R. A. Vigna, A. M. Nigen, J. S. Morrow, and F. R. N. Gurd, *J. Biol. Chem.*, 1974, **249**, 4149.
[174] H. Nimz, I. Mogharab, and H.-D. Lüdemann, *Makromol. Chem.*, 1974, **175**, 2563.
[175] A. K. Bhattacharjee, H. J. Jennings, C. P. Kenny, A. Martin, and I. C. P. Smith, *J. Biol. Chem.*, 1975, **250**, 1926.
[176] P. A. J. Gorin, *Carbohydrate Res.*, 1975, **39**, 3.
[177] M. T. Covacevich and G. N. Richards, *Proc. Queensland Soc. Sugar Cane Technol.*, 1974, **41**, 171.
[178] P. Colson, H. J. Jennings, and I. C. P. Smith, *J. Amer. Chem. Soc.*, 1974, **96**, 8081.
[179] H. Nimz, *Z. Pflanzenphysiol.*, 1974, **73**, 226.
[180] J. Bode, M. Blumenstein, and M. A. Raftery, *Biochemistry*, 1975, **14**, 1153.
[181] W. E. Hull and B. D. Sykes, *Biochemistry*, 1974, **13**, 3431.

insensitive to rotation about the C_β-aromatic ring bond, and were consistent with a model in which there was isotropic motion of the protein with a correlation time of about 7×10^{-8} s. As part of a general biochemical study[182] of gene-5 protein from bacteriophage fd, m-fluorotyrosine was substituted for the five normal tyrosine residues of the protein. The ^{19}F spectrum showed five resonances, three ca. 58 p.p.m. to low frequency of TFA and thought to be from nuclei exposed to the solution, and another two from nuclei buried in the interior of the protein, occurring at 56 p.p.m. N.m.r. and sedimentation techniques were used to study[183] several trifluoroacetyl derivatives of insulin at pH 2.0 and 6.8. From the linewidths it was suggested that the derivatives might be in dimers at pH 2.0 and larger aggregates at pH 6.8, while chemical shift results indicated some conformational changes at pH 8.7. Both ^{19}F and ^{31}P n.m.r. were used to study[184] the binding of ligands, particularly 2,3-diphosphoglycerate (DPG) to human haemoglobin. ^{19}F Chemical shifts of the trifluoroacetonylated protein showed the presence of species, intermediate in the ligand-binding process. ^{31}P Resonances showed that DPG, which exchanged rapidly between protein and solution, bound to these intermediate protein species. A model consistent with the n.m.r. data was proposed. From the ^{1}H and ^{31}P spectra[185] of coenzyme A, chemical shifts and coupling constants for nuclei in the pantetheine and adenosine parts were calculated. These suggested that although the enzyme was flexible several bonds had preferred orientations, and some of these were discussed. The ^{1}H and ^{31}P T_1 values of polyriboadenylic acid in D_2O were measured[186] between 287 and 355 K. Minima in T_1 between 308 and 318 K were used to calculate the correlation times and activation energies for the sugar residues.

5 Small Molecules

The papers in this section are concerned with effects of macromolecules on the spectra of small molecules and, as for the previous review, are arranged according to the nucleus studied, with a separate section on hydration.

A. ^{1}H: Hydration.—The water proton T_1 values in a solution of ferric haemoglobin and myoglobin were measured[187] between 278 and 308 K and for frequencies between 1.5 and 60 MHz. The correlation time for the relaxation was thought to be the electron relaxation time, with exchange between solution and bound sites becoming important at higher temperatures. Spin–lattice relaxation times of water protons in solutions of normal and sickle-cell haemoglobin, normal red blood cells, and whole blood were measured[188] as a function of frequency and pH. Correlation times for the water molecules associated with the proteins under various conditions indicated that, whether entrained within the cells or interstices of a gel of low protein concentration, both water–haemoglobin interactions and haemoglobin dynamics were essentially unaltered.

[182] R. A. Anderson, Y. Nakashima, and J. E. Coleman, *Biochemistry*, 1975, **14**, 907.
[183] R. A. Paselk and D. Levy, *Biochemistry*, 1974, **13**, 3340.
[184] W. H. Huestis and M. A. Raftery, *Biochemistry*, 1975, **14**, 1886.
[185] C.-H. Lee and R. H. Sarma, *J. Amer. Chem. Soc.*, 1975, **97**, 1225.
[186] K. Akasaka, *Biopolymers*, 1974, **13**, 2273.
[187] G. Lahajnar, I. Zupančič, R. Blinc, G. Pifat, and S. Maričič, *Biopolymers*, 1974, **13**, 1187.
[188] T. R. Lindstrom and S. H. Koenig, *J. Magn. Resonance*, 1974, **15**, 344.

Macromolecules 225

The T_1 of water protons in solutions of ferricytochrome c was measured[189] as a function of pH, and pK values of 2.5 and 4.5 were obtained in aqueous and 6M urea solutions, respectively. Addition of different anions, including methane sulphonate, decreased the water T_1 by 85%, though not through binding of the anions to the haem iron, as shown by the lack of chemical shift or linewidth increase of the methyl resonance. Gupta and Mildvan[190] have measured water relaxation times in solutions of human methaemoglobin and fluoromethaemoglobin containing varying amounts of inositol hexaphosphate and at different pH. From the temperature and frequency dependence of the T_1 values, the dominant relaxation processes such as exchange were deduced, and iron–water or iron–distal histidine distances calculated. Water proton T_1 values have been measured[191] in solutions of haemoglobins containing carbon monoxide and inositol hexaphosphate (IHP). IHP caused an increase in the solvent T_1^{-1} which was attributed to an increase in the exchange between the haeme group and the bulk solvent, and inferences were made about the influence of IHP on the structure of the haeme group.

The temperature dependence of T_1 and T_2 at two frequencies for water in oxygenated and deoxygenated erythrocytes from normal adults, and with sickle-cell disease, was interpreted[192] by assuming a three-phase model for water. These phases were bulk water, weakly-bound and strongly-bound to the protein; as a detailed assignment of the phases is not possible, conclusions reached concerning protein reactions from water relaxation times must be speculative. The pH dependence of the T_1 of water protons in solutions of horseradish peroxidase showed[193] no difference between the high-spin and low-spin states, indicating that water was not coordinated to the ferric centre of the enzyme. Darke and Finer[194] studied the hydration of poly-L-lysine hydrobromide using ^{1}H, ^{2}H, ^{13}C, and ^{81}Br resonances from the polymer and solvent. A gel–liquid transition occurred at about 13 water molecules per polymer molecule, about seven of which were associated with helical residues and the others with the disordered part of the macromolecule. Gel systems giving a singlet ^{1}H spectrum of water produced a doublet ^{2}H water spectrum, which was attributed to rapid hydrogen or deuteron exchange between water molecules in the bulk and adjacent to the polymer. The T_1 of water in solutions containing gelatin or cotton was found[195] to increase with increasing water content in the normal way, showing the difficulty of using water T_1 measurements as a diagnostic tool for cancerous mammalian tissue. Westover and Dresden[196] have measured T_1, T_2, and the diffusion coefficient D for water in solutions and gels of native and denatured collagen. The decrease in T_1 and T_2 and the constancy of D, with increasing collagen concentration were attributed to the presence of bound water, with possible long-range order, though this terminology is potentially quite misleading. The accessibility of cellulose samples has been estimated[197] by D_2O exchange. The intensity of

[189] A. Lanir and I. Aviram, *Arch. Biochem. Biophys.*, 1975, **166**, 439.
[190] R. K. Gupta and A. S. Mildvan, *J. Biol. Chem.*, 1975, **250**, 246.
[191] G. Pifat, B. Benko, S. Maričič, and S. Vuk-Pavlović, *Croat. Chim. Acta*, 1974, **46**, 145.
[192] B. C. Thompson, M. R. Waterman, and G. L. Cottam, *Arch. Biochem. Biophys.*, 1975, **166**, 193.
[193] A. Lanir and A. Schejter, *Biochem. Biophys. Res. Comm.*, 1975, **62**, 199.
[194] A. Darke and E. G. Finer, *Biopolymers*, 1975, **14**, 441.
[195] W. R. Inch, J. A. McCredie, C. Geiger, and Y. Boctor, *J. Nat. Cancer Inst.*, 1974, **53**, 689.
[196] C. J. Westover and M. H. Dresden, *Biochim. Biophys. Acta*, 1974, **365**, 389.
[197] E. K. H. Wittich, *Progr. Colloid Polymer Sci.*, 1975, **57**, 141.

the HOD resonance was used as a measure of the accessibility and exchange over a period of several weeks. Spin–lattice relaxation measurements, made[198] at different frequencies, of the non-freezing component of water in agarose gels between 263 and 193 K showed that the correlation time for water in the 'bound' phase might be described by a log–normal distribution. Relaxation times (T_1 and T_2) of water in aqueous solutions of dextrans[199] of different molecular weights showed slight changes if solutions were kept at 373 K for 30 min. These changes were attributed to aggregation of dextrans, though the unknown purity of the samples precludes the general acceptance of the results. Similar T_1 and T_2 measurements[200] of water (0—35%) in different starches were used to estimate proportions of bound and unbound water, using a simple two-site model. The dependence on temperature and frequency of the water proton T_1 and T_2 values in solutions of manganese carbonic anhydrase have been measured[201] by pulse techniques. From the frequency dependence of T_1 and the T_1/T_2 ratio at a high frequency, it was estimated that one water molecule was bound to the metal at high pH, but none were bound to the metal at low pH.

The reaction between concanavalin A and Mn^{2+}, Ca^{2+}, or α-methyl-D-mannoside has been studied[202] by observing the signal height of the H_2O resonance as a function of time, after rapid stopped-flow mixing of the protein–ligand solutions in tubes situated inside an n.m.r. probe. Binding of the Mn^{2+} to the protein caused broadening of the water resonance, and kinetic constants for possible non-protein reactions were calculated. In a brief note[203] the linewidth of the methyl resonance of ethanol in ethanol–water mixtures in calcium alginate gels was used as a monitor of the alcohol binding to the gels. At 95% ethanol, linewidths comparable to pure ethanol were obtained whereas for 100% ethanol in the gel, broadening to several Hz was observed, indicating preferential solvation of the gel by the water.

Proton Relaxation Enhancement (PRE). Papers describing the use of PRE by paramagnetic ions to obtain details of macromolecule-small molecule interactions are listed in Table 3.

Table 3

Macromolecule	Comments	Ref.
Aspartate transcarbamylase	Zn^{2+} replaced Mn^{2+}; PRE of bound succinate protons or cytidine 5'-triphosphate used to calculate proton correlation times (dipolar interaction) and Mn^{2+}—H distances.	204

[198] I. D. Duff and W. Derbyshire, *J. Magn. Resonance*, 1975, **17**, 89.
[199] E. I. Fedin, V. G. Tsitsishvili, V. Ya. Grinberg, T. I. Bakari, and V. B. Tolstoguzov, *Carbohydrate Res.*, 1975, **39**, 193.
[200] H. J. Hennig and H. Lechert, *Stärke*, 1974, **26**, 232.
[201] A. Lanir, S. Gradstajn, and G. Navon, *Biochemistry*, 1975, **14**, 242.
[202] J. J. Grimaldi and B. D. Sykes, *J. Biol. Chem.*, 1975, **250**, 1618.
[203] H. Grasdalen, I. Svare, and O. Smidsrod, *Acta Chem. Scand. (B)*, 1974, **28**, 966.
[204] S. Fan, L. W. Harrison, and G. G. Hammes, *Biochemistry*, 1975, **14**, 2219.

Macromolecule	Comments	Ref.
Alcohol dehydrogenase	Water proton enhancement with Zn^{2+} partially replaced by Co^{2+} showed rapid H_2O exchange with about half of Co^{2+}; no exchange with other half. Effects of added ligands such as ethanol or isobutyramide studied.	205
Bovine superoxide dismutase	Maximum in water proton relaxation rates between pH 6 and 13 attributed to changes in Cu^{2+}-enzyme internuclear distance.	206
Superoxide dismutase	Temperature and frequency dependence of water proton T_1 and T_2 of Mn^{2+} and Fe^{3+}-enzyme complex provided information about the hydration of the metal site.	207
Arginine kinase	Water proton enhancement of Mn–ADP–enzyme complex was 18; dissociation constant of $Mn-ADP^-$ from enzyme found to be μmol l^{-1}.	208
Arginine kinase	Frequency dependence of PRE used to estimate number of water ligands on Mn^{2+} in enzyme–Mn^{2+}–ADP–arginine complexes.	209
Arginine kinase	Structure of quaternary arginine kinase Mn–ADP–arginine complex and influence of anions on kinetics of enzyme reaction investigated.	210
Fully adenylated glutamine synthetase	Frequency, temperature, and substrate dependence of water PRE of enzyme–Mn^{2+}–substrate complexes indicated whether Mn^{2+} is directly involved in binding.	211
Alkaline phosphatase	Water proton $T_1:T_2$ ratios of Mn^{2+}-, Cu^{2+}-, and Co^{2+}-enzyme complexes used to calculate metal–water distances and correlation times. Added phosphate ions did not displace water from metal.	212
Pyruvate carboxylase	Relaxation times of methyl protons on pyruvate in pyruvate–enzyme complex, measured at 60 and 220 MHz, showed Mn^{2+}–methyl distance ca. 0.6—0.7 nm.	213
Phosphoenol pyruvate-gadolinium-pyruvate kinase	Analysis of enhancement of water and enolpyruvate proton relaxation indicated Gd^{3+} electron relaxation provided the dominant correlation time; structure of the complex uncertain.	214
Pyruvate kinase	Formation of ternary complex Mn^{2+}–enzyme–oxalate investigated and compared to enzyme–enolpyruvate interaction.	215

[205] D. L. Sloan, J. M. Young, and A. S. Mildvan, *Biochemistry*, 1975, **14**, 1998.
[206] M. Terenzi, A. Rigo, C. Franconi, B. Mondovi, L. Calabrese, and G. Rotilio, *Biochim. Biophys. Acta*, 1974, **351**, 230.
[207] J. J. Villafranca, F. J. Yost, jun., and I. Fridovich, *J. Biol. Chem.*, 1974, **249**, 3532.
[208] W. J. O'Sullivan, E. Smith, B. E. Chapman, and K. H. Marsden, *Biochim. Biophys. Acta*, 1974, **370**, 153.
[209] D. H. Buttlaire and M. Cohn, *J. Biol. Chem.*, 1974, **249**, 5741.
[210] D. H. Buttlaire and M. Cohn, *J. Biol. Chem.*, 1974, **249**, 5733.
[211] J. J. Villafranca and F. C. Wedler, *Biochemistry*, 1974, **13**, 3286.
[212] R. S. Zukin and D. P. Hollis, *J. Biol. Chem.*, 1975, **250**, 835.
[213] G. H. Reed and M. C. Scrutton, *J. Biol. Chem.*, 1974, **249**, 6156.
[214] G. L. Cottam, K. M. Valentine, B. C. Thompson, and A. D. Sherry, *Biochemistry*, 1974, **13**, 3532.
[215] G. H. Reed and S. D. Morgan, *Biochemistry*, 1974, **13**, 3537.

Macromolecule	Comments	Ref.
Pyruvate kinase	From frequency and temperature dependence of relaxation times of pyruvate protons in complex enzyme–Mn^{2+}–pyruvate (with and without ADP), correlation times and Mn^{2+}–pyruvate distances calculated.	216
Phosphofructokinase	Temperature and frequency dependence of water PRE in Mn^{2+}–ATP–enzyme complex allowed various contributions to overall τ_c to be estimated.	217
Concanavalin A (Con A)	Water PRE of Con A–Mn^{2+}–saccharide systems related to binding affinity of Con A to saccharide; contributions of exchange and electron relaxation to τ_c determined from temperature and frequency dependence of water T_1 and T_2.	218
Concanavalin A (Con A)	Addition of Ca^{2+} to Con A-Mn^{2+} complex caused longer residence times for water on Mn^{2+}.	219
Formyltetrahydrofolate synthetase	Water PRE used to calculate dissociation constants for enzyme–Mn ATP or ADP complexes; competitive binding of tetrahydrofolate studied.	220
Formyltetrahydrofolate synthetase	Frequency dependence of water PRE with enzyme–Mn^{2+}–ADP–tetrahydrofolate complex allowed correlation times for relaxation to be calculated; results correlated with extensive e.s.r. data.	221
Lysozyme	Temperature and frequency dependence of water PRE for enzyme–Gd^{3+} complex measured; co-ordination number and correlation times for Gd^{3+} estimated.	222
Mandelate racemase	Enhancement of water T_1 measured for enzyme–Mn^{2+} plus inhibitors. Correlation times and number of metal–water ligands determined. Distances between Mn^{2+} and some ^{13}C (enriched) nuclei of enzyme calculated.	223
Rat liver mitochondria	Mitochondria–Mn^{2+} or –Gd^{3+} complexes produced enhancements of 10; ligands such as phosphates or succinates reduced this.	224
Carboxypeptidase B	From water T_1 and T_2 values of Mn^{2+}–enzyme complex, it was calculated that there is one H_2O molecule bound to Mn^{2+}. Effects of inhibitor investigated.	225

B. ^1H N.M.R.: General Studies.—The papers in this section are mainly concerned with the binding of small molecules to macromolecules, particularly proteins, and

[216] T. L. James and M. Cohn, *J. Biol. Chem.*, 1974, **249**, 3519.
[217] R. Jones, R. A. Dwek, and I. O. Walker, *European J. Biochem.*, 1974, **47**, 285.
[218] J. J. Villafranca and R. E. Viola, *Arch. Biochem. Biophys.*, 1974, **165**, 51.
[219] B. H. Barber and J. P. Carver, *Canad. J. Biochem.*, 1975, **53**, 371.
[220] D. H. Buttlaire, G. H. Reed, and R. H. Himes, *J. Biol. Chem.*, 1975, **250**, 254.
[221] D. H. Buttlaire, G. H. Reed, and R. H. Himes, *J. Biol. Chem.*, 1975, **250**, 261.
[222] R. Jones, R. A. Dwek, and S. Forsén, *European J. Biochem.*, 1974, **47**, 271.
[223] E. T. Maggio, G. L. Kenyon, A. S. Mildvan, and G. D. Hegeman, *Biochemistry*, 1975, **14**, 1131.
[224] G. D. Case, *Biochim. Biophys. Acta*, 1975, **375**, 69.
[225] N. Zisapel, G. Navon, and M. Sokolovsky, *European J. Biochem.*, 1975, **52**, 487.

the resultant changes in mobility or environment of the bound species. The proton T_1 values of the solvent in the liquid crystal system CH_2Cl_2 plus poly(γ-benzyl-L-glutamate) were found[226] to be frequency dependent on passing from isotropic solution to the liquid crystal. This frequency dependence was attributed to solvent orientation fluctuation of several types, with some being clearly identified as the most likely. The temperature and concentration dependence of the self-diffusion coefficient of acetone and benzene in solutions of poly(methyl methacrylate) were measured[227] by the usual pulsed methods and the results interpreted using Cohen–Turnbull's free volume theory. The addition of poly(glutamic acid) to sonicated lecithin vesicles caused[228] a decrease in the intensity of the lipid resonances, which was interpreted as introducing lateral tightening of the bilayer membrane. Davis has studied[229] the binding of NH_4^+ to the antibiotics valinomycin and nonactin in $CDCl_3$ by measuring the linewidths of the triplet proton resonance. The difference between the linewidths for the central and outer lines was used to separate the dipolar and quadrupolar contributions to relaxation and, by making the unjustified assumption that the correlation times for these two processes are the same, the authors calculated a quadrupole coupling constant for the ^{14}N nucleus. The interaction between bovine neurophysin and oxytocin was studied[230] from changes in the resonances of the C-2 proton of the histidine residue on oxytocin. The negligible chemical shift of the resonance on complex formation indicated no π–π interactions which occur during the stacking of aromatic rings, whereas the splitting of the histidine resonance at equimolar protein:hormone ratios suggested protein conformational changes in the complex. The T_1 and T_2 values of tyrosyl protons of oxytocin and [8-lysine]vasopressin bound separately to bovine neurophysin II were measured[231] as a function of frequency and temperature. Whereas T_2 decreased with increasing neurophysin II concentration, T_1 was unaffected, suggesting fast exchange between the free hormone and hormone-neurophysin II complex.

Oakes[232] measured the chemical shifts and linewidths of alkyl resonances of sodium dodecyl sulphate (SDS) bound to bovine serum albumin and deductions concerning the bound SDS state were made, assuming a simple two-state model. These were that above 10—20 SDS molecules per protein molecule, the surfactants were in a micelle-like environment which included apolar groups of the protein, and that both the surfactant head groups and alkyl chains were associated with the protein at the initial 10 binding sites. The relative binding affinities of atropine acetylcholine and physostigmine to acetylcholinesterase were estimated[233] from changes in the linewidth of the N-methyl resonances of the ligands. By assuming a two-state model and that exchange or susceptibility contributions were negligible, the authors calculated binding constants for the ligands to the enzyme. In the binding[234] of 3′- and 5′-cytidine monophosphate to ribonuclease A, the contribution of exchange

---

[226] W. A. Hines and E. T. Samulsky, *J. Polymer Sci., Symposia*, 1974, No. 44, 11.
[227] J. Schlegel and R. Kosfeld, *Progr. Colloid Polymer Sci.*, 1975, **57**, 69.
[228] C.-A. Chang and S. I. Chan, *Biochemistry*, 1974, **13**, 4381.
[229] D. G. Davis, *Biochem. Biophys. Res. Comm.*, 1975, **63**, 786.
[230] J. H. Griffin, J. S. Cohen, P. Cohen, and M. Camier, *J. Pharm. Sci.*, 1975, **64**, 507.
[231] R. Alazard, P. Cohen, J. S. Cohen, and J. H. Griffin, *J. Biol. Chem.*, 1974, **249**, 6895.
[232] J. Oakes, *J.C.S. Faraday I*, 1974, **70**, 2200.
[233] G. Kato, *J. Pharm. Sci.*, 1975, **64**, 488.
[234] D. G. Gorenstein and A. M. Wyrwicz, *Biochem. Biophys. Res. Comm.*, 1974, **59**, 718.

between solution and enzyme to the line-broadening of the proton resonances from CMP was estimated. For increasing ratios of enzyme to CMP a non-linear increase in line-broadening of H-6 and H-5 resonances from CMP was observed, this non-linearity possibly arising from the exchange contribution. Correlation times for motion of the pyrimidine (4.4×10^{-9} s) and ribose rings (1.2×10^{-8} s) were calculated. As part of a general study[235] of the peptide interactions with DNA, Gourévitch et al. used the linewidth changes of the N-Me and C-Me resonances of malouetine to obtain the isotherm of its adsorption onto DNA. From the pH dependence[236] of the chemical shift of the H-5, H-6, and H-1' resonances of 3'- and 5'-CMP–ribonuclease A complexes, pK_a values for these protons were determined. These pK_a values and absolute chemical shifts were used to indicate detailed differences between the two nucleotide–enzyme structures. The binding of the inhibitor 4-(N-acetylaminoglucosyl)-N-acetylglucosamine to lysozyme was investigated[237] using deuteriated and normal samples of the inhibitor. By subtracting spectra of the complexes containing deuteriated and protonated forms, a spectrum of the deuteriated positions in the complex was obtained for the binding process. In the interaction[238] between lysozyme and (NAG-NAM)$_2$ [(2-acetamido-2-deoxy-D-glucose-2-acetamido-2-deoxy-D-muramic acid)$_2$] changes in the coupling constant between the anomeric ring proton H-1 and the ring proton H-2 indicated a distortion of the reducing ring of the (NAG-NAM)$_2$ to the half-chair conformation in the bound state.

The binding of [^{2}H]propan-2-ol to horse liver alcohol dehydrogenase was studied[239] by the unusual method of monitoring indirectly the deuterium relaxation rate by means of the proton spin-coupled to the deuterium nucleus. The resonance of this proton changed from a triplet to a singlet with changing deuterium relaxation rates and this was used to detect binding of the alcohol to the enzyme. The exchange of the α-hydrogen of alanine for water protons is accelerated by the enzyme alanine racemase.[240] The activity of the enzyme in D$_2$O was estimated by observing the reduction in intensity of the doublet methyl resonance (when the α-carbon has a proton) and the appearance of a single methyl resonance (when the C_α has a deuteron). An interesting method for following the hydrolysis of L-phenylalanine t-butyl ester by α-chymotrypsin was described by Grimaldi and Sykes.[241] Stopped-flow FT difference spectra of the t-butyl resonance taken after as short a time as 1 s allowed the reaction kinetics to be followed. Using this technique weak signals can be detected in the presence of strong solvent signals.

C. ^{13}C N.M.R. Studies.—One application of ^{13}C n.m.r. is using enriched materials in the study of small molecules bound to macromolecules. N-Acetyl-L-tyrosine semicarbazide (I) is hydrolysed by chymotrypsin to N-acetyl-L-tyrosine and semicarbazide. Using N-acetyl-L-[^{13}C]tyrosine enriched at the carboxy carbon, the enzyme–

---
[235] M. Gourévitch, P. Puigdoménech, A. Cavé, G. Etienne, J. Méry, and J. Parello, Biochimie, 1974, 56, 967.
[236] D. G. Gorenstein and A. Wyrwicz, Biochemistry, 1974, 13, 3828.
[237] J. H. Baldo, S. E. Halford, S. L. Patt, and B. D. Sykes, Biochemistry, 1975, 14, 1893.
[238] S. L. Patt, D. Dolphin, and B. D. Sykes, Proceedings of the Lysozyme Conference, 1974, p. 229.
[239] T. Drakenberg, J. Magn. Resonance, 1974, 15, 354.
[240] U. M. Babu, R. B. Johnston, and L. C. McNeff, Analyt. Biochem., 1975, 63, 208.
[241] J. J. Grimaldi and B. D. Sykes, J. Amer. Chem. Soc., 1975, 97, 273.

substrate (I) interaction was studied[242] by high-resolution methods. As no chemical shift was observed when the fraction of substrate bound varied from 0.17 to 0.70, it was concluded that negligible strain occurred in this complex at the site of enzyme attack. The relaxation times (T_1 and T_2) of ^{13}C-enriched HCO_3^- bound to bovine cobalt or zinc carbonic anhydrase were measured[243] in the presence and absence of the inhibitor acetazolamide. Large paramagnetic effects indicated a binding site at the metal, and a broadening of 4 Hz in the presence of the fully inhibited enzyme was taken as evidence for a second, weaker binding site. Assuming a correlation time of the cobalt electron spin relaxation, a ^{13}C—Co^{2+} distance was calculated. The reverse hydration of CO_2, catalysed by carbonic anhydrase, was followed[244] by monitoring the linewidths of enriched $^{13}CO_2$ and $H^{13}CO_3^-$ in solutions of the native human B enzyme. The residence lifetimes on the enzymes is long compared with T_2 of $H^{13}CO_3^-$, thus allowing the reaction to be followed from linewidth changes. The rate constants were found to be in general agreement with those from stopped-flow techniques. The broadening of the resonance from $^{13}CO_3^{2-}$ bound to Fe^{3+}-human transferrin was used[245] to estimate an upper limit of 0.9 nm for the Fe^{3+}-CO_3^{2-} distance, the lower limit depending on the assumptions made about the relaxation processes. The chemical shifts of ^{13}CO bound to the α- and β-chains of various haemoglobins were found[246] to be different, though independent of whether the chains were incorporated in one molecule or as isolated units. The relative affinity of oxygen as compared to ^{13}CO for the α-chain was shown to be different from that of the β-chain. Phosphatidylcholines and sphingomyelins were enriched[247] with ^{13}C in the choline moiety, in different carbon atoms of oleic and linoleic acid chains of lecithins and sphingomyelin, and in cholesterol esters. These were recombined with total apoproteins of human high-density lipoprotein and the separated apolipoproteins apo A-I and apo A-II. From T_1 values, the general conclusion reached was that the phospholipids bind through their hydrophobic group and not their hydrophilic group. Assignments have been made[248] of ^{13}C spectra from high, low, and very low density lipoproteins isolated from human plasma. Relaxation times of the lipid moieties indicated that although they were in a liquid-like state, their segmental motion was significantly restricted compared with that in organic solvents. Relaxation times and chemical shifts of nuclei in the polar head-group of phospholipids, both in solvents and combined with lipid–apoprotein complexes showed[249] that the head-group was in similar hydrophilic environments in the complexes and in sonicated lipid particles. The relaxation times also suggested that binding in the lipid–apoprotein was primarily hydrophobic. In a similar paper,[250]

[242] G. Robillard, E. Shaw, and R. G. Shulman, *Proc. Nat. Acad. Sci. U.S.A.*, 1974, **71**, 2623.
[243] P. L. Yeagle, C. H. Lochmüller, and R. W. Henkens, *Proc. Nat. Acad. Sci. U.S.A.*, 1975, **72**, 454.
[244] S. H. Koenig and R. D. Brown, *Pure Appl. Chem.*, 1974, **40**, 103.
[245] D. C. Harris, G. A. Gray, and P. Aisen, *J. Biol. Chem.*, 1974, **249**, 5261.
[246] R. B. Moon and J. H. Richards, *Biochemistry*, 1974, **13**, 3437.
[247] W. Stoffel, O. Zierenberg, B. D. Tunggal, and E. Schreiber, *Z. physiol. Chem.*, 1974, **355**, 1381.
[248] J. A. Hamilton, C. Talkowski, R. F. Childers, E. Williams, A. Allerhand, and E. H. Cordes, *J. Biol. Chem.*, 1974, **249**, 4872.
[249] G. Assmann, R. J. Highet, E. A. Sokoloski, and H. B. Brewer, jun., *Proc. Nat. Acad. Sci. U.S.A.*, 1974, **71**, 3701.
[250] W. Stoffel, O. Zierenberg, B. Tunggal, and E. Schreiber, *Proc. Nat. Acad. Sci. U.S.A.*, 1974, **71**, 3696.

Stoffel et al. showed from relaxation times of phospholipids and cholesteric lipids that their binding to human high-density apolipoprotein was probably through hydrophobic rather than hydrophilic groups. Stejskal and Schaefer[251] have noticed a simple correlation between the ability of a dense polymer film to separate styrene from ethylbenzene and the observed linewidths of the adsorbed species, though no conclusive explanation was given.

D. Other Nuclei.—The papers in this section are concerned with studies using fluorine, phosphorus, and deuteron resonance. The high-resolution ^{19}F spectra[252] of N-trifluoroacetylated chitotriose in the presence of turkey lysozyme showed no marked pH dependence, in contrast to earlier studies with hen lysozyme, where at low pH, Asp-101 is protonated. The replacement in turkey lysozyme of Asp-101 by glycine was thought to account for this difference, and a pK_a of 4.2 for Asp-101 in the hen enzyme was calculated. A brief study[253] has been made of the binding of a fluorinated phosphonate to alkaline phosphatase by following ^{19}F linewidth changes. In the absence of metal ions no binding was detected, but in the presence of Mn^{2+} some broadening was observed, and with Zn^{2+} two resonances appeared, indicating slow exchange between two sites. The interaction of fluorinated acetates with Co^{2+} human carbonic anhydrase C was investigated[254] by measuring both ^{1}H and ^{19}F relaxation as a function of temperature and at two frequencies. Differences in the relaxation rates in the presence and absence of sulphonamides indicated a complex proton transfer between the acetates and sulphonamide near the active site of the enzyme. The binding of N-trifluoroacetyl-α- and -β-D-glucosamine to dimeric and tetrameric concanavalin A was studied[255] using ^{19}F resonance. The lack of chemical shift on binding suggested that the site of attachment to the protein was not a hydrophobic one as previously thought. This was also implied from the ^{19}F resonance broadening when the sugar was bound to concanavalin A containing Mn^{2+} in the transition-metal site, the broadening being used to calculate a metal-^{19}F distance of 1.2 to 1.4 nm. The interaction of small fluorine-containing molecules with polyethyleneimine was investigated[256] by ^{19}F n.m.r. in both aqueous and organic solvents. From the chemical shifts of the polymer–ligand complex, deductions were made about the conformation of the complex, though in view of the assumptions involved these must be quite tentative. Cozzone et al.[257] investigated the binding of ATP to rabbit muscle G-actin by ^{31}P FT techniques at 40.5 MHz. A single average resonance was observed for bound and free molecules, showing rapid exchange with the protein. From linewidth measurements, P_β and P_γ were thought to be more involved in the binding than P_α. The structure of 2′,5′-thymidine diphosphate bound to the active site of the calcium-binding enzyme staphylococcal nuclease was investigated[258] by measuring the enhanced ^{1}H and ^{31}P relaxation

---

[251] E. O. Stejskal and J. Schaefer, *Macromolecules*, 1974, **7**, 767.
[252] N. Arnheim, F. Millett, and M. A. Raftery, *Arch. Biochem. Biophys.*, 1974, **165**, 281.
[253] H. Lilja, H. Csopak, B. Lindman, and G. Fölsch, *Biochim. Biophys. Acta*, 1975, **384**, 277.
[254] P. Taylor, *J. Pharm. Sci.*, 1975, **64**, 501.
[255] G. M. Alter and J. A. Magnuson, *Biochemistry*, 1974, **13**, 4038.
[256] T. W. Johnson and I. M. Klotz, *Macrolecules*, 1974, **7**, 618.
[257] P. J. Cozzone, D. J. Nelson, and O. Jardetzky, *Biochem. Biophys. Res. Comm.*, 1974, **60**, 341.
[258] B. Furie, J. H. Griffin, R. J. Feldmann, E. A. Sokoloski, and A. N. Schechter, *Proc. Nat. Acad. Sci. U.S.A.*, 1974, **71**, 2833.

caused by substituting Gd^{3+} for Ca^{2+}. By taking a correlation time of 2.6×10^{-9} s from earlier work on similar systems, and neglecting scalar contributions, Gd^{3+}–^{31}P and Gd^{3+}–thymidine proton distances were calculated, which were in reasonable agreement with X-ray data. ^{31}P and ^{1}H T_1 and T_2 values were measured[259] to study the interaction between glucose 1- or 6-phosphates and manganese phosphoglucomutase. From these relaxation times, the distances between the manganese and phosphorus or protons of the bound glucose phosphates were calculated. The ^{31}P spectrum[260] of phosphate in free pyridoxal (I) or pyridoxamine phosphate (II) is a single resonance, if proton noise-decoupled, and shows a pH-dependent chemical shift. Deprotonation resulted in a shift to higher frequencies. When (I) or (II) was bound to aspartate transaminase the signal corresponded to a fully ionized phosphate monoester in the pH range 5.6—9.2. Deuterium relaxation times and diffusion coefficients of $CDCl_3$ have been measured[261] by pulsed methods in solutions of poly(vinyl pyrrolidone) (PVP) and polystyrene consisting of mixed solvents. The lifetime of the PVP–$CDCl_3$ hydrogen bond was estimated to be ca. 10^{-10} s, and correlation times for motions of solvent molecules about various axis were calculated. Andrasko and Forsén[262] studied the interaction between [$^{2}H_8$]edta and several proteins by measuring deuteron T_1 and T_2 times. The fraction of bound edta increased with decreasing pH, and from the frequency dependence of T_1 and T_2 correlation times for the macromolecular rotation were calculated.

Counter-ions. The T_1 values of $^{23}Na^+$ were measured[263] by pulsed techniques in solutions of beef brain and dogfish (sodium + potassium)-activated adenosine triphosphatase containing various amounts of ATP. In the absence of ATP there was almost no difference between the T_1 values, but this increased with increasing ATP concentration, suggesting binding of the Na^+ at specific sites on the enzyme. The variation[264] of the T_1 of $^{23}Na^+$ with degree of ionization in aqueous solutions of poly(styrene sulphonic acid) neutralized by both sodium and barium hydroxide suggested that the majority of the Ba^{2+} ions were closely associated with the polymer. The T_1 of $^{23}Na^+$ and $^{87}Rb^+$ in aqueous DNA solutions were shown[265] to increase with DNA concentration, this increase being interpreted in terms of a simple two-site (bound and free) model. Shporer and Frommer[266] measured the ^{17}O ($H_2^{17}O$) and $^{23}Na^+$ resonance of electrolyte solutions in contact with cellulose acetate membranes. Single-component relaxation times and a decreasing T_1 with decreasing water content led to the simple conclusion that water exchanged rapidly between solvent and membrane. The linewidth of $^{35}Cl^-$ bound to carboxypeptidase A was measured[267] as a function of pH and temperature. A binding site, associated with the glutamic acid-270 residue, was identified, and chemical exchange was

[259] D. G. Gadian, G. K. Radda, and R. E. Richards, *Biochim. Biophys. Acta*, 1974, **358**, 57.
[260] M. Martinez-Carrion, *European J. Biochem.*, 1975, **54**, 39.
[261] H. Rockelmann and H. Sillescu, *Z. phys. Chem. (Frankfurt)*, 1974, **92**, 263.
[262] J. Andrasko and S. Forsén, *Chimica Scripta*, 1974, **6**, 163.
[263] F. Ostroy, T. L. James, J. H. Noggle, A. Sarrif, and L. E. Hokin, *Arch. Biochem. Biophys.*, 1974, **162**, 421.
[264] J. J. van der Klink, D. Y. H. Prins, S. Zwolle, F. van der Touw, and J. C. Leyte, *Chem. Phys. Letters*, 1975, **32**, 287.
[265] J. Reuben, M. Shporer, and E. J. Gabbay, *Proc. Nat. Acad. Sci. U.S.A.*, 1975, **72**, 245.
[266] M. Shporer and M. A. Frommer, *J. Macromol. Sci., Phys.*, 1974, **B10**, 529.
[267] R. S. Stephens, J. E. Jentoft, and R. G. Bryant, *J. Amer. Chem. Soc.*, 1974, **96**, 8041.

found to dominate the linewidth. The linewidths of $^{35}Cl^-$ measured[268] in the presence of alkaline phosphatase and varying amounts of Zn^{2+} showed little change below Zn^{2+} : enzyme ratios of 2:1, but decreased at higher ratios, though no unique explanation for this decrease could be determined. Addition of orthophosphate did not change the $^{35}Cl^-$ linewidth in the absence of Zn^{2+}, but decreased it in the presence of Zn^{2+}. Marwedel et al.[269] have reported some brief results of $^{19}F\ T_1^{-1}$, in the presence of galactose oxidase, that indicated binding of F^- to the enzyme. The binding of Tl^+ to ATP in the presence of Mn^{2+} has been studied by ^{205}Tl resonance.[270] Increases in the relaxation rate of the enzyme–Mn^{2+}–Tl^+ complex were used to calculate $Mn^{2+} - Tl^+$ distances, and the results were compared with X-ray data.

[268] J.-E. Norne, H. Csopak, and B. Lindman, *Arch. Biochem. Biophys.*, 1974, **162**, 552.
[269] B. J. Marwedel, R. J. Kurland, D. J. Kosman, and M. J. Ettinger, *Biochem. Biophys. Res. Comm.*, 1975, **63**, 773.
[270] C. M. Grisham, R. K. Gupta, R. E. Barnett, and A. S. Mildvan, *J. Biol. Chem.*, 1974, **249**, 6738.

9
The Solid State

BY P. S. ALLEN

1 General Introduction

In the current volume the practice, begun in Volume 4, of including both spectral and relaxation studies in a single chapter on the Solid State is continued. By contrast with Volume 4, however, surface phenomena are covered in a separate chapter; otherwise the organization of this chapter bears a strong resemblance to that of Chapter 8 of Volume 4 on the Solid State, being based on the bulk magnetic properties of the materials under investigation, since these properties are usually determined by the same solid-state features which influence the nuclear resonance behaviour.

While not claiming to be completely exhaustive, the aim of this Report is to be comprehensive by including the overwhelming majority of papers published in the mainstream languages; dissertations have been excluded.

During the time-span of this Report two editions of the proceedings for the 18th Ampère Congress, held in Nottingham, U.K., in September 1974, have appeared. The commercially available edition of these proceedings[1] appeared in 1975, whereas the participants edition, published by the local organizing committee, is dated 1974. The papers contained in these proceedings, while not referred to individually in this Report, constitute almost 300 publications of relevance to the scope of this chapter. Textbooks oriented to the nuclear resonance of the solid state appear only infrequently. Nevertheless the current catchment period has spawned 'Nuclear Magnetic Resonance' by Rushworth and Tunstall.[2] This book is aimed at an elementary post-graduate level.

It must be recorded that numerical values quoted in this Report may not be in the same units as those used by the parent author. The reason for this is an effort to achieve overall consistency of units in the interests of comparison and standardization. For example, activation energy units have, where necessary, been consistently converted into $kJ\ mol^{-1}$.

2 Half-integral Spins in Diamagnetic Solids

A. Introduction.—The behaviour of nuclei of spin $\frac{1}{2}$ in diamagnetic solids is usually dominated by the internuclear dipole–dipole interaction. The effect which this

---

[1] 'Proceedings of the 18th Ampère Congress', ed. P. S. Allen, E. R. Andrew, and C. A. Bates, North-Holland, Amsterdam, 1975.
[2] F. A. Rushworth and D. P. Tunstall, 'Nuclear Magnetic Resonance', Gordon and Breach, London and New York, 1973.

interaction has on the nuclear resonance observables reflects both the motional and structural properties of the solid under investigation. As a result, its exploitation forms the basis of a large proportion of published nuclear resonance work. Papers which deal with the theoretical problems rather than individual materials are covered in subsection 2B, whereas those where the prime interest is in a specific material are covered in subsections 2C—2G.

Publications on image formation are somewhat sparse this year, and for convenience have been transferred to Section 6 on techniques. Moreover, though some semiconductors are diamagnetic, others display a weak paramagnetism. Rather than divide this group of materials into separate parts, the diamagnetic cases have been collected together with all conducting materials in Section 4. The number of papers on ^{3}He has increased markedly owing to the significant contributions nuclear resonance can make to the elucidation of its superfluid phases. Helium has accordingly been given a subsection to itself.

B. Theoretical Aspects.—Nuclear magnetic resonance in solids has been profoundly influenced by the spin-temperature theory.[3] To a large extent the applications of this theory have been limited to the high-temperature approximation, which is valid under most experimental conditions. This approximation leads to a linear theory, where one evaluates physical quantities to first order with respect to the inverse temperature. Goldman *et al.*[4] have now extended the theory, and have subsequently analysed the situation where the temperature is too low to justify the use of the linear (high-temperature) approximation. Their study is limited to systems containing only one species of spin $\frac{1}{2}$, in high external magnetic fields. Their Hamiltonian consists of a Zeeman interaction plus spin–spin interactions which commute with the Zeeman term and which they suppose to be purely dipolar. Their calculation yields expressions for physical quantities which are exact and valid to all orders with respect to the Zeeman interaction, but which are power expansions in inverse temperature with respect to the spin–spin interaction.

A general source of confusion with all theories is the employment of equivalent, but seemingly different, formalisms. A classic case of such confusion[5] was the turmoil produced by the alternative formalisms of Quantum Mechanics, before the reconciling work of Von Neumann. The theory of nuclear resonance is no exception to this syndrome. Over the past several years a variety of statistical mechanical formalisms have been applied to spin–lattice relaxation theory. For example the projection operator formalism has been used by Argyres and Kelly[6] and by Albers and Deutch,[7] whereas Kubo[8] and Freed[9] have employed generalized cumulant expansions. Two recent publications,[10,11] however, attempt to remove any confusion by showing the equivalence of the various methods. The more substantial of

[3] M. Goldman, 'Spin Temperature and Nuclear Magnetic Resonance in Solids', Oxford, 1970.
[4] M. Goldman, J. F. Jacquinot, M. Chapellier, and Vu Hoang Chan, *J. Magn. Resonance*, 1975, **18**, 22.
[5] M. Jammer, 'The Philosophy of Quantum Mechanics', Wiley, New York, 1974.
[6] P. N. Argyres and P. L. Kelley, *Phys. Rev. (A)*, 1964, **134**, 98.
[7] J. Albers and J. M. Deutch, *J. Chem. Phys.*, 1971, **55**, 2613.
[8] R. Kubo, *J. Phys. Soc. (Japan)*, 1962, **17**, 1100.
[9] J. H. Freed, *J. Chem. Phys.*, 1968, **49**, 376.
[10] B. Yoon, J. M. Deutch, and J. H. Freed, *J. Chem. Phys.*, 1975, **62**, 4687.
[11] S. Emid, *Physica*, 1974, **78**, 563.

these two papers, by Yoon, Deutch, and Freed, first establishes the equivalence of the expressions for the spin density matrix obtained by the two projection operator methods. It then develops two variants of the cumulant expansion method: first, a 'total time-ordered cumulant', which is found to be equivalent to the projection operator methods and more convenient to use in the frequency domain (*i.e.* for calculating spectra), and second, a 'partial time-ordered cumulant' which has different characteristics from the projection operator methods and is more convenient to use in the time domain. It is noted by these authors, however, that when the lattice model is that of a classical stationary Markov process, the stochastic Liouville equation turns out to be more convenient to use than either the projection operator or the cumulant expansion methods.

With regard to spectra alone, Dugdale[12] uses quantum field theory to present a generalized treatment of exchange averaging, which reduces to the perturbation results of Kubo and Tomita[13] in the limiting cases of strong and weak averaging. Although this paper is primarily intended for consumers in the area of e.s.r., it cannot be neglected by nuclear spectroscopists. A similar problem of the bandshape of a system of two nuclear species coupled by the scalar interaction $h J\mathbf{I}_1 . \mathbf{I}_2$, but where the spin–lattice relaxation rates of each species may be very different, is evaluated by Hughes *et al.*[14] They compute the bandshapes as a function of the relative magnitude of the exchange frequency and the respective relaxation rates.

It is often convenient to describe the shape of a nuclear resonance spectrum in a solid in terms of its moments. Indeed, a number of papers on techniques in Section 6 are specifically related to the accurate determination of such parameters. For some time, moment data have been used to gain structural information in rigid lattices and motional information in the presence of diffusion or rotation. In a recent paper Polak *et al.*[15] have developed the mathematical apparatus needed to correct the intermolecular second moment for molecular vibrations. They illustrate their mathematics with calculations for *trans-trans*-muconodinitrile, which exhibits a large (17°) amplitude of libration, pyrene, and anthracene. It is to be noted, however, that highly accurate moment data are required to reveal these small vibrational effects.

C. Structural Studies (Broadline Techniques).—The nuclear resonance spectrum is extensively employed to investigate the water structure in crystalline hydrates. By far the dominating contribution to the proton spectrum from hydrates is the intra-proton-pair contribution. However, if accurate data are to be obtained the inter-pair contribution must also be known, and this was, for some time, difficult to obtain. Boden, Levine, and co-workers have, by developing the experimental technique (see, *e.g.*, Section 6A), shown empirically[16] that in polycrystalline samples the inter-pair contribution to the second moment, M_2(inter), is five-sixths of that [$^{vv}M_2$ (inter)] calculated on the basis of Van Vleck's equation.[17] In a more recent paper,[18] they adopt a rectangular model of two proton pairs and evaluate the inter-

---

[12] D. E. Dugdale, *J. Phys. (C)*, 1974, **7**, 3758.
[13] R. Kubo and K. Tomita, *J. Phys. Soc. (Japan)*, 1954, **9**, 888.
[14] D. G. Hughes, M. R. Smith, and D. A. M. Switzer, *J. Chem. Phys.*, 1974, **60**, 4890.
[15] M. Polak, M. Sheinblatt, and U. Shmueli, *J. Magn. Resonance*, 1974, **16**, 252.
[16] N. Boden, Y. K. Levine, M. Mortimer, and R. T. Squires, *Phys. Letters*, 1974, **46A**, 329.
[17] J. H. Van Vleck, *Phys. Rev.*, 1948, **74**, 1168.
[18] N. Boden and Y. K. Levine, *Mol. Phys.*, 1975, **29**, 1221.

pair second moment as a function of d/a, where a and d are respectively the intra- and inter-pair proton separations. For a rigid rectangle they find that M_2 (inter) approaches 5 $^{vv}M_2$ (inter)/6 asymptotically as d becomes much greater than a. Furthermore, for rapidly flipping pairs, they find the M_2 (inter) to $^{vv}M_2$ (inter) ratio fixed at 5/6, independent of d/a. The origin of this 'loose coupling' reduction factor of 5/6 is in the quenching of non-secular terms in the inter-pair dipolar Hamiltonian. The same group have employed the same model[19, 20] to account for the proton echo response to $90 - \tau - \beta_\theta$ pulse sequences, where τ is the pulse separation, β is the length of the second pulse, and θ is the r.f. phase difference between pulses. They establish which features of the echo response are directly related to the inter-pair interactions and independent of the intra-pair interactions. Moreover, their analysis is applied to experimental data on $CaSO_4,2H_2O$; $K_2C_2O_4,H_2O$; $AlCl_3,6H_2O$; and $BaBr_2,2H_2O$ to confirm the predicted behaviour.

During an investigation of the possibility of solid–solid phase transitions in the homologous series of n-alkylammonium halides, Tsau and Gilson[21] observed a non-reversible behaviour of the proton n.m.r. spectrum with temperature during the initial heating cycle but not for subsequent cycles. They conclude that this behaviour is due to a change in the crystal structure on initial heating (as the sample is taken above a transition temperature) which does not revert on cooling. From n.m.r. and differential scanning calorimetry results they suggest that this transition temperature will be above room temperature if the chain length is greater than 10 carbon atoms.

Structural studies making use of other spin-$\frac{1}{2}$ nuclei have been reported for amorphous BeF_2,[22] selenium and As_2Se_3,[23] and for a series of lead compounds.[24] O'Reilly and Peterson[22] used the second moment of the ^{19}F resonance to help clarify the proportion of vacancies in the β-quartz structure of amorphous BeF_2. Their data are consistent with a recently suggested[25] structure for BeF_2 glass containing 6% randomly distributed ionic vacancies. Bishop and Taylor[23] employed the ^{77}Se spectrum to establish that markedly different chemical-shift behaviours occur in the glassy states of selenium and As_2Se_3. An isotropic shift of 0.064% relative to a H_2SeO_2 standard is obtained for the former, while the latter displays a 0.111% shift. Bishop and Taylor postulate that this effect may be used to evaluate structures in mixed chalcogenide glass systems. The ^{207}Pb resonance was used by Kim and Bray[24] to gain information about the bonding character and site symmetry of lead atoms in $Pb(NO_3)_2$, $PbCl_2$, Pb_3O_4, PbO_2, and PbO in its orthorhombic and tetragonal forms.

D. High Resolution in Solids.—Experiments exploiting the currently developing field of high-resolution techniques in solids have in essence been structural investigations, rather than enquiries into the dynamical properties of materials. Here, 'structural' should be interpreted in its wider sense of incorporating electronic structure, rather than in the narrower crystallographic sense. For this reason, this

[19] N. Boden, Y. K. Levine, and R. T. Squires, *Chem. Phys. Letters*, 1974, **28**, 523.
[20] N. Boden, Y. K. Levine, D. Lightowlers, and R. T. Squires, *Mol. Phys.*, 1975, **29**, 1877.
[21] J. Tsau and D. F. R. Gilson, *Canad. J. Chem.*, 1974, **52**, 2421.
[22] D. E. O'Reilly and E. M. Peterson, *J. Chem. Phys.*, 1975, **62**, 2512.
[23] S. G. Bishop and P. C. Taylor, 'Proceedings of the 11th International Conference on the Physics of Semiconductors', 1972, p. 576.
[24] K. S. Kim and P. J. Bray, *J. Magn. Resonance*, 1974, **16**, 334.
[25] A. H. Narten, *J. Chem. Phys.*, 1972, **56**, 1905.

section is now placed adjacent to the structural studies carried out by the more traditional techniques. The papers are grouped according to their method of line narrowing. High-resolution techniques in solids have been reviewed recently by Vaughan.[26]

Multiple-pulse Methods. The frequency resolution achievable in multiple-pulse experiments is limited by residual interactions not fully removed by the multiple-pulse sequence. If this residual broadening were independent of frequency over the extent of the high-resolution spectrum to be observed, then the experimental spectrum would be uniformly broadened, and the true spectrum obtained by simple deconvolution. Unfortunately it is not! For example, the line-narrowing ability varies by about a factor of three on going from resonance to 3 kHz off-resonance. Garroway, Mansfield, and Stalker,[27] in a paper devoted to the limits to resolution in multiple-pulse n.m.r., identify three mechanisms of frequency-dependent broadening and suggest several new pulse sequences which reduce the broadening caused by these sources and extend the frequency range of highest resolution. Moreover, the spectrometer alignment for multiple-pulse-sequence experiments is discussed in detail, and the effect of a particular phase alignment error is considered in depth.

Grosescu et al.[28] have applied the multiple-pulse technique to a single crystal of maleic acid and have made independent determinations of the magnetic shielding tensors for the four magnetically inequivalent protons forming hydrogen bonds. They furthermore exclude the existence of a bifurcated hydrogen bond in maleic acid. By analysing their data in conjunction with those for protons in hydrogen bonds in malonic acid and oxalic acid, they are able to make the following propositions: (i) in single-bond systems there is a direct correspondence between the bond directions and the principal directions of the shielding tensors. The dominant contribution to the shielding anisotropy stems from the quadrupolar ground-state term; (ii) the double-bond principal directions are reflected in those of the shielding tensors of the olefinic protons, indicating a predominant contribution from excited configurations; (iii) the magnetic shielding tensor is a transferable physical index of intermolecular hydrogen bonds. An example of a non-hydrogen-bonded proton is the hydroxyl proton in $Ca(OH)_2$. The chemical shift tensor of this proton has been measured by Schreiber and Vaughan,[29] using an eight-pulse cycle, and found to be axially symmetric with components $\sigma_\perp = -9.3 \pm 1$ p.p.m. and $\sigma_\parallel = +4.7 \pm 1$ p.p.m. relative to TMS. By means of the WAHUHA four-pulse cycle, Burghoff et al.[30,31] have resolved the proton–phosphorus dipolar splitting in KH_2PO_4.

Multiple-pulse methods have been applied to polymers by Lind and Ames[32] and by Garroway et al.[33] The former group of authors were able to resolve the aromatic from the methyl protons in a 50 wt % polydimethylsiloxane/bisphenol-A polycarbonate copolymer at room temperature, but were unable to detect an anisotropy of the chemical shielding. The latter group observed the ^{19}F resonance

[26] R. W. Vaughan, *Ann. Rev. Mater. Sci.*, 1974, **4**, 21.
[27] A. N. Garroway, P. Mansfield, and D. C. Stalker, *Phys. Rev. (B)*, 1975, **11**, 121.
[28] R. Grosescu, A. M. Achlama, U. Haeberlen, and H. W. Spiess, *Chem. Phys.*, 1974, **5**, 119.
[29] L. B. Schreiber and R. W. Vaughan, *Chem. Phys. Letters*, 1974, **28**, 586.
[30] U. Burghoff, G. Scheler, and R. Muller, *Phys. Status Solidi (A)*, 1974, **25**, K31.
[31] U. Burghoff, H. Rosenberger, and R. Zeiss, *Phys. Status Solidi (A)*, 1974, **26**, K171.
[32] A. C. Lind and D. P. Ames, *J. Polymer Sci. (B)*, 1974, **12**, 339.
[33] A. N. Garroway, D. C. Stalker, and P. Mansfield, *Polymer*, 1975, **16**, 161.

from an aligned specimen of PTFE fibres at 77 K, from which they deduced that the most screened component of the ^{19}F chemical shielding tensor lies along the C—F bond and that the least screened is aligned at *ca.* 20° to the chain axis. At room temperature, molecular rotation about the chain axis yields an axially symmetric shielding tensor.

Double-resonance Method. Waugh and co-workers[34] have extended their analysis of durene[35] to single crystals of hexamethylbenzene, hexaethylbenzene, and pentamethylbenzene, for which they quote detailed ^{13}C shielding tensors of the ring and methyl carbons, the ring, methylene, and methyl carbons, and the ring carbons, respectively. They concluded that although substitution in the benzene ring substantially affects the shielding for ring carbons, the effect is essentially localized on the ring carbon at which the substitution occurs and is insensitive to substitution on neighbouring carbon atoms. For a given substituent at a ring carbon, the shielding at this carbon does not differ much from molecule to molecule. The orientations of the eigenvectors are consistent, within experimental error, with the point-group symmetry expected for the free molecule, and there is no positive evidence of any intermolecular perturbations of the shielding. In all the cases studied, the largest shielding for ring carbons occurs when the field is closely normal to the ring and the smallest for a field directed along the bond joining the ring carbon and the substituent or hydrogen. The methyl carbons exhibit small anisotropies, and their principal axes are less correlated to bond directions.

The search for trends in the anisotropies of the ^{13}C chemical shielding tensors of various functional groups has also been focused on the carboxy-group by investigations on solid ammonium tartrate,[36] solid dimethyl oxalate,[37] and solid ammonium hydrogen malonate.[38] It appears that the greatest shielding for the carboxyl carbon occurs when B_0 is perpendicular to the ^{13}COO$^-$ plane. The smallest shielding always occurs when B_0 is in the carboxyl plane, but its direction depends on the molecular structure. For example, in ammonium tartrate it bisects the C—O bonds, whereas in ammonium hydrogen malonate it makes an angle of 25° to this bisector and is rotated towards the long C—O bond. The other functional groups studied as a matter of completeness in these papers are the hydroxy-group,[35] the carbonyl group,[36] and the methylene group.[37]

An important consideration in the design and efficiency of double-resonance experiments is the dynamics of polarization transfer from the *I* to the *S* spin system, since this determines the sensitivity enhancement and the time-scale accessible, and therefore the range of materials which can be studied. Pines and Shattuck[39] have illustrated this problem by considering differing methods of polarization transfer to establish which is most efficient in solid adamantane, and they recorded the cross-polarization times as a function of the r.f. field intensity, which is in turn related to the expenditure of r.f. power.

[34] S. Pausak, J. Tegenfeldt, and J. S. Waugh, *J. Chem. Phys.*, 1974, **61**, 1338.
[35] S. Pausak, A. Pines, and J. S. Waugh, *J. Chem. Phys.*, 1973, **59**, 591.
[36] A. Pines, J. J. Chang, and R. G. Griffin, *J. Chem. Phys.*, 1974, **61**, 1021.
[37] A. Pines and F. Abramson, *J. Chem. Phys.*, 1974, **60**, 5130.
[38] J. J. Chang, R. G. Griffin, and A. Pines, *J. Chem. Phys.*, 1975, **62**, 4923.
[39] A. Pines and T. W. Shattuck, *J. Chem. Phys.*, 1974, **61**, 1255.

Other Papers. Papers which do not fall neatly into either the muitiple-pulse or the double-resonance category include a study of the ^{19}F chemical shifts in the alkali-metal fluorides by Sears,[40] who makes use of the experimental technique of Kunitomo. Sears attempts to correlate his ^{19}F chemical shifts with the overlap repulsion energy between ions in units of the mean excitation energy of F^- in the individual fluorides. Schmiedl[41] has devoted himself not to the experimental evaluation of chemical shielding anisotropies, but to the problems occurring in their interpretation. He presents a possible interpretation of the chemical shift, using only the electron density of the system in the absence of an external field.

E. Motional Studies (Classical Regime).—In this subsection on motional studies, the majority of which make use of the temperature dependence of relaxation rates, the papers are organized, by and large, in order of increasing complexity of molecular motion. Diffusion is dealt with first, before graduating through the various forms of molecular reorientation. It is to be noted, however, that plastic crystals, clathrates, ferroelectrics, and polymer systems have been filtered out and are considered separately towards the end of the subsection.

When Slichter and Ailion[42] introduced their strong-collision theory of nuclear spin relaxation, they extended to very slow motions the technical range of motional studies using n.m.r. by several orders of magnitude. Their theory has been utilized most to study slow translational diffusion in solids, and the application to slow reorientation has been somewhat neglected. Goren and Pintar,[43] therefore, present a calculation of the Slichter–Ailion order parameter for three cases relevant to molecular reorientation. They are (i) two and four tetrahedrally co-ordinated protons, (ii) the two water protons, and (iii) the three protons formed by a water molecule adjacent to a hydroxy-group.

Diffusion. In a series of papers Wolf has progressively extended the isotropic random-walk diffusion model of Torrey to anisotropic diffusion in single crystals. Using the random-walk model, he first of all calculates[44] the correlation functions, which govern the nuclear relaxation times, in a representation involving a series expansion in increasing powers of time, the leading term of which is the simple exponential correlation function. This model is further extended[45] to correlated diffusion, to take account of the fact that in a crystal lattice successive jumps of atoms caused by the migration of vacancies are not independent of each other. The anisotropy of the diffusion is expected to affect T_1 on the low-temperature side of its minimum, while the shape and width of the T_1 and $T_{1\rho}$ minima are expected to be sensitive to the details of the diffusion mechanism. For example, diffusion *via* monovacancies in monoatomic b.c.c. or f.c.c. crystals should produce considerable differences in shape and width from those produced by random-walk diffusion. Wolf then proceeds[46] to apply this model to the strong-collision theory of Slichter and Ailion for low-field rotating-frame relaxation.

[40] R. E. J. Sears, *J. Chem. Phys.*, 1974, **61**, 4368.
[41] H. Schmiedl, *Phys. Status Solidi (B)*, 1975, **67**, K27.
[42] C. P. Slichter and D. C. Ailion, *Phys. Rev. (A)*, 1964, **135**, 1099.
[43] S. Goren and M. M. Pintar, *Fizika (Zagreb)*, 1974, **6**, 91.
[44] D. Wolf, *J. Magn. Resonance*, 1975, **17**, 1.
[45] D. Wolf, *Phys. Rev. (B)*, 1974, **10**, 2710.
[46] D. Wolf, *Phys. Rev. (B)*, 1974, **10**, 2724.

Two-spin Reorientation. The dynamics of water molecules in natrolite ($Na_2Al_2Si_3O_{10}, 2H_2O$) have been studied by Thompson *et al.*,[47] who employed both spectral and relaxation techniques over a temperature range 4.2—600 K. Over such a wide temperature range the water motion takes on many forms, ranging from infrequent 180° flips to diffusion out of the material altogether.

Three-spin Reorientation. Andrew *et al.*[48] have now given a fuller account of their work on 10 polycrystalline amino-acids $\overset{+}{N}H_3CHRCOO^-$. The activation energies for the NH_3 groups remain as cited last year, but the methyl-group rotation activation energies shown in Table 1 have now been included. By observing the temperature dependence of T_1, $T_{1\rho}$, and T_{1D}, Thompson and Kydon[49] have been able to

Table 1 *Methyl-group rotation activation energies for some amino-acids*

Amino-acid	R	E_a/kJ mol^{-1}
L-Alanine	CH_3	21
L-Isoleucine	$CH(CH_3)CH_2CH_3$	13
DL-Leucine	$CH_2CH(CH_3)_2$	14
L-Valine	$CH(CH_3)_2$	12

confirm the existence of the OH_3^+ group in hydrated gallium sulphate. They therefore support the formula $(H_3O)Ga_3(OH)_6(SO_4)_2$ for this compound. Moreover, they quote activation energies of (7.5 ± 1) kJ mol^{-1} for OH_3^+ reorientation and (9 ± 1) kJ mol^{-1} for OH_3^+ tumbling. Kadaba *et al.*[50] have detected in AsF_3 a bandwidth transition of the ^{19}F spectrum below 200 K. The change in the ^{19}F second moment is too large to be due to three-fold reorientation of the fluorine triangles; the nature of the motion causing the bandwidth transition is not therefore identified.

Other work on the triangular configuration of three spins appears in the next subsection on quantum effects, and in Section 3 on $I > \frac{1}{2}$ nuclei.

Four-spin Reorientation. Activation energies for some tetrahedrally co-ordinated four-spin systems are quoted in Table 2, together with their corresponding references. Smith[51] has attempted to rationalize the activation energies of certain four-spin tetrahedra in terms of their torsional energy states. His suggestion (that the activation energy corresponds to the energy difference between the ground harmonic level and the energy levels just above the highest anharmonic level) is more precise than, and therefore supersedes, the earlier notion that it corresponds to the barrier amplitude minus the zero-point energy. The data on NH_4Cl are taken from a series of four papers by Mandema and Trappeniers,[52-55] which constitutes a 'tour de force' on the proton spin–lattice relaxation in that material as a function of temperature and pressure. They begin by obtaining an equation for T_1 which is the sum of four exponentials, prior to considering two different models for the reorienta-

[47] R. T. Thompson, R. R. Knispel, and H. E. Petch, *Canad. J. Phys.*, 1974, **52**, 2164.
[48] E. R. Andrew, W. S. Hinshaw, and M. G. Hutchings, *J. Magn. Resonance*, 1974, **15**, 196.
[49] R. T. Thompson and D. W. Kydon, *J. Chem. Phys.*, 1974, **61**, 1813.
[50] P. K. Kadaba, J. Slivnik, R. Blinc, J. Pirnat, and Z. Trontelj, *Mol. Phys.*, 1975, **29**, 1485.
[51] D. Smith, *J. Chem. Phys.*, 1975, **62**, 4497.
[52] W. Mandema and N. J. Trappeniers, *Physica*, 1974, **76**, 73.
[53] W. Mandema and N. J. Trappeniers, *Physica*, 1974, **76**, 85.
[54] W. Mandema and N. J. Trappeniers, *Physica*, 1974, **76**, 102.
[55] W. Mandema and N. J. Trappeniers, *Physica*, 1974, **76**, 123.

tion. Although it is not possible to discriminate between these models from the temperature dependence of T_1, they succeed (from the angular dependence of T_1 in a single crystal) in highlighting, as the most probable mechanism, one which allows direct reorientation to all other equilibrium orientations. The rotational activation energies for the NH_4 group, above and below the order–disorder transition in NH_4Cl, which are quoted in Table 2 correspond to a pressure of 1 atm. As the pressure increases, this transition becomes less sharp, and at 950 atm it changes its character altogether. Yet another investigation of NH_4Cl in the region of its λ transition has been made by Shimomura et al.[56]

Table 2 Activation energies for some tetrahedrally co-ordinated four-spin systems

Material	Four-spin system	E_a/kJ mol^{-1}	Ref.
NH_4Cl ($T < T_c$)	NH_4	21.2 ± 0.1	54
NH_4Cl ($T > T_c$)	NH_4	17.64 ± 0.06	54
$(NH_4)_2RuCl_6$	NH_4	5.1	57
$(NH_4)_2PdCl_6$	NH_4	1.8	57
$(NH_4)_2PtCl_6$	NH_4	1.6	57
GeF_4	GeF_4	80.0	58

More work on four-spin tetrahedra will be found in Section 3 on investigations employing nuclei of $I > \frac{1}{2}$.

Six-spin Reorientation. The six-spin group, in its octahedral configuration and containing ^{19}F nuclei, is a highly symmetric and mobile rotor, as evidenced by the data quoted in Table 3.[59, 61, 62] Skjaeveland and Svare,[59] who measured the temperature dependent linewidths in the metal fluorosilicates, employed the approximate

Table 3 Activation energies for some octahedral six-spin systems containing ^{19}F

Material	Six-spin group E_a/kJ mol^{-1}	Other motions studied; E_a/kJ mol^{-1}	Ref.
$MnSiF_6,6H_2O$	35	$Mn(H_2O)_6^{2+}$; 35	59
$FeSiF_6,6H_2O$	28	$Fe(H_2O)_6^{2+}$; 34	59
$CoSiF_6,6H_2O$	31	$Co(H_2O)_6^{2+}$; 52	59
$NiSiF_6,6H_2O$	23	—	
$CuSiF_6,6H_2O$	19	$Cu(H_2O)_6^{2+}$; 54	59
$ZnSiF_6,6H_2O$	23	$Zn(H_2O)_6^{2+}$; 52	59
$(ND_4)_2GeF_6$	56 ± 7	ND_4^+; 4.2 ± 0.4	61
C_6H_6	17.2		62

relation proposed by Waugh and Fedin[60] to obtain their activation energies, viz. E_a/cal mol$^{-1} \simeq 37 T_c$, where T_c is the temperature at which the resonance line has been narrowed by half the maximum. In addition, they observed the proton reson-

[56] K. Shimomura, R. Sugimoto, and H. Negita, *J. Sci. Hiroshima University (A)*, 1974, **38**, 71.
[57] M. Bonori and M. Terenzi, *Chem. Phys. Letters*, 1974, **27**, 281.
[58] L. Niemela and J. Makela, *Phys. Letters*, 1974, **50A**, 67.
[59] S. M. Skjaeveland and I. Svare, *Phys. Scripta*, 1974, **10**, 273.
[60] J. S. Waugh and E. I. Fedin, *Soviet Phys. Solid State*, 1963, **4**, 1633.
[61] J. E. Tuohi, E. E. Ylinen, and L. K. E. Niemela, *Chem. Phys. Letters*, 1974, **28**, 35.
[62] J. Wendt and F. Noack, *Z. Naturforsch.*, 1974, **29a**, 1660.

ance in order to monitor the motion of the $X(H_2O)_6^{2+}$ octahedral complex. The activation energies of the water complexes were found to correlate with the densities of the corresponding fluorosilicates, whereas those of the SiF_6^{2-} groups did not. All data on other six-spin systems, including the planar configuration of benzene,[62] were obtained from spin–lattice relaxation measurements. These new benzene data provide evidence of correlated motions of the C_6H_6 molecules.

Miscellaneous Molecular Motion. In order to elucidate the correct structure for the trigonal phase of sodium hydrosulphide, Jeffrey[63] carried out both bandshape and relaxation measurements. He obtained the best agreement for his data if he assumed that the SH^- ion underwent motion in which the proton flipped back and forth between two equilibrium sites on the trigonal axis. Andrew *et al.*[64] have illustrated that P_4S_3 is an exceptionally mobile molecule in the solid state. As a result the dipolar interaction plays only a minor role in the spectral and relaxational behaviour of ^{31}P, and this material provides an example, unusual for a solid, where the relaxation is dominated by the chemical shielding anisotropy and the spin–rotation interaction. Molecular motion in solid complexes is a subject of study undertaken by Fyfe, who used bandshape measurements to investigate the dynamic structure of a series of π–π molecular complexes of the acceptor molecules tetracyanoethylene, pyromellitic dianhydride, tetracyanobenzene, and hexafluorobenzene with the donor molecules pyrene[65] and naphthalene.[66]

Motion in Plastic Crystals. The globular molecule adamantane continues to attract attention because of its high degree of molecular mobility. Graham and Choi[67] have investigated the plastic phase of both adamantane and its twist-boat isomer, twistane (tricyclo[4,4,0,0]decane). Bearing in mind the uncertainty of activation energies from bandshape transitions, their results on adamantane are reasonably consistent with the previous work of Resing.[68] Their new data on twistane suggest that reorientation and diffusion (which affects the n.m.r. observables as much as 100 K below the melting point) have activation energies of 7.2 and 67 kJ mol^{-1}, respectively. In order to explain the proton relaxation data from PH_3 in its plastic crystalline phase, Boden and Folland[69] found it necessary to employ the *M*-diffusion limit of rotational diffusion (only the orientation of the angular momentum is randomized on collision), rather than the *J*-diffusion limit (both the magnitude and orientation of the angular momentum are randomized), which has been successful for small spherical and symmetric-top molecules in liquids. In addition they found a discontinuous change in the reorientational activation energy from 2.5 to 1.3 kJ mol^{-1} at the melting point. Both nuclear resonance and radiotracer techniques have been used to evaluate self-diffusion in the plastic solid hexamethyldisilane.[70]

---

[63] R. Jeffrey, *Canad. J. Phys.*, 1974, **52**, 2370.
[64] E. R. Andrew, W. S. Hinshaw, and M. G. Hutchings, *Chem. Phys. Letters*, 1974, **27**, 96.
[65] C. A. Fyfe, *J. C. S. Faraday II*, 1974, **70**, 1633.
[66] C. A. Fyfe, *J. C. S. Faraday II*, 1974, **70**, 1642.
[67] J. D. Graham and J. K. Choi, *J. Chem. Phys.*, 1975, **62**, 2509.
[68] H. A. Resing, *Mol. Crystals Liquid Crystals*, 1969, **9**, 101.
[69] N. Boden and R. Folland, *Chem. Phys. Letters*, 1975, **32**, 127.
[70] A. V. Chadwick, J. M. Chezeau, R. Folland, J. W. Forrest, and J. H. Stange, *J.C.S. Faraday I*, 1975, **71**, 1610.

Motional Studies in Ferroelectrics. A review of n.m.r. studies of collective atomic motion near the ferroelectric phase transition has been published recently by Bjorkstam.[71] He has also contributed[72] to the literature on KDP-type ferroelectrics by analysing previously published data both in terms of the 'collective model–fluctuation dissipation theorem' method and in terms of critical indices. KDP-type ferroelectrics more recently privileged by public discussion include ammonium-doped RbH_2PO_4,[73] KH_2AsO_4,[74] and KH_2PO_4 itself.[75]

The effect of the reorientational motion of ammonium ions on the onset of ferroelectricity is the subject of a comprehensive proton resonance study of $(NH_4)_2Cd_2(SO_4)_3$, the ammonium analogue of langbeinite, by McDowell *et al.*,[76] and of sodium ammonium tartrate by Morimoto.[77, 78] McDowell and co-workers have provided the first experimental evidence of the order–disorder nature of the ferroelectric transition in the former material. The hydrogen-bonded trihydrogen selenite ferroelectric $RbH_3(SeO_3)_2$ has provided evidence[79] that its phase transition below 113 K is triggered by the hydrogen atoms. However, the same authors cannot confirm that the antiferroelectric phase-transition in $CsH_3(SO_3)_2$ is governed by the hydrogen dynamics. An additional antiferroelectric in which the decisive proton dynamics have recently been investigated[80, 81] is copper formate, for which Zumer and Pirs conclude that the correct description is one of a simple compressible Ising model, which gives a temperature dependence of the molecular reorientation time and of the spontaneous sub-lattice polarization.

Motional Studies in Polymer Systems. McBrierty[82] has recently reviewed the application of nuclear resonance to solid polymers, including both the structural and the motional aspects. Together with Wardell and Douglass,[83] he has also compared the motional behaviour in the homopolymers polystyrene and polybutadiene with that in the block copolymer styrene–butadiene–styrene. Their results show that the vigorous motions in the polybutadiene phase of the copolymer instigate motion in the polystyrene phase prematurely, and that the motion in polybutadiene is more hindered in the copolymer than in the pure polymer as a result of the cross-linking effect of the polystyrene domains. The conclusions of reference 83 were drawn from relaxation-time data. Bandshape data are, on the other hand, made use of by Zachmann[84] to investigate the glass transition in polymers, and by Sobottka *et al.*[85] to investigate molecular motion in the ethylene–vinyl acetate copolymer. Dybowski and Vaughan[86] employed the powerful multiple-pulse technique to study the non-

[71] J. L. Bjorkstam, *Adv. Magn. Resonance*, 1974, **7**, 1.
[72] J. L. Bjorkstam and C. Wei, *Ferroelectrics*, 1974, **7**, 127.
[73] R. Blinc, G. Lahajnar, A. Levstik, R. Osredkar, and L. A. Shuvalov, *Phys. Status Solidi (B)*, 1974, **65**, 397.
[74] T. Takoshima, I. Tatsuzaki, and T. Hikita, *J. Phys. Soc. (Japan)*, 1974, **37**, 574.
[75] L. C. Gupta and R. Vijayaraghavan, *Solid State Chem.*, 1974, 703.
[76] C. A. McDowell, P. Raghunathan and R. Srinivasan, *Mol. Phys.*, 1975, **29**, 815.
[77] K. Morimoto, *J. Phys. Soc. (Japan)*, 1974, **37**, 1714.
[78] K. Morimoto, *J. Phys. Soc. (Japan)*, 1974, **37**, 1574.
[79] A. A. Silvidi and T. D. Workman, *Ferroelectrics*, 1974, **6**, 183.
[80] C. I. Massara and E. R. Mognaschi, *Ferroelectrics*, 1974, **7**, 115.
[81] S. Zumer and J. Pirs, *Ferroelectrics*, 1974, **7**, 119.
[82] V. J. McBrierty, *Polymer*, 1974, **15**, 503.
[83] G. E. Wardell, V. J. McBrierty, and D. C. Douglass, *J. Appl. Phys.*, 1974, **45**, 3441.
[84] H. G. Zachmann, *J. Polymer Sci., Polymer Symposia*, 1973, **43**, 111.
[85] J. Sobottka, F. Keller, and K. Wunderlich, *Faberforsch. Textiltech*, 1974, **25**, 352.
[86] C. R. Dybowski and R. W. Vaughan, *Macromolecules*, 1975, **8**, 50.

isotropic motion in *cis*-polyisoprene in both a natural gum sample and in a carbon-filled, sulphur-vulcanized sample. They accordingly found it possible to characterize the secular dipolar broadening as being due to anisotropic or restricted motion, and to associate an order parameter of 0.006 to characterize the degree of anisotropy within the natural rubber sample. The size of the chemical-shift anisotropy could also be determined.

Motional Studies in Clathrates. Since they are more relevant to a surface-study enquiry, experiments involving zeolites are covered in the chapter devoted to surfaces. In contrast, the motion of guest molecules encaged in the clathrate hydrates, when compared with their motion in their parent material, gives valuable insight into the effects on this motion of changing the crystalline environment. Enclathration in D_2O ice also provides a useful method of isolating guest molecules from intermolecular magnetic interactions and thereby allows the corresponding intramolecular interactions to be studied in more detail. A survey of the bandshape characteristics due to magnetic dipolar interactions for many guest molecules in clathrate deuteriates has been provided by Garg and Davidson.[87] The same group have also illustrated the great freedom of movement of the guest molecules isoxazole[88] and tetrahydrofuran[89] which, due to an activation energy of 3.85 kJ mol^{-1}, only exhibit a rigid-lattice bandshape below 10 K.

F. Quantum Effects in Molecular Solids.—Nuclear resonance has been applied[90] to the task of continuously monitoring the relative ortho and para populations of solid hydrogen. It has been shown that the integral of the nuclear resonance absorption curve can be calibrated and used to serve as a measure of the ortho concentration at a fixed temperature. Different calibration curves are obtained at each temperature.

The quantum behaviour of the three spin-$\frac{1}{2}$ triangle has been the subject of most attention this year. Beginning with the conventional technique of bandshape studies, Mottley and Johnson[91] have reported the bandshapes of CH_3 groups in the matrix-isolated, partially deuteriated molecules CH_3CD_2OD, CH_3CD_2Cl, CH_3CD_2Br, and CH_3CD_2I at 4.2 K. They attempt to interpret these spectra in terms of theoretical ground torsional state spectra, and find that in order to make experiment and theory agree, it is necessary to superpose the theoretical spectra corresponding to two different barrier heights in equal weights. They therefore conclude that in each of their samples there must be two environmentally inequivalent methyl groups, one of which is identical with the neat parent compound and the other of which corresponds to a gas-phase molecule. Since these data were obtained at 4.2 K, the interpretation of Johnson and Mottley in terms of ground-state spectra is reasonable. However, at higher temperatures, though still below those at which a classical calculation might be valid, Allen and Taylor[92] have established that the bandshape will not resemble a ground-state spectrum because of temperature-dependent

[87] S. K. Garg and D. W. Davidson, in 'Physical Chemistry of Ice', ed. E. Whalley, S. J. Jones, and L. W. Gold, 1973, p. 56.
[88] S. R. Gough, S. K. Garg, and J. A. Ripmeester, *Canad. J. Chem.*, 1974, **52**, 3193.
[89] S. K. Garg, D. W. Davidson, and J. A. Ripmeester, *J. Magn. Resonance*, 1974, **15**, 295.
[90] F. Schmidt, *Rev. Sci. Instr.*, 1974, **45**, 1430.
[91] C. Mottley and C. S. Johnson, jun., *J. Chem. Phys.*, 1974, **61**, 1078.
[92] P. S. Allen and D. G. Taylor, *J. Phys. (C)*, 1975, **8**, 3036.

thermal excitations to the first excited torsional state. They calculate methyl-group spectra at a series of temperatures for a number of barrier heights within the range 14—21 kJ mol^{-1}, where the temperature-dependent bandshape changes are most marked. Such temperature-dependent effects on the bandwidth have been observed in tetramethylsilane by Hasebe et al.[93]

Tunnelling-affected spin relaxation has received somewhat more attention than the spectral investigations. A phenomenon which has a bearing on a number of high-temperature relaxation studies was brought to light initially by Emid and Wind.[94] They pointed out that spin diffusion between adjacent methyl groups is limited by symmetry restrictions on the inter-group transitions. These restrictions lead, via a master-equation approach, to the prediction for the nuclear magnetization recovery which, while still being the sum of three exponentials, is intermediate between the theory of Hilt and Hubbard[95] and the single exponential of the spin-temperature theory. Notwithstanding the idea of symmetry-restricted spin diffusion, it is still possible to bring about equilibrium within the Zeeman and dipolar systems in an assembly of methyl groups on a time-scale shorter than T_1. However, transitions which maintain these spin temperatures do not maintain a Boltzmann equilibrium distribution between the overall populations of the different symmetry species. Emid and co-workers[96] therefore adopted a different point of view; they argued in terms of population differences, and produced a double-exponential prediction for the magnetization recovery. These ideas were further refined,[97] until they produced[98] a description of spin–lattice relaxation for methyl groups in solids in terms of four coupled parameters describing the Zeeman, the dipolar, the tunnelling, and the rotational polarization systems. The region of rapid motional narrowing ($\omega_0\tau_c \ll 1$) is specified in all these papers. Experimental demonstrations of the coupling between the Zeeman system and the rotational polarization system have been given by Wind et al.[97, 99] The relaxation data of Punkkinen et al.[100] illustrate non-exponential relaxation in the different regime of $\omega_0\tau_c > 1$, and show that in such circumstances rotational polarization decays in a time very much shorter than the Zeeman relaxation time.

Clough and co-workers have modified their previous attempt to explain how a sudden temperature change can induce proton dipolar polarization by making[101] 'more plausible assumptions' about symmetry conversion transitions. This work is also based on the assumptions that spin diffusion can maintain common Zeeman and dipolar temperatures but cannot equalize the populations of different symmetry species. The occurrence of symmetry-conversion transitions themselves has been detected thermally[102] as a rise in temperature of the crystal lattice when the electronic Larmor frequency is tuned into the tunnelling frequency to promote the transitions. The use of an electronic spin system, as well as a nuclear spin system,

---

[93] T. Hasebe, G. Soda, and H. Chihara, *Proc. Japan Acad.*, 1975, **51**, 168.
[94] S. Emid and R. A. Wind, *Chem. Phys. Letters*, 1974, **27**, 312.
[95] R. L. Hilt and P. S. Hubbard, *Phys. Rev. (A)*, 1964, **134**, 392.
[96] S. Emid, R. A. Wind, and S. Clough, *Phys. Rev. Letters*, 1974, **33**, 769.
[97] R. A. Wind and S. Emid, *Phys. Rev. Letters*, 1974, **33**, 1422.
[98] S. Emid and R. A. Wind, *Chem. Phys. Letters*, 1975, **33**, 269.
[99] R. A. Wind, S. Emid, and J. F. J. M. Pourquie, *Phys. Letters*, 1975, **53A**, 310.
[100] M. Punkkinen, L. P. Ingman, D. G. Taylor, and P. S. Allen, *J. Phys. (C)*, 1975, **8**, 2159.
[101] S. Clough and J. R. Hill, *Phys. Letters*, 1974, **49A**, 461.
[102] S. Clough, J. R. Hill, and T. Hobson, *Phys. Rev. Letters*, 1974, **33**, 1257

constitutes a powerful extension of resonance techniques to the study of molecular group tunnelling.

G. N.M.R. Studies of Helium.—The application of n.m.r. spectroscopy to the properties of ^{3}He is expanding. The proceedings of the 13th international conference on low-temperature physics[103] contain a number of such papers. Work on adsorbed ^{3}He published either in ref. 103 or at the symposium on monolayer and sub-monolayer helium films,[104] or published independently, is covered in Chapter 10 on Heterogeneous Systems.

Anderson[105] has presented an overview of the situation regarding the low-temperature superfluid phases of ^{3}He, concluding with the remark that 'n.m.r. probably offers a unique opportunity for the detailed confirmation of the nature of the low-temperature phase of ^{3}He'. The general problem of spin dynamics in anisotropic phases of a Fermi superfluid is discussed at length by Leggett.[106, 107] The longitudinal and transverse resonances have been investigated in the B phase by Osheroff[108] and by Brinkmann.[109] The first nuclear resonance measurement from the A_1 region of the A phase, produced by high fields at melting pressures, is presented by Osheroff and Anderson,[110] who conclude that the A phase is to be identified as the Anderson–Brinkmann–Morel state, whereas the B phase is to be identified as the Balian–Werthamer state. Ahonen *et al.*[111] have, on the other hand, observed both the A and the B phases away from the melting plane. The longitudinal n.m.r. in superfluid ^{3}He is considered as an internal Josephson effect by Maki and Tsuneto,[112] and the self-diffusion coefficient of ^{3}He is measured by Barbe *et al.*[113] Additional n.m.r. shift and linewidth data in the superfluid state are presented by Combescot and Ebisawa.[114]

Solid helium presents other features of interest. For example, the activation energies and tunnelling frequencies of vacancies in both the b.c.c. and h.c.p. phases of solid ^{3}He are given by Sullivan *et al.*,[115] the effects on nuclear spin ordering under the influence of magnetic fields up to 1.2 T have been observed by Kummer *et al.*,[116] while the influence of ^{4}He impurities on ^{3}He spin–lattice relaxation has been investigated by Bernier and Deville.[117]

3 Nuclei of Spin $> \frac{1}{2}$ in Diamagnetic Solids

A. Introduction.—This section is restricted to the situation where the interaction between the nuclear electric quadrupole moment and the electric field gradient is

[103] Proceedings of 13th International Conference on Low Temperature Physics, ed. K. D. Timmerhaus, W. J. Sullivan, and E. F. Hammel, Plenum, New York, 1974.
[104] Monolayer and Submonolayer Helium Films, ed. J. G. Daunt and E. Lerner, Plenum, New York, 1973.
[105] P. W. Anderson, Proc. 24th Nobel Symposium, 1974, p. 103.
[106] A. J. Leggett, Proc. 24th Nobel Symposium, 1974, p. 109.
[107] A. J. Leggett, *Ann. Phys. (New York)*, 1974, **85**, 11.
[108] D. D. Osheroff, *Phys. Rev. Letters*, 1974, **33**, 1009.
[109] W. F. Brinkman, *Phys. Letters*, 1974, **49A**, 411.
[110] D. D. Osheroff and P. W. Anderson, *Phys. Rev. Letters*, 1974, **33**, 686.
[111] A. J. Ahonen, M. T. Haikala, and M. Krustus, *Phys. Rev. Letters*, 1974, **33**, 1595.
[112] K. Maki and T. Tsuneto, *Progr. Theor. Phys.*, 1974, **52**, 773.
[113] R. Barbe, M. Leduc, and F. Laloe, *J. Phys. (Paris)*, 1974, **35**, 935.
[114] R. Combescot and H. Ebisawa, *Phys. Rev. Letters*, 1974, **33**, 810.
[115] N. Sullivan, G. Deville, and A. Landesman, *Phys. Rev. (B)*, 1975, **11**, 1858.
[116] R. B. Kummer, E. O. Adams, and W. P. Kirk, *Phys. Rev. Letters*, 1975, **34**, 517.
[117] M. Bernier and G. Deville, *J. Low Temp. Phys.*, 1974, **16**, 349.

less than that between the nuclear magnetic moment and the externally applied static magnetic field. This restriction follows from the fact that the prime interest of this volume is nuclear magnetic resonance and not pure n.q.r. Moreover, this particular section is devoted to diamagnetic homogeneous insulators. Quadrupolar effects in systems with unpaired electrons are discussed in later sections of this chapter.

An accurate knowledge of the nuclear magnetic moments is of fundamental importance to all applications of nuclear resonance. Using CW methods in CuCl, CuI, and $CuRh_2Se_4$, Locher[118] has determined a new value for the nuclear moment ratio of the isotopes ^{63}Cu and ^{65}Cu. Beckmann et al.[119] have employed the high-precision atomic beam magnetic resonance method to determine the nuclear magnetic dipole moments of ^6Li, ^7Li, ^{23}Na, ^{39}K, and ^{41}K to seven significant figures.

A statistical technique has been applied by Peterson et al.[120] to the study of n.m.r. bandshapes in amorphous materials, where the major broadening agent is the nuclear quadrupole interaction. They take account of the randomness of sites by writing both the Larmor frequency and the NQCC as simultaneously random and uncorrelated variables, and represent them both by probability density functions. Once the density function is determined, the bandshape may be calculated by mapping into a co-ordinate frame in which one of the axes represents frequency. An important point about this method is that the calculations are exact; the bandshapes are given by closed mathematical expressions.

The intermediate region, where the Zeeman energy and the quadrupole coupling are comparable, is often a difficult area to treat. Brooker and Creel[121] have solved exactly for the combined Zeeman and quadrupolar Hamiltonian in the special case of an applied field parallel to one of the quadrupolar principal axes, and they have presented the resonance bandshape for powders of spins $I = 3/2$. For nuclei of $I = 1$, Scheinbeim and Schemp[122] have also solved exactly for the combined Zeeman and quadrupolar Hamiltonian in order to devise a geometrical method for obtaining the asymmetry parameter η and the principal axes of the electric field gradient when $\chi < 2\gamma B_o/\pi$.

B. ^2H N.M.R.—As an aid to uncovering the dynamic processes in ice, n.m.r. is most sensitive to the changes in local environment of the resonant nuclei. Deuteron n.m.r. offers the advantage over its proton counterpart that lines from different sites may be separated by different quadrupole couplings. Schmidt[123] has therefore made an n.m.r. study of a zone-refined single crystal of D_2O ice between 167 and 273 K. He finds that the diffusion process, which enables deuterons originally in bonds parallel to the c-axis to mix with deuterons in bonds not parallel to c, is thermally activated, with an activation energy (58 ± 3) kJ mol^{-1} above 197 K and (12 ± 3) kJ mol^{-1} below that temperature.

The ferroelectric phase transition in a single crystal of triglycine sulphate can

---

[118] P. R. Locher, *Phys. Rev. (B)*, 1974, **10**, 801.
[119] A. Beckmann, K. D. Boeklers, and D. Elke, *Z. Phys.*, 1974, **270**, 173.
[120] G. E. Peterson, C. R. Kurkijian and A. Carnevale, *Phys. and Chem. Glasses*, 1974, **15**, 59.
[121] H. R. Brooker and R. B. Creel, *J. Chem. Phys.*, 1974, **61**, 3658.
[122] J. Scheinbeim and E. Schemp, *J. Magn. Resonance*, 1974, **15**, 40.
[123] H. V. Schmidt, *Phys. Chem. Ice*, 1973, 212.

also be more advantageously studied by ^{2}H n.m.r. than by ^{1}H n.m.r. By doing just that and observing the ND_3 groups, Stepisnik and Slak[124] confirmed the proton disorder motion in the hydrogen bond between the glycinium II ions and the glycine III zwitterions. At the same time they cast doubt upon the hypothesis concerning the critical flipping of the glycinium I ions. They also present evidence indicating a faster hindered rotation of the glycinium I ND_3 group, compared with that of such groups at the other glycine ions. The antiferroelectric phase of $ND_4D_2PO_4$ has been studied by Blinc et al.,[125] who have measured its O—D$\cdots$O deuteron quadrupole coupling tensors. They find that these data fully support the Nagamiya model of proton ordering in the antiferroelectric phase of ammonium dihydrogen phosphate; that the chemical equivalence of the deuteron e.f.g. tensors shows that the distortions of PO_4 tetrahedra below T_c are not significantly larger in ammonium dihydrogen phosphate than in potassium dihydrogen phosphate; and finally that the antiferroelectric phase transition in ammonium dihydrogen phosphate is connected with an order–disorder transition of the O—H$\cdots$O hydrogens, as in potassium dihydrogen phosphate.

The deuteron NQCC and a modified asymmetry parameter for the CD_3 deuterons in fully deuteriated toluene are presented by Barnes et al.,[126] who also reiterate the corresponding values for *ortho*- and *para*-xylene. The NQCC's are equal to within experimental uncertainty at about 52 kHz, but the asymmetry parameters vary by a factor of three. This large variation is assigned to the range of the rigid bond-coupling constants in each of the three materials. Asymmetry parameters and NQCC's for deuterons in the chlorobenzenes are given by Ragle and Mokarram.[127] Again the coupling constants are very similar at (181.9 ± 0.3) kHz for 1,3,5-trichlorobenzene and (180.7 ± 0.2) kHz for 1,2,3,4-tetrachlorobenzene, but the asymmetry parameters (between 4 and 6%) are substantial. These large asymmetry parameters were not unexpected in view of the low local symmetry around the carbon atom to which the deuteron is bonded, and they are in line with the qualitative description of the non-axial component of the e.f.g. surrounding a C—D moiety given by Bersohn.[128] Ragle et al.[129] have also observed the deuteron NQCC in $CDBrCl_2,(C_2D_5)_2O$ in order to justify the assumption made previously by him[130] that the hydrogen bonding of $CDCl_3$ with ketones and ethers produces structures similar to that reported for $CHBrCl_2,(C_2H_5)_2O$.

In contrast to the experimental determination of NQCC's, Goren[131] makes use of the experimental deuteron NQCC's as an empirical constraint in a systematic study of a bond-orbital method for evaluating orbital wavefunctions in a series of MD_4 molecules, where M = B, C, N, Al, Si, or P. He demonstrates that a minimal basis set may well be sufficient to describe the first-row deuteriates, but it is insufficient for deuteriates of the second-row elements, where double zeta orbitals are required.

[124] J. Stepisnik and J. Slak, *J. Chem. Phys.*, 1975, **62**, 34.
[125] R. Blinc, J. Slak, and I. Zupanic, *J. Chem. Phys.*, 1974, **61**, 988.
[126] R. G. Barnes, W. C. Harper, and D. R. Torgeson, *J. Chem. Phys.*, 1975, **62**, 4572.
[127] J. L. Ragle and M. Mokarram, *J. Chem. Phys.*, 1975, **62**, 3361.
[128] R. Bersohn, *Mol. Phys.*, 1974, **27**, 605.
[129] D. Schwartz and J. L. Ragle, *J. Chem. Phys.*, 1974, **61**, 429.
[130] J. L. Ragle, C. Minott, and M. Mokarram, *J. Chem. Phys.*, 1974, **60**, 3184.
[131] S. D. Goren, *Theor. Chim. Acta*, 1974, **34**, 39.

C. **The Halogen Nuclei.**—^{20}F is a fluorine isotope with a spin $I = 2$. Its nuclear quadrupole moment (0.064 × 10^{-28} m^2) and its NQCC (5.77 MHz) and asymmetry parameter (0.317) in MgF$_2$ have been determined by Stoeckmann et al.,[132] who observed the effect of n.m.r. transitions on the asymmetric β decay of ^{20}F to ^{20}Ne. They establish that at room temperature they are observing ^{20}F ions on normal lattice sites, in an environment undisturbed by the capture of the polarized neutrons which were required to produce the isotope.

Because halogen atoms often occupy terminal positions in molecules, they are exposed to a considerable range of intermolecular interactions. Brill[133] considers how those interactions occurring within 0.02 nm either side of the van der Waals contact distance affect halogen e.f.g.'s. He illustrates his paper with a discussion of a number of chlorine compounds (mostly ionic). Cooke and Jeffrey[134] have also employed the ^{35}Cl isotope. They have investigated the order–disorder transition at 248.9 K in ND$_4$Cl. In the region of this phase transition they find that the ^{35}Cl relaxation rate is dominated by the quadrupole interaction and is very similar to that found in NH$_4$Cl. The time dependence of the field gradient responsible for the quadrupolar relaxation is due to the thermally activated flipping of the ND$_4$ tetrahedra.

The temperature dependence of the ^{81}Br n.m.r. linewidth between 173 K and the melting point of AgBr enabled Becker and Richtering[135] to monitor the diffusion of foreign ions in that material. They assign their linewidth variations to the changes in the dynamic quadrupole interactions between Br nuclei and the diffusing excess charges, and they quote activation energies for the mean jump frequency for Li$^+$, Na$^+$, K$^+$, Cu$^+$, Cd^{2+}, and Mg^{2+} ions. All foreign ions except Cu$^+$ appear to migrate *via* cation vacancies.

The room-temperature and low-temperature phases of Ag$_2$H$_3$IO$_6$ have been investigated by Ross and Kind,[136] who made use of a ^{1}H–{^{127}I} double-resonance technique. In the paraelectric phase all the iodine sites are chemically equivalent, whereas in the antiferroelectric phase two chemically inequivalent iodine sites were found, and domains were observed which are related by a triad axis in the hexagonal c-direction of the paraelectric phase. The structural model derived from these data results in four pairwise antipolar sublattices and a new unit cell. Straightforward ^{127}I n.m.r. intensity measurements have been employed by Gerhart and Hooper[137] to study dislocation effects in CsI single crystals. The intensity variations, which are dependent on the angle between the static magnetic field and the crystal axes through the first-order quadrupole broadening interaction, suggest that preferentially oriented edge and screw dislocations exist, which have a [100] slip direction. By submitting their sample to uniaxial stress, a gradient elastic tensor $C_{44} = 16 \times 10^4$ statvolt dyn^{-1} was obtained. Hackeloer and Kanert[138] have evaluated both the C_{11} and C_{44} components of the gradient elastic tensor in a single crystal of RbI by fitting their spin-echo data to their calculations of the effect on the bandshape of an elastic stress in a solid containing a large number of defects.

---

[132] H. J. Stoeckmann, H. Ackermann, and D. Dubbers, *Z. Phys.*, 1974, **269**, 47.
[133] T. B. Brill, *J. Chem. Phys.*, 1974, **61**, 424.
[134] D. F. Cooke and K. R. Jeffrey, *J. Magn. Resonance*, 1975, **18**, 455.
[135] K. D. Becker and H. Richtering, *Ber. Bunsengesellschaft phys. Chem.*, 1974, **78**, 461.
[136] J. Roos and R. Kind, *Ferroelectrics*, 1974, **8**, 553.
[137] G. R. Gerhart and H. O. Hooper, *Phys. Rev. (B)*, 1974, **10**, 4413.
[138] H. J. Hackeloer and O. Kanert, *J. Magn. Resonance*, 1975, **17**, 367.

D. The Alkali-metal Nuclei.—^{7}Li has been used[139] as a probe of ionic motion in four lithium and lithium–potassium silicate glasses of composition $R_2O,4SiO_2$ and $R_2O,2SiO_2$. Moreover, the authors reanalyse previous motional data from some alkali-metal borate glasses of composition $3R_2O,7B_2O_3$. Although all glasses with higher alkali contents appear to be homogeneous, both the borates and silicates with lower alkali contents appear to be heterogeneous, and it is shown how the heterogeneities form barriers to the alkali-metal ion motion. These heterogeneities appear to be on a smaller scale in the borates than in the silicates. In addition to the long-range ^{7}Li ionic motion observed in the homogeneous glasses (and the intermediate-range motion in the heterogeneous glasses), which proceeds by an interstitial process, there is evidence of short-range motion within severely restricted regions in nearly all the glasses reported. Another lithium isotope, ^{8}Li, has had its quadrupole moment, $(32 \pm 6) \times 10^{-31}$ m^2, and its NQCC in LiNbO$_3$, (43 ± 3) kHz, measured[140] at room temperature by means of the quadrupolar split n.m.r. spectrum detected via ^{8}Li β radiation.

Sodium nitrite is a ferroelectric material at room temperature and as such represents one of the few ferroelectric materials for which the gradient elastic tensors have been evaluated. Kanashiro et al.[141, 142] employed acoustic excitation to saturate specific n.m.r. transitions of the ^{23}Na spectrum, while monitoring the change in the nuclear resonance bandshape with a frequency-swept marginal oscillator. They found that the gradient elastic tensors extracted from their data fit better to a point-multipole model of the system rather than a point-charge model. Gradient elastic tensors at ^{87}Rb sites in RbI are reported in the previously mentioned paper by Hackeloer and Kanert.[138] The quadrupolar distortions around ^{23}Na sites in NaCl single crystals, caused by 0.01% of bivalent Ca^{2+} impurities, were evaluated by Meissner and Kanert,[143] who employed a pulsed n.m.r. technique.

E. Transition-element Nuclei.—The electronic properties of diamagnetic compounds of the transition elements are readily amenable to study by nuclear resonance techniques. For example, the observation of the ^{59}Co resonance in Co(NH$_3$)$_6$Cl$_3$ single crystals has enabled Reynhardt[144] to extract the chemical-shift tensors at the five non-equivalent cobalt sites. He finds that the principal axes of these tensors almost coincide with the principal directions of the e.f.g. tensors, which he suggests are possibly reorientation axes for the five non-equivalent hexammine octahedra. From a similar study[145] in Co(C$_5$O$_2$H$_7$)$_3$, the same author provides evidence that to assume a D_3 symmetry for the crystal field at ^{59}Co sites is to oversimplify the real situation. The T_{1g} excited state is split into three non-degenerate levels, resulting in a chemical-shift tensor which deviates from axial symmetry.

F. Other Nuclei.—The ^{11}B nuclear resonance bandshape in glasses of the system MgO–Na$_2$O–B$_2$O$_3$ is readily decomposed into a narrow component from boron

[139] J. R. Hendrickson and P. J. Bray, *J. Chem. Phys.*, 1974, **61**, 2754.
[140] H. Ackermann, D. Dubbers, M. Grupp, P. Heitjans, and H. J. Stockmann, *Phys. Letters*, 1974, **52B**, 54.
[141] T. Kanashiro, T. Ohno, and M. Satoh, *J. Phys. Soc. (Japan)*, 1974, **37**, 1179.
[142] T. Kanashiro, T. Ohno, and M. Satoh, *J. Phys. Soc (Japan)*, 1975, **38**, 1293.
[143] P. Meissner and O. Kanert, *Phys. Status Solidi (B)*, 1974, **65**, K85.
[144] E. E. Reynhardt, *Solid State Comm.*, 1974, **15**, 643.
[145] E. C. Reynhardt, *J. Phys. (C)*, 1974, **7**, 4135.

nuclei in BO_4 tetrahedra and a broad component from ^{11}B nuclei in the BO_3 triangles. This has allowed Kim and Bray[146] to investigate the relative fraction of boron atoms in BO_4 units, and to evaluate the ^{11}B quadrupolar coupling parameters for comparison with other glasses. The composition dependence of the fraction of BO_4 units is interpreted in terms of the role of Na_2O and MgO as network modifiers.

The temperature dependence of the ^{14}N NQCC in sodium nitrate has been of interest for some time. In the most recent report on this material, Gourdji et al.[147] explore the temperature region from 77 K up to and including the phase transition at 548 K. They discuss the inter-ionic contribution to the nitrogen NQCC in the nitrate ion in terms of the σ- and π-orbital populations in the nitrogen and of the electric charges in the oxygen atoms. The decrease in the ^{14}N NQCC with increasing temperature does not follow the Bayer–Kushida theory well above room temperature but is consistent with the postulated high-amplitude internal motions of the NO_3^- ions as the transition temperature is approached. The ^{14}N nuclei were also used in a previously mentioned paper[134] concerning the investigation of the order–disorder transition in ND_4Cl. The quadrupolar mechanism governs the nitrogen relaxation in the region of the phase transition, and it is time dependent due to the thermal flipping of the ND_4 tetrahedra.

4 Systems with Unpaired Electrons in the Absence of Magnetic Ordering

A. Non-metallic Paramagnetic Systems.—An extension of Redfield's relaxation theory has been made by Bosse et al.[148] in order to explain the dynamics of nuclear spins in the presence of electronic relaxation. A strong motivating influence in this work was to draw attention to the inter-relationship between the various spin-relaxation parameters measured in different types of experiment, such as n.m.r., Mössbauer spectroscopy, or perturbed angular distribution of γ-rays. Not only are the limiting cases of fast and slow electronic relaxation considered but the transition region between them is also dealt with.

The nuclear resonance study of specific paramagnetic systems continues to provide data on exchange mechanisms, their temperature dependence, and the mechanics of phase transitions. For example, the ^{19}F relaxation in paramagnetic O_2AsF_6 is dominated by its coupling with the dioxygenyl unpaired electrons. The relative reorientational motion of the diatomic anion and the octahedral cation modulates this coupling, and it is the source of its temperature dependence. Consequently the ^{19}F nuclear resonance behaviour can be used[149] to correlate the temperature dependence of the inter-ionic motions with the approach to, and occurrence of, the phase transition at (255 ± 3) K. Cerous magnesium nitrate hydrate, $Ce_2Mg_3(NO_3)_{12}, 24H_2O$, is a salt which remains paramagnetic down to temperatures in the milliKelvin range. For this reason it is of great interest in magnetic thermometry. A proton spin–lattice relaxation study of this material has shown[150] that, above 200 K, the dominant relaxation mechanism is the internuclear dipole–

---

[146] K. S. Kim and P. J. Bray, Phys. and Chem. Glasses, 1974, **15**, 47.
[147] M. Gourdji, L. Guibe, and A. Peneau, J. Phys. (Paris), 1974, **35**, 497.
[148] J. Bosse, H. Gabriel, and W. Vollmann, Phys. Status Solidi (B), 1975, **68**, 81.
[149] P. Rigny and W. E. Falconer, J. Chem. Phys., 1975, **62**, 2581.
[150] D. B. Utton, J. Chem. Phys., 1975, **62**, 670.

dipole interaction modulated by 180° flips of the water molecules. The two types of water molecule have activation energies of 28 and 40 kJ mol^{-1}. That the internuclear dipolar interaction should dominate at all is due in part to the large separation of Ce^{3+} ions, a fact which determines the large paramagnetic temperature range of this material. However, below 200 K, as the flipping slows down, the paramagnetic moments of Ce^{3+} begin to govern the proton relaxation rate, and this makes it possible to estimate a value of k (200 ± 18 K) for the energy of the second excited Ce^{3+} doublet.

Other paramagnetic systems reported this year include: (i) copper acetate monohydrate,[151] where ^{63}Cu n.m.r. transitions of bivalent copper ions have been detected for the first time and where quadrupole coupling parameters together with indirect nuclear exchange parameters are determined at the ^{63}Cu sites; (ii) copper complexes of pyrazine,[152] where proton spectra are evaluated in terms of the exchange interactions between the copper ions; (iii) tetramethylammonium manganese chloride,[153, 154] where proton T_1's have been used to clarify spin diffusion along the linear chains; (iv) cadmium fluoride strongly doped with ErF_3 and YbF_3,[155] for which the presence of (rare earth^{3+} − F$^-$)$_2$ dimers has been established; (v) gadolinium phosphide,[156]—the authors reanalysed earlier data of Myers and Narath;[157] and (vi) DPPH,[158] for which the hyperfine coupling to protons at helium temperatures has been elucidated.

Dynamic nuclear polarization in systems containing paramagnetic ions enables either or both of the electric and magnetic hyperfine interactions to be investigated. By means of this technique Brodbeck et al.[159] examined the mutually correlated n.m.r. and e.s.r. strain broadenings in several ruby crystals. They showed that the n.m.r. quadrupolar structure at the ^{27}Al sites and the e.s.r. fine structure constants at the Cr^{3+} ions simultaneously deviate in the same manner from their perfect crystalline values. An experimental and theoretical appraisal of dynamic nuclear polarization in samarium-doped lanthanum magnesium nitrate has been presented by Byvik and Wollan,[160] and its use in preparing highly polarized proton targets has been applied to glycerol and 2,3-butyleneglycol complexes of chromium(v) by Svoboda.[161]

B. Metals.—Measurements of the Knight shift have become, over the years, a standard test of the electronic structure models for individual metals. Das and Pant[162] have established that a density matrix approach can be used to calculate the Knightshift, and they illustrate their communication by evaluating the shift in Mg, Al, and In using a model local pseudopotential. The pressure dependence of the Knight shift, when coupled with the pressure dependence of the Pauli spin

[151] A. Kawamori and G. Soda, *Mol. Phys.*, 1975, **29**, 1085.
[152] M. Inoue, S. Emori, K.-I. Hara, and M. Kubo, *J. Magn. Resonance*, 1975, **17**, 212.
[153] Y. H. Tchao and S. Clement, *Phys. Letters*, 1974, **48A**, 295.
[154] Y. H. Tchao and S. Clement, *J. Phys. (Paris)*, 1974, **35**, 861.
[155] M. Greenblatt and E. Banks, *J. Chem. Phys.*, 1975, **62**, 2700.
[156] J. W. Tucker, *Solid State Comm.*, 1974, **15**, 1679.
[157] S. M. Myers and A. Narath, *Phys. Rev. (B)*, 1973, **7**, 4776.
[158] R. Verlinden, P. Grobet, and L. Van Gerven, *Chem. Phys. Letters*, 1974, **27**, 535.
[159] C. M. Brodbeck, S. Lee, and H. H. Niebuhr, *Phys. Rev. (B)*, 1974, **10**, 844.
[160] C. E. Byvik and D. S. Wollan, *Phys. Rev. (B)*, 1974, **10**, 791.
[161] J. Svoboda, *Czech. J. Phys. (B)*, 1975, **25**, 340.
[162] A. K. Das and M. M. Pant, *J. Phys. Soc. (Japan)*, 1974, **37**, 1259.

susceptibility, gives directly the pressure dependence of the spin density at the nucleus. Mahanti[163] has collected together both sets of data[164, 165] for lithium, and he concludes that the expected increase in spin density due to the volume decrease is substantially modified by the corresponding increase in the negative core polarization contribution to the spin density.

Electric field gradients in metals are also sensitive to the electronic structure. Using pseudopotential theory and taking account only of conduction electrons external to the ion cores, Lodge and Sholl[166] calculate the e.f.g.'s in Be, Mg, Zn, Cd, and In, but only obtain good agreement with experiment for Be. For the other metals the calculated values are smaller than the experimental values, and it is therefore concluded that in these metals conduction electrons within the ion core make a significant contribution to the e.f.g. Sholl[167] has also developed a theory of quadrupolar relaxation in liquid metals. Nuclear spin relaxation in metals due to the vacancy-type self-diffusion mechanism is discussed in two papers by Cavelius, who concentrates on dipolar interactions in one[168] and quadrupolar interactions in the other.[169]

The experimental investigations on specific metals which were reported during the current catchment period include those summarized in Table 4.[170-177]

Table 4 N.m.r. studies in metals

Nucleus	Metallic state	Temperature range	Type of investigation	Ref.
^{27}Al	powder	room	Quadrupolar second moment	170
	powder	room	Knight shift	171
	powder	< 4.2 K	Fine particle size effects	172
^{51}V	single crystal	4.2 K	Dipolar second moment	173
^{63}Cu	powder	< 4.2 K	Fine particle size effects	172
^{69}Ga	liquid		Relaxation theory	174
^{93}Nb	single crystal	77 and 300 K	Dipolar and indirect exchange contributions to the second moment	175
^{109}Ag	powder and liquid	room to 1358 K	Knight shift and relaxation	176
^{111}Cd ^{113}Cd	powder	room	Knight shift	171
^{137}Ba	powder	4.2 to 530 K	Knight shift	177

[163] S. D. Mahanti, Solid State Comm., 1974, **15**, 779.
[164] T. Kushida and M. Hanabusa, Bull. Amer. Phys. Soc., 1974, **19**, 64.
[165] G. B. Benedek and T. Kushida, J. Phys. and Chem. Solids, 1958, **5**, 241.
[166] K. W. Lodge and C. A. Sholl, J. Phys. (F), 1974, **4**, 2073.
[167] C. A. Sholl, J. Phys. (F), 1974, **4**, 1556.
[168] E. Cavelius, Phys. Status Solidi (B), 1974, **65**, 181.
[169] E. Cavelius, Phys. Status Solidi (B), 1974, **66**, 321.
[170] E. R. Andrew, W. S. Hinshaw, and R. S. Tiffen, J. Phys. (F), 1974, **4**, L215.
[171] E. R. Andrew, W. S. Hinshaw, and R. S. Tiffen, J. Magn. Resonance, 1974, **15**, 191.
[172] S. Kobayashi, ref. 103, vol. 4, p. 315.
[173] D. Hechtfischer, R. Karcher, and K. Lueders, Phys. Status Solidi (B), 1974, **65**, K89.
[174] N. C. Halder, Phys. Rev. (B), 1974, **10**, 2333.
[175] J. Pellisson and J. Buttet, Phys. Rev. (B), 1975, **11**, 48.
[176] V. A. Hanay, M. Shaham, and D. Zamir, Phys. Rev. (B), 1974, **10**, 2343.
[177] G. A. Styles and G. Tranfield, Phys. Letters, 1974, **48A**, 471

C. Alloys and Intermetallic Systems.—Much current interest in dilute alloy systems centres on the electronic charge and spin density oscillations set up around an impurity site in the host matrix. The particular case of dilute transition-metal impurities, which may give rise to a Kondo effect at low enough temperatures, is of special interest. Nuclear resonance provides a sensitive means of probing these oscillations because they affect both the magnetic hyperfine interactions (*e.g.* Knight shift) and the electric quadrupole interactions at the host nuclei neighbouring the impurity sites. A calculation of the effect of the charge and spin density oscillations in noble- or transition-metal alloys with transition-metal impurities has been given by Parlebas,[178] whereas Alloul[179] has evaluated the spatial dependence of the spin polarization in dilute alloys for which both cases of magnetic and non-magnetic impurities are considered.

Table 5[180-192] lists a number of specific metallic state binary and tertiary systems which have received attention during the past one-year period. The table attempts to highlight the salient features of these investigations, but space availability prohibits a detailed discussion of those findings which relate to oscillations of electronic spin and charge densities.

D. Hydrides.—The mobility of hydrogen atoms in metal lattices has been the subject of a small number of reported investigations this year. Fedders[193] has developed a theory of dynamic quadrupolar spin relaxation applicable to Ta nuclei, when the field gradients are governed by mobile hydrogen impurities in the parent metal. In contrast, the studies of $LaNiH$,[194] the palladium–hydrogen system,[195] and hafnium hydride[196] all concentrate on observing the hydrogen nucleus itself.

E. Semiconductors.—The elemental semiconductors selenium and tellurium are the subject of a chemical-shift tensor calculation by Bensoussan,[197] who concludes that the local electronic configuration of selenium is made up of two bonding orbitals within the trigonal chain and two equivalent lone pairs oriented such that the overlap between second neighbours is a minimum; whereas for tellurium the two

[178] J. C. Parlebas, *J. Phys. (F)*, 1974, **4**, 1392.
[179] H. Alloul, *J. Phys. (F)*, 1974, **4**, 1501.
[180] A. C. Gossard and J. P. Remeika, *Solid State Comm.*, 1974, **15**, 609.
[181] G. K. Schoep, H. J. van der Valk, G. A. M. Frijters, H. B. Kok, and N. J. Poulis, *Physica*, 1974, **77**, 449.
[182] M. Bose and S. Ganguli, *Phys. Letters*, 1974, **48A**, 357.
[183] M. Matsumura and S. Wada, *J. Phys. Soc. (Japan)*, 1974, **37**, 876.
[184] H. Alloul, J. Darville, and P. Bernier, *J. Phys. (F)*, 1974, **4**, 2050.
[185] G. Serforzo, K. Tompa, and A. Lovas, *J. Phys. and Chem. Solids*, 1974, **35**, 1303.
[186] K. Tompa, *Pure Appl. Chem.*, 1974, **40**, 61.
[187] J. Hesse and Z. Szyecs, *Appl. Phys.*, 1975, **6**, 55.
[188] H. T. Weaver, J. E. Schirber, and A. Narath, ref. 103, vol. 4, p. 157.
[189] H. T. Weaver and R. K. Quinn, *Phys. Rev. (B)*, 1974, **10**, 1816.
[190] H. Bertschat, H. Haas, and F. Pleiter, *Z. Phys.*, 1974, **270**, 203.
[191] S. K. Malik, R. Vijayaraghavan, and S. K. Garg, *Phys. Status Solidi (B)*, 1975, **68**, 399.
[192] I. D. Weisman, L. H. Bennett, and A. J. McAlister, *Phys. Rev. (B)*, 1975, **11**, 82.
[193] P. A. Fedders, *Phys. Rev. (B)*, 1974, **10**, 4510.
[194] T. K. Halstead, *J. Solid State Chem.*, 1974, **11**, 114.
[195] G. K. Schoep, N. J. Poulis, and R. R. Arons, *Physica*, 1974, **75**, 299.
[196] H. T. Weaver, *J. Magn. Resonance*, 1974, **15**, 84.
[197] M. Bensoussan, *J. Phys. and Chem. Solids*, 1974, **35**, 1661.

Table 5 N.m.r. studies of alloys and intermetallic systems

Probe nucleus	System investigated	Type of investigation	Ref.
^{17}O	V_2O_3	Spectral and spin relaxation studies of vanadium–oxygen covalency	180
^{51}V	(a) V_3X (X = Co, Rh, Ir, or Au) (b) V_xX_{1-x} (X = Ga, Ir, or Pt and $0 \leqslant x \leqslant 1$)	Magnetic and quadrupolar interactions studied by spin-echo spectra	181
^{51}V	XVO_4 (X = Ce, Pr, Nd, Sm, Gd, Tb, Dy, or Yb)	Knight shift and quadrupolar interactions	182
^{51}V	$(Mo_{1-x}Re_x)_{0.96}Co_{0.01}V_{0.03}$ where $0 < x < 1$	Knight shift and spin–lattice relaxation	183
^{59}Co	$(Mo_{1-x}Re_x)_{0.96}Co_{0.01}V_{0.03}$ where $0 < x < 1$	Knight shift and spin–lattice relaxation	183
^{63}Cu	CuCr, CuMn, CuFe	Concentration dependence of second moment	184
^{63}Cu	CuZn, CuGa, CuGe, CuAg	Concentration dependence of quadrupole spectra	185
^{63}Cu	copper-based alloys	Survey	186
^{63}Cu	CuFe	Spectral temperature dependence	187
^{71}Ga	$AuGa_2$	Knight shift pressure dependence	188
^{103}Rh	$Pt_{1-x}Rh_x$ ($0 \leqslant x \leqslant 1$)	Knight shift and spin–lattice relaxation	189
^{111}Cd	CdPd	Knight shift temperature dependence	190
^{119}Sn	CeSn	Spectra	191
^{195}Pt	$Pt_{1-x}Rh_x$ ($0 \leqslant x \leqslant 1$)	Knight shift and spin–lattice relaxation	189
^{195}Pt	$LaNi_{5-x}Pt_x$ ($0 \leqslant x \leqslant 5$)	Knight shift and spin–lattice relaxation	192

lone pairs are non-equivalent, and one is oriented in such a way as to make the overlap between second neighbours a maximum. The vanadium–oxygen system in its different forms lends itself to many applications. The V_2O_3 system already mentioned[180] facilitates the study of metal–insulator transitions. The study of impurity-doped VO_2, on the other hand, is important because it is expected that it will lead to a better understanding of the semiconductor-to-metal transition, and for this reason Reyes et al.[198] have observed the concentration and temperature dependence of the ^{51}V spectra in $V_{1-x}Al_xO_2$ with $0 \leqslant x \leqslant 0.038$. The V_2O_5–MoO_3 and V_2O_5–WO_3 systems are also semiconductors, and the deformation of the V_2O_5 lattice by MoO_3 and WO_3 is the subject of a short paper by Coldea et al.,[199] who observe the concentration-dependence of the ^{51}V NQCC. N.m.r. has been used to detect charge defects in GaAs,[200] by observing the second moments of ^{69}Ga, ^{71}Ga, and ^{75}As as a function of static magnetic field.

[198] J. M. Reyes, S. L. Segel, and M. Sayer, *J. Solid State Chem.*, 1975, **12**, 298.
[199] M. Coldea, L. Stanescu, and I. Ardelean, *Phys. Status Solidi (A)*, 1974, **26**, K145.
[200] R. K. Hester, A. Sher, J. F. Soest, and G. Weisz, *Phys. Rev. (B)*, 1974, **10**, 4262.

F. **One-dimensional Conductors.**—The TCNQ (tetracyanoquinodimethanide) salts[201, 202] have been replaced this year as the most studied type of one-dimensional conductor by the square-planar platinum cyanide complexes. Butler and Guggenheim[203] have investigated the importance of the water content and motion on the conductivity of $K_2Pt(CN)_4Cl_{0.3},nH_2O$. They suggest that the importance of the water molecules arises from their electric dipole moments, which provide significant shielding of the platinum chains from potentials generated by the random occupation of K^+ and Cl^- sites. The analogous Br salt, with three waters of crystallization, has enabled Brinkmann and Keller to propose that the motion of all the water molecules is not the same[204] and to study the metal–insulator phase transition.[205] Moreover, it has been used by Muller et al.[206] to explore the effects of pressure on the electronic structure surrounding the ^{195}Pt nuclei.

5 Ordered Magnetic Materials

A. **Ferromagnetic Materials.**—Materials which order magnetically lend themselves to nuclear resonance evaluation because the time-average hyperfine interaction is a sensitive indicator of their electronic structure, and because the time-dependent hyperfine interaction reveals valuable information about dynamic processes occurring within them. Attention on ferromagnets this year has been focused away from pure metals, mainly on to binary systems. This is illustrated by the range of ferromagnetic systems listed in Table 6,[207–220] which begins with one pure metal but includes a number of doped metals and alloy systems. Masuda et al.,[211] for example, have given a comprehensive survey of the electronic properties, hyperfine field components, and relaxation parameters of various doping transition elements in Fe. Table 6 also includes some ferromagnetic chalcogenides, in which interest is still strong,[217, 218] and penultimately makes note of an extensive account published by the Kamerlingh Onnes group[219] of the magnetic interactions in six three-dimensional b.c.c. Heisenberg ferromagnets, all of which have structures isomorphous with $K_2CuCl_4,2H_2O$.

---

[201] J. Alizon, G. Berthet, and J. P. Blanc, *Phys. Status Solidi (B)*, 1974, **65**, 577.
[202] F. Devreux and M. Nechtschein, *Solid State Comm.*, 1975, **16**, 275.
[203] M. A. Butler and H. J. Guggenheim, *Phys. Rev. (B)*, 1974, **10**, 1778.
[204] D. Brinkmann and H. U. Keller, *Phys. Status Solidi (B)*, 1975, **67**, K43.
[205] D. Brinkmann and H. U. Keller, *Helv. Phys. Acta*, 1974, **47**, 387.
[206] W. H. G. Muller, H. Niedoba, and D. Jerome, *Solid State Comm.*, 1975, **16**, 655.
[207] M. N. Ghatikar and J. Itoh, *J. Phys. Soc. (Japan)*, 1975, **38**, 283.
[208] P. K. James, N. J. Stone, and H. R. Foster, *Phys. Letters*, 1974, **48A**, 237.
[209] R. Laurenz, E. Klein, and W. D. Brewer, *Z. Phys.*, 1975, **270**, 233.
[210] M. Kawakami and Y. Koi, *J. Phys. Soc. (Japan)*, 1974, **37**, 1257.
[211] Y. Masuda, T. Hioka, and M. Kontani, *Internat. J. Magnetism*, 1974, **6**, 143.
[212] J. Englich, M. Rotter, and J. Rijacek, *Czech. J. Phys. (B)*, 1974, **24**, 1069.
[213] M. Nomura and H. Fujiwara, *J. Phys. Soc. (Japan)*, 1975, **38**, 55.
[214] R. L. Streever, *Phys. Rev. (B)*, 1974, **10**, 3969.
[215] F. Y. Fradin and J. P. Williamson, *Phys. Rev. (B)*, 1974, **10**, 2803.
[216] K. H. J. Bushaw and A. Oppelt, *J. Phys. (F)*, 1974, **4**, 1246.
[217] H. Yasuoka and Y. Nakamura, *J. Phys. Soc. (Japan)*, 1975, **38**, 81.
[218] N. Karnezos, L. B. Welsh, and M. W. Shafer, *Phys. Rev. (B)*, 1975, **11**, 1808.
[219] T. O. Klaassen, W. J. Looyestijn, and N. J. Poulis, *Physica*, 1974, **77**, 43.
[220] L. C. Gupta, R. Vijayaraghavan, S. D. Damle, U. R. K. Rao, Le Dang Khoi, and P. Veillet, *J. Magn. Resonance*, 1975, **17**, 41.

The Solid State

Table 6 N.m.r. studies of ferromagnetic materials

System	Probe nucleus	Type of investigation	Ref.
Gd Metal	Gd	Resonance properties as a function of thermal and mechanical history	207
Fe + $10^{-4}\%$ I	^{131}I, ^{132}I	I hyperfine fields at 15 mK	208
Fe + 0.01% Co	^{57}Co	Nuclear moment of ^{57}Co	209
Co + 0.8% Mn	^{55}Mn	Temperature dependence of Mn hyperfine field between 4.2 and 296 K	210
Fe X, where $X = \begin{cases} \text{V, Mn, Co, Cu,} \\ \text{Nb, Ta, Ir, Au} \\ \text{at 1\%} \\ \text{Mo, Ru, Pd} \\ \text{at 2\%} \\ \text{Rh, W} \\ \text{at 3\%} \end{cases}$	X	Relaxation behaviour of the impurity nucleus as a function of its electronic structure and external magnetic field at 4.2 K	211
Fe$_4$N	^{57}Fe	Fe hyperfine fields at different sites and 4.2 K	212
Fe$_{1-x}$V$_x$ where $0 \leq x \leq 1$	^{51}V, ^{57}Fe	x dependence of hyperfine field at 4.2 K	213
Ni$_{1-x}$Mn$_x$ ($0.01 \leq x \leq 0.14$)	^{55}Mn	x dependence of spin–lattice relaxation at 4.2 and 77 K	214
V$_3$Ga$_{1-x}$Si$_x$ ($0 \leq x \leq 1$)	^{51}V	Spin–lattice relaxation and quadrupole coupling as a function of x	215
GdMg$_{1-x}$Zn$_x$ ($0 \leq x \leq 1$)	^{67}Zn, ^{155}Gd, ^{157}Gd	x dependence of hyperfine field at 4.2 K and T_c	216
CoS$_2$	^{59}Co	Hyperfine field between 4.2 and 77 K	217
NbSe$_2$ intercalated with Ga, In, Tl, Sn, and Pb	^{93}Nb	Effect of intercalate nuclei on metallic behaviour between 1.5 and 4.2 K	218
NbS$_2$ intercalated with In	^{93}Nb	Effect of intercalate nuclei on metallic behaviour between 1.5 and 4.2 K	218
M$_2$CuX$_4$,2H$_2$O (M = K, NH$_4$, Rb, or Cs, and X = Cl or Br)	^{35}Cl, ^{63}Cu, ^{79}Br	Analysis of covalency effects and resonance observables below 2K	219
Rb$_2$CuF$_4$	^{19}F, ^{63}Cu	Hyperfine fields	220

B. Antiferromagnetic Materials.—Table 7[221–235] is a listing of antiferromagnetic materials upon which reports have appeared during the past year. The majority of

[221] T. Kohara and K. Asayama, *J. Phys. Soc. (Japan)*, 1974, **37**, 393.
[222] T. Kohara and K. Asayama, *J. Phys. Soc. (Japan)*, 1974, **37**, 401.
[223] M. Coldea, M. Crisan, A. Neda, and I. Pop, *J. Phys. and Chem. Solids*, 1974, **35**, 1095.
[224] K. Okada and H. Yasuoka, *J. Phys. Soc. (Japan)*, 1974, **37**, 1711.
[225] K. Okada, *Phys. Rep. Kumamoho University*, 1974, **1**, 57.
[226] D. Paquette, A. R. King, and V. Jaccarino, *Phys. Rev. (B)*, 1975, **11**, 1193.
[227] A. R. King, D. Paquette, and V. Jaccarino, *Phys. Letters*, 1974, **50A**, 229.
[228] T. Horiguchi and E. R. Hunt, *Phys. Letters*, 1974, **49A**, 227.
[229] T. Jimbo and C. Elbaum, *Phys. Rev. (B)*, 1974, **10**, 2131.
[230] E. R. Hunt and T. Horiguchi, *Phys. Rev. (B)*, 1974, **11**, 1804.
[231] W. J. M. de Jonge and C. H. W. Swuste, *J. Chem. Phys.*, 1974, **61**, 4981.

Table 7 N.m.r. studies of antiferromagnetic materials

System	Probe nucleus	Type of investigation	Ref.
α-Mn alloys with 9 different metallic impurities up to ~10%	^{55}Mn	Néel temperature as a function of impurity and hyperfine field as a function of concentration	221
β-Mn alloys with 17 different metallic impurities up to ~40%	^{55}Mn	Hyperfine field as a function of type and concentration of impurity	222
CrAl$_7$	^{27}Al	Temperature dependence of Knight shift	223
CoO	^{59}Co	Hyperfine field at 4.2 and 77 K	224
CoCl$_2$,2H$_2$O	^{35}Cl	Field dependence of spectra and relaxation times	225
MnF$_2$	^{19}F, ^{55}Mn	Field dependence of spin–lattice relaxation	226
MnF$_2$	^{55}Mn	Field-dependent n.m.r. enhancement	227
RbMnF$_3$	^{19}F	Linewidth data	228
RbMnF$_3$	^{55}Mn	Acoustic excitation near T_N	229
RbMnF$_3$	^{87}Rb	Temperature dependence of spin–lattice relaxation	230
α-RbMnCl$_3$,2H$_2$O	^{35}Cl, ^{87}Rb	Establishment of antiferromagnetic state	231
CsMnBr$_3$,2H$_2$O, Cs$_2$MnBr$_4$,2H$_2$O, Rb$_2$MnBr$_4$,2H$_2$O	^{1}H, ^{79}Br ^{81}Br, ^{85}Rb ^{87}Rb, ^{133}Cs	Temperature dependence of the magnetic interaction parameters and magnetic phase diagrams	232
Mn(CO$_2$Me)$_2$,4H$_2$O	^{1}H	Verified the magnetic transition	233
(CH$_3$)$_4$NMnCl$_3$ CuCl$_2$,2NC$_5$H$_5$	^{1}H	Temperature dependence of spin–lattice relaxation	234
CaF$_2$	^{19}F	Susceptibility measurements in nuclear antiferromagnetic state	235

these are insulators, and (CH$_3$)$_4$NMnCl$_3$[234] has been referred to previously in its paramagnetic state.[154] Reference 232 is a 'tour de force' and cannot be ignored by anyone concerned with magnetic salts, while the letter on CaF$_2$[235] is somewhat different from the rest, because it deals with an ordered nuclear spin system rather than an ordered system of electronic spins.

C. Ferrimagnetic Materials.—The nuclear resonances of ^{53}Cr and ^{59}Co have been studied by Tsuda *et al.*[236] in the ferrimagnetic phase of CoCr$_2$O$_4$ in order to obtain, from the field dependence of the resonance frequency, the half-cone angles of the ferrimagnetic spiral. In yttrium iron garnet doped with Mn ions, on the other hand.

[232] C. H. W. Swuste and W. J. M. de Jonge, *Physica*, 1974, **76**, 21.
[233] R. D. Spence, *J. Chem. Phys.*, 1975, **62**, 3659.
[234] P. M. Richards and F. Borsa, *Solid State Com.*, 1974, **15**, 135.
[235] J. F. Jacqinot, M. Chapellier, and M. Goldman, *Phys. Letters*, 1974, **48A**, 303.
[236] T. Tsuda, H. Abe, and A. Hirai, *J. Phys. Soc. (Japan)*, 1975, **38**, 72.

Le Dang Khoi et al.[237] have measured the hyperfine fields at ^{55}Mn nuclei at various crystalline sites and in different valence states, by observing their spin-echo signals. The hyperfine fields have also been reported for ^{59}Co in Co^{2+} ions in Co ferrite[238] and for tervalent diamagnetic cation nuclei in nickel ferrite.[239] The diamagnetic ion nuclei (gallium) have also been investigated in lithium ferrite gallate, where hyperfine fields and quadrupolar interactions have been obtained from nuclear resonance data.[240]

6 New Techniques and Apparatus

The discussion in this section is brief and restricted to those developments which are relevant to the studies reviewed in this chapter.

A. Techniques.—A method of measuring the secular dipolar energy, which makes use of the first moment of the nuclear absorption signal, has been described by Goldman.[241] The method is valid at all temperatures and is applicable even in the presence of magnetic ordering. It will be useful for investigating nuclear spin dynamics either at a fundamental level or as a tool for investigating the solid state. The potential of free induction decay (FID) measurements, which in turn lead to the evaluation of the moments of the nuclear resonance line, has been greatly increased recently by the improvement of zero-time resolution techniques by Lowe and co-workers.[242] In a more recent paper[243] this group record the FID shapes over four orders of magnitude in time and claim to evaluate from these shapes all even moments of the ^{19}F resonance in CaF_2 from the second to the fourteenth. Their values agree with the theoretical predictions so far available for the second, fourth, sixth, and eighth moments. The use of dipolar echoes has also been extended recently by Boden and co-workers,[244] who develop the echo method to separate the homo- and hetero-nuclear contributions to the moments of n.m.r. absorption spectra in solids containing two dipolar-coupled spin-½ species.

The application of double-resonance techniques to the observation of quadrupole effects which are difficult to observe by single resonance techniques is being advanced by the Ljubljana group. Blinc and Zumar,[245] for example, have described a method of using the proton dipolar T_{1D} to give a measure of the relaxation rates of non-resonant quadrupolar nuclei which are very difficult to detect individually. In addition, Seliger et al.[246, 247] have developed a double-resonance technique for obtaining the quadrupole spectra of integral spins, based on coupling between spin systems induced by a strong r.f. magnetic field. The sensitivity of this method is higher than the previously used level-crossing technique.

---

[237] Le Dang Khoi, P. Veillet, and R. Krishnan, *Compt. rend.*, 1974, **278**, B, 1047.
[238] T. Tsuda and K. Okada, *J. Phys. Soc. (Japan)*, 1974, **37**, 1713.
[239] Y. Miyahara and S. Iida, *J. Phys. Soc. (Japan)*, 1974, **37**, 1248.
[240] V. D. Doroshev, V. A. Klochan, and N. M. Kovton, *Phys. Status Solidi (A)*, 1974, **26**, 77.
[241] M. Goldman, *J. Magn. Resonance*, 1975, **17**, 393.
[242] I. J. Lowe, K. W. Vollmers, and M. Punkkinen, Proceedings of the First Specialized Colloque Ampère, ed. J. W. Hennel, Krakow, 1973, p. 70.
[243] M. Englesberg and I. J. Lowe, *Phys. Rev. (B)*, 1974, **10**, 822.
[244] N. Boden, M. Gibb, Y. K. Levine, and M. Mortimer, *J. Magn. Resonance*, 1974, **16**, 471.
[245] R. Blinc and S. Zumar, *J. Chem. Phys.*, 1975, **62**, 3118.
[246] J. Seliger, R. Blinc, M. Mali, and R. Osredkar, *Phys. Status Solidi (A)*, 1974, **25**, K121.
[247] J. Seliger, R. Blinc, M. Mali, and R. Osredkar, *Phys. Rev. (B)*, 1975, **11**, 27.

A new method of nuclear magnetic resonance image formation has been published by Garroway et al.,[248] which is based on selective r.f. irradiation of the sample in a switched magnetic field gradient. In other words, the sample is prepared by irradiating it with r.f. while it is subject to a field gradient in one direction, before switching the field gradient direction and applying a 'read' pulse of r.f. A crucial feature of the preparatory irradiation is that it ensures that only those parts of the spin system which have been specially selected are irradiated. The method may be applied to both two- and three-dimensional spin-density maps. Van Putte and van den Enden[249] have developed a quick, accurate, and fully automatic method for the determination of the solid fat content in partially crystallized fats. This is important because the solid fat content is used for process control by the fat industry. The characteristics of the pulsed n.m.r. method of determining solid fat content are also compared with those of dilatometry.

The coupling of n.m.r. spectroscopy with other solid-state methods can produce a combined technique which has some advantages over both the component methods. Such is the situation with Mössbauer spectroscopy[250] and with nuclear orientation,[251] where the development of the combined techniques continues to grow.

B. Data Analysis.—The modification of the width of n.m.r. lines by an unsuitable choice of experimental conditions has long been a subject for discussion, and this year is no exception to that rule. Both saturation broadening and narrowing are discussed by Emid and Wind,[252] who in addition[253] develop a method for correcting the modulation broadening of resonance signals by means of their Fourier transform. The errors obtained in T_1 and T_2 evaluations when using Carr–Purcell–Meiboom–Gill pulse sequences which exhibit baseline drift also continue to be a source of controversy, with the various protagonists still exchanging papers.[254]

C. Apparatus.—New circuits and experimental developments are most concisely summarized in tabular form by means of Table 8.[255-261]

[248] A. N. Garroway, P. K. Grannell, and P. Mansfield, *J. Phys.* (C), 1974, **7**, L457.
[249] K. Van Putte and J. van den Enden, *J. Amer. Oil Chemists' Soc.*, 1974, **51**, 316.
[250] B. Krishnamurthy and K. P. Sinha, *J. Magn. Resonance*, 1975, **17**, 189.
[251] P. T. Callaghan, P. D. Johnson, and N. J. Stone, *J. Phys.* (C), 1975, **7**, 3161.
[252] S. Emid and R. A. Wind, *J. Magn. Resonance*, 1975, **17**, 125.
[253] R. A. Wind and S. Emid, *J. Phys.* (E), 1975, **8**, 281.
[254] T. E. Bull, *J. Magn. Resonance*, 1974, **16**, 192.
[255] Z. Pajak, K. Jurga, and S. Jurga, *Postepy Fiz.*, 1974, **25**, 193.
[256] I. J. Lowe and M. Englesberg, *Rev. Sci. Instr.*, 1974, **45**, 631.
[257] P. A. Speight, K. R. Jeffrey, and J. A. Courteney, *J. Phys.* (E), 1974, **7**, 801.
[258] B. Ramadan, T. C. Ng, and E. Tward, *Rev. Sci. Instr.*, 1974, **45**, 1174.
[259] G. Odberg and L. Odberg, *J. Magn. Resonance*, 1974, **16**, 342.
[260] A. Briguet, J. L. Culty, and J. C. Duplan, *J. Phys.* (E), 1974, **7**, 71.
[261] T. Merrick, S. Smiriga, and A. Pines, *J. Magn. Resonance*, 1974, **14**, 270.

Table 8 N.m.r. apparatus development

Development	Ref.
New pulse spectrometer for work on solids	255
Fast-recovery pulsed n.m.r. sample probe	256
Modification to pulsed n.m.r. sample probe to eliminate ringing	257
Fast-recovery post-amplifier for pulsed n.m.r.	258
Quadrupole coil system for field gradient generation	259
A device permitting selective relaxation studies	260
A data aquisition system for proton-enhanced n.m.r. spectroscopy	261

10
Heterogeneous Systems

BY W. DERBYSHIRE

1 Introduction

This chapter is an innovation, but it is primarily based upon the section on surface phenomena included in the chapters on the solid state in previous Volumes. In studies on such phenomena, *e.g.* the adsorption of small molecules onto surfaces such as silica or alumina, there is interest in the nature of the complex formed between the surface and the adsorbed molecule, in any modification to the structure of either species, and in any motion occurring in the adsorbed state. In practice, different types of binding site exist, with different binding energies, and the adsorbed molecules exhibit exchange between the various binding sites, and with any bulk phase, if present. The range of timescales available in the various n.m.r. experiments is such that signals from the separate regions are sometimes resolvable and sometimes averaged, but often the intermediate situation occurs when analysis of the data provides information on the kinetics of the exchange processes. The relevant theory is reasonably well documented, and general reviews[1-4] exist, although none was published in the period of this literature survey.

The factors determining behaviour in the classical surface systems are also operable in the binding of small molecules, or ions, to macromolecules, and in systems such as zeolites, clays, and exchange resins. For the clays and resins the bulk phase may be extensive, but the relaxation behaviour is often determined by a small fraction of bound species, reorienting at a reduced rate, and hence relaxing at an enhanced rate, in rapid exchange with a major bulk phase. The enhancement of the bound-phase relaxation rate is such that, even though present at low concentrations, the bound phase determines the population-weighted average relaxation rate. The same factors are operable in solutions of macromolecules, in gels, and in cellular systems. Particularly detailed information can be obtained from solutions of metallo-proteins when the metal atom is paramagnetic, and when H_2O exchange between the primary hydration shell and the bulk phase is rapid. Analysis of the frequency dependence allows the number of H_2O molecules in the hydration shell and the mean distance from the metal atom to be determined. Changes in these factors as a result of changes of pH or temperature, substrate binding, or the incorporation of denaturing agent have offered fertile fields of study. Inclusion of this material on solvent relaxation would make literature extensive, and it has been necessary to limit the discussion. This chapter is more concerned with the n.m.r.

---

[1] K. J. Packer, *Progr. N.M.R. Spectroscopy*, 1967, **3**, 86.
[2] H. A. Resing, *Adv. Mol. Relaxation Processes*, 1967, **1**, 109.
[3] H. A. Resing, *Adv. Mol. Relaxation Processes*, 1972, **3**, 199.
[4] H. Pfeifer, *N.M.R. Basic Principles Progr.*, 1972, **7**, 53.

properties of the solvent in macromolecular solutions than in the macromolecular conformation or changes in conformation resulting from changes in concentration, pH, ionic strength *etc.*; *e.g.*, changes in protein conformation as a result of substrate binding are not discussed here. Essentially the chapter is limited to a brief discussion, scarcely more than a listing, of typical or more important papers concerned with solvent or substrate molecule behaviour in macromolecular solutions.

In cellular systems (Section 5) the bulk phase is water, and papers fall into two groups; one group interpreting observations in terms of the model outlined above, involving rapid molecular exchange between bulk water and a bound phase, which may itself be complex, and the second group of papers adopting the minority view[5] that all the cellular water has a structure different from that of bulk water. An observation by Damadian[6] which has attracted considerable interest is that water proton spin–lattice relaxation times are enhanced in cellular systems incorporating a growing tumour. There is interest in determining the origin of this effect and/or developing a diagnostic procedure. Experiments involving alkali-metal ions have also been interpreted from both viewpoints, *i.e.* involving a rapid exchange between bound and free phases, or alternatively invoking a long-range structural modification. Interest in these ions originates, at least in part, in attempts to understand the mechanism of ion transport through cell membranes, and to prove the presence or absence of ion pumps embedded in the cell membrane. Studies of membrane systems, intact and model, are reviewed in Section 6. These studies include investigations of the mobility (rotational, vibrational, and translational) of the membrane constituents, primarily phospholipid molecules, and of the effect of the incorporation of cholesterol, lipoproteins, drug molecules *etc.* Model systems often include lamellar liquid crystals, and in Section 6 there is an interface with the liquid-crystal chapter, included biennially in this series of Volumes. The literature on membranes and model membranes is also very extensive, and the papers selected for comment mostly involve fundamental membrane processes.

It should not be forgotten that the properties of a surface species are different to bulk properties, even in the supposed absence of an adsorbed layer. This is best exemplified by a system with a degree of co-operativity and/or spatially extensive waveforms. Ido and Hoshino[7] provide an example concerning fine copper particles with diameters between 7 and 60 nm. They have found a line broadening resulting from a distribution of Knight shifts caused by fluctuations of the electronic spin polarization in the surface region; this is a direct consequence of oscillations in the electron wavefunctions as a result of the requirement that the wavefunctions vanish at the boundary. The susceptibility χ was also dependent upon whether the number of conduction electrons was odd or even. In general, however, papers like this, which are concerned with specific surface effects, are not discussed in this chapter, which is concerned with adsorbate–adsorbent interactions.

2 Adsorption Studies

A. General.—This section is concerned with studies where the bulk phase is either

---

[5] C. F. Hazlewood, *Ann. N.Y. Acad. Sci.*, 1973, Vol. 204.
[6] R. Damadian, *Science*, 1971, **171**, 1151.
[7] M. Ido and R. Hoshino, *J. Phys. Soc. Japan*, 1974, **36**, 1325.

absent or present in only a small amount. Resing[8] has provided an introduction to the use of n.m.r. in conventional surface studies, and he has shown that (a) for water molecules a single correlation time, τ_c, can be used to specify molecular rotation and diffusion, whereas two times are required for benzene, and that (b) in general it is necessary to invoke a distribution of jump times. Water molecules adsorbed on charcoal, in zeolites, and on bacterial cell membranes have a mobility intermediate between ice and water and do not freeze readily. The mobility of a surface species is in general inhibited compared with the bulk phase, and spectral widths are wider. However, the application of pulse-narrowing sequences on benzene adsorbed on charcoal showed a narrow central ^{13}C peak together with an axially symmetric powder pattern with chemical shift components similar to those reported for solid benzene, features corresponding to isotropically reorientating and to bound benzene molecules, respectively. In a study typifying what can be achieved, Staudte[9] determined the ^{1}H linewidth of the surface OH groups on silica gel from 148 to 423 K in the absence of any adsorbed species, determining the packing and density of surface OH groups. At low temperatures rotation occurred about the Si—O bond, followed at higher temperatures by OH group diffusion. The presence of an adsorbed species permitted OH group diffusion at a lower temperature by reducing the activation energy required. In a study of ^{1}H signal intensity and T_2 values as a function of temperature (153—293 K), coverage, and isotopic composition in a mixture of C_6H_6–C_6D_6 of benzene adsorbed on a deuteriomethylated silica at $\frac{1}{2}$, 1, and 2 monolayers, Boddenberg[10,11] determined that 0.3 monolayers were located between surface methoxy-groups. These molecules exhibited anisotropic motion, with a correlation time $\tau_c < 10^{-9}$ s at 163 K, for rotation about the hexad axis. Rotation about the diad axis was considerably slower, and was describable by a log normal distribution of correlation times. Excess benzene was present as a capillary condensate, which froze at temperatures well below the bulk triple point, the size of the freezing-point depression being a function of pore size.

In a continuing investigation of the use of ^{13}C n.m.r. in surface studies and the use of pulse narrowing, proton decoupling, and signal enhancement, Chang et al.[12] investigated the trimethylsilyl-, vinyl-, and allyl-derivatives of chrysotile asbestos; in the first two cases the authors confirmed, and in the last denied, the validity of reported synthetic procedures. All three cases showed a sharp line corresponding to methyl groups bound to silicon, with chemical shift anisotropy as a major source of linewidth. There is also a discussion of the problems of this type of study, and some speculation on future development.[12]

It has previously been assumed that excess water adsorbed onto boehmite $Al_2O_3,1H_2O$ is H-bonded between boehmite layers, giving an increased lattice spacing and a composite n.m.r. spectrum comprising superposed broad and narrow lines. Baker and Pearson[13] have now proposed that water is co-ordinated to the aluminium atom. If this is so then the amount of adsorbed water should be a func-

---

[8] H. A. Resing, in 'Mass Spectrometry and N.M.R. Spectroscopy in Pesticide Chemistry', ed. R. Haque and F. J. Biros, Plenum, New York, 1974, p. 273.
[9] B. Staudte, Z. phys. Chem. (Leipzig), 1974, **255**, 158.
[10] B. Boddenberg, Ber. Bunsengesellschaft phys. Chem., 1974, **78**, 361.
[11] B. Boddenberg, Ber. Bunsengesellschaft phys. Chem., 1974, **78**, 354.
[12] J. J. Chang, A. Pines, J. J. Fripiat, and H. A. Resing, Surface Sci., 1975, **47**, 661.
[13] R. R. Baker and R. M. Pearson, J. Catalysis, 1974, **33**, 265.

tion of crystallite size; some experimental data are cited in support. Thompson et al.[14] have analysed lineshapes and relaxation times of H_2O molecules in the hydrated form of the mineral natrolite, again finding superposed broad and narrow lines with temperature-dependent relative intensities. Resing[15] has suggested that these data provide an example of his apparent phase transition, occurring when there is a continuous distribution of correlation times;[16] he proposes that the temperature at which the apparent intensities are both 50% can be related to the activation energy by a relationship similar to that of Waugh and Fedin.[17] Woessner[18] has reviewed some earlier work on the use of $90-\tau-180$ and $90-\tau-90_{90}$ pulse sequences on H_2O molecules with a non-random average orientation at an interface; he selects hectorite and a series of montmorillonite clays as examples, and introduces the use of T_2 and T_3 decays. [The T_3 decay involves the observation of echo amplitudes obtained in a $90_0-\tau-90_{90}$ pulse sequence as a function of pulse spacing, τ.] Comparison allowed both the interaction constant and the correlation time to be derived. In a subsequent paper[19] the theory was developed in greater detail, and applied to several cation-exchanged chabazites. Proton-exchange lifetimes decreased in the cation order $Ca^{2+} \approx Na^+ > K^+ > Li^+$ with a temperature dependence which was consistent with the Arrhenius equation and was a function of particle size. The opportunity was taken to use the theory to analyse previous measurements on adsorbed water on various montmorillonite-type clays.[20]

The interactions of water and organic molecules with minerals are of practical significance in the estimation of potential oil yields from oil shales. Mikinis et al.[21] have extended earlier work[22] by using CW techniques and relaxation studies to distinguish between adsorbed water and organic constituents. Particular attention was given to the minerals illite, dawsonite, nordstrandite, and analcite nacholite. The time for the n.m.r. assay was one tenth of that for the traditional Fischer assay and correlated to the order of 0.9.

B. Cryogenic Studies.—The use of very low temperatures introduces some additional special features into adsorption studies, arising from quantum effects. There has been a recent flurry of papers on the properties of adsorbed ^3He; the earlier studies were on heterogeneous systems, such as Vycor glass and the zeolites, but recently the more homogeneous systems of graphite and grafoil sheets have received attention. In general, published papers have been directed towards a readership composed of workers in low-temperature physics. Early work was consistent with specific-heat data, but recently the situation has become more complex, and the data from different groups are not always consistent. Rollefson[23] measured the susceptibility and

[14] R. T. Thompson, R. R. Knispel, and H. E. Petch, *Canad. J. Phys.*, 1974, **52**, 2164.
[15] H. A. Resing, *Canad. J. Phys.*, to be published.
[16] H. A. Resing, *J. Chem. Phys.*, 1965, **43**, 669.
[17] J. S. Waugh and I. Fedin, *Soviet Phys. Solid State*, 1963, **4**, 1633.
[18] D. E. Woessner, in ref. 8, p. 279.
[19] D. E. Woessner, *J. Magn. Resonance*, 1974, **16**, 483.
[20] D. E. Woessner and B. S. Snowden, jun., *J. Colloid and Interface Sci.*, 1969, **30**, 54.
[21] F. P. Mikinis, A. W. Decora, and G. L. Cook, U.S. Bureau of Mines, Report of Investigations No. 7984, 1974.
[22] A. W. Decora, F. R. McDonald, and G. L. Cook, U.S. Bureau of Mines, Report of Investigations No. 7523, 1971.
[23] R. J. Rollefson, in 'Monolayer and Submonolayer Helium Films,' ed. J. G. Daunt and E. Lerner, Plenum, New York, 1974, p. 115.

linewidth as a function of temperature and coverage for ^{3}He adsorbed on graphite. The susceptibility obeyed the Curie Law. At low coverages, ≤ 0.3 monolayers, no temperature dependence of linewidth was observed, but at higher coverages (0.6 to 0.9) a dependence was present, the maximum width observed being consistent with a rigid-lattice calculation made assuming a uniform distribution. Daunt's group used CW,[24] and relaxation techniques[25] to investigate submonolayers (0.20 to 0.95) of ^{3}He on grafoil in the temperature range 0.07—4.2 K. At low coverages (less than 0.4 monolayers) the sample had a Curie susceptibility down to below 1 K, where a deviation occurred. The system could be treated as a two-dimensional Fermi gas, but with an apparent degeneracy temperature lower than the ideal. At higher coverages the degeneracy temperature was reduced, becoming consistent with bulk liquid ^{3}He. For coverages greater than 0.4 the Curie Law applied down to the lowest temperatures studied, and the ^{3}He could then be treated as a solid. This is not consistent with the paper by Grimmer and Luszczynski.[26] The CW linewidths and FID times did not show any large coverage or temperature dependence; the FID times were 150—200 µs at 4.2 K and 100—150 µs at 0.1 K, the lineshape varying from Lorentzian to Gaussian. The limiting linewidth was only one quarter of that calculated for the rigid lattice, a reduction attributed to exchange. It was suggested that the large ^{3}He linewidths on grafoil were caused by a large anisotropic diamagnetism; this would result in large field gradients over the small voids. Husa and Hickernell and their colleagues[27,28] have made measurements on grafoil sheets soaked in ethyl alcohol at 15, 30, and 40 MHz. At 30 MHz there was no shift from the bulk liquid value when the sheets were parallel to the magnetic field. When perpendicular the shift was *ca.* 800 p.p.m. and dependent on sample geometry. The gradients of graphs of T_2^{-1} against B_0^2 were dependent upon the sheet orientation. The data were interpreted in terms of a model invoking a boundary limited diffusion consistent with the void size. By making a comparison with silica gel systems the authors speculated that exchange between chemically shifted sites was negligible. In an earlier paper Creswell *et al.*[29] considered the nature of an adsorbed ^{3}He monolayer and whether it was a liquid, solid, or a collection of atoms isolated by the potentials of adsorbent sites on Vycor glass substrate. The linewidth decreased with increasing coverage up to 1 monolayer, where the M_2 value was consistent with triangular packing. At lower coverages the observed M_2 values were too large for the same correlation time, which was assumed to be single. Below a 0.5 monolayer coverage M_2 was proportional to the coverage, and packing was random. The relaxation times had no strong temperature dependence and were thus indicative of a tunnelling type of motion. In the succeeding paper Brewer and Rolt[30] determined the susceptibility and spin–lattice relaxation time of ^{3}He and ^{3}He–^{4}He films on Vycor from 0.4 to 1.5 K; χ either obeyed the Curie Law or showed a small Fermi degeneracy, and was consistent with the statistical layer model in which the first two layers were magnetically independent of each other and the rest of the

[24] S. G. Hedge, E. Lerner, and J. G. Daunt, *Phys. Letters (A)*, 1974, **49**, 437.
[25] D. C. Hickernell, D. L. Husa, and J. G. Daunt, *Phys. Letters (A)*, 1974, **49**, 435.
[26] D. P. Grimmer and K. Luszczynski, in ref. 23, p. 123.
[27] D. L. Husa, D. C. Hickernell, and J. E. Piott, in ref. 23, p. 133.
[28] D. C. Hickernell, D. L. Husa, J. G. Daunt, and J. E. Piott, *J. Low Temp. Phys.*, 1974, **15**, 29.
[29] D. J. Creswell, D. F. Brewer, and A. L. Thomson, *Proc. Low Temp. Phys.*, 1973, **13**, 195.
[30] D. F. Brewer and J. S. Rolt, *Proc. Low Temp. Phys.*, 1973, **13**, 200.

film. The value of χ for the film was higher than that for the bulk liquid, and it followed the Curie Law down to lower temperatures. Subsequently Brewer et al.[31] studied the frequency dependence of ^{3}He films on Vycor between 0.65 and 7.1 MHz, for submonolayer coverages between 0.3 and 2.5 K, and for higher coverages down to 45 mK; χ was proportional to coverage up to 2 monolayers. Departures from the Curie Law were expected at lower temperatures, but for 1 monolayer the parameter Δ in the expression $\chi = \frac{C}{T+\Delta}$ had an upper limit of 5 mK. A tendency towards parallel alignment in filled pores indicated the occurrence of an exchange interaction enhanced over that occurring in bulk liquid. The value of T_1 exceeded that of T_2 and had an opposite and negative coverage dependence, a reversal of the dependences found by Monod and Cowen in zeolites. For the lowest coverage (0.1 monolayer) the value found for T_2 was 70 μs; this corresponds to an M_2 value $[\approx (1/T_2)^2]$ similar to that estimated for a monolayer, suggesting that clustering occurs. Brewer and Rolt[32, 33] determined values of the self-diffusion coefficient D of ^{3}He, inside and outside the pores of average diameter 7 nm in Vycor, finding values reduced from the bulk liquid value, and temperature independent. This is in contrast with a T^{-2} dependence observed in the bulk liquid, which provided a temperature calibration. The ^{3}He was generally in the collisionless regime, but mean-free-path effects were observed between 35 and 200 mK. The magnitude of the diffusion rate was consistent with the modified Landau theory. From the CW spectrum Brewer et al.[34] determined the susceptibility of ^{3}He adsorbed on the free surface of a ^{4}He film forming a two-dimensional Fermi system. This was attained by the use of isotopic mixtures adsorbed in a Vycor glass with a pore diameter of 6 nm, the ^{4}He being adsorbed at the wall preferentially. The Fermi temperature was independent of coverage, again implying a constant atomic spacing, i.e. atomic aggregation. Barbe et al.[35] have found very long ^{3}He T_1 relaxation times, e.g. greater than 2 d at 4.2 K, with solid H_2 coatings on cell walls; they have also discussed[35] the use of other cryogenic coatings e.g. D_2 and N_2, in reducing the effects of collisions with the walls. Finally, Dubault and Legrand[36] have described a method of determining the anisotropic part of the adsorption potential of H_2 on a surface at low temperature where ortho hydrogen is adsorbed preferentially, faujasite being taken as an example. Several theoretical models have been prepared based mainly upon a hindrance of the motion. Dubault and Legrand considered the modifications to the rotational energy-level pattern produced by electrostatic quadrupolar, induced dipolar, and Stockmayer interactions.

C. Zeolites.—Zeolites are traditionally included in reviews of surface phenomena. The n.m.r. literature on zeolite systems is extensive. A leading group is that in Leipzig headed by Pfeifer, who has provided a survey of recent work on the state of water and cyclic hydrocarbons[37] in zeolites, considering the extraction of the

[31] D. F. Brewer, D. J. Creswell, Y. Goto, M. G. Richards, J. Rolt, and A. L. Thomson, in ref. 23, p. 101.
[32] D. F. Brewer and J. S. Rolt, Phys. Letters (A), 1974, 48, 141.
[33] D. F. Brewer and J. S. Rolt, Proc. Low Temp. Phys., 1973, 13, 177.
[34] D. F. Brewer, D. J. Creswell, and A. L. Thomson, Troc. Low Temp. Phys., 1973, 13, 163.
[35] R. Barbe, F. Laloe, and J. Brossel, Phys. Rev. Letters, 1974, 34, 1488.
[36] A. Dubault and A. P. Legrand, J. Chem. Phys., 1974, 61, 1000.
[37] H. Pfeifer, 'Proceedings 18th Colloque Ampère', ed. P. S. Allen, E. R. Andrew, and C. A. Bates, University of Nottingham Press, Nottingham, 1974, p. 51.

exchange rate from the temperature dependence of T_2, and the effect of decationization *etc.* As a further example of this kind of study Nagel *et al.*[38] examined ^{1}H spin–lattice relaxation in the series cyclohexane, cyclohexene, cyclohexadiene, and benzene, finding a shift in the T_1 minimum to higher temperatures. For cyclohexane a distribution of correlation times was found. In all examples τ_c had an exponential temperature dependence, differences in activation energy being small (*ca.* 50%), whereas the pre-exponential factor differed by a factor of 200. It was considered that there was a specific adsorption to Na^+ ions through the π-bond system.

In an extension of an earlier paper[39] describing procedures for deriving the kinetics of surface reactions from relaxation times, Resing and his colleagues have investigated the reaction of water with the surface complex in zeolite 13-X,[40] determining ^{1}H relaxation as a function of temperature, coverage, deuteriation level, and degree of hydrolysis. Hydrolysis produces a new surface proton site which exchanges protons with adsorbed water, the exchange reaction being half-order with respect to water and first-order with respect to the surface complex. Proton M_2 data indicated an $Al(OH)_n$ structure for the surface complex, and ^{23}Na intensity measurements indicated that a previous model, in which the removal of Na^+ ions and their replacement by protons occurred during the hydrolysis of structural aluminium, was incorrect. The authors proposed that the broad distribution of correlation times arose because of non-uniformities in the zeolite structure, and in cavity occupancy by hydrolysis products. This was investigated by various heat treatments. The Al OH proton rotated quickly, and water had a maximum mobility at a pore-filling factor of 0.6. Oehme *et al.*[41] have considered the number and arrangement of OH groups in ion-exchanged and decationized zeolites of types A, X, and Y, and have found them readily distinguishable from water protons. For A-type zeolites the number of hydroxy-groups was proportional to the metal-cation concentration, the constant of proportionality being dependent upon the pretreatment temperature. However, the statistical arrangement of OH groups was different to that on decationated zeolites. In Y-type zeolites the H—Al distance decreased at higher degrees of decationization. Karger and Renner[42] extracted a value of 2×10^{-12} m^2 s^{-1} at 293 K for the diffusion coefficient of butane in NaCaA zeolite with pore-filling factors in excess of unity.

Stejskal *et al.*[43] used sample rotation at the magic angle to reduce the linewidth of ^{13}C spectra of CO_2 adsorbed in a variety of channel and cavity sieves at temperatures between 200 and 300 K. Spectra, asymmetric due to local susceptibility heterogeneities, were narrowed by rotation at 3 kHz. For small cage dimensions, less than 0.8 nm, T_1 was determined by paramagnetic impurities, and in larger cages by spin–rotation interactions. For CO_2 in Na^+ mordenite a single symmetric line was observed, consistent with CO_2 being physisorbed instead of chemisorbed. In zeolites of types 4 A and Y an asymmetric line was observed with a sense dependent on coverage. On spinning, this became symmetric and centred at the normal

---

[38] M. Nagel, H. Pfeifer, and H. Winkler, *Z. phys. Chem. (Leipzig)*, 1974, **255**, 283.
[39] H. A. Resing, *J. Phys. Chem.*, 1974, **78**, 1279.
[40] J. S. Murday, R. L. Patterson, H. A. Resing, J. K. Thompson, and N. H. Turner, *J. Phys. Chem.*, 1975, **79**, 2674.
[41] W. Oehme, D. Freude, and H. Schmiedel, *Z. phys. Chem. (Leipzig)*, 1974, **255**, 566.
[42] J. Karger and E. Renner, *Z. phys. Chem. (Leipzig)*, 1974, **255**, 357.
[43] E. O. Stejskal, J. Schaefer, J. M. S. Henis, and M. K. Tripodi, *J. Chem. Phys.*, 1974, **61**, 2351.

bulk position. Deininger et al.[44] have recorded the ^{13}C chemical shifts of toluene adsorbed on NaY and NaMgY zeolites, and compared them with estimates based on a simple adsorption model and semi-empirical CNDO calculations. Interactions with active sites on the zeolite surface were thought to cause changes in electronic structure. Karger and Walter[45] measured the self-diffusion of alkanes in NaX zeolites as a function of crystallite size, pore-filling factor, and number of CH_2 groups. Between 150 and 500 K the activation energy for diffusion of butane, pentane, hexane, and heptane increased regularly from 31.4 to 109 kJ mol^{-1}. This represents the energy required to remove the alkane to the outer surface. Labisch et al.[46] used n.m.r. and gas chromatography to study the adsorption of various butenes and n-butane in modified NaCaA molecular sieves, with the object of gas separation. Different behaviour was explained by electrostatic interactions at the opening to the supercages; *trans*-2-butene had a peculiar adsorption behaviour on the 4A sieve. There was a strong interaction with 1-butene and *cis*-2-butene, whereby the *cis*-compound could penetrate the large cavities only if 1-butene were present. In a series of papers Lechert and his colleagues investigated zeolites of the faujasite type.[47-50] A model was assumed, whereby for zeolites with low Si:Al ratios there were many shallow potential holes, whereas at high Si:Al ratios there were a relatively few deep potential wells near the cations of the large cavities. Proton n.m.r. demonstrated that the model was applicable for molecules with permanent electric dipoles, *e.g.* diethyl ether or THF, or for these with easily induced dipoles, *e.g.* benzene. The effect of a limited number of vacancies at high coverages on cation mobility was investigated. At low temperatures and coverages benzene molecules rotated about an axis in the molecular plane, whereas at higher coverages rotation was hindered. At higher temperatures benzene molecules jumped between different sorption sites. The mobilities of dioxan molecules in Ca forms of two faujasite samples were compared with those in the Na forms. Differences were explained in terms of the depths of the potential wells at different Si:Al ratios, and the number of cations in the large cages. Strong restrictions on mobility occurred at high dioxan concentrations. The ^{23}Na linewidth in synthetic faujasite zeolites at low Si:Al ratio was investigated as a function of water content, and showed an initial steep increase followed by a short range at constant width. Calculations of the electric field gradient and asymmetry factor showed that both of two proposed models had to be considered, *i.e.* with H attached to oxygen ions connecting octahedra, and with H_2O attached to Na^+ ions at four-membered rings at the cage walls.

3 Polymers, Exchange Resins, and Gels

Doskocilova et al.[51] have used magic-angle sample rotation to investigate internal motions in cross-linked poly(methyl methacrylate) gels swollen in $CHCl_3$ as a

[44] D. Deininger, D. Geschke, and W. D. Hoffman, *Z. phys. Chem. (Leipzig)*, 1974, **255**, 273.
[45] J. Karger and A. Walter, *Z. phys. Chem. (Leipzig)*, 1974, **255**, 142.
[46] L. Labisch, R. Schollner, D. Michel, V. Robiger, and H. Pfeifer, *Z. phys. Chem. (Leipzig)*, 1974, **255**, 581.
[47] H. Lechert, H. J. Hennig, and S. Mirtsch, *Surface Sci.*, 1974, **43**, 88.
[48] H. Lechert and H. J. Hennig, *Z. Naturforsch.*, 1974, **29a**, 1065.
[49] H. Lechert, K. P. Wittern, and H. J. Hennig, *Z. Naturforsch.*, 1975, **30a**, 690.
[50] H. Lechert and H. Henneke, *Ber. Bunsengesellschaft phys. Chem.*, 1974, **78**, 347.
[51] D. Doskocilova, B. Schneider, and J. Trekoval, *Coll. Czech. Chem. Comm.*, 1974, **39**, 2943.

function of cross-linking agent. The authors were interested in the temperature dependences of the limiting linewidths of the OMe, α-Me, and skeletal CH_2 groups. Tao et al.[52] note that in resins composed of relatively large microparticles, microns in size, the internal water line is broadened by kinetic processes in the resin phase. In small particles (ca. 10 nm) inter-particle heterogeneity was shown by multiple water peaks. In previous papers concerned with exchange between aqueous and resin phases interest had resided in chemical shifts and intensities, and not in the linewidths, which depended upon local susceptibilities. However, as in the example of ^{13}C spectra, this effect can be removed by sample spinning at the magic angle. In a bead copolymer of styrene and technical divinylbenzene two signals were observed with a chemical-shift difference which was a function of the internal molality of H^+. On spinning, the linewidth of the external water was less than that of the internal water, which differed in different types of resin.

A number of papers have been concerned with 'technical applications'. For example Smith and Robb[53] have used n.m.r. to investigate the effect of water, which acts as a plasticizer, on the glass transition temperature of polyethylene and a polystyrene latex. By incorporating Mn^{2+} ions, varying the particle diameter from 30 to 210 nm, and applying the standard theories of relaxation and spin-diffusion, ion-penetration depths of 1 nm were established. Mouse and Aziz[54] have investigated the adsorption of water in graft copolymers of cotton and methyl methacrylate. They claim that a distinction can be made between true graft copolymers and mixtures by studying the dependence of linewidth upon water content. The nature of the adsorbed water can be modified by the introduction of hydrophobic components. Lastly, Anderson[55] has noted that addition of water caused the CO_2H peak of acetic acid to shift upfield by an amount proportional to the amount of water added. Thus, by using a solvent containing acetic acid, the water content could be determined and the spinning process controlled. Studies of this type overlap with solution studies: e.g., by studying the ^2H and ^{35}Cl relaxations, and the frequency dependence of ^1H relaxation in solutions of poly(vinylpyrrolidone) and polystyrene in $CHCl_3$–$CDCl_3$, Rockelmann and Sillescu[56] obtained information on molecular motion, preferential solvation, H-bonding etc. van der Klink et al.[57] examined solvent relaxation also, in aqueous solutions of poly(acrylic acid) and poly(methacrylic acid). Proton exchange with the acid-group affected ^1H and ^2H relaxations differently, the latter being dominated by acid-group relaxation. The situation was complicated by changes in the number of protons incorporated in the acid group with degree of neutralization, and by changes in the interaction, which in general is anisotropic, as a result of reorientation of the acid group and/or chemical exchange. It was necessary to use the frequency dependence of ^2H relaxation to remove some ambiguities in interpretation. This paper provides a number of references to other work on solvent relaxation.

Lechert and his colleagues have published three papers concerned with water

---

[52] D. D. Tao, D. Doskocilova, and J. Stamberg, *Angew. Makromol. Chem.*, 1974, **38**, 129.
[53] E. G. Smith and I. D. Robb, *Polymer*, 1974, **15**, 713.
[54] A. M. Mouse and A. Aziz, *J. Appl. Polymer Sci.*, 1975, **19**, 1495.
[55] C. E. Anderson, *Analyt. Chem.*, 1975, **47**, 918.
[56] H. Rockelmann and H. Sillescu, *Z. phys. Chem. (Frankfurt)*, 1974, **92**, 263.
[57] J. J. van der Klink, J. Schriever, and J. Leyte, *Ber. Bunsengesellschaft phys. Chem.*, 1974, **78**, 369.

binding to starch, and water mobility in starch gels. Water in excess of 15% by weight was subject to capillary condensation, and gave two-component spin–spin relaxation, whereas spin–lattice relaxation had a single component, with T_1 exceeding the T_2 values by a factor of 100, but with an opposite dependence upon water content, T_1 decreasing to 12% and both T_2 values increasing. Starches from different sources (maize, rice, *etc.*) were used.[58] Previous measurements had indicated an increased quantity of bound water in industrial potato starches compared with samples prepared under controlled conditions. Investigations[59] showed that this did not arise as a result of different drying conditions, although the amount was a function of granule size. Instead the difference was related to the ion content of the water used during the isolation process, although the mechanism itself was not clear. Lechert's group[60] also measured the diffusion rate of water in starch gels from 50 to 95% H_2O and 274 to 320 K, finding $D_{gel} \approx D_{H_2O} \times [H_2O \text{ fraction}]^2$. The diffusion rate had the same activation energy as bulk water and was uniform and unrestricted, indicating that the boundaries of the swollen starch granules presented no barrier to diffusion.

Measurements such as this, giving diffusion coefficients and activation energies similar to bulk water, provide strong support for the two-phase model for water in macromolecular solutions and gels, and by inference in cellular systems, implying that only a small fraction is bound and has modified properties. There is still the problem of identifying the extent of the bound phase and determining its relaxation behaviour. One expedient is to identify the bound phase with water that does not freeze.[61] Ramirez *et al.*[62] have criticized this procedure on the basis of CW measurements on frozen aqueous solutions of acids, bases, salts, amino-acids, polypeptides, and proteins. They claimed that the data could be interpreted as a eutectic phenomenon in terms of the appropriate phase diagrams. Polypeptide solutions behaved similarly to the constituent amino-acids. The non-freezing component in bovine serum albumin (BSA) solution decreased with decreasing pH, although low pH values would be expected to result in chain unfolding and an increase in hydration. In the opinion of the Reporter this eutectic effect must certainly be operative, and in experiments where the amount of non-freezing water is strongly temperature dependent and for which $T_1 \approx T_2$, the non-freezing fraction probably cannot be identified as a bound component. In an attempt to investigate the validity of the two-phase model and this method of identifying the bound component, Derbyshire and Duff[63] measured ^1H T_1 and T_2 relaxation and diffusion in agarose gels as a function of gel concentration and temperature, the bound fraction corresponding to 0.59 g H_2O (g dry agarose)$^{-1}$. The T_1 data were completely consistent with the model, but the T_2 data presented some strange features. An investigation of the frequency and temperature dependence of relaxation of the non-freezing fraction[64] showed that $T_1 \gg T_2$, and it was necessary to invoke a distribution of correlation times to explain the T_1 frequency dispersion. The T_1 data between 10 and 90 MHz

---

[58] H. J. Hennig and H. Lechert, *Stärke*, 1974, **26**, 232.
[59] H. J. Hennig, H. Lechert, and B. Krische, *Stärke*, 1974, **27**, 151.
[60] W. Basler and H. Lechert, *Stärke*, 1974, **26**, 39.
[61] I. D. Kuntz, T. S. Brassfield, G. D. Low, and G. V. Purcell, *Science*, 1969, **163**, 1969.
[62] J. E. Ramirez, J. R. Cavanaugh, and J. M. Purcell, *J. Phys. Chem.*, 1974, **78**, 807.
[63] W. Derbyshire and I. D. Duff, *Discuss. Faraday. Soc.*, 1974, **57**, 243.
[64] I. D. Duff and W. Derbyshire, *J. Magn. Resonance*, 1975, **17**, 89.

were fitted to a log normal distribution, but with a low interaction constant. The T_2 relaxation was an order of magnitude faster than calculated. It was concluded that there were some additional components of motion at low frequencies, possibly arising from anisotropic rotation. Andrasko[65] produced additional support for the applicability of the two-phase model in agarose gels. He observed non-exponential $T_2(^{23}Na)$ relaxation in agarose gels from 1% concentration, although T_1 only became non-exponential at higher values. The data were fitted to a theory developed by Bull[66] for nuclei of spin $I = 3/2$ involved in exchange, and gave a τ_c value of 13 ns.

4 Protein Hydration

These studies involve direct adsorption, or gels or solutions. The n.m.r. of cellular systems is discussed in Section 5. As indicated in the introduction, the literature is extensive, particularly on protein solutions, and as this section interfaces with other chapters of this Volume, discussion is brief. Lynch and Webster[67] have provided an introductory review of the application of diffusion and relaxation techniques to protein hydration from the adsorption viewpoint, discussing the orientation dependence and relaxation in multiphase systems.

For the purposes of this review, polypeptides are discussed with proteins. Darke and Finer[68] used ^1H, ^2H, ^{13}C, and ^{81}Br n.m.r. to investigate the polylysine hydrobromide–water system (in the gel and liquid-crystal phases and briefly in the solid) as a function of temperature and water content. The polymer molecules were flexible in the liquid-crystal phase but less so than in trifluoroacetic acid solutions. In the gel phase the presence of α-helices caused water molecules to exhibit a preferential orientation. Discontinuities in the dependence of the ^2H quadrupolar splitting on water content indicated that there were seven water molecules bound per lysine residue (a single doublet indicated that D_2O molecules experienced both helical and disordered regions on an n.m.r. timescale; the ND_3 group gave an overlapping quadrupolar pattern and singlet in the liquid-crystalline phase). A sample cooled slowly to the gel phase in the magnetic field became ordered, giving two doublets, only one of which had an angular dependence.[69] Similar phases and behaviour have been observed in aqueous dispersions of phospholipid molecules. The authors make comparison with similar studies on phosphatidylethanolamine.[70]

Adsorption of water onto collagen has been a particularly rewarding, if a sometimes confusing, exercise in studies of protein hydration, and the system continues to receive attention. Westover and Dresden[71] determined the T_1, T_2, and D values of water protons in solutions and gels of native and denatured collagen *in vitro*. On heat denaturation of collagen gels T_1 and T_2 increased irreversibly, while D remained constant, implying a loss of bound water and/or an increase in bound-water relaxation time or, in terms of the structural theory of water, a decrease in long-range order. The authors discussed their data in terms of both theories, noting that

[65] J. Andrasko, *J. Magn. Resonance*, 1974, **16**, 502.
[66] T. E. Bull *J. Magn. Resonance*, 1972, **8**, 344.
[67] L. J. Lynch and D. S. Webster, *J. Polymer Sci., Part C, Polymer Symposia*, 1975, **43**, 49.
[68] A. Darke and E. G. Finer, *Biopolymers*, 1975, **14**, 441.
[69] E. G. Finer and A. Darke, *J.C.S. Faraday I*, 1975, **71**, 984.
[70] E. G. Finer and A. Darke, *Chem. and Phys. Lipids*, 1974, **12**, 1.
[71] C. J. Westover and M. H. Dresden, *Biochim. Biophys. Acta*, 1974, **365**, 389.

Heterogeneous Systems

T_2 had a different temperature dependence to bulk water. Fung et al.[72] recorded the frequency dependence of T_1 as a function of temperature, 183 to 293 K, for H_2O adsorbed on collagen at different surface coverages. Water in excess of 0.5 to 0.55 g H_2O (g collagen)$^{-1}$ behaved as bulk water and froze at 263 K. To interpret the T_1 data it was again necessary to invoke a log normal distribution of correlation times, but it was significant that the parameters changed with water content. Hsi and Bryant[73] examined the ^1H n.m.r. relaxation in frozen lysozyme solutions and in crystals, where 80% of the water of crystallization remained unfrozen. The ^1H T_1 relaxation was multi-component, but the T_2 decay was single. The data were fitted to a log normal distribution of correlation times with a low frequency cut-off. Eley et al.[74] recorded the H_2O ^1H relaxation in solutions of BSA modified by amidination, acetylation, or succinylation, as a function of lithium and sodium ion concentrations. These ions modified the dependence of relaxation rate upon protein concentration. For the unmodified protein the solvent-relaxation enhancement produced by the protein was unaffected by Na and reduced by Li, whereas for the chemically modified protein, lithium salts had no effect and sodium salts caused an enhancement. As stated earlier, the literature on solvent relaxation in protein solutions, and particularly metallo-protein solutions, is extensive. The observed relaxation is frequently determined by that occurring in the primary hydration sphere of the metal (often paramagnetic), transmitted to the bulk by molecular exchange, depending upon the electron spin relaxation rate or the molecular exchange rate. The basic theory has been developed by Bloembergen[75] and Solomon,[76] and extended by Swift and Connick[77] and Luz and Meiboom.[78] Campbell et al.[79] have provided a useful description of the procedure. Studies on diamagnetic derivatives are often included so that the diamagnetic contribution to relaxation can be excluded. A list of such papers is given in Table 1.

Table 1 *Solvent proton relaxation studies incorporating an investigation of the frequency dependence*

Study	Ref.
Ferri and CO-bovine haemoglobin and equine myoglobin	80
Haemoglobin, methaemoglobin, and diamagnetic oxyhaemoglobin	81
Horse-heart metmyoglobin fluoride (relaxation not due to protons in primary co-ordination sphere)	82
Complex of Mn—fully adenylated glutamine synthetase	83
Mn^{II}-concanavalin A	84
Mn^{II}-bovine carbonic anhydrase	85
Gd^{III}-lysozyme	86
(This paper contains a useful discussion of procedure)	
Mn ADP- or ATP-arginine kinase	87
Mn-Formyl tetrahydrofolate synthetase	88
E. coli alkaline phosphate	89

[72] B. M, Fung, J. Witschel, jun., and L. L. McAmis, *Biopolymers*, 1974, **13**, 1767.
[73] E. Hsi and R. G. Bryant, *J. Amer. Chem. Soc.*, 1975, **97**, 2397.
[74] D. D. Eley, M. J. Hey, and A. J. I. Ward, *J.C.S. Faraday I*, 1975, **71**, 1106.
[75] N. Bloembergen, *J. Chem. Phys.*, 1957, **27**, 572.
[76] I. Solomon, *Phys. Rev.*, 1957, **99**, 559.
[77] T. J. Swift and R. E. Connick, *J. Chem. Phys.*, 1962, **37**, 307.
[78] Z. Luz and S. Meiboom, *J. Chem. Phys.*, 1964, **40**, 2686.
[79] I. D. Campbell, J. P. Carver, R. A. Dwek, A. J. Nummelin, and R. E. Richards, *Mol. Phys.* 1971, **20**, 913.

Table 2 *Less complete solvent proton relaxation studies*

Study	Ref.
Human methaemoglobin	90
Carbonic anhydrase—monovalent anions	91
—aromatic sulphonamides	
Horseradish peroxidase (water not co-ordinated to Fe^{3+})	92
Bovine superoxide dismutase	93
Mn^{2+}–Arginine kinase	94
Ferricytochrome C	95
Mn^{2+}–Formyl tetrahydrofolate synthetase	96

A number of papers involving solvent proton relaxation do not report a full frequency dependence, the authors often being interested in some particular feature. A listing of such papers is made in Table 2.

A complete coverage of literature concerning the observation of the n.m.r. of a substrate during substrate–enzyme binding has not been attempted but a few papers of more than usual interest are considered below. Moon and Richards[97] have used ^{13}C n.m.r. to follow the binding of ^{13}C-labelled CO to haemoglobin and replacement by oxygen, separate signals being resolved for CO bound to α- and β subunits, whereas Brewer *et al.*,[98] also using ^{13}C n.m.r., observed the binding of α- and β-D-glucopyranosides to concanavalin A and obtained the binding kinetics, for different occupancies of the metal binding site, by using the Woessner–Zimmerman equations.[99] Drakenberg,[100] interested in the binding of 2-propanol to horse liver alcohol dehydrogenase and the rate of hydroxy-exchange, used the ^1H lineshape of nuclei with a spin–spin coupling to the O^2H site to find the ^2H exchange rate. Ostroy *et al.*[101] have used ^{23}Na n.m.r. to observe sodium binding to adenosinetriphosphatase, whereas Arnheim *et al.*[102] used ^{19}F n.m.r. to observe the binding of

[80] G. Lahajnar, I. Zupancic, R. Blinc, G. Pifat, and S. Maricic, *Biopolymers*, 1974, **13**, 1187.
[81] R. Blicharska and J. I. Dmitriew,'Proceedings 18th Colloque Ampère', ed. P. S. Allen, E. R. Andrew, and C. A. Bates, University of Nottingham Press, Nottingham, 1974, p. 281.
[82] M. E. Fabry and M. Eisenstadt, *J. Biol. Chem.*, 1974, **249**, 2915.
[83] J. J. Villafranca and F. C. Wedler, *Biochemistry*, 1974, **13**, 3286.
[84] J. J. Villafranca and R. E. Viola, *Arch. Biochem. Biophys.*, 1974, **165**, 51.
[85] A. Lanir, S. Gradstajn, and G. Navon, *Biochemistry*, 1975, **14**, 242.
[86] R. Jones, R. A. Dwek, and S. Forsén, *European J. Biochem.*, 1974, **47**, 271.
[87] D. H. Buttlaire and M. Cohn, *J. Biol. Chem.*, 1974, **249**, 5741.
[88] D. H. Buttlaire, G. H. Reed, and R. Himes, *J. Biol. Chem.*, 1975, **250**, 261.
[89] R. S. Zukin and D. P. Hollis, *J. Biol. Chem.*, 1975, **250**, 835.
[90] R. K. Gupta, and A. S. Mildvan, *J. Biol. Chem.*, 1975, **250**, 246.
[91] P. Taylor, *J. Pharm. Sci.*, 1975, **64**, 501.
[92] A. Lanir and A. Schejter, *Biochem. Biophys. Res. Comm.*, 1975, **62**, 199.
[93] M. Terenzi, A. Rigo, C. Franconi, B. Mondovi, L. Calabrese, and G. Rotilio, *Biochim. Biophys. Acta*, 1974, **351**, 230.
[94] D. H. Buttlaire and M. Cohn, *J. Biol. Chem.*, 1974, **249**, 5733.
[95] A. Lanir and J. Aviram, *Arch. Biochem. Biophys.*, 1975, **166**, 439.
[96] D. H. Buttlaire, G. H. Reed, and R. H. Himes, *J. Biol. Chem.*, 1975, **250**, 254.
[97] R. B. Moon and J. H. Richards, *Biochemistry*, 1974, **13**, 3437.
[98] C. F. Brewer, D. Marcus, A. P. Grollman, and H. Sternlicht, 'Lysozyme', Academic Press, New York and London, 1974, p. 239.
[99] D. E. Woessner and J. P. Zimmerman, *J. Phys. Chem.*, 1963, **67**, 1590.
[100] T. Drakenberg, *J. Magn. Resonance*, 1974, **15**, 354.
[101] F. Ostroy, T. L. James, J. H. Noggle, A. Sarrif, and L. E. Hokin, *Arch. Biochem. Biophys.*, 1974, **162**, 421.
[102] N. Arnheim, F. Millett, and M. A. Raftery, *Arch. Biochem. Biophys.*, 1974, **165**, 281.

chitotriose [Glu(F Ac)$_3$] to lysozyme. Horne et al.[103] used ^{35}Cl relaxation to investigate binding to Zn atoms in E. coli alkaline phosphatase—a rapid exchange with a bulk phase ^{35}Cl gave an averaged relaxation rate which could be related to the amount and type of zinc binding. Fischer[104] has provided a review of n.m.r. studies of the binding of drugs to macromolecules, giving a number of examples. A more recent example of what can be achieved is the observation of the linewidth of the NMe group in acetylcholine on binding to acetylcholinesterase.[105]

Although they are not proteins, the nucleotides also exhibit binding. Reuben et al.[106] observed the frequency dependence of the ^{23}Na T_1 values with some support from ^{87}Rb measurements. Concentration plots indicated that a two-phase model was appropriate, with rapid exchange between bound and free species. The bound-phase ^{23}Na relaxation rate was similar to that reported for ^{23}Na bound to soluble RNA, polyphosphates, and poly(acrylic acid), suggesting that a similar mechanism was operative.

5 Cellular Systems

The interpretation of n.m.r. measurements on cellular systems has for some time been controversial. Most workers interpret their observations in terms of the model exploited in the papers reported in the earlier sections, based on a rapid exchange of water molecules or ions between a small bound region and a majority bulk phase. Other workers assume that the structure of cellular water is modified from bulk water, and that all the ions are complexed and have a reduced mobility, instead of only a small fraction being affected. The proceedings of a conference on the subject provide summaries of both viewpoints.[5] In the period of this review there have been relatively few papers concerned with this controversy, although Hazlewood[107] has provided a general paper discussing the origins of the model involving extensive structuring. Cope and Damadian,[108] supporters of this minority viewpoint, have noted that the T_1 and T_2 values of ^{39}K in rat muscle and brain were reduced to values comparable to those observed in exchange resins. The authors found that for solutions of KCl, introduction of 0.1 mol l^{-1} of paramagnetic ions or of acetate or phosphate at physiological concentrations was ineffective in reducing the relaxation times to the values observed in cellular systems, and concluded that these data supported their interpretation. An odd feature was that the free induction decay of the KCl solution was Gaussian, implying, perhaps, that field inhomogeneities were responsible. It has been claimed by exponents of this viewpoint that because of binding and consequent line-broadening, part of the water or alkali-metal n.m.r. signal might not be observable. Monoi[109] has examined the ^{23}Na n.m.r. of an homogenate of rat liver consisting of various particulate fractions, and concluded that there was no missing signal. He discusses the lineshape expected in such systems for different exchange rates between the bound and free fractions. The controversy on the absence or presence of long-range ordering is not restricted to cellular systems and even

---

[103] J. E. Horne, H. Csopak, and B. Lindman. *Arch. Biochem. Biophys.*, 1974, **162**, 552.
[104] J. J. Fischer, *Mod. Pharmacol.*, 1973, **1**, 583.
[105] G. Kato, *J. Pharm. Sci.*, 1975, **64**, 488.
[106] J. Reuben, M. Shporer, and E. J. Gabbay, *Proc. Nat. Acad. Sci. U.S.A.*, 1975, **72**, 245.
[107] C. F. Hazlewood, *Bull. Texas Heart Inst.*, 1975, **2**, 83.
[108] F. W. Cope and R. Damadian, *Physiol. Chem. Phys.*, 1974, **6**, 17.
[109] H. Monoi, *Biophys. J.*, 1974, **14**, 645.

arises in a system as 'inert' as a porous glass. Roberts and Northey[110] found a H_2O proton line-broadening as a function of pore size, and interpreted it in terms of long-range ordering; this was criticized by Belfort,[111] who applied the standard two-phase model.

Neville et al.[112] examined the H_2O ^1H n.m.r. of the lens of the eye from frogs and rabbits, finding $T_2^* < T_1$, with the T_2^* decay multi-component. The values depended upon the H_2O content, which was changed by soaking the lens in hypo- or hyper-osmotic Ringer solution. The data were not consistent with a simple two-phase rapid-exchange model. Treatment with microwaves or with a 5% CCl_3CO_2H solution rendered the lens opaque, decreasing T_1 and T_2^* without affecting the water content. Hazlewood et al.[113] examined the T_2 relaxation of water protons in rat gastrocnemius muscle, finding three components, in good agreement with Belton et al.[114] The three components were identified with three regions, water associated with macromolecules (8%), with the myoplasm (82%), and with the extracellular space (10%). The decay was corrected for diffusion, which was determined at different pulse spacings. On the basis of measurements at two frequencies the following suggestion was rejected, viz. that the reduction in T_2 values was caused by diffusion through field gradients produced by sample heterogeneities. Comparison with ^2H studies suggested that paramagnetic impurities were not responsible. Belton and Packer[115] considered the effect of dehydration on the three T_2 components in frog gastrocnemius muscle, and the effect of dehydration and subsequent rehydration on the freezing properties. Dehydration occurred as a two-stage process, and difficulty was experienced in removing the last 10 to 20% of the water. The three T_2 components were removed sequentially, but this was a kinetic study. Mahendroo et al.[116] found that on allowing re-equilibration to occur after partial dehydration, some redistribution of water occurred amongst the three components. The values of the three relaxation times were decreased, as was the fraction of freezable water. In porcine muscle Pearson et al.[117] observed a single T_1 and two T_2 components whose relative intensities changed during the *rigor* process. The observations were interpreted in terms of a model invoking two separated regions within each of which bound and free H_2O occurred in rapid exchange, but exchange between regions was, on an n.m.r. timescale, slow. Over the period of the *rigor* process there was a net transfer of free water from one region to the other, presumably intracellular to extracellular. The bound water was again identified with the non-freezing fraction. Again, the relaxation behaviour of the non-freezing fraction was complex,[118] and it was necessary to invoke a log normal distribution of correlation times to fit the frequency and temperature dependences of the spin–lattice relaxation data. As in frozen agarose gels,[64] the fitted interaction constant was anomalously low and

[110] N. K. Roberts and H. L. Northey, *Nature*, 1972, **237**, 144.
[111] G. Belfort, *Nature*, 1974, **249**, 593.
[112] M. C. Neville, C. A. Paterson, J. L. Rae, and D. E. Woessner, *Science*, 1974, **184**, 1072.
[113] C. F. Hazlewood, D. C. Chang, B. L. Nichols, and D. E. Woessner, *Biophys. J.*, 1974, **14**, 583.
[114] P. S. Belton, R. R. Jackson, and K. J. Packer, *Biochim. Biophys. Acta*, 1972, **286**, 16.
[115] P. S. Belton and K. J. Packer, *Biochim. Biophys. Acta*, 1974, **354**, 305.
[116] P. P. Mahendroo, D. R. Woodhouse, W. Derbyshire, and S. M. A. Baghdadi, in ref. 81, p. 285.
[117] R. T. Pearson, I. D. Duff, W. Derbyshire, and J. M. V. Blanshard, *Biochim. Biophys. Acta*, 1974, **362**, 188.
[118] I. D. Duff and W. Derbyshire, *J. Magn. Resonance*, 1974, **15**, 310.

this, together with an enhanced ^{1}H $T_{1\rho}$ and spin–spin relaxation rate, indicated that mobility at low frequencies was greater than that obtained using the calculated log normal distribution. These results and conclusions were consistent with those of Fung and McGaughy[119] who fitted the spin–lattice relaxation data above and below the freezing point, obtaining separate values for the activation energy. The authors incorporated deuteriated DMSO solutions, which had the effect of depressing the freezing point and increasing the fraction of non-freezable water.

This is part of an extended study of cellular systems by Fung's group. In a subsequent investigation[120] using ^{2}H n.m.r. it was found that there was no n.m.r.-invisible water. Proton T_1 data were fitted to a log normal distribution in gastrocnemius muscle and liver of mice, and in egg white. It was claimed that because ^{2}H relaxation is determined almost entirely by intramolecular terms there are fewer ambiguities in the assignment of an interaction constant. The ^{1}H T_1 decays were single, except for the egg-white sample. The potassium and sodium ion concentrations were obtained for the three systems but did not correlate with the water proton relaxation. Finch and Homer[121] also examined ^{1}H relaxation (T_1, T_2, and $T_{1\rho}$) in frog muscle as a function of frequency and temperature. The application of a regression analysis showed a distribution skewed towards free water. Knipsel et al.[122] recorded the T_1 and $T_{1\rho}$ as a function of frequency (17 to 45 MHz for the former and 10^3 to 10^5 Hz for the latter), for healthy and tumorous spleen and muscle of mice, and attempted an interpretation based on the two-phase model, but found that a minimum of three relaxation mechanisms was required.

Civan and Shporer[123] examined the ^{17}O relaxation from $H_2{}^{17}$O in frog striated muscle. For fresh samples T_1 was non-exponential, the relative fractions of the two components being temperature independent. As sample deterioration occurred the non-exponentiality became less marked. Evacuation also improved the exponentiality; T_1 was a function of water content and was consistent with a two-phase model. T_1 was reduced by a factor of 3 and T_2 by a factor of 4 or 5 from bulk water values. Non-exponentiality of ^{17}O T_1 relaxation was also found in red cell suspensions, packed cells, and the supernatant of human erythrocytes.[124] Because ^{17}O relaxation times are three orders of magnitude shorter than those for protons, separate signals could be obtained from intra- and extra-cellular regions, and the exchange process monitored. The signal from the extra-cellular regions could be removed by the introduction of Mn^{2+} ions. By incubating in a low concentration K^+ Ringer solution Shporer and Civan[125] enhanced the Na^+ concentration in frog striated muscle and studied the temperature and frequency dependence of the ^{23}Na relaxation, observing a single T_1 decay and two-component T_2; T_1 and T_2^* were a factor of 4 or 5 less than in aqueous solutions of sodium halides and had a different temperature dependence. It was concluded that the fraction of bound ions subject to a severe restriction in mobility could not exceed a few percent. Hoult et al.[126] observed ^{31}P

[119] B. M. Fung and T. W. McGaughy, *Biochim. Biophys. Acta*, 1974, **343**, 663.
[120] B. M. Fung, D. L. Durham, and D. A. Wassil, *Biochim. Biophys. Acta*, 1975, **399**, 191.
[121] E. D. Finch and L. D. Homer, *Biophys. J.*, 1974, **14**, 907.
[122] R. R. Knipsel, R. T. Thompson, and M. M. Pintar, *J. Magn. Resonance*, 1974, **14**, 44.
[123] M. M. Civan and M. Shporer, *Biochim. Biophys. Acta*, 1974, **343**, 399.
[124] M. Shporer and M. M. Civan, *Biochim. Biophys. Acta*, 1975, **385**, 81.
[125] M. Shporer and M. M. Civan, *Biochim. Biophys. Acta*, 1974, **354**, 291.
[126] D. I. Hoult, S. J. W. Busby, D. G. Gadian, G. K. Radda, R. E. Richards, and P. J. Seeley, *Nature*, 1974, **252**, 5481.

n.m.r. in intact biological tissue, particularly muscle at 7.5 T, resolving peaks and monitoring concentrations of many sugar phosphates and glycolytic intermediates. The state of ionization and interactions with metal ions also affected the resonance positions. The time dependences of the concentrations of ATP, ADP, AMP, glucose-1-phosphate, glucose-6-phosphate, and inorganic phosphate were followed during the 'flash' activation of phosphorylase in glycogen particles. In intact rat muscle the positions of the β and γ ATP peaks were consistent with the complexation of ATP with Mg^{2+} ions. Whilst most muscle phosphate lines were broader than in normal solution, the inorganic phosphate peak width was greatest. Having established that this was not attributable to field or susceptibility inhomogeneities, it was proposed that a distribution of pH values within the muscle was responsible. The intensity of the creatine phosphate signal gave a measure of the degree of stimulation before death.

Diffusion of water in cellular systems is of interest because of the occurrence of obstacles and barriers. Neuman[127] discussed the echo signals of spins diffusing in a bounded medium, considering planar, spherical, and cylindrical boundaries. For a pulse spacing τ, either long or short compared with the diffusion time between boundaries, the distribution of phase shifts was Gaussian. Neuman derived the phase distribution at all values of τ. Cooper et al.[128] used the pulsed field-gradient experiment to observe H_2O diffusion in human blood plasma and red cells, and in rat and rabbit heart and liver. The packed red blood cells showed a barrier spacing at 2.3 μM, presumably the red cell membrane, and a permeability of 1.4×10^{-4} m s^{-1}. For the rat liver the boundaries occurred at 2 μM, probably mitochondria, whilst none of the remaining systems had a dependence upon the field-gradient pulse-spacing.

Measurements of the type discussed in this section have been applied to cellular systems subject to tumour growth, and to erythrocyte cells subject to sickling. The number of published papers in the first group is particularly extensive, mostly concerned with the enhancement of H_2O ^1H spin–lattice relaxation times in tumorous tissues, usually being greater for malignant tumours than for benign. The situation is somewhat complicated in that the interpretation of observations depends upon the model of intracellular water adopted; e.g. Inch et al.[129] determined the T_1 values in regenerating liver and foetal tissue, which have a higher H_2O content than most cells, and compared them with those obtained from normal and neoplastic tissues. The data were fitted to a hyperbolic expression relating T_1 to water content, and it was concluded that the enhancement of T_1 in tumorous tissue was related to water content and not to any long-range ordering. This procedure was disputed by Hazlewood,[130] who applied a regression analysis, finding a correlation between T_1 and water content in liver only and not in spleen and kidney. Even within the framework of the two-phase model, Block and Maxwell[131] came to different conclusions than Inch and his colleagues. Block and Maxwell examined the ^1H relaxa-

---

[127] C. H. Neuman, J. Chem. Phys., 1974, 60, 4508.
[128] R. L. Cooper, D. B. Chang, A. C. Young, C. J. Martin, and B. Ancker-Johnson, Biophys. J., 1974, 14, 161.
[129] W. R. Inch, J. A. McCredie, R. R Knispel, R T. Thompson, and M. M. Pintar, J. Nat. Cancer. Inst., 1974, 32, 353.
[130] C. F. Hazlewood, J. Nat. Cancer Inst., 1974, 52, 625.
[131] R. E. Block and G. P. Maxwell, J. Magn. Resonance, 1974, 14, 329.

tion in normal and tumorous rat tissue at two frequencies. They applied the two-phase model, assuming a single correlation time for the bound phase, and iteratively fitted this and the amount of bound water, finding the latter to be reduced in tumorous tissue. Chang and Hazlewood[132] were unhappy with this interpretation, referring to their own work on diffusion in cardiac muscle as a function of pulse spacing[133] and stating that it was normally considered necessary to invoke a distribution of correlation times, and that in any event the model did not fit the T_2 data. In two papers Fung and his colleagues[134, 135] determined that, whilst the total water content was higher in tumorous tissue, the amount of bound water relative to the solid was unchanged at 0.34 g H_2O (g solid)$^{-1}$, the K^+ concentration was markedly lower, whilst the Na^+ concentration was slightly higher. These papers reported studies on the relaxation behaviour over wide limits of frequency and temperature. The literature is extensive, and as a book has been published recently[136] devoted specifically to n.m.r. investigations of cancer, published papers are merely listed (Table 3) together with a few short comments.

Table 3 *Investigations of tumour detection*[a]

Study	Ref.
Regenerating liver, foetal tissue, and normal and neoplastic tissue compared with gelatine and cotton	(−) 137
White and grey matter in human and canine brain	(−) 138
T_{1p} frequency dispersion and T_2–Meiboom dispersion independent of presence of tumour	(−) 139
Normal and tumorous tissue of rats and mice, H_2O and Na^+ concentration higher, K^+ similar. ^{39}K decay in 90–τ–90 sequence oscillatory	(+) 140
Mice mammary nodule and neoplasms distinguishable. T_1 in other organs affected. T_1 increase not directly correlated with water content	(+) 141
Human breast dysplasias and neoplasms. Distinguish non-neoplastic from neoplastic. T_2 single, T_1 multicomponent	(+) 142
Reduction in D different in normal, preneoplastic, and neoplastic tissue	(+) 143
Mouse liver and kidney T_1 related to H_2O concentration. Bathed in NaCl solution to modify H_2O content	(+) 144
Appearance and development of tumour systemic effect in serum, liver, kidneys, spleen. T_1 and H_2O content; variability too big in skeletal muscle	(+) 145
Yoshida ascites tumour and mouse fibrosarcoma	(+) 146
Human cells *in vitro*. Benign tumours intermediate T_1	(+) 147
Time courses of ^1H T_1 in mice developing ascites tumours, blood serum, liver homogenate	(+) 148
Nine-day embryos, immature tissues, neoplasms. T_1 related to H_2O concentration, T_2 related to pH	(+) 149
Rat tissue T_1 related to H_2O concentration. T_1 of healthy tissues in same animal affected	(+) 150

The (−) signs indicate that the authors were pessimistic and the (+) signs that they were optimistic concerning the use of n.m.r. as a diagnostic probe. In many cases the optimism or pessimism had to be qualified.

[132] D. C. Chang and C. F. Hazlewood, *J. Magn. Resonance*, 1975, **18**, 550.
[133] D. C. Chang, *Biophys. J.*, 1974, **14**, 161.
[134] B. M. Fung, *Biochim. Biophys. Acta*, 1974, **362**, 209.

The use of solvent molecule relaxation in studies of sickle-cell haemoglobin is neither as extensive, nor as controversial. Zipp et al.[151] determined the linewidth at two frequencies, 60 and 100 MHz, and found a linear dependence upon percentage sickling in the deoxygenated samples that was reversed on oxygenation. They considered that the sickled cells suffered a 10% loss in water. Lindstrom and Koenig[152] examined the frequency dependence of the ^{1}H spin–lattice relaxation of normal and sickle-cell haemoglobin solutions, normal red blood cells, and whole blood. The value of τ_c for the haemoglobin molecules was unchanged on incorporation into the cells. At 283 K, where the deoxygenated sickle-cell haemoglobin forms a gel, 90% of the haemoglobin molecules remained in solution, but at 308 K all the haemoglobin was incorporated into the gel structure. The most detailed studies were performed by Cottam and Waterman and their colleagues.[153–155] They again found that the T_2 values of blood, washed erythrocytes, and haemoglobin with sickle cell were decreased on deoxygenation, whereas for normal adults the decrease in T_2 was minor; thus in effect they were monitoring gelation. Reducing temperature reversed the gelation and caused T_2 to have an opposite temperature dependence to that observed in oxyhaemoglobin S and oxy- or deoxy-normal haemoglobin. To explain the frequency dependence it was necessary to invoke three environments, whereby 0.2% of the H_2O was very tightly bound (with $\tau_c \geqslant 10^{-7}$s), 1.3 to 1.5% was in the hydration sphere of the haemoglobin (with $\tau_c \approx 2$ to 4 ns), independent of the erythrocyte, and the remaining 98% was bulk water. The T_1 values were not sensitive to sickling. In the last papers the authors observed the pH dependence of the T_2

---

[135] B. M. Fung, D. A. Wassil, D. L. Durham, R. W. Chestnut, N. N. Durham, and K. D. Berlin, *Biochim. Biophys. Acta*, 1975, **385**, 180.
[136] 'The Nuclear Resonance Effect in Cancer', ed. R. Damadian, Pacific Publishing, 1974.
[137] W. R. Inch, J. A. McCredie, C. Geiger, and Y. Bocter, *J. Nat. Cancer Inst.*, 1974, **53**, 689.
[138] R. G. Parrish, R. J. Kurland, W. W. Janese, and L. Bakay, *Science*, 1974, **183**, 438.
[139] M. M. Pintar, in ref. 136, Chap. 31.
[140] R. Damadian and F. W. Cope, *Physiol. Chem. Phys.*, 1974, **6**, 309.
[141] C. F. Hazlewood, G. Cleveland, and D. Medina, *J. Nat. Cancer Inst.*, 1974, **52**, 1849.
[142] D. Medina, C. F. Hazlewood, G. G. Cleveland, D. C. Chang, and H. J. Spjut, *J. Nat. Cancer Inst.*, 1975, **54**, 813.
[143] D. C. Chang, H. E. Rorschach, G. G. Cleveland, and C. F. Hazlewood, in ref. 136, Chap. 15.
[144] L. A. Saryan, D. P. Hollis, J. S. Economou, and J. C. Eggleston, *J. Nat. Cancer Inst.*, 1974, **52**, 599.
[145] D. P. Hollis, L. A. Saryan, J. S. Economou, J. C. Eggleston, J. L. Czeisler, and H. P. Morris, *J. Nat. Cancer Inst.*, 1974, **53**, 807.
[146] R. S. Chaughule, S. R. Kasturi, R. Vijayaraghavan, and S. S. Ranade, *Indian J. Biochem. and Biophys.*, 1974, **11**, 256.
[147] S. S. Renade, R. S. Chaughule, S. R. Kasturi, and J. S. Nadkarni, *Indian J. Biochem. and Biophys.*, 1975, 12.
[148] R. A. Floyd, J. S. Leigh, jun., B. Chance, and M. Miko, *Cancer Res.*, 1974, **34**, 89.
[149] I. C. Kiricuta, V. Simplaceanu, and D. Demco, 'Proceedings 18th Colloque Ampère', ed. P. S. Allen, E. R. Andrew, and C. A. Bates, University of Nottingham Press, Nottingham, 1974, p. 287.
[150] W. Bovee, P. Huisman, and J. Smidt, *J. Nat. Cancer Inst.*, 1974, **52**, 595.
[151] A. Zipp, I. D. Kuntz, S. J. Rehfeld, and S. B. Shohet, *F.E.B.S. Letters*, 1974, **43**, 9.
[152] T. R. Lindstrom and S. H. Koenig, *J. Magn. Resonance*, 1974, **15**, 344.
[153] G. L. Cottam, K. M. Valentine, K. Yamaoka, and M. R. Waterman, *Arch. Biochem. Biophys.*, 1974, **162**, 487.
[154] B. C. Thompson, M. R. Waterman, and G. L. Cottam, *Arch. Biochem. Biophys.*, 1975, **166**, 193.
[155] A. H. Chuang, M. R. Waterman, K. Yamaoka, and G. L. Cottam, *Arch. Biochem. Biophys.*, 1975, **167**, 145.

relaxation. In the deoxy-form the spin–spin relaxation rate was maximum in the middle of the pH range, whereas it increased with pH in the oxy-form.

So far all discussion has related to animal cells. However, a few papers have been concerned with plant tissues. Macdowell and Buchanan[156] determined that the amount of non-freezing water was greater in frost-resistant wheat leaves than in the normal leaf sheath. Nanassy[157] found overlapping wide and narrow lines in green and remoistened wood from white birch and white spruce. On dehydration and rehydration the intensity and linewidth of the narrow component showed hysteresis; the intensity and width of the broad component also exhibited a dependence upon water content, but no hysteresis. Only 40% of the H_2O molecules were exchangeable with D_2O. Finally, Burke et al.[158] examined cold-acclimatized and non-acclimatized red osier dogwood stem, finding again that the T_2 decay was multi-component, but in this instance an orientation dependence was established. The amount of non-freezable water as g H_2O (g solid)$^{-1}$ was found to be independent of acclimatization, but the amount of freezable water was found to be reduced.

In the future it appears probable that the technique variously named zeugmatography, n.m.r. diffraction, or spin mapping will prove useful in studies of heterogeneous systems. Although in its infancy, spatial resolution is of the order of 1 mm. Lauterbur is a leading exponent and has examined the water distributions inside a small clam, a pine branch, and the thoracic cavity of a mouse.[159,160] He examined the transport of the H_2O through a parsley stem immersed in D_2O, but found no evidence of transport when a clam was immersed in D_2O. Other examples on mice, onions, fingers etc. have been provided by Hutchison et al.,[161] Mansfield et al.,[162] and Hinshaw.[163] Garroway has discussed the effects of flow and diffusion.[164]

6 Membranes

As discussed in the Introduction, the study of membranes and model membranes is very popular, interest focusing on the arrangement and structure of phospholipid molecules that are the major constituents, and the interactions with cholesterol lipoproteins, ions, water, and drug molecules. This section interfaces with some other chapters and inevitably there is a certain degree of arbitrariness in the division of material. One reason is that surfactant molecules resemble phospholipids and the lamellar liquid-crystal phase resembles the membrane bilayer. Much of this material is covered in the chapter on liquid crystals included biennially. Even so, a large number of papers remain, and a selection has had to be made. The basis for this has been to choose the ones which seem to the Reporter to be of greater significance, or to be typical of the type of study being discussed. In general, this section is not concerned with the structure and/or mobility of membrane molecules in isolation. An

---

[156] F. D. H. Macdowell and G. W. Buchanan, Canad. J. Biochem., 1974, 52, 652.
[157] A. J. Nanassy, Wood Science, 1974, 7, 61.
[158] M. J. Burke, R. G. Bryant, and C. J. Weiser, Plant Physiol., 1974, 54, 392.
[159] P. C. Lauterbur, Pure Appl. Chem., 1974, 40, 149.
[160] P. C. Lauterbur, 'Proceedings 1st International Conference on the Use of Stable Isotopes in Chemistry, Biology and Medicine', U.S. Atomic Energy Commission, 1974, p. 255.
[161] J. M. S. Hutchison, J. R. Mallard, and G. C. Goll, in ref. 149, p. 283.
[162] P. Mansfield, P. K. Grannell, and A. A. Maudsley, in ref. 149, p. 431.
[163] W. S. Hinshaw, in ref. 149, p. 433.
[164] A. N. Garroway, in ref. 149, p. 435.

example of this is an investigation[165] of the conformation of phosphatidylethanolamine in solutions of $CDCl_3$ and CD_3OD and comparison with results for phosphoethanolamine.

Seelig and his colleagues employed ^2H n.m.r. to investigate the mobility at different points along the acyl chains, using selectively deuteriated compounds. The system was first tested on a lamellar phase comprising sodium decanoate, decanol, and water.[166] The quadrupolar interactions of the CD deuterons were partially averaged as a result of chain motion, the degree of averaging being a function of position along the chain. An order parameter S was derived; it was constant at 0.5 except for the three positions nearest the terminal methyl group. The averaging was explained by the presence of kink defects, of the order of one per chain, with a lifetime of 1 μs. Seelig and Seelig[167] extended this type of measurement to selectively deuteriated dipalmitoyl-lecithins (DPL). They found that the two acyl chains were not completely equivalent. At 314 K the order parameter S was constant at 0.45 and then decreased. At higher temperatures S was reduced as a consequence of chain shortening resulting from bent configurations. The variation of order parameter with chain position was not completely consistent with e.s.r. spin-label studies, which showed a regular decrease in order parameter along the chain. Seelig and Niederberger[168] made a more detailed comparison and concluded that the introduction of nitroxide spin labels caused perturbation. They also compared[169] order parameters calculated from powder spectra (as described earlier) with those calculated from spectra obtained from selectively deuteriated lipids oriented between glass slides. Previous models of mobility in phospholipid molecules were based on those used for free polymer chains. Marcelja[170,171] derived a statistical model based on the probability of *trans* and *gauche* configurations, and in turn on more specific defects, *e.g.* kinks and jogs. Schindler and Seelig[172] used this model to calculate segmental order parameters. Fujiwara *et al.*[173] used techniques similar to those of Seelig's group in studies of mobility in the lipophilic region. Stockton *et al.*[174] applied similar techniques, but used deuteriated phospholipid and fatty acid probe molecules intercalated in the liquid-crystal phase of aqueous dispersions of egg phosphatidylcholine. The incorporation of cholesterol was found to affect the quadrupolar splittings of the hydrocarbon chain, but not those of the choline methyl group. Arvidson *et al.*[175] fed rats with deuteriated cholines. On examining the membrane ^2H resonance they found a smaller quadrupolar splitting than in model liquid-crystalline phases, and they concluded that the choline group interacted with other membrane molecules, probably proteins.

Lamellar multilayers of the various lecithin molecules are often used as a model

[165] H. Richard, J. Dufourcq, and C. Lussau, *F.E.B.S. Letters*, 1974, **45**, 136.
[166] V. W. Niederberger and J. Seelig, *Ber. Bunsengesellschaft phys. Chem.*, 1974, **78**, 947.
[167] A. Seelig and J. Seelig, *Biochemistry*, 1974, **13**, 4839.
[168] J. Seelig and W. Niederberger, *Biochemistry*, 1974, **13**, 1585.
[169] J. Seelig and W. Niederberger, *J. Amer. Chem. Soc.*, 1974, **96**, 2069.
[170] S. Marcelja, *Biochim. Biophys. Acta*, 1974, **367**, 165.
[171] S. Marcelja, *J. Chem. Phys.*, 1974, **60**, 3599.
[172] H. Schindler and J. Seelig, *Biochemistry*, 1975, **14**, 2283.
[173] F. Fujiwara, L. W. Reeves, A. S. Tracey, and L. A. Wilson, *J. Amer. Chem. Soc.*, 1974, **96**, 5249.
[174] G. W. Stockton, C. F. Polnaszek, L. C. Leitch, A. P. Tulloch, and I. C. P. Smith, *Biochem Biophys. Res. Comm.*, 1974, **60**, 844.
[175] G. Arvidson, G. Lindblom, and T. Drakenberg, *F.E.B.S. Letters*, 1975, **54**, 249.

system. Feigenson and Chan[176] examined the field dependence of the ^{1}H linewidth in lecithin multilayers, which was attributed to chemical-shift differences: T_1 was measured as a function of temperature and frequency and interpreted in terms of segmental motion of the methylene chains and choline head-groups. It was considered that the terminal CH_3 groups, together with a few mobile CH_2 groups at the chain ends, provided a relaxation sink for the remaining CH_2 groups by means of spin diffusion; the choline methyl groups were considered to be uncoupled. Kink formation was invoked to explain relaxation of the methylene chains. Urbina and Waugh[177] used the double-resonance technique developed by Pines et al.[178] to measure the ^{13}C n.m.r. of selectively labelled ^{13}C sites in dimyristoyl- and dipalmitoyl-lecithins (DML and DPL), above and below the temperatures of the gel to liquid-crystal transitions, finding that the linewidths were determined by chemical-shift anisotropies, the efficiency of the cross-relaxation process being strongly dependent upon the motion of the group involved. Phillips and Finer[179] compared the immobilization of lecithin chains by cholesterol and by 5α-androstan-3β-ol, confirming a 1:1 stoicheiometry of the complex between lecithin and cholesterol. The amplitude of the CH_2 signal was reduced, the reduction corresponding to binding to cholesterol. There was also some discussion of the co-operative motion of the hydrocarbon chain. Ulmius et al.[180] undertook a related study, examining lamellar liquid crystals and gel phases containing lecithins and cholesterol. Spectra were consistent with the molecules possessing cylindrical symmetry, implying that the lateral diffusion was sufficiently rapid to remove the intermolecular dipolar interaction. Lineshapes in the lamellar and gel phases were super-Lorentzian, corresponding to 3 and 2 components, respectively. An order parameter was estimated from the rigid-lattice linewidth calculated for geminal methylene protons. This was consistent with Seeligs' value[181] but larger than that of Seiter and Chan.[182] Incorporation of cholesterol enhanced lateral diffusion in the gel phase, but produced effects which were not consistent with the 1:1 complex proposed by Phillips and Finer.[179] There was evidence of phase separation in the mixed DPL–DML system. Dijkema and Berendsen[183] oriented a model membrane system, octylammonium chloride and potassium oleate, between glass slides. By rotating the glass slides relative to the magnetic field they could study the orientation dependence. With the slides at an angle of 54°44′ relative to the magnetic field the dipolar interactions were reduced to a small level, and well-resolved spectra of Lorentzian lineshape were obtained. The sample alignment was calculated as being correct to 15′. At small deviations there was evidence of splittings of the methyl and methylene peaks. Peaks were broadened sequentially in the order α-CH_2, CH_3, CH_2, and finally the H_2O protons. At orientations of both 0° and 90° spectra consisted of a well-resolved doublet, with a superposed central peak only partially due to the water protons. It was concluded that in both systems the 5 or 6 atoms in the hydrocarbon chains near

[176] G. W. Feigenson and S. I. Chan, J. Amer. Chem. Soc., 1974, 96, 1312.
[177] J. Urbina and J. S. Waugh, Proc. Nat. Acad. Sci. U.S.A., 1974, 71, 5062.
[178] A. Pines, M. G. Gibby, and J. S. Waugh, J. Chem. Phys., 1973, 59, 569.
[179] M. C. Phillips and E. G. Finer, Biochim. Biophys. Acta, 1974, 356, 199.
[180] J. Ulmius, H. Wennerstrom, G. Lindblom, and G. Arvidson, Biochim. Biophys. Acta, 1975, 389, 197.
[181] J. Seelig and A. Seelig, Biochem. Biophys. Res. Comm., 1974, 57, 406.
[182] C. H. A. Seiter and S. I. Chan, J. Amer. Chem. Soc., 1973, 95, 7541.
[183] C. Dijkema and H. J. C. Berendsen, J. Magn. Resonance, 1974, 14, 251.

the polar head had a uniform degree of order, whilst the molecule rotated rapidly about the optical axis. The authors also discussed relaxation in the system, and the origin of the order parameter.

In general, spectra from intact membranes or the lamellar liquid-crystal phase of model membranes are wide, although some narrow signals from different chemical groups are observed superposed on a broad background. Valic et al.[184] claimed, on the evidence of lineshape analysis, that the narrow peaks arose from singularities in the powder pattern corresponding to regions of membrane oriented at the magic angle. A commonly used expedient to obtain high-resolution n.m.r. spectra is to sonicate the intact system. This produces spherical vesicles, 20—30 nm in diameter, and yields high-resolution spectra. The reason for the success of this procedure has been a subject of debate for the past five or six years, one reason being that the vesicle tumbling rate is insufficient to narrow statically broadened lines. Finer[185] claimed that as the molecular motion was anisotropic it was only necessary for the vesicle tumbling rate to be rapid relative to the partially reduced linewidth to produce further narrowing. Chan's group have for some time adopted the position that sonication does in fact disrupt the regular packing of the phospholipid molecules, presumably as a consequence of the high surface curvature of vesicles 30 nm in diameter. Chan and his colleagues[186] disputed Finer's conclusions, claiming that he had used an inappropriate value for the reduced linewidth in his analysis. They also disputed the use of the Gutowsky–Pake formula for inhomogeneously broadened lines. On the evidence of measurements above and below the thermal phase transition they again claimed that sonication does produce disruption. Sears[187] determined the ^{13}C T_1 values and linewidths in sonicated and unsonicated forms of egg phosphalidylcholine: T_1 was sensitive to segmental motions, whereas the linewidth reflected small changes in membrane structure. These interpretations are not consistent with those of Lee et al.[188] He also examined phosphatidylcholine in both H_2O and methanol, and concluded that sonication produced minor but significant changes in structure. Godici and Landsberger[189] determined the ^{13}C T_1 values of sonicated aqueous dispersions of egg yolk lecithin, finding that T_1 values increased from the glycerol backbone towards the choline moiety and terminal methyl groups. The authors also considered the effect of stearic acid derivative spin labels on the T_1 values. They concluded that lateral motion was accompanied by a whipping motion of the acyl chains. Brulet and McConnell[190] determined the spin-label-enhanced ^{13}C relaxation in unsonicated dispersions of phosphatidylcholine. The theory for a two-dimensional lateral diffusion was developed and applied to an analysis of ^1H and ^{13}C relaxation in sonicated vesicles, suggesting that the lateral diffusion rate exceeded $2 \times 10^{-12} m^2 s^{-1}$. Stoffel et al.[191] determined the ^{13}C T_1 and

[184] M. I. Valic, E. Enga, E. E. Burnell, and M. Bloom, in ref. 149, p. 573.
[185] E. G. Finer, J. Magn. Resonance, 1974, **13**, 76.
[186] D. Lichtenberg, N. O. Peterson, J. L. Girardet, M. Kainosho, P. A. Kroon, C. H. A. Seiter, G. W. Feigenson, and S. I. Chan, Biochim. Biophys. Acta, 1975, **382**, 10.
[187] B. Sears, J. Membrane Biol., 1975, **20**, 59.
[188] A. G. Lee, N. J. M. Birdsall, Y. K. Levine, and J. C. Metcalfe, Biochim. Biophys. Acta, 1972, **255**, 43.
[189] D. E. Godici and F. R. Landsberger, Biochemistry, 1974, **13**, 362.
[190] P. Brulet and H. M. McConnell, Proc. Nat. Acad. Sci. U.S.A., 1975, **72**, 1451.
[191] W. Stoffel, B. D. Tunggal, O. Zierenberg, E. Schreiber, and E. Binczek, Z. physiol. Chem., 1974, **355**, 1367.

NOE of mixed vesicles of phosphatidylcholine and sphingomyelins labelled at specific positions of their hydrophilic and hydrophobic regions, and considered the influence of cholesterol and the effect of medium viscosity. Barden et al.[192] used ^{1}H and ^{31}P n.m.r., together with ferricyanide shift reagents, to determine the outside/inside distribution of phospholipids in mixed vesicles. The preference was also pH-dependent. The significant factors were the charge and packing properties of the head groups, the charged groups tending to be on the outside. The effect of vesicle size on the ^{31}P linewidth was also investigated, leading to the conclusion, consistent with previous studies, that chemical-shift anisotropy was the dominant feature. In previous studies Barden and his colleagues[193] examined the ^{31}P linewidth as a function of frequency in DPL, brain sphingomyelin, and egg lecithin, finding it strongly temperature dependent in DPL below 314 K, but temperature independent above. At 323 K two signals were resolved, one of which, assigned to phospholipids on the outside of the bilayer vesicle, disappeared on addition of $PdCl_2$. The outside/inside ratios were calculated to be 1.53 for lecithin and 1.94 for sphingomyelin. Heterogeneity of chemical shift was rejected as a possible mechanism for the production of two signals, as this would be removed by the rapid lateral diffusion. Gent and Prestegard[194] investigated the relationship between vesicle size and ^{1}H linewidth in cholesterol–phosphatidylcholine (PC) systems of varying composition. With mixed PC–phosphatidylethanolamine (PE) vesicles the CH_2 linewidth increased with increase in PE content and vesicle size. With the PC–cholesterol system the behaviour was more complex. Up to 30% cholesterol, the linewidth increased; above 30% the dependence became more pronounced, the spectrum becoming two-component, with the intensity of the narrow component decreasing. Finally, Prestegard and Fellmeth[195] used ^{1}H intensities and linewidths to monitor the fusion of DML vesicles, finding a two-step process, and the appearance of a bimodal distribution of larger vesicles. The fusion rate increased drastically near the hydrocarbon phase-transition temperature at 293 K. This system was intended as a model to the study of cell–cell interactions and cell fusion. Lange et al.[196] examined the amino-proton resonance in phosphatidyl ethanolamine in sonicated mixed phospholipid vesicles, finding it 3.3 p.p.m. to high frequency of H_2O. Above pH 5 it broadened and became non-observable as a result of exchange with H_2O protons. On increasing pH, the signal from groups inside the vesicle remains and can be used to assay for inside/outside ratios and proton transfer across and along vesicular membranes etc.

Cornell et al.[197] performed a detailed frequency-dependent study of the ^{1}H and ^{2}H T_1 and $T_{1\rho}$ of H_2O/D_2O adsorbed on DPL as a function of water content, from 253 to 338 K, supplemented where necessary by the $90–\tau–90_{90}$ pulse sequence. Results on the liquid-crystal phase were consistent with four molecules of water being bound to the lecithin head-group and exhibiting anisotropic motion, with differential tumbling rates of 10^{10} and 10^6 Hz. Relaxation in the gel phase was a

---

[192] J. A. Barden, R. W. Barker, and G. K. Radda, *Biochim. Biophys. Acta*, 1975, **375**, 186.
[193] J. A. Barden, P. R. Cullis, D. I. Hoult, A. C. McLaughlin, G. K. Radda, and R. E. Richards, *F.E.B.S. Letters*, 1974, **46**, 55.
[194] M. P. N. Gent and J. H. Prestegard, *Biochemistry*, 1974, **13**, 4027.
[195] J. H. Prestegard and B. Fellmeth, *Biochemistry*, 1974, **13**, 1122.
[196] Y. Lange, E. K. Ralph, and A. G. Redfield, *Biochem. Biophys. Res. Comm.*, 1975, **62**, 891.
[197] B. A. Cornell, J. M. Pope, and G. J. F. Troup, *Chem. and Phys. Lipids*, 1974, **13**, 183.

complex function of thermal history. Prestegard and Wilkinson[198] examined the H_2O ^1H relaxation in concentrated aqueous DML solutions (from 19 to 39 molecules of water per lecithin). The T_1 and T_2 values corresponded to an increase in mobility at the hydrocarbon phase transition. The onset of an exchange process with a low activation energy was interpreted as involving diffusion over large distances. Andrasko and Forsen[199] inserted paramagnetic ions into vesicles, and measured ^1H relaxation times as a function of temperature, obtaining values for water diffusion and the permeability of the DPL vesicles. Diffusion was discussed as a two-site chemical exchange between inside and outside regions. Hayter et al.[200] also recorded the rate of H_2O diffusion through model membranes, using a combination of n.m.r. relaxation studies and neutron scattering.

The group of Lindman is very active in the study of the binding of alkali-metal ions to membranes and model membranes. Wennerstrom et al.[201] have published a review of the n.m.r. of alkali-metal ions and halide counter-ions in such systems, considering the effect of anisotropies, macroscopic and microscopic (in effect a rapid local anisotropic motion together with a slow isotropic one over the dimensions of the aggregates). Gustavsson et al.[202] considered the interaction of Na^+ ions with anionic surfactants, recording ^{23}Na quadrupolar splittings, plus T_1 and T_2 times, counter-ion chemical shifts, H_2O ^2H splittings, and ^{14}N ammonium quadrupolar splittings. The water motion was deemed responsible for ^{23}Na relaxation, and the mode of binding was dependent upon the surfactant end-group. The same group[203] also considered the binding of Na^+ and H_2O to lecithin in lecithin–NaCl–D_2O and lecithin–cholesterol–NaCl–D_2O systems as a function of composition and temperature. The numbers of Na^+ and H_2O binding sites were observed to increase with temperature and electrolyte concentration, the proposed explanation invoking a change in orientation of the choline group. The ^{23}Na and ^2H quadrupolar splittings were strongly dependent upon both cholesterol and water content. Initial addition of cholesterol caused a partial release of Na^+ ions from the lamellae, but at higher cholesterol concentrations the Na^+ splitting increased. Deuteron splittings were smaller with K^+ than with other cations, but increased with increasing temperature as a result of exchange between water and choline and/or phosphate groups. Persson et al.[204] confirmed the previous measurements, and attributed broadening of ^2H spectra to deuteron exchange between water and cholesterol.

It is sometimes considered that the binding of alkali-metal ions to ionophores is the rate-determining step in ion transport through membranes. Shporer et al.[205] tested this by examining the binding of Na^+ to valinomycin in methanol, using ^{23}Na T_1 measurements from 173 to 303 K. The observation of a maximum and a minimum

---

[198] J. H. Prestegard and A. Wilkinson, *Biochim. Biophys. Acta*, 1974, **345**, 439.
[199] J. Andrasko and S. Forsen, *Biochem. Biophys. Res. Comm.*, 1974, **60**, 813.
[200] J. B. Hayter, A. M. Hecht, J. W. White, and G. J. T. Tiddy, *Discuss. Faraday Soc.*, 1974, **57**, 130.
[201] H. Wennerstrom, G. Lindblom, and B. Lindman, *Chemica Scripta*, 1974, **6**, 97.
[202] H. Gustavsson, G. Lindblom, B. Lindman, N. O. Persson, and H. Wennerstrom, in 'Liquid Crystals and Ordered Fluids', ed. J. F. Johnson, and R. S. Porter, 1974, p. 161.
[203] G. Lindblom, N. O. Persson, and B. Lindman, *Ber. Bunsengesellschaft phys. Chem.*, 1974, **78**, 955.
[204] N. O. Persson, G. Lindblom, B. Lindman, and G. Arvidson, *Chem. and Phys. Lipids*, 1974, **12**, 261.
[205] M. Shporer, H. Zemel, and Z. Luz, *F.E.B.S. Letters*, 1974, **40**, 357.

demonstrated the existence of an exchange process. In a later paper[206] these measurements were extended to the binding of K^+ with dibenzo-18-crown-6 (^{39}K) and to Rb^+ in RbCNS (^{87}Rb).

Perhaps the most difficult problem to investigate in membranology is the interaction between membrane constituents, e.g. phospholipids and lipoproteins. Chang and Chan[207] considered the interactions between sonicated lecithin bilayers and poly-(L-glutamic acid). On association, the choline methyl groups showed a large reduction in intensity. The ^1H n.m.r. showed that the poly-(L-glutamic acid) remained in a random-coil configuration. Subsequently Lau and Chan[208] extended the measurements to the surface-active agent alamethicin, which interacted primarily with the polar choline head group. Alamethicin induced aggregation and subsequent fusion of small bilayer vesicles. Nicolau et al.[209] also examined a model system, the interaction between DPL and alanylalanine and alanylglycine, obtaining a single ^{13}C T_1 for the choline methyl group which correlated with chemical shifts, parallel to changes in T_1 observed in membrane systems. Stoffel et al.[210] used ^{13}C n.m.r. in lipid human high-density lipoproteins. In reassembled high-density lipoprotein the phospholipid molecules phosphatidylcholine and sphingomyelin bound to the apoproteins with the hydrophobic chains, and not with the hydrophilic zwitterionic groups. The effect of the incorporation of cholesterol was also investigated. Finally, Chapman[211] has provided an introductory discussion to the effect of ion binding and drug incorporation on membranes, citing antidepressants and tranquillizers.

Reeves's group is currently making a series of measurements on model membrane systems; a number of papers have been published in the period of this review, of which three have been selected. Deuteron n.m.r. has been used[212] to determine the orientation of acetate ion and its mobility in a model based on the cationic detergent decylammonium chloride. The distortions of the BF_4^- and NH_4^+ ions in the electric field of the double layer of a model membrane system have been determined,[213] NH_4^+ ions being more distorted in cationic than in anionic systems. In a later paper[214] the measurements were extended to alkali-metal ions. These replaced D_2O molecules in the more ordered regions of the detergent surface. In other systems, lithium dodecylsulphate and sodium dodecylsulphate, the opposite effect occurred.

7 Micellar Systems

Most of the pioneering work establishing the dominant factors affecting the n.m.r. behaviour of micellar systems was done several years ago, and by and large, current studies appear to be aimed at a better understanding of the particular system under investigation, or to the use of a micellar system as a convenient tool to investigate a particular interaction. However, Nakagawa[215] provided a critical analysis of

[206] M. Shporer and Z. Luz, *J. Amer. Chem. Soc.*, 1975, **97**, 665.
[207] C. A. Chang and S. I. Chan, *Biochemistry*, 1974, **13**, 4381.
[208] A. L. Y. Lau and S. I. Chan, *Biochemistry*, 1974, **13**, 4942.
[209] C. I. Nicolau, H. Dreeskamp, and D. Schulte-Frohlinde, *F.E.B.S. Letters*, 1974, **43**, 148.
[210] W. Stoffel, O. Zierenberg, B. D. Tunggal, and E. Schreiber, *Z. physiol. Chem.*, 1974, **355**, 1381.
[211] D. Chapman, *Chem. and Ind.*, 1975, 98.
[212] D. M. Chen, L. W. Reeves, A. S. Tracey, and M. M. Tracey, *J. Amer. Chem. Soc.*, 1974, **96**, 5349.
[213] F. Fujiwara, L. W. Reeves, and A. S. Tracey, *J. Amer. Chem. Soc.*, 1974, **96**, 5250.
[214] D. M. Chen, K. Radley, and L. W. Reeves, *J. Amer. Chem. Soc.*, 1974, **96**, 5251.
[215] T. Nakagawa, *Colloid and Polymer Sci.*, 1974, **252**, 56.

published data on the rate of micellar dissociation, employing a variety of techniques: stop-flow, pressure jump, temperature jump, ultrasonic absorption, and signal-shape analysis for both n.m.r. and e.s.r. Different models of micelle-monomer conversion have been employed, and a new interpretation has been proposed. Tiddy's group has published two papers on mixed surfactant systems. In the first[216] they investigated the phase diagram of octyltrimethylammonium bromide and sodium dodecylsulphate (SDS) in water at 298 K. The surfactant n.m.r. linewidths were found to correlate with the rheological properties. A possible explanation invoked the coexistence of both spherical and cylindrical micelles, although the occurrence of micro-emulsions could not be excluded. Both the ^{1}H T_1 and T_2 relaxations were non-exponential, the latter being subject to wide variation. In the viscoelastic region close to the SDS-rich boundary, differential peak-broadening occurred. In the second paper,[217] the same authors performed a similar study on a mixed zwitterionic anionic surfactant system, hexadecyldimethylammoniopropane sulphonate-SDS-water. Again viscoelasticity was interpreted by the coexistence of spherical and cylindrical micelles. Menger and Lynn[218] determined the rate of proton transfer at the surfaces of micelles composed of long-chain amine salts. Fendler et al.[219] investigated the effect of surfactants on ligand-exchange reactions of vitamin B_{12a} in water and in benzene, considering the effect of aqueous micelles and of solvent restrictions. The same group[220] also studied the environment of acetophenone and benzophenone solubilized in aqueous micelles of hexadecyltrimethylammonium bromide, hexadecyl pyridinium chloride, and SDS, together with others. The solubilization site was found to depend upon the solubilizate, the surfactant, and the extent of water penetration into the micelle. Marked differences in solubilization indicated a dependence upon the nature of the head groups. On solubilization the ^{1}H n.m.r. showed considerable differences in the CH_3 and CH_2 signals, in addition to those from the surfactant head groups and substrates.

Robb and Smith[221] determined the T_1 relaxation of alkali-metal counterions at micellar interfaces, and in electrolyte solutions also. For alkyl sulphate micelles the relaxation rate of the adsorbed species was in the order Cs^+, Rb^+, Na^+, Li^+. Proton relaxation-rate enhancements due to the incorporation of manganese ions into the Stern layer indicated that the Mn^{2+} rotation rate decreased by a factor of 30%. As this decrease was expected to be smaller for univalent ions, it was concluded that the major effect is a change in field gradient produced by changes in hydration, and this would be expected to be greater for the larger ions. Smith and Muller[222] reported ^{19}F chemical shifts in complexes of sodium trifluoroalkyl sulphate with a range of reduced proteins, finding that the chemical shift was a sensitive indication of bound detergent ions. The behaviour was very similar to that found for micelles, implying that the bound ions form micelle-like aggregates.

[216] C. A. Barker, D. Saul, G. J. T. Tiddy, B. A. Wheeler, and E. Willis, *J. C. S. Faraday I*, 1974, **70**, 154.
[217] D. Saul, G. J. T. Tiddy, B. A. Wheeler, P. A. Wheeler, and E. Willis, *J. C. S. Faraday I*, 1974, **70**, 163.
[218] F. M. Menger and J. L. Lynn, *J. Amer. Chem. Soc.*, 1975, **97**, 948.
[219] J. H. Fendler, F. Nome, and H. C. Van Woert, *J. Amer. Chem. Soc.*, 1974, **96**, 6745.
[220] J. H. Fendler, E. J. Fendler, G. A. Infante, P. S. Shih, and L. K. Patterson, *J. Amer. Chem. Soc.*, 1975, **97**, 89.
[221] I. D. Robb and R. Smith, *J. C. S. Faraday I*, 1974, **70**, 287.
[222] M. L. Smith and N. Muller, *Biochem. Biophys. Res. Comm.*, 1975, **62**, 723.

Turning to inverted micelles, Wells[223] found no evidence for two forms of bound water in micelles of phosphatidylcholine in diethyl ether, presumably because of rapid proton exchange. He also investigated the effect of micelle size. Klose and Stelzner[224] investigated the interaction of water with lecithin in benzene solution, using $^{31}P\ T_1$ relaxation, plus the ^1H linewidth and chemical shift. The authors found that water interacted with the phosphate group in two regions, and that exchange between them was rapid. Increased water concentrations produced a third region and enhanced the mobility of the head group. McKinney[225] also investigated the lecithin–benzene system in the presence of water, but considered interactions with 4,4′-dichlorodiphenylacetic acid and kelthane.

Fendler's group has produced a series of papers concerned with the facts that reversed micelles can act as solubilizers, and can provide a system where relaxation rates are often enhanced. In paper four of the series, El Seoud et al.[226] were concerned with the effects of DMSO, imidazole, methanol, pyrazole, 2-pyridone, and tetrabutylammonium perchlorate in dodecylammonium propionate (DAP) in benzene, $CDCl_3$, and dichloromethane. The ^1H chemical shift of the $\overset{+}{N}H_3$ peak had a linear dependence upon concentration for all combinations of solubilizate and solvent. In a subsequent paper[227] the authors determined the chemical shifts of the solubilizates as functions of DAP concentration. Equilibrium constants and the average lifetimes of solubilizate in the micellar environment were determined from linewidths of the solubilizate and DAP protons. Addition of solubilizate produced relatively small changes in the linewidths of the DAP protons, changes which were different for magnetically distinguishable protons, and were solvent- and solubilizate-dependent. Fendler's group extended these measurements[228] to study the aggregation of hexylammonium propionate (HAP) in DMSO–benzene and DMSO–water mixed solvents, using ^1H chemical shifts of HAP; they essentially mapped the phase diagram. El Seoud and Fendler[229] investigated the interactions of imidazole, methanol, and pyrazole with micellar sodium bis(2-ethylhexyl)sulphosuccinate (aerosol OT) in CCl_4, which can solubilize large quantities of water. Traces of water caused the collapse of the methanol doublet. The residence lifetimes in micelles were determined from the dependences of the shifts and linewidths on aerosol concentration.

[223] M. A. Wells, *Biochemistry*, 1974, **13**, 4937.
[224] G. Klose and F. Stelzner, *Biochim. Biophys. Acta*, 1974, **363**, 1.
[225] J. D. McKinney, *Chem. and Phys. Lipids*, 1974, **13**, 249.
[226] O. A. El Seoud, E. J. Fendler, and J. H. Fendler, *J. C. S. Faraday I*, 1974, **70**, 450.
[227] O. A. El Seoud, E. J. Fendler, and J. H. Fendler, *J. C. S. Faraday I*, 1974, **70**, 459.
[228] E. J. Fendler, V. G. Constein, and J. H. Fendler, *J. Phys. Chem.*, 1975, **79**, 917.
[229] O. A. El Seoud and J. H. Fendler, *J. C. S. Faraday I*, 1975, **71**, 452.

11
Medium Effects on Chemical Shifts and Coupling Constants

BY M. I. FOREMAN

1 Introduction

This chapter follows the general trend set in previous volumes in covering specific and non-specific effects of the medium, the latter topic receiving rather less attention within this reporting period. The literature on shift reagents has also again been covered. No attempt has been made at an exhaustive coverage; rather, the general trend of research in these areas is outlined and an effort has been made to draw attention to topics which have perhaps been neglected.

2 Non-specific Solvent Effects

The distinction traditionally made between specific and non-specific effects of the medium is, of necessity, arbitrary and relies to a large extent on intuition. It is clear for example, that a molecule in a condensed phase will interact with its environment to a greater or lesser degree. In some cases the interaction may be dominated by a form of chemical binding, such as occurs in hydrogen-bonded systems, or with respect to the type of molecular complexation responsible for the ASIS effect. These interactions are loosely referred to as specific effects,[1] and arise alongside the non-specific effects, discussed in this section, which have been classified in the classic equation of Buckingham, Schaefer, and Schneider.[2] This attempts to express the solute–solvent interaction energy as an additive sum of independent contributions. There has, to date, been no serious attempt to assess the validity of this type of approach. Most studies relate to methods by which one or more of the individual contributions may be calculated, or to discussion of the validity of the models employed for such calculations. As an example, one contribution to the chemical shift of polar solutes is felt to arise from the 'reaction field' term σ_E, whereby the solute dipole polarizes the solvent medium, and nuclei within the solute experience a nett electric field.[3] Early attempts to calculate the chemical-shift contribution from this effect were based on the Onsager model[4] in which the solute is regarded as a sphere, with a central point dipole, in a continuous medium. Various refinements

---

[1] T. Schaefer and W. G. Schneider, *J. Chem. Phys.*, 1960, **32**, 1218, 1224.
[2] A. D. Buckingham, T. Schaefer, and W. G. Schneider, *J. Chem. Phys.*, 1960, **32**, 1227.
[3] J. W. Emsley, J. Feeney, and L. H. Sutcliffe, 'High Resolution Nuclear Magnetic Resonance Spectroscopy', Pergamon Press, Oxford, 1965, Vol. 1; 1966, Vol. 2, p. 88.
[4] Ref. 3, Vol. 1, p. 56.

have been made to this basic idea, to include non-spherical solute 'cavities' and localized dipolar regions in molecules having zero nett dipole moment. The model is, however, limited, as has recently been demonstrated. It has been shown that the spectrum of a non-polar solute in a polar solvent may be perturbed by the application of an external electric field, which will impart to the solvent molecules a nett orientation parallel to the field, the molecular alignment of the solvent being transferred in some degree to the non-polar solute molecules,[5] an occurrence for which there is no explanation in terms of the Onsager 'reaction field' model. This phenomenon has been described in detail for the case of perdeuteriobenzene dissolved in nitrobenzene.[5] In such a case, the presence of an external field causes the ^{2}H resonance of the perdeuteriobenzene to split into two, the peak separation (in Hz) being given by equation (1),[5] where $\chi = e^2qQ/h$ is the quadrupole coupling constant of the

$$\Delta v = \tfrac{3}{2}\chi \, \langle \tfrac{3}{2}\cos^2\theta - \tfrac{1}{2} \rangle_E \, (\tfrac{3}{2}\cos^2\gamma - \tfrac{1}{2}) \tag{1}$$

deuterium nuclei; θ is the angle between the axis perpendicular to the plane of the [^{2}H$_6$]benzene ring and the static magnetic field B_o, γ is the angle between this axis and the principal axis of the electric field gradient at the ^{2}H nucleus for which axial symmetry along the C—^{2}H bond has been assumed ($\gamma = 90°$ therefore), $\langle \tfrac{3}{2}\cos^2\theta - \tfrac{1}{2} \rangle_E$ is the alignment average of [^{2}H$_6$]benzene in the presence of the field E within the medium. It has been demonstrated for this system that there is a reasonably good linear correlation between $\Delta\delta$ and the square of the applied field.[6] In this instance therefore, intermolecular interactions are of importance, at least some of which would normally be regarded as specific solvent effects, and treated separately from non-specific effects, of which the reaction field term is a part. (Nitrobenzene and benzene form a charge-transfer complex[7] of the type responsible for the ASIS phenomenon). It is in this sense, therefore, that the expression of the medium shift as a series of discrete additive terms may break down.

Tapia and Goscinski,[8] in developing an extension of Onsager's model in terms of a self-consistent quantum mechanical formalism, also stress the need ultimately to encompass molecular associations and dispersion forces within the model. With respect to dispersion forces the comments of Amos and Burroughs,[9] although not specifically directed towards n.m.r. solvent effects, are of interest. Further evidence of the failure of the Onsager model, due to its neglect of intermolecular interactions, has been proposed from a study of the temperature dependence of electric field effects on the ^{14}N spectra of nitromethane and nitrobenzene.[10]

In more general terms, the non-specific effects, of which dipolar interactions form a part, are frequently smaller, in terms of their contribution to the overall chemical shift, than those arising from specific effects. In many instances, therefore, these factors tend to be overlooked. One such example of this arises with regard to evaluations of ^{13}C substituent shift parameters in hydrocarbons. These shifts are generally measured for the neat liquid, which neglects entirely the effect of non-

[5] C. W. Hilbers, J. Biemond, and C. Maclean, *Pure Appl. Chem.*, 1972, **32**, 197.
[6] J. Biemond and C. Maclean, *Mol. Phys.*, 1974, **28**, 571.
[7] R. Foster, 'Organic Charge-Transfer Complexes', Academic Press, London and New York, 1969.
[8] O. Tapia and O. Goscinski, *Mol. Phys.*, 1975, **29**, 1653.
[9] A. T. Amos and B. L. Burrows, *Theor. Chim. Acta*, 1973, **29**, 139.
[10] J. Biemond, S. van der Groot, and C. Maclean, *Chem. Phys. Letters*, 1975, **32**, 390.

specific solvent interactions, since each hydrocarbon is effectively acting as its own solvent. Wilson, van der Wen, and de Haan[11] have stressed this point by studying ^{13}C shifts in binary mixtures of n-alkanes extrapolated to infinite dilution. From this work it is clear that the non-specific solvent effect is different for each of the n-alkanes, and existing ^{13}C substituent shifts are criticized for not taking proper account of this point. The authors[11] have therefore re-evaluated certain of these parameters in the light of their data.

One recurrent problem which complicates more exact studies of solvent effects is that of a suitable reference standard. For nuclei other than protons, internal referencing is particularly suspect, and an external standard should ideally be used. This, however, introduces the bulk diamagnetic susceptibility of the medium into consideration.[12] This being the case, the method described by Ward, Morris, and Healey[13] for the determination of the bulk susceptibility of a given sample is particularly valuable. It is asserted, perhaps a little optimistically, that the bulk susceptibility may be determined by this method to an accuracy of *ca.* 0.002 p.p.m.[13]

3 Specific Interactions with the Medium

A. General.—In this section, interactions between the solute and the surrounding solvent molecules are discussed in cases where one or more molecular species are formed which are definable entities. Stated in this way, the distinction between specific and non-specific effects seems fairly clear cut; in practice however, this is not so. The ASIS effect, as one example, may be considered as being due to the formation of one or more charge-transfer or hydrogen-bonded complexes which have definable geometries and stoicheiometries. In such a case the ASIS effect would be classified as a specific effect. Another view would regard the solute as being perturbed by collisions with the solvent in which so-called 'collision complexes'[7] are formed. In this approach the ASIS effect is not so easily classified. However, in general, such systems are considered to contain a number of definable solvent–solute or solute–solute associates, each having appropriate chemical shifts $\delta_{\text{obs},i}$; the observed shift is then given by equation (2), where P_i is the population fraction of species i, and site exchange is fast on the n.m.r. timescale. It should not be forgotten that the individual δ_i are subject to non-specific effects of the medium.

$$\delta_{\text{obs},i} = \sum_i \delta_i P_i \qquad (2)$$

B. Solvent Effects on Conformational Equilibria.—One slightly special case arises when site exchange occurs, not between different associated species but between different conformations of the same molecule. Where the energy barriers between the conformations are sufficiently low, the observed chemical shift for a given nucleus is an exchange average of all of the nuclear environments. The chemical shift appropriate to each conformation will be affected by the specific or non-specific solvent effects discussed elsewhere. In addition, the energy barriers between conformations may also be solvent dependent, in which case the exchange-averaged

---

[11] A. R. N. Wilson, L. J. M. van der Wen, and J. W. de Haan, *Org. Magn. Resonance*, 1974, **6**, 601.
[12] Ref. 3, Vol. 1, p. 260.
[13] A. J. I. Ward, A. Morris, and M. A. Healey, *J. Magn. Resonance*, 1975, **16**, 357.

shift will also be solvent dependent. Abraham and co-workers have investigated this type of behaviour in tetrahalogenoethanes[14] and discuss the results in terms of calculated solvation energies,[15] from which rotamer energies and coupling constants may be inferred. Solvent-dependent barriers to rotation about P—N bonds in $R^1P(X)NR^2$, where R^1 and R^2 are alkyl groups and X = F or Cl, have also been identified from this type of study.[16]

Bertran and Rodriguez[17] have measured the solvent dependence of the aldehydic proton of furfural, thiophenaldehyde, and benzaldehyde in order to determine whether solvent shifts are due predominantly to a direct effect or to solvent-dependent conformational equilibria. Benzaldehyde is taken as an example of the situation for which only direct solvent effects on the observed shift are important, since only a single minimum energy conformation is possible. The solvent shift behaviour of thiophenaldehyde and benzaldehyde is very similar, which is taken to imply that, whilst two conformations of thiophenaldehyde are possible in principle, only one is populated. Furfural behaves quite differently, which suggests that the conformer populations are solvent dependent.[17]

Most discussions of this type assume that only those conformations which correspond to potential-energy minima effectively contribute to the observed chemical shift. Emsley and Tabony[18] have attempted to improve on this approach by using probability distribution functions of the relevant dihedral angles to allow non-equilibrium conformations to be included in the calculations.

C. The ASIS Effect.—A number of applications of the ASIS effect have been reported, ranging from strictly empirical correlations which may prove useful in the identification of the site of methoxy substitution in coumarin trimethylsilyl ethers,[19] to structural assignments in dihydrobenzo[c]thiophen oxides based on an inferred geometry for the benzene–solute complex,[21] arising from benzene binding to the sulphinate group.[20, 21] There has, however, been a disappointing lack of discussion of the ASIS effect in terms of the approach by Engler and Laszlo,[22] which is probably the only really novel treatment of the ASIS effect to appear within the past few years.

The ASIS effect has been tabulated in some detail for methyl groups of monoethylquinolizidines[23] and for the substituent groups of compounds of the general type of structure (1).[24] In both instances the observed ASIS effect may be useful for assigning the site of substitution[23] or the molecular configuration.[24]

Of perhaps more general importance to an understanding of the ASIS effect, there has been a brief report of benzene-induced shifts in substituted ethylenes.[25] For

[14] R. J. Abraham, M. A. Cooper, T. M. Siverns, P. F. Swinton, H. G. Weder, and L. Cavalli, *Org. Magn. Resonance*, 1974, **6**, 331.
[15] R. J. Abraham and Z. L. Rossetti, *J. C. S. Perkin II*, 1973, 582.
[16] S. Di Stefano, H. Goldwhite, and E. Mazzola, *Org. Magn. Resonance*, 1974, **6**, 1.
[17] J. F. Bertran and M. Rodriguez, *Org. Magn. Resonance*, 1974, **6**, 525.
[18] J. W. Emsley and J. M. Tabony, *Mol. Phys.*, 1974, **28**, 423.
[19] D. Brown, R. O. Asplund, and V. A. McMahon, *Phytochemistry*, 1974, **13**, 1923.
[20] I. Sataty, *Org. Magn. Resonance*, 1974, **6**, 8.
[21] R. V. Norton and I. B. Douglass, *Org. Magn. Resonance*, 1974, **6**, 89.
[22] E. M. Engler and P. Laszlo, *J. Amer. Chem. Soc.*, 1971, **93**, 1317.
[23] R. T. Lalonde, T. N. Donvito, A. I.-M. Tsai, and C. Wong, *Org. Magn. Resonance*, 1975, **7**, 191.
[24] J. A. Lepoivre, H. O. Desseyn, and F. C. Alderweireldt, *Org. Magn. Resonance*, 1974, **6**, 279.
[25] D. Bryce-Smith, A. Gilbert, and H. M. Tyrrell, *J.C.S. Chem. Comm.*, 1974, 699.

(1) X = O or S

ethylenes which are electron donors relative to benzene, there is a slight benzene-induced shift to high frequency for the vinyl protons, whilst allylic or alkoxy protons shift slightly to low frequency. This is taken to suggest a solvent–solute complex having the time-average geometry of structure (2), whilst electron-acceptor ethylenes form complexes of structure (3), the vinyl protons showing a very marked low-frequency shift. It is interesting that these structures are in general agreement with the stereospecificity observed for 1,2-photoaddition reactions of these compounds.[25] The *endo* association possibly arises from a small hydrogen-bonding interaction with the π-electron system of the benzene ring [structure (4)].[25]

donor ethylenes (2)   acceptor ethylenes (3)   (4)

An ASIS effect induced by pyridine on protons of benzylacetone, reported earlier[26] and rationalized in terms of the averaged structure (5), has been re-examined in the opposite sense by studying the shifts induced in nitrogen heterocycles by acetone as the solvent.[27] Structure (5) assumes an interaction between the positive end of the carbonyl group dipole and the nitrogen lone pair, and should therefore affect the pyridine proton resonances in a manner similar to N-protonation. This has in fact been shown to be the case, although the magnitudes of the shifts are much reduced.[27]

(5)

Inferred geometries of the above type arise from a qualitative understanding of the anisotropy of the magnetic field surrounding molecules such as benzene. Quantita-

[26] S. S. Danyluk, *Canad. J. Chem.*, 1963, **41**, 387.
[27] G. V. Rao, M. Balakrishnan, and N. V. Subramanian, *Org. Magn. Resonance*, 1974, **6**, 515.

tive mappings of the magnetic field about benzene are therefore of considerable interest. To the early calculation[28,29] of this field based on the electron ring-current model has been added a more general analytical approach based on magnetic multipolar expansions[30,31] which has recently been elaborated by Stiles.[28,32-34] This approach has the advantage that it may be used to treat the case of the macrocyclic shift reagents described, for example, by Maskasky and Kenney.[35]

D. Hydrogen Bonding.—Previous articles in this series have discussed the general features of chemical-shift perturbations which may be used to identify the presence and type of hydrogen-bonded aggregates in solution. Von Dreele and Stenhouse have also recently outlined the various methods for identifying the occurrence of hydrogen bonding based on deuterium substitution experiments and studies of chemical-shift variations with temperature; they discuss problems arising in the analysis of the results.[36]

There is a general observation that the magnitude of the change in the proton chemical shift which occurs when a hydrogen bond is formed is related to the strength of the hydrogen bond. This can prove useful in a diagnostic sense. For epimeric cyclanols[37] in DMSO solution, where proton exchange at the hydroxy-group is diminished and the hydrogen-bonded proton shift is therefore readily determined, it has been found that the resonance of the equatorial hydroxyl proton at infinite dilution occurs at higher frequencies than that of the axial hydroxyl. This is thought to reflect stronger binding to the equatorial group, which is less sterically shielded than the axial group.[37]

Martin and Fujiwara[38] have further studied the hydrogen bonding of HF using carefully purified aprotic basic solvents, in which self aggregation and chemical exchange are thought to be minimized. In these circumstances the HF coupling has been observed and its dependence on the solvent measured. It has been found that the HF coupling is considerably less in such solvents than in the gas phase, and decreases with the probable strength of the hydrogen bond. This behaviour is not observed for N—H or C—H fragments which may be involved in hydrogen bonding; the observed behaviour of HF is therefore attributed to the high polarizability of this bond.[38] It has also been proposed that the proton shielding correlates with the strength of the hydrogen-bonding interaction; this does not seem to be true for the shielding of the fluorine nucleus, however.[38] A slightly different dependence of coupling constants on solvent hydrogen-bonding strength has been reported for the $^3J(F,H)$ coupling in $CF_2BrCHFCl$,[39] which may arise from solvent-induced changes in the rotamer populations or from an intrinsic effect of hydrogen-bond

---

[28] P. J. Stiles and R. M. Wing, *J. Magn. Resonance*, 1974, **15**, 510.
[29] C. W. Haigh and R. B. Mallion, *Org. Magn. Resonance*, 1972, **4**, 203.
[30] A. D. Buckingham and P. J. Stiles, *Mol. Phys.*, 1972, **24**, 99.
[31] P. J. Stiles, *Proc. Roy. Soc.*, 1974, **A336**, 251.
[32] P. J. Stiles, *Proc. Roy. Soc.*, 1974, **A336**, 251.
[33] P. J. Stiles, *Mol. Phys.*, 1975, **29**, 1271.
[34] P. J. Stiles, *Chem. Phys. Letters*, 1975, **30**, 259.
[35] J. E. Maskasky and M. E. Kenny, *J. Amer. Chem. Soc.*, 1973, **95**, 1443.
[36] P. H. von Dreele and I. A. Stenhouse, *J. Amer. Chem. Soc.*, 1974, **96**, 7546.
[37] R. K. Sehgal, R. U. Koenigsberger, and T. J. Howard, *Tetrahedron Letters*, 1974, 4173.
[38] J. S. Martin and F. Y. Fujiwara, *J. Amer. Chem. Soc.*, 1974, **96**, 7632.
[39] S. Ng, *J. Magn. Resonance*, 1975, **17**, 244.

formation on the relevant couplings, such as that reported above or that proposed for CF_3CHCl_2.[40]

One further, simple application of hydrogen-bond shifts is in the distinction between inter- and intra-molecular hydrogen bonding. The two types of bond behave fairly differently, in that intramolecularly bonded X—H groups show very high-frequency proton shifts which are independent of concentration and solvent. In 2-fluoroethanol, for example, the prfeerred *gauche* conformation has been attributed to an intramolecular hydrogen bond.[41] The hydroxyl proton shift, however, is markedly concentration dependent in $CDCl_3$, where the *gauche* conformation persists, and the infinite-dilution shift is in fact similar to that of normal alcohols.[42] This behaviour is wholly incompatible with an intramolecular hydrogen bond.[42] Conversely, it has been reported that methyl-N-benzoyl-L-valyl-L-valinate and methyl-N-benzoyl-D-valyl-L-valinate have magnetically non-equivalent methyl ester groups in $CDCl_3$.[43] The non-equivalence is not, however, evident in DMSO solution. These observations are rationalized in terms of structure (6), where the methyl ester groups are fixed in position with respect to the anisotropic field of the benzene ring by a weak intramolecular bond. Chloroform is presumably not a sufficiently good proton donor to compete for the acceptor site, whilst DMSO competes sufficiently strongly to disrupt the bond, leading to time-averaged equivalence of the methyl ester groups.[43]

(6)

The relative ability of halogen groups to act as proton acceptors in intramolecular hydrogen bonds has been inferred from the preferred conformations of species such as (7) and (8), the conformation being deduced from the magnitude of the ^5J-(HOCCCH) coupling.[44] Similar considerations have been applied to the spectra of substituted naphthaquinones.[45] In compound (9) the hydroxyl proton resonance occurs at an unusually high frequency, characteristic of an intramolecular bond. The

[40] S. Ng, *J. Magn. Resonance*, 1972, **7**, 370.
[41] K. Hagen and K. Hedberg, *J. Amer. Chem. Soc.*, 1973, **95**, 8263.
[42] R. C. Griffith and J. D. Roberts, *Tetrahedron Letters*, 1974, 3499.
[43] J. S. Davies, R. J. Thomas, and M. K. Williams, *J.C.S. Chem. Comm.*, 1975, 76.
[44] J. B. Rowbotham and T. Schaefer, *Canad. J. Chem.*, 1974, **52**, 3037.
[45] S. B. Padhye and B. A. Kulkarni, *J. Magn. Resonance*, 1974, **16**, 150.

same is not, however, true for compound (10).[45] Other systems studied include α-alkyl- and αα- dialkyl-o-methoxybenzyl alcohols[46] and N-alkyl-substituted pyruvamides.[47]

Rather more quantitative studies have been reported, including some which investigate the above-mentioned correlation between hydrogen-bond shifts and some property which reflects the hydrogen-bond strength in more detail. For example, for the hydroxyl proton shift of t-butyl alcohol in both benzene and also carbon tetrachloride it has been found that[48]

$$\delta_{\infty,i} - \delta_{x,i} \propto H_{v,\infty} - H_{v,x} \tag{3}$$

where $\delta_{\infty,i}$ and $\delta_{x,i}$ are the hydroxyl proton shifts at infinite dilution and mole fraction x, respectively, $H_{v,\infty}$ and $H_{v,x}$ being the corresponding heats of dilution. For hydrogen bonding between 4-methylpyridine N-oxide and the carboxylic acids RCO_2H (R = CF_3, CCl_3, CCl_2H, or CH_2Cl) in various solvents the hydrogen-bond shifts correlate with ΔpK_a, the difference between the dissociation constants of the N-oxide and the appropriate acid in aqueous solution.[49] Wong and Ng[50] have carried out a rather more extensive study involving chloroform binding to aliphatic tertiary amines and ethers in cyclohexane. The hydrogen-bond shift may be correlated with the enthalpy change ΔH^0 for each series of proton acceptors, and also with Taft σ^* values,[51] at least for the ethers. These authors also discuss the temperature dependence of the shift in terms of the model of Muller and Reiter,[52] which

---

[46] A. Yamashita, K. Hara, S. Aizawa, and M. Hirota, *Bull. Chem. Soc. Japan*, 1974, **47**, 2508.
[47] G. Fischer and A. Schellenberger, *Tetrahedron Letters*, 1974, 3307.
[48] R. Radeglia and W. Storek, *Z. Chem.*, 1975, **15**, 78.
[49] Z. Dega-Szafran, M. Z. Nascret-Barciszewska, and M. Szafran, *J.C.S. Perkin II*, 1974, 763.
[50] K. F. Wong and S. Ng, *J.C.S. Faraday II*, 1975, 622.
[51] R. W. Taft, in 'Steric Effects in Organic Chemistry', ed. M. S. Newman, Wiley, New York, 1956, Chapter 13.
[52] N. Muller and R. C. Reiter, *J. Chem. Phys.*, 1965, **42**, 3265.

proposes that the hydrogen-bond length increases with temperature and the hydrogen-bond shift therefore diminishes.[52, 53] For many of the cases examined there is a linear temperature dependence of the shift over the range studied, with the exception of sterically hindered amines. These results are discussed in terms of the potential-energy configurations of the hydrogen bonds formed.[50] It is suggested that a requirement for a uniform dependence of the hydrogen-bond shift with temperature within a series of similar bases is that the hydrogen bonds formed must be in the minimum potential-energy configuration. If steric interactions in a given molecule prevent this configuration from being adopted, the temperature dependence of the hydrogen-bond shift does not then follow the pattern set by the non-hindered molecules of the series.[50] It has also been argued that a similar shift *versus* $\Delta H°$ correlation exists for sterically hindered phenols bonded to alicyclic ethers and pyridines;[54] the experimental support for this is, however, very limited.

Generally, correlations of the above type require the evaluation of some thermodynamic property, such as $\Delta H°$, for the interaction, which reflects the strength of the hydrogen bond formed. Measurement of the equilibrium constant as a function of temperature is the method most commonly encountered, using the hydrogen-bond shift dependence on the relative amounts of proton donor and acceptor present in the system. Problems which may arise in determinations of this kind have already been outlined in this series.[55] One difficulty arises with respect to the estimation of the chemical shift of the free monomeric species in the solution. Other complications have been discussed by Ng and co-workers. It has been pointed out,[56] for example, that a base which hydrogen-bonds at two independent sites has an effective concentration which is in excess of the molecular concentration. The effect of using the molecular concentration in such situations in the conventional manner (in this case using a derivative of the Benesi–Hildebrand method)[57] has been assessed by a study of chloroform binding to bases having one or more acceptor sites. The equilibrium constants for certain twin-site bases are rather higher, relative to the single-site analogue, than is consistent with two sites of equal strength acting independently. It is felt that the proton-donor binding to such molecules preferentially exchanges between the two sites of the same molecule rather than between sites of different molecules.[56] Conversely, for a proton donor having more than one equivalent donor site the observed shift is the exchange-average over all such sites.[58] For the species XH_2 the predominant complex is that having one hydrogen involved in a hydrogen bond, in which case equation (4) applies, where $P_{AB,i}$ is the population fraction of

$$\delta_{obs,i} = \frac{P_{AB,i}}{2}(\delta_{AB,i}+\delta_{AB',i})+P_{m,i}\,\delta_{m,i} \qquad (4)$$

hydrogen-bonded species, $\delta_{AB,i}$ is the shift of the hydrogen-bonded proton in the complex, $\delta_{AB',i}$ the shift of the non-involved hydrogen in the complex, while $P_{m,i}$

[53] R. E. Cramer, P. L. Dahlstrom, and H. Heya, *J. Phys. Chem.*, 1975, **79**, 376.
[54] T. S. Pang and S. Ng, *Spectroscopy Letters*, 1974, **7**, 377.
[55] M. I. Foreman, in 'Nuclear Magnetic Resonance', ed. R. K. Harris, (Specialist Periodical Reports), The Chemical Society, London, 1974, Vol. 3, p. 352.
[56] K. F. Wong, T. S. Pang, and S. Ng, *Chem. Phys. Letters*, 1975, **30**, 309.
[57] I. D. Kuntz, jun., F. P. Gasparro, M. D. Johnston, jun., and R. P. Taylor, *J. Amer. Chem. Soc.*, 1968, **90**, 4778.
[58] T. S. Pang and S. Ng, *Spectroscopy Letters*, 1974, **7**, 511.

and $\delta_{m,i}$ are the population fraction and shift of the non-hydrogen-bonded proton donor, respectively. This treatment is in fact very similar to that proposed for isomeric 1:1 charge-transfer complexes.[7]

Other quantitative studies of hydrogen bonding include the very strong hydrogen bonds formed by CF_3CO_2H and its anion,[59] and of other α-halogen-substituted carboxylic acids,[60,61] and also of the binding of chloroform to organosulphur compounds.[62]

Foster et al.[63] have investigated in some detail the binding of DDT to the π-donors benzene and hexamethylbenzene, to estimate the involvement of possible hydrogen bonding between the doubly benzylic proton of DDT and the aromatic ring of the π-donors in the formation of the complex. This work has the merit that some effort has been expended in substantiating the assumption that the complexes formed do indeed have a 1:1 stoicheiometry by studying the chemical-shift data over wide ranges of the saturation fraction;[64,65] the authors adopted the very useful practice of quoting the range of saturation fraction involved in each measurement. Other systems studied in this work include hydrogen bonding between chloroform and acetone, evaluated from both proton and ^{13}C shift measurements; $\Delta\delta_{AB}$ for the ^{13}C nucleus is greater (in p.p.m.) than that for the proton. For the chloroform–benzene complex, the ^{13}C shift is rather smaller and of opposite sense. Chloroform binding to pyridine has also been studied.[66] In this instance chloroform may bind to the nitrogen lone-pair electrons, or to the π-system (11) or (12); shift data suggests that binding is predominantly to the nitrogen atom. As before,[56] the error resulting from neglect of the π-bonded case is considered.

(11)   (12)

One other study of interest involves the behaviour of phenol in cyclohexane, carbon tetrachloride, and benzene.[67] Extensive thermodynamic, dielectric, and n.m.r. data are used to show that there is strong self-association of phenol in cyclohexane (where cyclic associates are of major importance), less in carbon tetrachloride, and very little in benzene, possibly owing to competitive binding to the solvent in the latter case. The effect of the medium on hydrogen-bond shifts has also been studied

---

[59] J. H. Clark and J. W. Emsley, *J.C.S. Dalton II*, 1974, 1125.
[60] L. Kimtys, *Org. Magn. Resonance*, 1975, 7, 179.
[61] Z. Dega-Szafran, E. Grach, M. Z. Nascret-Barciszewska, and M. Szafran, *J.C.S. Perkin II*, 1975, 250.
[62] K. W. Jolley, L. M. Hughes, and I. D. Watson, *Austral. J. Chem.*, 1974, 27, 287.
[63] A. A. S. Bright, R. Foster, and J. A. Chudek, *J.C.S. Perkin II*, 1975, 64.
[64] D. A. Deranleau, *J. Amer. Chem. Soc.*, 1969, 91, 4044.
[65] D. A. Deranleau, *J. Amer. Chem. Soc.*, 1969, 91, 4050.
[66] T. S. Pang and S. Ng, *J. Magn. Resonance*, 1975, 17, 166.
[67] F. Kohler, E. Liebermann, R. Schano, H. E. Affsprung, J. K. Morrow, K. Sosnkowska-Kehiaian, and H. Kehiaian, *J. Chem. Thermodynamics*, 1975, 7, 241.

in great detail for phenol hydrogen-bonding,[68] where it is also demonstrated that 2:1 as well as 1:1 binding of the proton acceptor to the phenol must be considered as a possibility.

E. **Ionic Solutions.**—As in previous years, n.m.r. studies of ionic solutions are being fairly extensively used to probe the nature of such systems. It is generally recognized in such work that the chemical shifts observed are determined by a complex composite of simultaneous interactions, so that any conclusions which may be drawn rely heavily on inference. In certain cases where the situation is simplified to some extent it may be reasonable to assume that one contribution dominates all others. In general, however, it is important that as much data as possible be obtained for any given system before conclusions may be drawn with any confidence. As an example of the difficulties which may arise, there have been several reports of interactions between Ca^{2+}, Sr^{2+}, and Ba^{2+} chlorides and guanosine.[69-72] Chang and Marzilli[73] have pointed out that such complexation is unprecedented, and, from n.m.r. evidence, propose that it is the chloride anion which is involved in the binding rather than the cation, since the >NH and —NH_2 group protons are both perturbed by addition of the salt, and the same effect is observed for NEt_4Cl, but not for the alkaline-earth-metal nitrates or perchlorates. The proposed interaction is shown in (13).[73]

(13)

It is often an advantage in work of this kind, therefore, to complement chemical shift studies with other data. Nuclear relaxation measurements, such as those relating to contact ion-pairing involving aqueous ^{35}Cl, can provide[74] valuable additional information, as, of course, can alternative spectroscopic techniques. It is surprisingly rare, however, to find instances of ionic systems simultaneously studied by more than one method; most n.m.r. studies rely exclusively on the interpretation of chemical-shift data. In certain cases, the nature of the nucleus being studied is such that the chemical shift is perturbed by one type of interaction in preference to all others. For $^{19}F^-$, for example, the chemical shift is mainly perturbed by molecules in the first co-ordination sphere of the ion. Consequently, in the first instance, it is possible to use the $^{19}F^-$ chemical shift as a probe of the structure and behaviour

[68] D. Baron and N. Lumbroso-Bader, *J. Phys. Chem.*, 1975, **79**, 479.
[69] F. Jordan and B. Y. Farquhar, *J. Amer. Chem. Soc.*, 1972, **94**, 6557.
[70] S. Shimokawa, H. Fukui, J. Soma, and K. Hotta, *J. Amer. Chem. Soc.*, 1973, **95**, 1777.
[71] S. M. Wang and N. C. Li, *J. Amer. Chem. Soc.*, 1968, **90**, 5069.
[72] L. S. Kan and N. C. Li, *J. Amer. Chem. Soc.*, 1970, **92**, 281, 4823.
[73] C.-H. Chang and L. G. Marzilli, *J. Amer. Chem. Soc.*, 1974, **96**, 3656.
[74] H. A. Berman and T. R. Stengle, *J. Phys. Chem.*, 1975, **79**, 1001.

of this co-ordination sphere. Bulk solvent effects appear to be unimportant, to the extent that the same type of ^{19}F shift behaviour persists even at such high salt concentrations that it is no longer appropriate to think in terms of 'bulk' solvent. However, insofar as molecules in the second co-ordination sphere of the fluoride ion may perturb those in the first, such outer-sphere molecules may thereby induce a second-order perturbation of the $^{19}F^-$ ion shift.[75] This is thought[75] to account for the effect of the addition of NN-dimethylformamide, acetonitrile, acetone, or 1,4-dioxan on the ^{19}F shift of aqueous KF. With the addition of certain amines there is a similar effect, but with a marked discontinuity in the concentration dependence of the $^{19}F^-$ shift when, presumably, the amine begins to compete effectively with water for sites in the first co-ordination sphere.[75] More detailed analysis of this system has also shown that counter-ion effects are small, that the effect of the presence of anions is less than that of the counter cation, and that the anion effects are cation-dependent.[75] This implies that a further second-order effect is operative, in that direct anion–anion interactions are unimportant, but that added anions perturb the normal cation–$^{19}F^-$ interaction. A much more general quantitative account of the kinetics of solvent competiton for sites within the primary co-ordination sphere of a dissolved ion has appeared, which is of relevance to this type of study.[76]

With modern n.m.r. spectrometers it is becoming increasingly possible to study in detail the chemical shifts of less tractable nuclei, which is of great potential value in augmenting current understanding of ionic solutions. ^{23}Na shifts have been fairly extensively studied in this respect,[77] and interest is increasingly being turned towards ^{7}Li shifts. The dependence of the $^{7}Li^+$ shift on solvent and counter-ion has been described[78,79] and discussed in terms of equilibria between intimate and solvent-separated ion pairs.[78] The incorporation of the Li^+ ion into the hexaoxadiamine macrobicyclic ligands (crypts) of the type (14) has also been demonstrated from a study of the $^{7}Li^+$ shift in the presence of these species.[80] Generally, fast-exchange conditions prevail except for the case where $a = b = c = 1$, when separate resonances have been observed for the free and bound cation.[80]

(14)

Kostelnik and Bothner-By have reported a study of the solvent and counter-ion dependence of $^{113}Cd^{2+}$ shifts at 13.13 MHz, mainly in aqueous solution.[81] Shifts are quoted in p.p.m. relative to the extrapolated shift at infinite dilution. In certain cases, however, the shift *versus* concentration plots are markedly curved and

---

[75] J. P. K. Tong, C. H. Langford, and T. R. Stengle, *Canad. J. Chem.*, 1974, **52**, 1721.
[76] A. D. Covington and A. K. Covington, *J.C.S. Faraday I*, 1975, 831.
[77] J. P. Kintzinger and J. M. Lehn, *J. Amer. Chem. Soc.*, 1974, **96**, 3313.
[78] R. H. Cox and H. W. Terry, jun., *J. Magn. Resonance*, 1974, **14**, 317.
[79] Y. M. Cahen, P. R. Handy, E. T. Roach, and A. I. Popov, *J. Phys. Chem.*, 1975, **79**, 80.
[80] Y. M. Cahen, J. L. Dye, and A. I. Popov, *Inorg. Nuclear Chem. Letters*, 1974, **10**, 899.
[81] R. J. Kostelnik and A. A. Bothner-By, *J. Magn. Resonance*, 1974, **14**, 141.

extrapolation may therefore yield an imprecise reference. In some respects, the $^{113}Cd^{2+}$ shift appears[82] to behave in a similar manner to that of $^{205}Tl^{2+}$, and is markedly dependent upon cation–anion interactions and the presence of co-ordinating ligands. Shifts of $^{45}Sc^{3+}$ in aqueous solution have been reported[83] relative to uncorrected external 0.1M-$Sc(ClO_4)_3$ (aq.). The shifts are found to be markedly concentration dependent, those for the chloride, bromide, and perchlorate tending to the same infinite-dilution limit (supposedly the shift of the isolated $[Sc(OH)]^{2+}$ species). The nitrate and sulphate salts behave very differently and tend to a different infinite-dilution value. This may suggest that a strong ion-pair is formed which persists even at low concentrations.[83]

F. **Micellar Systems.**—Solubilizate chemical-shift behaviour in surfactant solutions is proving to be a sensitive probe of micelle interiors. Surprisingly few groups of authors seem, however, to be engaged on detailed studies of this kind. One system thus studied[84] comprises acetophenone and benzophenone solubilized by hexadecyltrimethylammonium bromide, hexadecylpyridinium chloride, sodium decyl sulphate, 3-(dimethyldodecylammonio)propane-1-sulphonate, and polyoxyethylene(15) nonylphenol (Igepal CO-730). Spectra obtained in D_2O show that, for a fixed concentration, the surfactant proton resonances shift to lower frequencies, the shift change being proportional to the concentration within certain ranges. Where acetophenone is the solubilizate, the surfactant proton resonances were resolved into two peaks, the low-frequency peak being markedly concentration dependent. The authors[84] argue that this type of behaviour suggests that the aromatic ring of acetophenone shields those protons of the surfactant which are close to the micelle surface. Increasing the concentration may cause deeper penetration of solubilizate molecules into the micelle core, which would be consistent with the observed chemical-shift dependence upon the concentration. A very loose description of the average micelle solubilizate is shown in Figure 1.

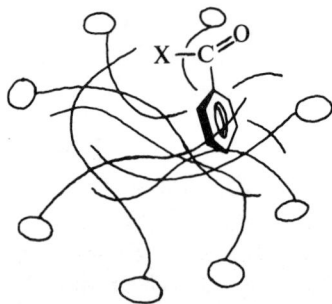

Figure 1 *Very schematic representation of the general solubilization sites of acetophenone and benzophenone in aqueous micellar solutions*
(Reproduced by permission from *J. Amer. Chem. Soc.*, 1975, **97**, 89)

[82] R. Freeman, R. P. H. Gasser, R. E. Richards, and D. H. Wheeler, *Mol. Phys.*, 1959, **2**, 75.
[83] G. A. Melson, D. J. Olszanski, and E. T. Roach, *J.C.S. Chem. Comm.*, 1974, 229.
[84] J. H. Fendler, E. J. Fendler, G. A. Infante, P.-S. Shih, and L. K. Patterson, *J. Amer. Chem. Soc.* 1975, **97**, 89.

In a similar way, the sites of occupancy of pyrene, pyrenebutyric acid (PBA), and pyrenesulphonic acid (PSA) in micelles of cetylmethylammonium bromide have been inferred.[85] Here the shift data are interpreted to mean that PSA spends most time in the region of the micelle surface, whilst pyrene itself tends to be randomly distributed throughout the micelle interior. PBA behaves in a manner which is roughly intermediate between these two extremes. There have been two further studies on reversed micelles. Here 'inverted' surfactant molecules form micelles in non-aqueous systems with the polar head-group now directed towards the micelle core. In one case it has been suggested that imidazole, methanol, and pyrazole, when solubilized by reversed micelles of sodium bis-(2-ethylhexyl)sulphosuccinate in nonpolar solvents, form hydrogen bonds to the SO_3^- group in the micelle core.[86] Such an inference is consistent with the chemical-shift behaviour of the surfactant protons, since only those very close to the micelle core are dependent on the solubilizate concentration.

One further interesting study has considered in detail the inversion of micelle structure on going from an aqueous to a non-polar organic solvent.[87] Hexylammonium propionate has magnetically discrete groups of protons which provide a useful probe of the system. Surfactant aggregation has been studied for this system in benzene–DMSO and water–DMSO systems of various compositions covering the whole range of solvent polarity, from aqueous to non-polar. Fairly detailed examination of the systems suggests that the micelle structure, c.m.c., aggregation number, and micelle association constants are all critically dependent upon solvent composition.

4 Shift Reagents

A. General.—Once again it is necessary to make rather arbitary definitions for the purpose of further discussion. Strictly, any reagent which induces chemical-shift changes in a dissolved substrate is a 'shift reagent'. Within this broad definition, therefore, fall almost all of the species discussed so far which induce chemical-shift changes by ASIS effects, hydrogen bonding, or other mechanisms. Such reagents include, for example, the BPh_4^- anion, which has been referred to in previous volumes and has recently been studied with respect to shifts induced in aminosulphonium cations,[88] and paraquat, the NN'-dimethyl-4,4',-dipyridylium cation, which is proving useful in the study of polypeptide and protein conformations.[89]

Arsenic trichloride has been mentioned as a potential shift reagent for compounds containing hydroxy-groups, although the main interest of this species is as a possible solvent for ^{13}C spectroscopy.[90] There has been rather less interest in the d-series transition metals, beyond their use as relaxation-, rather than shift-, reagents,[91] and relatively little interest in chiral solvents as selective shift reagents for mixtures of enantiomers,[14,92] or for inducing differential chemical shifts in otherwise

[85] M. Grätzel, K. Kalyanasundaram, and J. K. Thomas, *J. Amer. Chem. Soc.*, 1974, **96**, 7869.
[86] O. A. El Seoud and J. H. Fendler, *J.C.S. Faraday I*, 1975, **71**, 452.
[87] E. J. Fendler, V. G. Constien, and J. H. Fendler, *J. Phys. Chem.*, 1975, **79**, 917.
[88] G. P. Schiemenz and P. Klemm, *Org. Magn. Resonance*, 1974, **6**, 276.
[89] J. W. Verhoeven, A.-M. A. Verhoeven-Schoff, A. Masson, and R. Schwyzer, *Helv. Chim. Acta*, 1974, **57**, 2503.
[90] A. K. Bose, M. Sugiura, and P. R. Srinivasan, *Tetrahedron Letters*, 1975, 1251.
[91] J. G. Brassington, R. J. P. Williams, and P. E. Wright, *J.C.S. Chem. Comm.*, 1975, 338.
[92] W. H. Pirkle, S. D. Beare, and R. L. Muntz, *Tetrahedron Letters*, 1974, 2295.

isochronous nuclei in *meso* compounds.[14] One report has appeared of the use of uranyl-β-diketonates in enhancing the chemical-shift difference between methylene protons adjacent to sulphur in sulphoxides.[93] A further group of authors have described the use of a shift reagent which combines the advantages of the lanthanide chelates and the ASIS-type macrocyclic reagents[35] by synthesizing lanthanide complexes of *meso*-tetraphenylporphine.[94]

However, interest in the field of shift reagents is overwhelmingly concerned with the lanthanide chelates, to the extent that the term 'shift reagent' is widely used as an abbreviation for 'lanthanide shift reagent'. These compounds are of value with respect to their ability to modify nuclear chemical shift differences selectively (as opposed to their use as relaxation reagents) within a given molecule,[95] particularly in chiral and related systems.[96-98] Primarily, however, interest is centred about the possibility that shifts induced in a given substrate by association with a lanthanide chelate may be related to the detailed substrate geometry; to quote J. D. Roberts *et al.*,[99] who have, incidentally, provided a highly readable introduction to the topic of shift reagents, 'the lure of measuring interatomic distances by n.m.r. with naught but a pinch of lanthanide chelate has been irresistible'. The basic problem, however, is to decide exactly how this highly desirable objective may be attained. In what follows, therefore, emphasis is given to those articles which contribute to current understanding of the shift-reagent phenomenon, and which indicate the minimum requirements which must be met in order to obtain meaningful information relating to molecular structures in solution.

In this respect, the lanthanide chelates are basically used in the following way. The spectrum of a suitable substrate is obtained in the presence of varying relative amounts of a lanthanide chelate. The observed chemical-shift perturbations are taken to be mainly dipolar in origin, and are therefore used to calculate[100] the topology of the substrate molecule–lanthanide chelate complex. It is true that, for compounds of known structure, particularly where the molecule has a rigid framework, this simple approach can apparently be very successful. However, in previous volumes of this series a number of difficulties were anticipated which have still to be fully resolved.

B. The Stoicheiometry of the Shift-reagent–Substrate Complex.—In almost all early work relating to lanthanide shift reagents the following assumptions were made: (i) a rapidly exchanging 1:1 substrate–shift-reagent complex is formed; (ii) the complex has axial symmetry, with the major molecular axis coincident with the principal magnetic axis. These two assumptions are important in that, if true, it becomes relatively easy to obtain the chemical shifts for the nuclei of the substrate fully complexed to the shift reagent, and these shifts may then be related to the

[93] M. M. Dhingra and M. S. Subramanian, *Chem. Phys. Letters*, 1975, **30**, 83.
[94] C.-P. Wong, R. F. Venteicher, and W. DeW. Horrocks, jun., *J. Amer. Chem. Soc.*, 1974, **96**, 7149.
[95] J. W. Faller, M. A. Adams, and G. N. La Mar, *Tetrahedron Letters*, 1974, 699.
[96] H. M. Bell, *Org. Magn. Resonance*, 1975, **7**, 240.
[97] C. M. Deber and E. R. Blout, *J. Amer. Chem. Soc.*, 1974, **96**, 7566.
[98] A. Mannschreck, V. Jonas, H.-O. Bödecker, H.-L. Elbe, and G. Köbrich, *Tetrahedron Letters*, 1974, 2153.
[99] J. D. Roberts, G. E. Hawkes, J. Husar, A. W. Roberts, and D. W. Roberts, *Tetrahedron* 1974, **30**, 1833.
[100] H. M. McConnell and R. E. Robertson, *J. Chem. Phys.*, 1958, **29**, 1361.

spatial position of the nucleus relative to the paramagnetic lanthanide metal atom. There is, however, no *a priori* reason for supposing that only one complex with a 1:1 stoicheiometry should be formed.[101] The point is of rather fundamental importance since if more than one complex is formed, or if a single 1:2 complex is formed, it becomes more difficult to extract chemical shifts appropriate to each complex, and the validity of the application of the McConnell–Robertson equation[100] to such complexes becomes suspect.

Evidence has now been accumulated in sufficient quantity to show that at least two complexes LS and LS_2 (L = lanthanide chelate, S = substrate) will in general be formed in solution.[102] There has been one detailed analysis of the shifts induced by $Eu(fod)_3$ in cyclohexanones and cyclohexanols which illustrates the point extremely well.[102] Detailed bound shifts and binding constants were determined assuming that LS and LS_2 species are formed simultaneously in CCl_4 solution for the 28 substrates studied. In almost every case the analysis suggests that LS and LS_2 complexes are in fact formed, although no evidence was found to suggest that self-association of the lanthanide chelate itself occurred.[103,104] Bound shifts determined in this way for the 1:1 complex may then be used with the McConnell–Robertson equation.[100] The present authors have stressed this point by considering the effect that neglect of the LS_2 species would have. Where the ratio of L to S is high, the 1:1 complex would tend to predominate and the measured shifts might arguably be used in the McConnell–Robertson equation.[100] However, most frequently, shifts are determined for low L to S ratios, where the LS_2 species would predominate, and such a complex will not in general have the necessary axial symmetry.[102]

Further involvement of LS_2 species has been shown more directly from low-temperature studies, where the substrate exchange is sufficiently slow that the resonances of the various species may be separately observed.[105] This type of low-temperature study is, however, open to the objection that the data obtained may not be relevant to the situation at ambient temperatures.

There has been an attempt to evaluate the relative merits of alternative published methods[106,107] for the evaluation of binding constants and bound shifts from shift-reagent studies.[108] The criterion by which the results were judged was, however, so vague that no convincing conclusions may be drawn from the work. There can be little doubt, however, that the method described by Johnston *et al.*[102] represents a considerable advance over other, less rigorous, methods.

In conclusion, therefore, there can be no justification for the assumption that a 1:1 complex alone is formed. It is much more reasonable to assume that an LS_2 species *at least* will also be formed.

C. Dipolar, Contact, and Complexation Shifts.

Once the bound shift for the 1:1 complex has been identified it becomes possible to use the McConnell–Robertson

[101] Ref. 55, 1972, Vol. 1, p. 320.
[102] M. D. Johnston, jun., B. L. Shapiro, M. J. Shapiro, T. W. Proulx, A. D. Godwin, and H. L. Pearce, *J. Amer. Chem. Soc.*, 1975, **97**, 542.
[103] J. F. Desreaux, L. E. Fox, and C. N. Reilley, *Analyt. Chem.*, 1972, **44**, 2217.
[104] R. Porter, T. J. Marks, and D. F. Shriver, *J. Amer. Chem. Soc.*, 1973, **95**, 3548.
[105] D. F. Evans and M. Wyatt, *J.C.S. Dalton*, 1974, 765.
[106] B. L. Shapiro and M. D. Johnston, jun., *J. Amer. Chem. Soc.*, 1972, **94**, 8185.
[107] D. Kelsey, *J. Amer. Chem. Soc.*, 1972, **94**, 1764.
[108] J. J. Cawley and D. V. Petrocine, *Org. Magn. Resonance*, 1974, **6**, 544.

equation for the elucidation of the complex structure, provided (in the ideal case) that the observed shift is wholly dipolar in origin. Failing this, it becomes necessary to determine that fraction of the shift which is due to the dipolar term. Increasingly, authors are becoming aware that the simple assumption is invalid, and the observed shift is regarded to consist of three terms [equation (5)].[109]

$$\Delta\delta_{obs,i} = \Delta\delta_{dipolar,i} + \Delta\delta_{contact,i} + \Delta\delta_{diamag,i} \qquad (5)$$

The third term arises simply because the substrate has become bound to the shift reagent, and is, in principle, independent of the paramagnetism of the lanthanide. One method by which this term might be assessed is to determine [110] the bound shift in the presence of the appropriate diamagnetic La^{3+} and/or Lu^{3+} chelates. These usually induce very similar shifts,[111] so that the approximation that the $\Delta\delta_{diamag,i}$ term is the same throughout the series is not unreasonable. The relative magnitude of this term appears to vary widely. For carbonyl compounds it may be as high as 17% of the total shift induced by Eu^{3+} and Pr^{3+} chelates for carbons close to the site of complexation.[112] For alcohol substrates it is less important.[112] This has also been noted by Ajisaka and Kainosho[113] for alcohols and other substrates, and for aniline and p-toluidine.

The second term which must be determined is the contact term. There have been further reports that appreciable contact shifts may be induced by the commonly used Eu^{3+} or Pr^{3+} chelates, particularly for nuclei close to the site of complexation.[115,116] Attempts have been made to estimate contact shifts theoretically with little apparent success.[114] One technique, however, which is becoming popular is to use the appropriate Gd^{3+} chelate, which is thought not to induce dipolar shifts,[117,118] but for which induced shifts are exclusively due to $\Delta\delta_{contact,i} + \Delta\delta_{diamag,i}$. This method has been applied to a study of quinuclidine shifts, but without due correction for $\Delta\delta_{diamag,i}$.[119] In one other study, properly corrected values of the contact term have been estimated for n-octylamine[113] and the amounts of dipolar and contact terms evaluated across the lanthanide series (Figure 2). The dipolar terms evaluated in this way[113] also appear to agree acceptably well with theoretical predictions.[118] It seems, however, to be a feature of this type of work that those authors who take great care to separate the observed shift into the separate contributions fail to take proper account of the possible multiplicity of complexes present in the systems studied. In the final example discussed above,[113] the experimental conditions were such that the substrate was in large excess, and appreciable amounts of the LS_2 species may well have been present in some of the systems studied, and might invalidate the detailed conclusions drawn from this study.

[109] Ref. 55, 1975, Vol. 4, p. 313.
[110] B. F. G. Johnson, J. Lewis, P. McArdle, and J. R. Norton, *J.C.S. Dalton*, 1974, 1253.
[111] K. Tori, Y. Yoshimura, M. Kainosho, and K. Ajisaka, *Tetrahedron Letters*, 1973, 3217.
[112] D. J. Chadwick and D. H. Williams, *J.C.S. Perkin II*, 1974, 1202.
[113] K. Ajisaka and M. Kainosho, *J. Amer. Chem. Soc.*, 1975, **97**, 330.
[114] M. Hirayama and M. Sato, *Chem. Letters*, 1974, 725.
[115] J. A. Young and J. G. Grasselli, *J. Magn. Resonance*, 1974, **14**, 194.
[116] G. Beech and R. J. Morgan, *Tetrahedron Letters*, 1974, 973.
[117] B. Bleaney, C. M. Dobson, B. A. Levine, R. B. Martin, R. J. P. Williams, and A. V. Xavier, *J.C.S. Chem. Comm.*, 1972, 791.
[118] B. Bleaney, *J. Magn. Resonance*, 1972, **8**, 91.
[119] A. A. Chalmers and K. G. R. Pachler, *J.C.S. Perkin II*, 1974, 748.

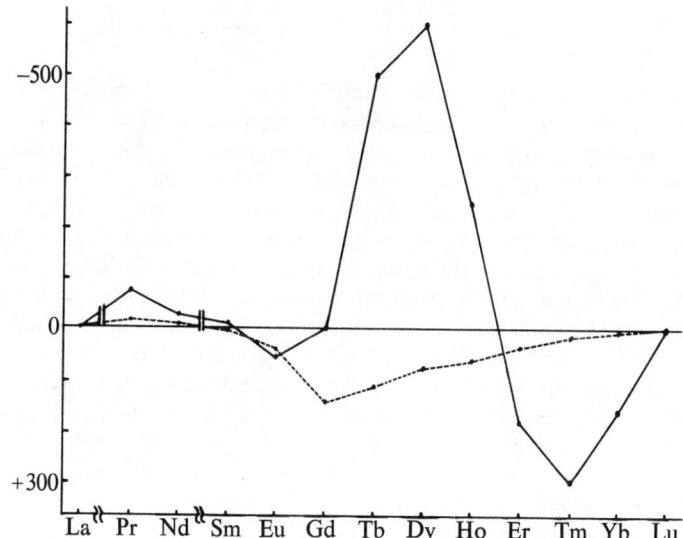

Figure 2 ^{13}C *Contact shifts (broken line) and pseudo-contact shifts (full line) for the α-carbon of n-octylamine plotted against the lanthanide metal (excluding Ce and Pm) in the* (fod)$_3$ *chelate. Shifts are in* p.p.m., *the positive sign denoting shifts to high frequency*

(Reproduced by permission from *J. Amer. Chem. Soc.*, 1975, **97**, 330)

D. Calculation of the Substrate–Shift-reagent Complex Geometry.—Almost certainly, the possibility that the detailed topology of substrate molecules in solution may be inferred from shift-reagent studies has been responsible for the massive interest in this type of work. It is probable, in fact, that techniques for the realization of this goal do now exist. Equally, it is certain that the simple approach used hitherto is not generally valid. In the optimum case, where bound shifts have been separately and uniquely determined for the LS and LS$_2$ complexes and where the dipolar contribution has been accurately determined, it remains to relate the observed dipolar shift to the relative spatial location of the lanthanide atom and the measured nucleus. The normal procedure is to assume an axially symmetric complex with coincident molecular and principal magnetic axes.

However, in general, for LS$_2$ complexes the above assumption cannot be valid. Cramer, Dubois, and Seff[120] have discussed this point in some detail for the case of Eu(tmhd)$_3$(py)$_2$ and Eu(tmhd)$_3$(3-picoline)$_2$. There is fairly good evidence that predominantly the LS$_2$ complex forms in solution although the experimental technique used here[120] does not uniquely determine the bound shifts for the LS$_2$ complex at ambient temperatures, and in fact, better agreement was found using the LS$_2$ shifts obtained directly at low temperatures. If the assumption of axial symmetry is discarded the dipolar shift is given by equation (6),[121] where r_i is the length of, and θ_i

---

[120] R. E. Cramer, R. Dubois, and K. Seff, *J. Amer. Chem. Soc.*, 1974, **96**, 4125.
[121] R. von Ammon and R. D. Fischer, *Angew. Chem. Internat. Edn.*, 1972, **11**, 675.

$$\Delta\delta_i = D_1\left(\frac{3\cos^2\theta_i - 1}{r_i^3}\right) + D_2\left(\frac{\sin^2\theta_i \cos 2\Omega_i}{r_i^3}\right) \qquad (6)$$

the angle subtended at the z magnetic axis by, the vector joining nucleus i and the paramagnetic centre, Ω_i is the angle between the projection of r_i onto the xy plane and the x magnetic axis, and D_1 and D_2 are constants determined by the magnetic anisotropy of the complex. For the pyridine and 3-picoline complexes with Eu(tmhd)$_3$ it is found[120] that the second term, normally discarded if axial symmetry is assumed, contributes 20—80% of the shift observed for each nucleus, and it is indeed necessary to consider the observed shifts in terms of the full equation.

Even for the LS species, complications may arise; probable structures have been calculated for monoadducts of lanthanide chelates on the basis of configurations of minimum energy, which suggests that three different configurations are possible.[122] However, it is probable that rapid exchange between different conformations of this type may give rise to an exchange-averaged structure having an effective axial symmetry,[123] in which case the McConnell–Robertson equation would be appropriate. There has in fact been one report of dipolar shifts, calculated on the basis of an X-ray crystal structure for a monoadduct of quinuclidine with Eu(tmhd)$_3$, which agree quite well with experiment.[124]

To date, no single report of shift-reagent studies is wholly rigorous in the sense of the above discussion, despite the fact that essentially these same points were stressed by Williams, Bleaney, and co-workers as early as 1972.[117]

E. **Structural Assignments.**—There have been a number of fairly simple applications of shift reagents in this respect, as, for example, the identification of the site of substitution in coumarins[125] or the assignment of ester configurations as being either *s-cis* or *s-trans*.[126,127] Shift reagents have also been used in the assignment of epoxide conformations,[128] and to determine the *syn–anti* ratio of oximes and 2,4-dinitrophenylhydrazones.[129] Some controversy exists with regard to the preferred site of co-ordination of oximes, *i.e.* whether it occurs at the oxygen or nitrogen atom. Evidence from shift-reagent studies had previously been presented to suggest that oxygen is the preferred site.[130] In the present work,[129] based on both shift-reagent and ASIS data (although the choice of CDCl$_3$ as a reference solvent is open to objection),[131,132] the preferred site is taken to be the nitrogen atom.

In contrast to studies of rigid molecules, there has been a continued interest in the study of molecules which exist in a state of conformational equilibrium. This topic has been discussed in detail by Servis *et al.*[133] with regard to cyclohexanones, by

[122] D. L. Kepert, *J.C.S. Dalton*, 1974, 617.
[123] Ref. 55, 1974, Vol. 3, p. 154.
[124] E. Bye, *Acta Chem. Scand. (A)*, 1974, **28**, 731.
[125] A. I. Gray, R. D. Waigh, and P. G. Waterman, *J.C.S. Chem. Comm.*, 1974, 632.
[126] N. Thoai and Trinh-Minh-Chan, *Canad. J. Chem.*, 1974, **52**, 1331.
[127] K. Sakamoto and M. Ōki, *Bull. Chem. Soc. Japan*, 1974, **47**, 2623.
[128] J.-P. Vidal, J. P. Girard, J.-C. Rossi, J. P. Chapat, and R. Granger, *Org. Magn. Resonance*, 1974, **6**, 522.
[129] M. Žinić, M. Štromar, M. Malnar, and D. Kolbah, *Croat. Chem. Acta*, 1974, **46**, 45.
[130] Z. W. Wolkowski, *Tetrahedron Letters*, 1971, 825.
[131] Ref. 55, 1972, Vol. 2, p. 374.
[132] J. Paasivirta, H. Häkli, and K. G. Widen, *Org. Magn. Resonance*, 1974, **6**, 380.
[133] K. L. Servis, D. J. Bowler, and C. Ishii, *J. Amer. Chem. Soc.*, 1975, **97**, 73.

Roberts et al.[99] with regard to menthone and isomenthone, and also in the context of N-nitrosopiperidines[134] and bicyclo[6,6,0]nona-2,4,6-trienes.[135]

In addition, in calculating structures corresponding to observed dipolar shifts, more authors are tending to include some degree of rotation about the lanthanide–substrate bond within their model.[99,136,137] This illustrates rather well the point made earlier, that the tendency is for workers in this field to concentrate their attention on one aspect of the problem to the exclusion of others, and thereby arrive at possibly erroneous conclusions. Armitage et al.,[138] for example, had earlier determined Eu(tmhd)$_3$-induced proton shifts in aniline which could not adequately be correlated with dipolar shifts using the simple model unless allowance was made for rotation about the lanthanide–substrate bond. Johnson and co-workers[110] have shown subsequently that, if proper correction is made for the contact contribution to this shift, a satisfactory correlation is obtained without the necessity of assuming internal rotation. However, since neither group has rigorously considered the stoicheiometry or multiplicity of the species in solution, the matter cannot be regarded as concluded.

F. Preferred Binding Sites.—In general, if shifted spectra are to be obtained, the substrate molecule must contain a suitable binding site in order to complex with the lanthanide chelate. It has not therefore been possible in the past to obtain shifted spectra for such compounds as (15) and (16). Evans, Tucker, and Villardi,[139] in an

(15) (16)

elegantly simple experiment, have produced lanthanide-shifted spectra of these compounds by adding to the substrate solution a mixture of $C_3F_7CO_2Ag$ and a lanthanide chelate. The silver salt binds initially to the unsaturated bonds of the substrate, the shift reagent then binds to the carbanion fragment, and shifted spectra are obtained. In a similar way, it has been observed that remarkably large lanthanide-induced shifts are found for N-benzylpyridinium salts, although there are no suitable functional groups present.[140] It is proposed, in this case, that an ion pair forms with the halide anion, which then binds to the lanthanide species.

For compounds having more than one functional group the situation becomes appreciably more complex. So far as relatively simple estimates of the actual site of binding are concerned for any specific case, the spatial proximity of the functional groups is important. For example, in aromatic compounds having COMe or CO_2Me groups ortho to OMe, both sites are simultaneously bound to the lanthan-

[134] T. F. Forrest, D. L. Hooper, and S. Ray, J. Amer. Chem. Soc., 1974, 96, 4286.
[135] C. J. Cheer and W. Rosen, Tetrahedron Letters, 1974, 4045.
[136] R. J. Abraham, S. M. Coppell, and R. Ramage, Org. Magn. Resonance, 1974, 6, 658.
[137] D. J. Chadwick, Tetrahedron Letters, 1974, 1375.
[138] I. M. Armitage, L. D. Hall, A. G. Marshall, and L. G. Werbelow, J. Amer. Chem. Soc., 1973, 95, 1437.
[139] D. F. Evans, J. N. Tucker, and G. C. de Villardi, J.C.S. Chem. Comm., 1975, 205.
[140] G. Montaudo, G. Kruk, and J. W. Verhoeven, Tetrahedron Letters, 1974, 1841.

ide.[141] The same appears to hold for certain diamagnetic nickel complexes which form structures of the type (17).[142]

(17)

Binding to anhydrides of the general type (18) and (19) also appears to involve a bidentate bond to the carbonyl groups from the metal atom, without, however, involving the ether oxygen.[143] Where the binding sites are too far separated for this type of interaction, binding may occur separately at the two sites. It has been observed, for example, that nuclei close to the two functional groups in 19α-methyl-17β-hydroxy-androstan-4-en-3-one, -dienone, -trienone, and -methanedienone all have induced shifts of similar magnitude.[144] This may indicate that the two binding sites in each molecule bind the shift reagent equally strongly, although this is probably a rather simplistic view. For (20) and (21) the problem is rather more complex. Here, surprisingly, binding is thought to occur predominantly at the

(18)   (19)

(20)   (21)

[141] N. Platzer, C. Lang, J.-J. Basselier, and P. Demerseman, *Bull. Soc. Chim. France*, 1975, 227
[142] L. F. Lindoy and W. E. Moody, *J. Amer. Chem. Soc.*, 1975, **97**, 2275.
[143] R. E. R. Craig, A. C. Craig, and G. D. Smith, *Tetrahedron Letters*, 1975, 1189.
[144] A. K. Lala and A. B. Kulkarni, *Indian J. Chem.*, 1974, **12**, 926.

5-alkoxy-group.[145] Shifts have also been reported for polyoxyalkylenes, which are thought to reflect the distribution of shift reagent at oxygen sites along the polymer chain.[146]

With regard to the binding of specific groups to shift reagents, there has been a further report of Eu(fod)$_3$-shifted spectra of phenols[147] which supports the earlier report of Shoffner[148] in removing the misapprehension that shift reagents generally are rapidly decomposed and rendered ineffective by organic acids. Shift-reagent binding to thiocyanates has been reported, but no shifts are observed with isothiocyanates.[149] This may prove to be a useful diagnostic test to distinguish between such compounds. Phosphorus-31 shifts have been observed for four-co-ordinate organophosphorus compounds with Yb(fod)$_3$ and Pr(fod)$_3$ binding to the phosphorus atom.[150] The effect seems to be rather selective, however; virtually no shifts are observed with Eu(fod)$_3$.[150]

[145] H. Cairns, D. Hunter, J. King, and N. H. Rogers, *Tetrahedron*, 1974, **30**, 79.
[146] A. Kh. Bulai, A. G. Gruznov, Ya. G. Urman, L. M. Romanov, and I. Y. Slonim, *Vysokomol. Soedineniya*, 1974, **16**, 2203.
[147] K.-T. Liu, M.-F. Hsu, and J. S. Chen, *Tetrahedron Letters*, 1974, 2179.
[148] J. P. Shoffner, *J. Amer. Chem. Soc.*, 1974, **96**, 1599.
[149] S. J. Anderson and A. H. Norbury, *J.C.S. Chem. Comm.*, 1975, 48.
[150] F. S. Mandel, R. H. Cox, and R. C. Taylor, *J. Magn. Resonance*, 1974, **14**, 235.

12
Oriented Molecules

BY P. DIEHL

1 Introduction

The basic format of this chapter is similar to that of Volumes 1 and 3. It deals with papers published within the two years from June 1973 to June 1975. Compared to Volume 1, two sections have been omitted and left to other specialized reports; these are 'relaxation', treated in Chapter 3 of this volume, and 'liquid crystals and micellar solutions', which may be found in Chapter 8 of Volume 4 as well as in forthcoming even-numbered volumes.

Only two reviews were published during this period. The first[1] deals with magnetic resonance studies of thermotropic liquid crystals. It discusses mainly pure liquid crystals but also contains some information on e.s.r. and n.m.r. spectra of molecules dissolved in liquid crystals. The second[2] covers the whole field of n.m.r. of oriented molecules for the period from 1963 to early 1975.

2 Structure Determination

The number of molecules for which structures have been determined by n.m.r. studies is steadily increasing. From about 15 molecules per year in the 'early days' of the method (1963–1969) and 25 per year in the period 1969–1971, the rate has increased to about 60 molecules per year. The total number of structures determined is now over 350. In parallel with this yearly increase there has been a trend towards bigger molecules. Only 4 structures of molecules with more than 6 magnetic nuclei were determined between 1963 and 1969, whereas over 20 such molecules were analysed during the two-year period of this Report. Furthermore, there has been an increase in the number of papers dealing with vibrational corrections (3 in Volume 3, 14 in Volume 5), with molecules dissolved in lyotropic liquid crystals (9 in Volume 3, 14 in Volume 5), and with molecules displaying internal motions (17 in Volume 3, 27 in Volume 5).

A. **Structures of Rigid Molecules.**—Within this subsection, as in previous articles, the molecules are arranged in order of increasing number of spins. This section also contains cases of internal motion where free rotation is assumed.

In order to provide any structural information, systems of 3 nuclei have to be highly symmetrical. The two cases studied recently were[3,4] the linear [1-^{13}C]-acetylene, and $P_3N_3Cl_6$, which was found to be planar with D_{3h} symmetry.

---

[1] G. R. Luckhurst, *Mol. Crystals Liquid Crystals*, 1973, **21**, 125.
[2] J. W. Emsley and J. C. Lindon, 'NMR Spectroscopy Using Liquid Crystal Solvents', Pergamon Press, Oxford, 1975.
[3] P. Diehl, S. Sýkora, W. Niederberger, and E. E. Burnell, *J. Magn. Resonance*, 1974, **14**, 260.
[4] N. Zumbulyadis and B. P. Dailey, *J. Magn. Resonance*, 1974, **13**, 189.

Approximately one third of all the molecules studied had 4 spins. Distance ratios were determined for o-, m-, and p-dicyanobenzene,[5] γ-hydroxypyridine,[6] ethylene,[7] and tellurophen.[8]

One of the easiest measurements is the determination of HCH bond angles in methyl groups or of corresponding angles in structurally equivalent groups. The following molecules were studied: $^{13}CH_3I$, $^{13}CH_3Br$, $^{13}CH_3Cl$,[9] CH_3F,[10] $^{13}CH_3OH$,[11] the [^{13}C]acetate ion,[12,13] phosphine,[14] phosphorus trifluoride,[15] phosphoryl fluoride,[16,17] cis-difluoroethylene,[18] and acetonitrile.[18a] In the methyl halides the HCH angles were found to vary with the liquid-crystal solvent. The variation was particularly large for methyl fluoride where the angle apparently changed by 2° between the solvents 4,4'-di-n-hexyloxyazoxybenzene and N-(p-ethoxybenzylidene)-p-n-butylaniline. The effect could not be attributed to the anisotropy of the indirect coupling. Vibrational corrections were not performed. Whereas the structures of the acetate ion, phosphine, and phosphorus trifluoride agreed with microwave (m.w.) and electron diffraction (e.d.) results, for phosphoryl fluoride the FPF angle was found to be 2° to 3° too small. One group attributed this result to an anisotropy in the (P, F) indirect coupling.[16] The other[17] discussed the following alternatives: solvent-dependent $J(P,F)$; difference between gas and liquid forms; vibrational corrections; dependence of geometry on orientation; or anisotropy in $J(F,F)$ or $J(P,F)$. The dependence of geometry on orientation seemed to be the best explanation, since a small 0.1° to 0.2° decrease of the OPF angle when the symmetry axis moves from a parallel to a perpendicular position with respect to the optic axis could account for the observation.

A number of molecular structures have been determined both in thermotropic and lyotropic liquid crystals and the results compared. The molecules were furan, thiophen, and p-dithiin[19] as well as pyrazine, pyrimidine, and pyridazine.[20] For furan and thiophen the results in the two media agreed. For pyrimidine, which originally showed the largest discrepancy,[20] the redetermination with improved scalar coupling constants[19] reduced the deviations to practically within twice the statistical error. It should be pointed out that when making such comparisons, the data obtained from lyotropic solvents may be, because of small degrees of order (dipolar coupling constants between 12 and 20 Hz), extremely sensitive to the indirect couplings (scalar coupling constants between 1 and 7 Hz).

[5] W. de Kievit and C. A. de Lange, *Chem. Phys. Letters*, 1973, **22**, 378.
[6] C. L. Khetrapal, A. C. Kunwar, and A. V. Patankar, *Org. Magn. Resonance*, 1974, **6**, 556.
[7] P. Diehl, S. Sýkora, and E. Wullschleger, *Mol. Phys.*, 1975, **29**, 305.
[8] A. d'Annibale, L. Lunazzi, F. Fringuelli, and A. Taticchi, *Mol. Phys.*, 1974, **27**, 257.
[9] P. K. Bhattacharyya and B. P. Dailey, *Mol. Phys.*, 1973, **26**, 1379.
[10] P. K. Bhattacharyya and B. P. Dailey, *J. Magn. Resonance*, 1974, **13**, 317.
[11] P. K. Bhattacharyya and B. P. Dailey, *J. Chem. Phys.*, 1973, **59**, 3737.
[12] D. M. Chen. L. W. Reeves, A. S. Tracey, and M. M. Tracey, *J. Amer. Chem. Soc.*, 1974, **96**, 5349.
[13] L. W. Reeves, A. S. Tracey, and M. M. Tracey, *J. Amer. Chem. Soc.*, 1973, **95**, 3799.
[14] N. Zumbulyadis and B. P. Dailey, *Mol. Phys.*, 1974, **27**, 633.
[15] N. Zumbulyadis and B. P. Dailey, *Mol. Phys.*, 1973, **26**, 777.
[16] P. K. Bhattacharyya and B. P. Dailey, *Mol. Phys.*, 1974, **28**, 209.
[17] J. Bulthuis and C. A. de Lange, *J. Magn. Resonance*, 1974, **14**, 13.
[18] J. W. Emsley and J. C. Lindon, *Mol. Phys.*, 1974, **28**, 1253.
[18a] P. K. Bhattacharyya and B. P. Dailey, *Chem. Phys. Letters*, 1975, **32**, 305.
[19] R. C. Long, jun., and J. H. Goldstein, *Mol. Crystals Liquid Crystals*, 1973, **23**, 137.
[20] R. C. Long, jun., K. R. Long, and J. H. Goldstein, *Mol. Crystals Liquid Crystals*, 1973, **21**, 299.

The four-spin molecule ethylene oxide[21] was used to demonstrate that the generalized NOE can be used to determine the energy level diagram by INDOR and to observe double-quantum transitions.

In an interesting study Haloui and Canet[22] compared two methods (^{13}C resonance at natural abundance, and ^{13}C satellites in H resonance) for the determination of molecular structures, with carbon positions included. They found that the satellite method utilizing FT proton spectroscopy was more accurate and much faster. In cyanopropyne the four carbon positions as well as the HCH angle were determined. The authors found a systematic discrepancy between their carbon–carbon bond lengths and m.w. data. This may be caused by their assumed CH bond length, without proper vibrational correction.

Again, tetrahedral species have been found to display apparent order in liquid crystals, in spite of their symmetry. The species studied were the ions NH_4^+, ND_4^+,[23] and BF_4^-.[24] The observed splittings were attributed to distortions of the tetrahedra through interaction with the anisotropic environment.[23a]

A surprising phenomenon was reported for the molecule [$^{13}C_2$]acetylene,[3] which in one liquid crystal showed an apparent change in geometry (distance ratio) by 30% over a temperature range of 20 °C. This change was attributed to an exchange of the acetylene between at least two sites in the liquid crystal that had opposite signs of the degree of order. A real change of geometry of the dissolved molecule between the sites is amplified considerably when the apparent degree of order is small. The authors demonstrated that a change in geometry of less than 1% together with an enthalpy difference of 800—1600 J mol^{-1} could account for the observed effect. It was later demonstrated that the same effect also explained the well-known discrepancies observed in the molecule $^{13}CH_3F$.[23b]

The following five-spin systems were studied: [^{13}C]methylphosphonate,[19] pyridine,[25, 26] $^{13}CH_3F$,[10] [^{13}C]ethylene,[7] trans-(py)(C_2H_4)$PtCl_2$,[27] BF_4^-,[24] benzonitrile,[28] pyridine N-oxide,[26] ethylene oxide and ethylene sulphide (^{13}C satellites),[29] and 1,4-dibromo-[30] and 1,4-dichloro-benzene (^{13}C satellites).[31] Of these, the ^{13}C satellites in ^1H-resonance cases are particularly interesting, since they allow the localization of the carbon atoms as well as the measurements of HCC and CCC angles. The separation of the effective couplings [$J(C,H) + 2D(C,H)$] was obtained by varying the degree of order, which only affects the $D(C,H)$. The deviations between n.m.r. and m.w. results for ethylene oxide and ethylene sulphide[29] may perhaps again be attributed to the CH bond length used as a basis and to a lack of vibrational corrections.

[21] J. Degelaen, E. Arte, and J. M. Dereppe, *J. Chem. Phys.*, 1974, **61**, 5295.
[22] E. Haloui and D. Canet, *Chem. Phys. Letters*, 1974, **26**, 261.
[23] L. W. Reeves and A. S. Tracey, *J. Amer. Chem. Soc.*, 1974, **96**, 365.
[23a] D. Bailey, A. D. Buckingham, F. Fujiwara, and L. W. Reeves, *J. Magn. Resonance*, 1975, **18**, 344.
[23b] E. E. Burnell, J. R. Council, and S. E. Ulrich, *Chem. Phys. Letters*, 1975, **31**, 395.
[24] F. Fujiwara, L. W. Reeves, and A. S. Tracey, *J. Amer. Chem. Soc.*, 1974, **96**, 5250.
[25] I. M. Armitage, E. E. Burnell, M. B. Dunn, L. D. Hall, and R. B. Malcolm, *J. Magn. Resonance*, 1974, **13**, 167.
[26] C. L. Khetrapal, A. C. Kunwar, and A. V. Pantankar, *J. Magn. Resonance*, 1974, **15**, 219.
[27] D. R. McMillin and R. S. Drago, *Inorg. Chem.*, 1974, **13**, 546.
[28] J. P. Jacobsen and K. Schaumburg, *Mol. Phys.*, 1974, **28**, 1505.
[29] E. Haloui and D. Canet, *J. Mol. Structure*, 1975, **24**, 85.
[30] E. E. Burnell and M. A. J. Sweeney, *Canad. J. Chem.*, 1974, **52**, 3565.
[31] E. E. Burnell, P. Diehl, and W. Niederberger, *Canad. J. Chem.*, 1974, **52**, 151.

The fact that during this two-year period three different groups studied the same six-spin molecules cyclopentadiene[32-34] and 2,5-dihydrofuran[35-37] indicates that the number of molecules with symmetry and with 6 or less spins which have not as yet been studied is rapidly diminishing. The results of the papers agreed with each other and with m.w. data (which were available only for cyclopentadiene). In one approach[33] the carbon positions were determined within the proton framework by minimizing the potential energy for C–C distances as well as CCC and CCH angles. The results of this study agreed with m.w. data within 0.001 nm for bond lengths and 1—2 degrees for bond angles. Other molecules containing 6 spins which have been studied are [^{15}N]pyridine,[38] benzene,[39] [^{13}C]benzonitriles,[28] [^{13}C$_2$]-ethylene,[7] coumarin,[40] H$_3$Ru$_3$(CO)$_9$CCH$_3$,[41,42] cyclobutanone,[43,44] 2-cyclopentenone,[44a] trimethylenesulphide,[43] m-difluorobenzene,[45] p-difluorobenzene,[46] 1,2,4,5-hexatetraene,[46a] the methyl-[23,47] and dimethyl-ammonium ions,[47] as well as the compounds dimethylmercury[48] and [^{15}N]pyrrole.[49] A conclusion reached by one group of authors[28] that 'n.m.r. spectroscopy is of questionable use for the determination of relative CH distances' is itself questionable since these authors did not perform the necessary vibrational corrections for their results.

The only seven-spin systems studied were 1-chloronaphthalene,[50] the dimethyltin ion,[47,51] [^{13}C]dimethylmercury,[48] and the dimethylthallium ion.[52]

For the eight-spin molecules naphthalene[53] and indene,[54] agreement between n.m.r. and electron-diffraction results was found. The spectrum of norbornadiene[55] was re-analysed in a different nematic phase. Apparently large deviations in distance ratios between the new and the old results were found. However, since the direct couplings being compared were small, it turned out that an error of 1 Hz in one coupling, direct or indirect, could account for the discrepancies. The structures of 2,5-dimethylpyrazine[56] and of [^{13}C$_2$]dimethylmercury[48] have been determined.

[32] J. W. Emsley, J. C. Lindon, D. S. Stephenson, and M. C. McIvor, *Mol. Phys.*, 1974, **28**, 93.
[33] H. Günther, W. Herrig, and J. B. Pawliczek, *Z. Naturforsch.*, 1974, **29b**, 104.
[34] C. A. Veracini, M. Guidi, M. Longeri, and A. M. Serra, *Chem. Phys. Letters*, 1974, **24**, 99.
[35] P. F. Swinton, *J. Mol. Structure*, 1974, **22**, 221.
[36] D. G. De Kowalewski and V. J. Kowalewski, *J. Mol. Structure*, 1974, **23**, 203.
[37] J. Courtieu and Y. Gounelle, *Mol. Phys.*, 1974, **28**, 161.
[38] C. Schumann and R. Price, *Angew. Chem.*, 1973, **85**, 989.
[39] R. C. Long, *J. Magn. Resonance*, 1973, **12**, 216.
[40] E. Capelli, A. Di Nola, and A. L. Segre, *Mol. Phys.*, 1974, **27**, 1385.
[41] A. D. Buckingham, J. P. Yesinowski, A. J. Canty, and A. J. Rest, *J. Amer. Chem. Soc.*, 1973, **95**, 2732.
[42] J. P. Yesinowsky and D. Bailey, *J. Organometallic Chem.*, 1974, **65**, C27.
[43] P. F. Swinton, *J. Magn. Resonance*, 1974, **13**, 304.
[44] A. D'Annibale, L. Lunazzi, G. Fronza, R. Mondelli, and S. Bradamante, *J.C.S. Perkin II*, 1973, 1908.
[44a] K. C. Cole and D. F. R. Gilson, *Canad. J. Spectroscopy*, 1975, **20**, 61.
[45] G. J. den Otter, J. Gerritsen, and C. MacLean, *J. Mol. Structure*, 1973, **16**, 379.
[46] G. J. den Otter, W. Heijser and C. MacLean, *J. Magn. Resonance*, 1974, **13**, 11.
[46a] B. Pedersen, J. H. Schaug, and H. Hopf, *Acta Chem. Scand. (A)*, 1974, **28**, 846.
[47] L. W. Reeves, J. Sanches de Cara, M. Suzuki, and A. S. Tracey, *Mol. Phys.*, 1973, **25**, 1481.
[48] A. Schumann, H. Dreeskamp, and K. Hildebrand, *J. Magn. Resonance*, 1975, **18**, 97.
[49] J. M. Briggs, E. J. Rahkamaa, and E. W. Randall, *J. Magn. Resonance*, 1975, **17**, 55.
[50] P. Diehl and J. Vogt, *Org. Magn. Resonance*, 1974, **6**, 33.
[51] L. W. Reeves, M. Suzuki, A. S. Tracey, and J. A. Vanin, *Inorg. Chem.*, 1974, **13**, 999.
[52] Y. Lee and L. W. Reeves, *Canad. J. Chem.*, 1975, **53**, 161.
[53] J. M. Dereppe, J. Degelaen, and M. van Meerssche, *J. Mol. Structure*, 1973, **17**, 225.
[54] P. Diehl and J. Vogt, *Org. Magn. Resonance*, 1975, **7**, 81.
[55] J. W. Emsley and J. C. Lindon, *Mol. Phys.*, 1975, **29**, 531.
[56] D. Canet, *Compt. rend.*, 1973, **276**, C, 315.

In the nine-spin system $P_3N_3F_6^4$ the structural information was obtained by simplification of the spectrum by means of $^{31}P-\{^{19}F\}$ decoupling. The spectrum of the dimethyl[^{15}N]ammonium ion was analysed and the structure determined.[57] Finally, the structure of the ten-spin system p-xylene was studied.[58]

B. Corrections for Vibrational Motion.—As pointed out above the effect of vibrational averaging on the geometries of molecules as determined by the n.m.r. of oriented molecules has been given as a possible explanation for observed deviations from m.w. data.[28, 43] Vibrational corrections were performed for the following molecules: cyclopentadienyl-compounds (cp)Ni(NO) and (cp)$_2$Hg,[59] cis-difluoroethylene,[18] cyclobutadiene,[60] cyclopentadiene,[32] norbornadiene,[55] CH_2DCH_2Br, CH_3CD_2Br,[61] 1,1-difluoroethene,[62] methylfluoride,[62] phosphorylfluoride,[17] [$^{13}C_2$]acetylene,[3] and [$^{13}C_2$]ethylene[7].

It has been found that often the vibration corrections do not account for the discrepancies[17, 59, 60, 62] or are small.[7, 18, 32, 55] On the other hand they seem to affect significantly the derived quadrupole coupling constants.[61] In acetylene the shrinking effects[3] were found to be in excellent agreement with i.r. data.

C. Molecules with Internal Motion.—Many papers have been published on molecules having internal motion, but in only a few has this motion been studied in any detail. Reasons for this are: for trimethylacetic acid[63] the spectrum turned out to be insensitive to the potential; for trimethylenemethane iron tricarbonyl[64] the broadened spectrum prevented a study of the motion; for methylsilane and methylgermane[65] as well as [$^2H_3$]methylsilane[66] the problem was underdetermined, so that staggered conformations or free rotation had to be assumed in order to derive further parameters. In tropolone[67] the exchanging hydroxyl proton was not observed and the spectrum had to be fitted with the assumption of a rather complex intramolecular motion; furthermore, a distance ratio apparently varied considerably with liquid-crystal solvent. Ethanol and its partially deuterated derivatives[68] associated with the liquid crystal, so that the direct coupling between the methyl and the hydroxy-groups could not be fitted with various models of internal motion, whereas the methylene–hydroxyl coupling agreed with free rotation. Evidence for intra: molecular rotation was obtained in π-cyclopentadienyliron tricarbonyl[69] and (cp)$_2$Hg.[59] The spectra of p-dioxene[70] and [^{15}N]formamide[71] indicated that the molecules were non-planar, with a twist angle of 28.4° for p-dioxene. In the methyl-

---

[57] L. W. Reeves and A. S. Tracey, *J. Amer. Chem. Soc.*, 1974, **96**, 7176.
[58] D. Canet and J. Barriol, *Mol. Phys.*, 1974, **27**, 1705.
[59] I. R. Beattie, J. W. Emsley, and R. M. Sabine, *J.C.S. Faraday II*, 1974, **70**, 1356.
[60] J. W. Emsley and J. C. Lindon, *Mol. Phys.*, 1974, **28**, 1373.
[61] J. W. Emsley and J. Tabony, *J. Magn. Resonance*, 1975, **17**, 233.
[62] J. Bulthuis and C. MacLean, *Chem. Phys. Letters*, 1973, **21**, 611.
[63] C. A. de Lange, *Chem. Phys. Letters*, 1974, **28**, 526.
[64] A. D. Buckingham, A. J. Rest, and J. P. Yesinowski, *Mol. Phys.*, 1973, **25**, 1457.
[65] R. Ader and A. Loewenstein, *J. Amer. Chem. Soc.*, 1974, **96**, 5336.
[66] R. Ader and A. Loewenstein, *Mol. Phys.*, 1974, **27**, 1113.
[67] J. W. Emsley, J. C. Lindon, and D. S. Stephenson, *Mol. Phys.*, 1974, **27**, 641.
[68] J. W. Emsley, J. C. Lindon, and J. Tabony, *Mol. Phys.*, 1973, **26**, 1485.
[69] D. Bailey, A. D. Buckingham, and A. J. Rest, *Mol. Phys.*, 1973, **26**, 233.
[70] C. A. de Lange and K. J. Peverelli, *J. Magn. Resonance*, 1974, **16**, 159.
[71] L. W. Reeves, J. M. Riveros, R. A. Spragg, and J. A. Vanin, *Mol. Phys.*, 1973, **25**, 9.

ammonium ion[72] practically free rotation of the methyl group was recorded, whereas the data on CH_2DCH_2Br and CH_3CD_2Br[61] indicated averaging over stable conformers.

Puckering motion in agreement with m.w. data was found for trimethylene oxide and trimethylene sulphide,[73] thietan,[4] and 2,5-dihydrofuran.[74] In a discussion of of this motion for cyclobutane[75] the authors used vibrational corrections for the benzene C—H bond in order to understand the deviations.

The structure and conformation of 4,4'-dichlorobiphenyl[76] were determined. Taking intramolecular potentials from theoretical predictions, the intermolecular contribution to the potential, and the equilibrium angle between the benzene planes were determined. In cyclo-octatetraene[77] a bond-shift process was observed which corresponds to a cyclic permutation of the carbon atoms along the ring. This process does not affect all the eigenstates, and consequently with variable-temperature experiments some transitions remain sharp while others are broadened. The activation energy of this process was determined.

A number of papers have dealt with molecules displaying intramolecular motion of 'large' groups with the corresponding well-known and serious problems of re-orientation and averaging. It is unfortunately still true that the direct couplings averaged over all conformations at a single temperature can never determine the isomeric ratio. Even if all the direct couplings can be obtained from the spectrum and the geometry of each conformer is known, it is still only possible to determine the quantities p_iS_{kl}, where p_i is the concentration of conformer i and S_{kl} are the various orientation parameters of this conformer. This limitation can only be surmounted by 'luck' (i.e. if only one conformer is important, as found for thiophen-2-carbaldehyde[78] and pyridine-2,6-dicarbaldehyde[79]), or by 'brute force', i.e. if the assumption is made that the various conformers orient very similarly (as was done for bithienyl derivatives[80] and [2-^{13}C]propene[81]), or if it is assumed that there exists an average degree of order (as for the cases of acroyl fluoride and acrolein,[82] 2,2'-bithienyl,[83] bipyridyl,[84] monofluoroacetone,[85] and 2-chloroethanol[86]). Both assumptions are difficult to prove, and there is consequently a continuum of equivalent solutions of the conformer ratio.

A second problem often overlooked in this field is that of having enough direct couplings measured to be over-determined. A typical case is 3,3',5,5-tetrachlorobiphenyl.[87] There is only one inter-ring coupling observed, which is

[72] L. W. Reeves and A. S. Tracey, *J. Amer. Chem. Soc.*, 1974, **96**, 1198.
[73] C. L. Khetrapal, A. C. Kunwar, and A. Saupe, *Mol. Phys.*, 1973, **25**, 1405.
[74] K. C. Cole and D. F. R. Gilson, *Canad. J. Chem.*, 1974, **52**, 281.
[75] K. C. Cole and D. F. R. Gilson, *J. Chem. Phys.*, 1974, **60**, 1191.
[76] W. Niederberger, P. Diehl, and L. Lunazzi, *Mol. Phys.*, 1973, **26**, 571.
[77] Z. Luz and S. Meiboom, *J. Chem. Phys.*, 1973, **59**, 1077.
[78] L. Lunazzi and C. A. Veracini, *J.C.S. Perkin II*, 1973, 1739.
[79] P. L. Barili, M. Longeri, and C. A. Veracini, *Mol. Phys.*, 1974, **28**, 1101.
[80] C. A. Veracini, D. Macciantelli, and L. Lunazzi, *J.C.S. Perkin II*, 1973, 751.
[81] L. F. Williams and A. A. Bothner-By, *J. Magn. Resonance*, 1973, **11**, 314.
[82] J. Courtieu, Y. Gounelle, P. Gonord, and S. K. Kan, *Org. Magn. Resonance*, 1974, **6**, 151.
[83] P. Bucci, M. Longeri, C. A. Veracini, and L. Lunazzi, *J. Amer. Chem. Soc.*, 1974, **96**, 1305.
[84] J. Courtieu, Y. Gounelle, J. Duret, P. Gonord, and S. K. Kan, *Org. Magn. Resonance*, 1974, **6**, 622.
[85] J. Courtieu and Y. Gounelle, *Org. Magn. Resonance*, 1974, **6**, 11.
[86] D. Pumpernik and A. Ažman, *Z. Naturforsch.*, 1974, **29a**, 527.
[87] A. D'Annibale, L. Lunazzi, A. C. Boicelli, and D. Macciantelli, *J.C.S. Perkin II*, 1973, 1396.

sensitive to the angle between the phenyl planes. This angle can consequently be determined with 'breath-taking' precision in a 'least squares' fit based on an assumed absolutely stable conformer. The molecule, however, displays large intramolecular motion which can only be detected if several inter-ring couplings are observed and if these are, because of short distances, sensitive to the motion. Similar remarks apply also to the molecule 2,2'-bithiophen.[88] It should be kept in mind that in just-determined or slightly over-determined systems the r.m.s. error of a least-squares fit approaching zero does not confirm the correctness of an assumed model but only means that the results are in agreement with both the assumed non-unique model and the observed spectra.

3 Determination of Indirect Couplings

The spectra of molecules oriented in lyotropic liquid crystals are often particularly sensitive to the magnitudes and signs of the indirect coupling constants when the direct couplings are small and of similar magnitude to the indirect ones. Also, because of sample rotation, the resolution of the spectra is better than for thermotropic solvents. These facts were used to determine magnitudes[89, 57, 72] as well as signs[89, 72] of indirect couplings in 1,2-difluoroethylene[89] and in dimethylthallium[57] and methylammonium[72] ions.

For the molecules cis-difluoroethylene,[18] trifluoroacetic acid,[90] and 1,4-dibromobenzene[30] the problem of separating the indirect from the direct contributions to effective coupling constants (caused by weak interaction) was solved by sample rotation, which only affected the direct couplings.

Similarly, the sign of $J(^{13}C,H)$ was found to be positive by varying the temperature for the species $^{13}CH_3X$, where $X = Cl$, I, or Br, since the wrong sign of J caused a temperature-dependent molecular geometry[9] when the degree of order was assumed to be positive.

4 Double-resonance Techniques

The theory of ^1H n.m.r. with deuteron decoupling was further developed[91] from the Bloom and Shoolery analysis by inclusion of direct and quadrupole couplings. In this theory the interaction between different nuclei was interpreted in terms of quantization in an effective field in a rotating co-ordinate system. Perturbation theory was used to include quadrupole coupling. The authors discussed the system HD in full detail as a function of the irradiation intensity. Furthermore, the special case in which two protons are attached to different carbon atoms, each of which also bears a deuteron, was discussed.

$^1H-\{^2H\}$ Decoupling experiments have been performed on $[^2H_{11}]$cyclohexane at the double-quantum deuteron frequency;[92] 30W of r.f. power were used for the decoupling experiment and the coils had to be cooled by water. The authors reached the following conclusion: 'whether the results of such an approach would justify the considerable effort required remains to be seen.'[92]

[88] C. L. Khetrapal and A. C. Kunwar, *Mol. Phys.*, 1974, **28**, 441.
[89] G. J. den Otter and C. MacLean, *Chem. Phys.*, 1974, **3**, 119.
[90] J. W. Emsley and J. C. Lindon, *Chem. Phys. Letters*, 1974, **26**, 361.
[91] L. C. Snyder and S. Meiboom, *J. Chem. Phys.*, 1973, **58**, 5096.
[92] R. C. Hewitt, S. Meiboom, and L. C. Snyder, *J. Chem. Phys.*, 1973, **58**, 5089.

$^1H-\{^{13}C\}$ and $^1H-\{^{15}N\}$ double-resonance experiments on methyl cyanide[93] provided data on ^{13}C as well as ^{15}N chemical-shift anisotropies.

$^1H-\{^2H\}$ INDOR measurements were made on the molecules CD_3CH_2OH and $CH_3CD_2OH^{68}$ and on CH_2DCH_2Br and $CH_3CD_2Br^{61}$ in order to obtain the quadrupole coupling constants as well as the relative signs of these with respect to the direct couplings.

A $^2H-\{^1H\}$ decoupling experiment allowed the observation of deuteron-deuteron direct couplings in n-[1,1-$^2H_2$]propanol and [1,1-$^2H_2$] decanol.[94] Such measurements are of importance for the determination of order parameters in biological systems, since they provide two degrees of order for different axes in the same CD_2 groups. It was confirmed that the usual assumption that one order parameter for each member of the hydrocarbon chain is fully sufficient to describe the order is essentially correct.

$^{31}P-\{^{19}F\}$ Decoupling was applied to the molecule $P_3N_3F_6$[4] in order to obtain structural information.

The general NOE recorded by INDOR has been used for the construction of an energy-level diagram;[21] it is, however, doubtful whether this method will be of much help in the analysis of complex spectra.

5 Order Parameters

Finding the degrees of order of molecules is usually considered a by-product of structure determination. On the other hand, more pre-knowledge of these parameters would certainly help the analysis of the often complex spectra.

Order parameters of dissolved molecules have been discussed in a review article[1] but not in great detail.

Particularly large degrees of order were detected for trimethylacetic acid[63] in liquid crystals of the benzoic acid type. On the other hand, a surprisingly low degree of order was detected for a certain axis in 2,2,2-trifluoroethanol[94a] where the coupling of the hydroxyl protons with the trifluoromethyl group was observable; however, the one with the CH_2 group was found to be zero.

In tellurophen[8] the ratio of degrees of order of the C_2 axis with respect to the axis perpendicular to it and in the molecular plane was found to be 6.4. This result may be compared with thiophen, for which the corresponding ratio is 2.4.

The asymmetry of the order tensor is small for coumarin,[40] the main long axis in the molecular plane being rotated towards the C=O side by 11.5°. Similarly in 1-chloronaphthalene[50] the orienting effect of the C—Cl group causes a rotation of the axis by 26.6° towards the substituent. In indene[54] the corresponding rotation away from the CH_2 group is only 2.4°.

A rule for degrees of order has been given for molecules possessing C_2 symmetry.[29] It states that $2S_{11} + S_{22} = A + B$, where S_{22} is the degree of order of the C_2 axis, S_{11} is that perpendicular to it and in the molecular plane, A is a constant for the unsubstituted compound, and B is a steric correction for the substituent. The formula is supposed to work for mono- and *para*-substituted benzenes as well as the pair ethylene oxide and ethylene sulphide, for which B is small. As two degrees of

---

[93] J. D. Kennedy and W. McFarlane, *Mol. Phys.*, 1975, **29**, 593.
[94] P. Diehl and W. Niederberger, *J. Magn. Resonance*, 1974, **15**, 391.
[94a] D. Pumpernik and A. Ažman, *J. Mol. Structure*, 1974, **22**, 463.

order fully characterize such molecules, the relation is not of much interest as soon as B is large.

With the increasing number of papers dealing with molecules dissolved in lyotropic liquid crystals there has been some interest in comparing degrees of order with those in thermotropic phases.[20] In such comparisons it should be kept in mind that the degree of order has a simpler meaning in thermotropic liquid crystals than in lyotropic ones, where the details of the liquid-crystal superstructure should be known in order to understand the solute distribution with respect to the applied magnetic field. This magnetic field may be at a certain angle α with respect to the liquid-crystal optic axis and there may also be, for example, a distribution of the superstructure axes in a cone (with angle β) around the magnetic field. As a consequence, the observed direct couplings are reduced by a factor of $\frac{1}{2}(3\cos^2\alpha - 1) \times \frac{1}{2}(3\cos^2\beta - 1)$. Such factors should be considered if degrees of order are compared.

Luz and Meiboom[95, 96] used biaxial smectic liquid crystals as solvents and, in terms of Wigner matrices, derived the corresponding formal theory of molecular order.

Alignment of non-polar molecules was obtained by first inducing order of polar molecules[97] with electric fields. This order is then partially transferred to the non-polar solute in the polar solvent. Results deduced from nitromethane and nitrobenzene ordered in electric fields disagreed with the Onsager liquid model.[97a]

The effect of sample rotation on the degree of order was again studied.[98]

6 Anisotropies in Indirect Couplings

The anisotropy of the indirect coupling constant (ΔJ) continues to be a rather controversial subject. Few measurements have been made and the results are sometimes contradictory. A typical case is phosphoryl fluoride,[16, 17] for which the FPF angle was found to deviate 2° to 3° from the m.w. value. Attributing the deviation to an anisotropy of $J(P,F)$, one group of authors obtained $\Delta J(P,F) = 3$ kHz,[16] and the other found a value of 12 kHz.[17] In both cases $\Delta J(F,F)$ was assumed as zero. Obviously, as also pointed out by Bulthuis and de Lange,[17] the anisotropy is the wrong explanation for this molecule.

For obtaining information on ΔJ, it is important to measure spectra at various ratios of order parameters. This variation may be obtained by combining data from lyotropic as well as thermotropic solvents. This technique was used for 1,2-difluoroethylene[89] and m-[45] and p-difluorobenzenes.[46] In 1,2-difluoroethylene the difference between the direct couplings $D(F,F)$ as obtained from an assumed e.d. geometry and from the experiment could be attributed to an anisotropic $J(F,F)$, in agreement with theoretical predictions. On the basis of no anisotropy in $J(H,H)$ and $J(H,F)$, a small resulting anisotropy was found in $J(F,F)$ for the molecules m-difluorobenzene[45] and p-difluorobenzene.[46]

In the determination of the vibrationally averaged structure of [1,2-^{13}C]ethylene,[7] it was found a small deviation in $D(C,C)$ could in principle be explained if an anisotropy of approximately 90 Hz was postulated for $J(C,C)$.

[95] Z. Luz and S. Meiboom, *J. Chem. Phys.*, 1973, **59**, 275.
[96] S. Meiboom and Z. Luz, *Mol. Crystals Liquid Crystals*, 1973, **22**, 143.
[97] J. Biemond and C. MacLean, *Mol. Phys.*, 1974, **28**, 571.
[97a] J. Biemond, S. van der Grot, and C. MacLean, *Chem. Phys. Letters*, 1975, **32**, 390.
[98] S. D. Goren, S. B. Marks, and R. Potashnik, *Chem. Phys. Letters*, 1974, **28**, 400.

Finally, in a study of dimethylmercury[48] the observed inconsistency in CH distances as calculated from the bonded or the non-bonded carbon–proton direct coupling was attributed exclusively to the neglect of vibrational correction to the bonded coupling. Subsequently the non-bonded coupling was used as basis for the determination of $\Delta J(Hg,C)$ and $\Delta J(C,C)$. Both turned out to be *ca.* 50% of the isotropic indirect coupling.

7 Anisotropies in Chemical Shifts

Appleman and Dailey[99] have reviewed the various methods used to determine chemical-shift anisotropies from spectra of oriented molecules. From the relation:

$$\sigma(\text{nematic}) = \sigma(\text{average}) + \tfrac{2}{3}\Delta\sigma\, S_{zz}$$

it can be seen that the anisotropy, $\Delta\sigma = \sigma_\| - \sigma_\perp$, may in principle be obtained at the phase transition. It turned out, however, that medium effects introduced errors of 100% or more. Better methods seem to be the ones which vary S_{zz} and determine $\Delta\sigma$ as the slope of $\sigma(\text{nematic})$ *versus* S_{zz}. The variation may be achieved by changing the temperature or the concentration of the solute, or by varying the sample spinning rate or the angle between the optic axis and the magnetic field in smectic solvents. Of these the first is the more suspect since $\sigma(\text{nematic})$ may itself be temperature-dependent. The authors[99] demonstrated the various methods using the molecule $CFCl_3$ and showed that the phase-transition and temperature-variation methods failed whereas the sample-rotation and the concentration-variation methods gave good results that were in agreement with each other. The smectic solvents produced slightly increased values.

A large amount of information on chemical-shift anisotropies has been collected during the past two years. Measurements have been made for the following nuclei and molecules:

(a) Protons: CH_3SiH_3,[63] CH_3SiD_3,[63] $CHCl_3$,[100] C_2H_2,[101] and PH_3.[14]
(b) Fluorine: $CFCl_3$,[99] POF_3,[16] CH_3F,[10] CF_3CO_2H,[90] CF_2CH_2,[102] and PF_3.[15]
(c) Carbon: CH_3I,[9] CH_3Br,[9] CH_3Cl,[9] CH_3F,[10] CH_3OH,[11] and CH_3CN.[18a, 93]
(d) Phosphorus: $P_3N_3Cl_6$,[4] $P_3N_3F_6$,[4] POF_3,[16] PF_3,[15] and PH_3.[14]
(e) Nitrogen: CH_3CN.[93]

Shift reagents were used in order to induce large chemical shifts in spectra of oriented methyl alcohol[103] and also for a study of the amount of isotropic and anisotropic contributions to the induced shifts in pyridine.[24]

In a theoretical paper[104] the effect of preferential molecular orientation on the paramagnetic shifts was discussed in general. Explicit formulae were given for axially symmetrical molecules.

---

[99] B. R. Appleman and B. P. Dailey, *J. Magn. Resonance*, 1974, **16**, 265.
[100] P. K. Bhattacharyya and B. P. Dailey, *J. Magn. Resonance*, 1973, **12**, 36.
[101] S. Mohanty, *Mol. Phys.*, 1973, **25**, 1173.
[102] G. J. den Otter and C. MacLean, *Chem. Phys. Letters*, 1973, **20**, 306.
[103] C. L. Khetrapal and A. C. Kunwar, *J. Magn. Resonance*, 1974, **15**, 389.
[104] G. Giacometti, P. L. Nordio, G. Rigatti, and U. Segre, *J.C.S. Faraday II*, 1973, **69**, 1815.

8 Nuclear Quadrupole Coupling Constants

Quadrupole coupling constants for deuterons have been determined for a number of molecules.

CH_3SiD_3,[65,66] CD_3SiH_3,[65] CD_3GeH_3,[65] CH_3GeD_3,[65] CD_3COO^-,[12] CH_3CD_2Br,[61] CH_3CD_2OH,[68] CD_3COCH_3,[105] CF_3CD_2OD,[94a] CD_3CO_2H,[106] $CH_3ND_3^+$,[23] $(CD_3)_2CO$,[107] $(CD_3)_2SO$,[107] and PD_3.[108]

Of these, the quadrupole coupling of the deuteron in a C—D bond is of particular interest since this constant is used in a rapidly growing number of biological studies in which the molecular degree of order in membranes is determined. The reported values are between 170[12] and 233 kHz,[94a] depending on the compound. The asymmetry parameter η for the electric field gradient was usually assumed to be zero. Its variation between $+0.03$ and -0.03 changed the coupling in CH_3CD_2Br[61] between 170 and 181 kHz. For the same molecule it was demonstrated[61] that the neglect of vibrational correction increased the coupling from 176 to 199 kHz.

Relative signs of nuclear quadrupole coupling constants with respect to direct coupling constants were derived by use of $^1H-\{^2H\}$ double resonance.[105,61] Tables of analytical expressions for line positions and intensities were presented for a group of three deuterons with D_3 symmetry.[105] Unfortunately, the tables contain several errors.

The author of this chapter is grateful to Dr A. S. Tracey for reading the manuscript.

[105] J. W. Emsley, J. C. Lindon, and J. Tabony, *Mol. Phys.*, 1973, **26**, 1499.
[106] D. Pumpernik and A. Ažman, *Z. Naturforsch.*, 1973, **28a**, 1746.
[107] M. Zaucer and A. Ažman, *J. Magn. Resonance*, 1973, **11**, 105.
[108] N. Zumbulyadis and B. P. Dailey, *J. Chem. Phys.*, 1974, **60**, 4223.

Author Index

Aalto, M. I., 129
Abdel-Alim, A. H., 206
Abe, H., 260
Abragam, A., 131
Abraham, R. J., 22, 44, 63, 77, 81, 156, 176, 184, 240, 295, 311
Abramson, E., 50
Abronin, I. A., 24
Achlama, A. M., 186, 239
Ackerman, J. L., 50
Ackermann, H., 251, 252
Adams, E., 248
Adams, M. A., 107, 306
Adams, R. D., 181
Adcock, W., 20, 28, 44, 54, 68, 78, 207
Ader, R., 82, 318
Affsprung, H. E., 301
Agova, M., 170
Ahamad, S. R., 210
Ahlbrecht, H., 67
Ahonen, A. J., 248
Ailion, D. C., 241
Aime, S., 70, 181
Aisen, P., 53, 231
Aizawa, S., 299
Ajisaka, K., 308
Akahori, Y., 154
Akasaka, K., 185, 224
Akhmetova, N.E., 157
Akiba, K., 55
Akitt, J. W., 55, 65, 190
Aksnes, D. W., 97, 154, 158, 161
Alazard, R., 229
Al'ber, S. I., 166
Albers, J., 236
Albrand, J. P., 71, 75, 157
Albrecht, H., 66
Albright, T. A., 69
Albriktsen, P., 155, 157, 158, 161
Alderweireldt, F. C., 295
Alekseev, N. V., 171
Alexander, S., 163
Alexandre, M., 53, 74, 104, 108
Alizon, J., 258
Allen, P. S., 145, 246, 247
Allerhand, A., 101, 193, 231
Alloul, H., 256
Alms, G. R., 113
Al'pert, M. L., 25
Al-Rawi, J. M. A., 48, 150, 189

Alscher, A., 40, 96
Alt, H., 180
Alter, G. M., 232
Ames, D. P., 239
Aminova, R. M., 16, 60
Amos, A. T., 37, 293
Amosova, S. V., 25, 27
Anand, N., 154, 157
Anastassiou, A. G., 179
Ancker-Johnson, B., 280
Anderson, C. E., 272
Anderson, C. R., 155
Anderson, J. M., 98, 159
Anderson, P. W., 248
Anderson, R. A., 224
Anderson, S. J., 313
Anderson, W. A., 188
Anderson, W. G., 166, 177
Ando, I., 16, 206
Andrasko, J., 233, 274, 288
Andrew, E. R., 242, 244, 255
Andruzzi, F., 214
Ane, W. P., 197
Anet, F. A. L., 86, 114, 175, 176, 177
Anteunis, M., 80, 88, 156, 158, 160, 215
Aoki, K., 98
Appleman, B. R., 49, 51, 183, 323
Apsimon, J. W., 35, 172
Araki, S., 73
Aranow, R. H., 115
Arata, Y., 106, 155
Arburov, B. A., 156
Ardelean, I., 257
Argyres, P. N., 236
Armitage, I. M., 115, 311, 316
Armitage, M., 136
Armstrong, J. A., 8
Armstrong, R. L., 124, 126, 144
Arnheim, N., 232, 276
Arnold, D. P., 18
Arnold, W., 18
Arnold, Z., 54
Arons, R. R., 256
Arte, E., 316
Arvidson, G., 284, 285, 288
Asai, T., 171
Asakura, T., 16, 206
Asayama, K., 259
Ashley-Smith, J., 180
Asplund, R. O., 295
Assink, R. A., 122, 135, 211

Assmann, G., 231
Atavin, A. S., 155
Atkins, P. W., 164
Aubry, A., 62
Aufderheide, B. E., 27, 35
Aviles, F. J., 217
Aviram, I., 225, 276
Axelson, D. E., 53
Axenrod, T., 28
Ayanbadejo, F. A. M., 161
Aycard, J. P., 157
Äyräs, P., 154, 160
Aziz, A., 272
Ažman, A., 110, 211, 319, 321, 324
Azoulay, M., 171
Azzaro, M., 26

Baas, J. M. A., 168
Babbitt, G. E., 212
Babiak, K. A., 52
Babu, S. V., 127, 135
Babu, U. M., 230
Bacchella, G. L., 131
Baghdadi, S. M. A., 278
Bailey, D., 316, 317, 318
Bailey, W. F., 42
Baitz-Gaes, E., 158
Bakari, T. I., 226
Bakay, L., 282
Baker, R. R., 266
Bakhmutov, V. I., 76, 79
Baklan, V. F., 27
Bakum, S. I., 64
Balakrishnan, M., 296
Baldo, J. H., 230
Balkan, F., 155
Ballardin, A., 219
Bamford, C. H., 210
Banck, J., 57, 73
Bangerter, B. W., 138, 202
Banks, E., 254
Baram, A., 163
Barbe, R., 248, 269
Barber, B. H., 228
Barbieri, G., 76
Barboiu, V., 56, 98
Barden, J. A., 287
Barfield, M., 61, 89
Barham, D. C., 145
Barieux, J.-J., 75
Barili, P. L., 319
Barker, C. A., 290
Barker, R. W., 287
Barnard, M., 31, 90
Barnes, R. G., 141, 250

Barnett, R. E., 234
Barojas, J., 108
Baron, D., 302
Barreau, J., 156
Barriol, J., 318
Bartenev, G. M., 209
Barth, R. C., 55
Bartholdi, E., 193, 196, 197
Bartle, K., 155
Baryshnikova, T. K., 56
Basch, H., 7
Basler, W., 273
Basselier, J. J., 25, 77, 312
Batiz-Hernandez, H., 45
Batts, B. D., 219
Bauer, D. R., 113
Bauta, E. B., 158
Bazzone, T. J., 155
Beak, P., 40
Beall, H., 65
Beare, S. D., 305
Beattie, I. R., 318
Becker, E. D., 28, 107, 114, 138, 199
Becker, K. D., 251
Becker, R. S., 52
Beckmann, A., 249
Beckmann, P. A., 127
Beech, G., 308
Begtrup, M., 67, 155
Beistel, D. W., 157
Belfort, G., 278
Bell, H. C., 85
Bell, H. M., 306
Bel'skii, V. E., 55
Belton, P. S., 278
Bembech, P., 157
Benassi, R., 84
Benedek, G. B., 255
Benessi, R., 157
Bender, H. J., 100
Benedek, G. B., 8
Benko, B., 225
Bennett, L. H., 256
Bennett, M. A., 159
Bensoussan, M., 256
Bentrude, W. G., 75, 157, 159
Benz, F. W., 194, 218
Berendsen, H. J. C., 285
Berezhnoi, V. G., 106
Berg, A., 19, 70
Berger, S., 52, 69, 114
Bergesen, K., 156, 158, 160
Berlin, K. D., 54, 282
Berman, H. A., 121, 302
Bernaert, E., 88
Bernheim, R. A., 45
Bernier, M., 248
Bernier, P., 118, 256
Bernstein, H. J., 49
Bersohn, R., 250
Berthet, G., 258
Bertran, J. R., 295
Bertrand, R., 42
Bertrand, R. D., 77
Bertschat, H., 256
Beveridge, D. L., 16, 33
Bezdadea, V. E., 206
Bhacca, N. S., 48, 70
Bhat, G., 166, 175
Bhattacharyya, P. K., 50, 51, 72, 223, 315, 323
Biemond, J., 189, 293, 322
Bigotto, A., 62

Bildsoe, H., 157, 158, 159
Bilofsky, H. S., 175
Binczek, E., 286
Binkley, R. W., 52
Binkley, W. W., 52
Binsch, G., 165
Birdsall, N. J. M., 99, 286
Bisanz, T., 154
Bishop, S. G., 238
Bjoergo, J., 24
Bjorkstam, J. L., 245
Bjoroy, M., 156, 158, 160
Blaauw, L. P., 212
Blackborow, J. R., 185
Blackwell, L. F., 18
Blais, P., 207
Blanc, J. P., 258
Blanshard, J. M. V., 278
Bleaney, B., 308
Bleich, H. E., 151, 194
Blicharska, R., 276
Blinc, R., 224, 242, 245, 250, 261, 276
Blinova, V. A., 25
Blizzard, A. C., 59
Bloch, F., 1
Block, R. E., 280
Bloembergen, N., 125, 275
Bloom, M., 124, 125, 126, 127, 286
Blout, E. R., 306
Bloxsidge, J. P., 48, 189
Bluenich, B., 200
Blumberg, W. E., 189
Blume, H., 73
Blumenstein, M., 223
Boccuzzi, F. C., 156
Bochkov, A. F., 81
Boctor, Y., 225, 282
Boddenberg, B., 266
Bode, J., 223
Boden, N., 237, 238, 244, 261
Bodner, G. M., 32, 55
Bodot, H., 81, 157
Bödecker, H.-O., 306
Boekelheide, V., 37, 38
Boeklers, K. D., 249
Bogdanov, V. S., 56, 67
Bohlmann, F., 27, 91
Boicelli, A. C., 173, 213. 319
Bomse, D. S., 167
Boni, R., 215
Bonori, M., 243
Booth, H., 48, 174
Booth, S., 145
Bopp, T. T., 102
Bordjabel, H., 158
Borgen, G., 156
Borsa, F., 260
Borudeaux-Pontier, M., 176
Bose, A. K., 305
Bose, M., 256
Bosse, J., 253
Bothner-By, A. A., 303, 319
Bottin-Strzalko, T., 89
Boublik, M., 217
Bovee, W., 282
Bovée, W. M. M. J., 111
Bovey, F. A., 39, 52, 205, 217, 218
Bowler, D. J., 74, 310
Boxer, S. G., 53
Boyd, D. R., 24

Bradamante, S., 74, 75, 157, 158, 317
Bradbury, A. F., 219
Bradbury, E. M., 217
Bradbury, J. H., 215, 216, 218
Bradley, C. H., 56
Bradley, R. B., 28, 118, 156
Braillon, B., 159
Bramley, R., 84
Brassfield, T. S., 273
Brassington, J. G., 305
Braterman, P. S., 72
Brauman, J. I., 113
Braun, D., 206
Braun, R. W., 73, 163
Bray, P. J., 57, 238, 252, 253
Breitmaier, E., 26, 222
Bremser, W., 25, 40, 96
Brewer, D. F., 268, 269, 276
Brewer, H. B., jun., 231
Brewer, W. D., 258
Briant, C., 176
Bridges, F., 124
Briggs, J. M., 158, 189, 317
Briggs, W. L., 55
Bright, A. A. S., 301
Briguet, A., 67, 105, 138, 262
Brill, T. B., 251
Brinkman, W. F., 248
Brinkmann, D., 258
Brodbeck, C. M., 254
Broecker, H. C., 212
Brooker, H. R., 249
Brookhart, M., 161
Brooks, A. A., 102
Brossel, J., 269
Brouant, P., 65
Brown, D., 295
Brown, I. M., 189
Brown, L. R., 215, 218
Brown, R. A., 219
Brown, R. D., 231
Brown, T. H., 8
Brownlee, R. T. C., 29
Brownstein, S. K., 172
Brulet, P., 286
Brune, A., 67
Brunelle, J. A., 175
Brunner, T. R., 136, 202
Bruno, G. V., 163
Brure, H. A., 155
Bryant, R. G., 186, 233, 275, 283
Bryce-Smith, D., 295
Bubnov, Yu. N., 56, 67
Bucci, P., 319
Buchachenko, A. L., 187
Buchanan, G. W., 24, 283
Buchner, W., 54, 55, 67, 70
Buckingham, A. D., 32, 33, 49, 64, 128, 292, 297, 316, 317, 318
Buckley, P. D., 18
Buckmaster, H. A., 137, 204
Buder, W., 55
Budnik, R. Z., 180
Buell, G. R., 30
Bürger, H., 66
Bugg, C. E., 172
Bukowska, M., 154
Bulai, A. Kh., 313

Author Index

Bull, T. E., 109, 136, 137, 262, 274
Bulthuis, J., 315, 318
Bundgaard, T., 25, 96, 201
Bunker, P. R., 74
Bunting, W. M., 178
Burgen, A. S. V., 219
Burghoff, U., 329
Burke, M. J., 283
Burke, T. E., 111
Burkey, D. L., 177
Burlingame, A. L., 149
Burnell, E. E., 94, 127, 286, 314, 316
Burnett, L. J., 66, 105, 142
Burnham, A. K., 40
Burrows, B. L., 293
Buruianǎ, E., 206
Busby, J. F., 63
Busby, S. J. W., 130, 279
Busch, M. A., 73
Bushaw, K. H. J., 258
Bushby, R. J., 173
Bushweller, C. H., 65, 84, 166, 175, 177
Buslaev, Yu. A., 53, 57, 72, 73, 79, 119
Butler, M. A., 258
Buttet, J., 255
Buttlaire, D. H., 227, 228, 276
Bye, E., 310
Byers-Brown, W., 2
Bystrov, V. F., 61
Byvik, C. E., 254
Bzhezovskii, V. M., 25, 27

Cabaret, F., 6
Çaccamese, S., 170
Cačković, H., 210
Caddy, D. E., 48
Cade, R. F., 6
Cahen, Y. M., 57, 303
Cairns, H., 313
Calabrese, L., 227, 276
Callaghan, P. T., 262
Callens, R., 215
Camier, M., 229
Campbell, I. D., 137, 151, 194, 218, 219, 275
Campbell, J. H., 117
Canet, D., 67, 98, 138, 203, 316, 317, 318
Canty, A. J., 317
Canuel, L., 48, 175
Canuel, L. L., 217, 218
Capen, G. L., 183
Cappelli, E., 95, 317
Caraculacu, A., 206
Cardin, A. D., 57, 67, 119
Carey, F. A., 35
Cargioli, J. D., 97, 109, 150, 203
Carlson, E. H., 176
Carlsson, D. J., 207
Carman, C. J., 213
Carnevale, A., 249
Carroll, F. I., 106
Carter, J. C., 27, 95, 155
Carter, R. E., 167
Carty, A. J., 55, 73
Carver, J. P., 228, 275
Cary, L. W., 55, 73
Cary, P. D., 217

Casanova, J., 44, 56, 70
Casaux, L., 156
Case, G. D., 228
Caulton, K. G., 181
Cavagna, F., 90, 159
Cavalli, L., 77, 159, 160, 295
Cavanaugh, J. R., 273
Cavé, A., 230
Cavelius, E., 255
Cavell, R. G., 73, 79
Cawley, J. J., 307
Cazaux, L., 81
Ceccarelli, G., 207, 214
Chachaty, C., 210
Chadaeva, N. A., 156
Chadwick, D. J., 173, 244, 308, 311
Chagin, U. I., 57, 73
Chaiken, I. M., 221
Chalmers, A. A., 88, 96, 98, 201, 308
Chamberlain, N. F., 213
Chan, L. H., 207
Chan, S. I., 111, 217, 229, 285, 286, 289
Chan, S. O., 101, 104, 137
Chance, B., 282
Chandler, D., 108
Chang, C.-A., 229, 289
Chang, C.-H., 302
Chang, D. B., 280
Chang, D. C., 278, 281, 282
Chang, J. J., 240, 266
Chang, S. H., 221
Chanon, M., 167
Chapat, J. P., 310
Chapellier, M., 236, 260
Chapman, B. E., 227
Chapman, D., 289
Chapman, G. E., 217
Chapman, R., 125
Charache, S., 216
Charlesby, A., 210
Chastellain, F., 186
Chaughule, R. S., 282
Chebotareva, E. G., 155
Cheer, C. J., 311
Chen, D. M., 289, 315
Chen, F. M., 123
Chen, H. E., 157
Chen, J. S., 313
Chen, L.-S., 156
Chen, R. M., 126
Cheney, B. V., 41
Cheng, A. K., 177
Chenon, M. T., 26
Chesnut, R. W., 282
Chewanhin, W., 156
Chezeau, J. M., 244
Chiabattoni, J., 179
Chidester, C. G., 176
Chihara, H., 247
Childers, R. F., 193, 231
Childs, R. F., 179
Chisholm, M. H., 32, 70
Chittenden, R. A., 54, 85
Chiu, J., 205
Chi-Yu Lee, 170
Chodosh, D. F., 181
Choi, J. K., 244
Chokki, Y., 207, 208
Chow, S. T., 78
Christensen, K. A., 18, 42, 202

Chuang, A. H., 282
Chudek, J. A., 301
Chuikova, T. V., 160
Chum, K., 62, 156
Chung, H. L., 184
Chung, M. T., 62
Chvalovsky, V., 56
Cistaro, C., 74, 75, 157, 158
Civan, M. M., 279
Clague, A. D. H., 212
Clare, P., 158
Clark, H. C., 32, 70
Clark, J. H., 301
Clark, P. W., 76
Clark, R. J., 70, 73
Clark, W. G., 142, 144
Clarke, D. E., 24
Clem, T. R., 114, 118
Clement, D. A., 54, 73
Clement, R. A., 22
Clement, S., 254
Cleveland, G., 282
Closs, G. L., 53
Clough, S., 247
Clouse, A. O., 190
Coates, R. M., 52
Coburn, W. C., 65
Cochoy, R. E., 93
Cochran, D. W., 42
Codrington, R., 1, 148
Cogne, A., 85, 158
Cohn, M., 227, 228, 276
Cohen, E. A., 71
Cohen, J. S., 118, 216, 221, 229
Cohen, M. S., 189
Cohen, P., 229
Cohen, S. M., 5, 8
Cohen-Addad, J. P., 209
Cohn-Sfectu, S., 204
Coldea, M., 257, 259
Cole, K. C., 319
Cole, P. E., 220
Cole, T., 50
Coleman, J. E., 224
Collan, H. K., 129
Colson, P., 223
Combescot, R., 248
Combrisson, S., 173
Comisarow, M. B., 189
Condon, F. E., 18
Connick, R. E., 185, 275
Constein, V. G., 291, 305
Conti, F., 213
Conway, E., 52
Cook, D. B., 6, 48
Cook, G. L., 267
Cook, M. J., 158
Cook, T. J., 130
Cooke, D. F., 251
Cooney, J. D., 172
Cooper, C., 196
Cooper, D. B., 85
Cooper, D. G., 32
Cooper, G. H., 54, 85
Cooper, M. A., 63, 77, 295
Cooper, R. L., 280
Cope, F. W., 277, 282
Coppell, S. M., 311
Cordes, E. H., 231
Cordle, J. E., 5, 64
Cornell, B. A., 138, 287
Corno, C., 213
Cornus, M., 156

Corradi, E., 36
Cotton, F. A., 52, 103, 180, 181, 187, 225, 227, 282
Council, J. R., 316
Courtieu, J., 317, 319
Courtney, J. A., 124, 144, 262
Covacevich, M. T., 223
Coville, N. J., 180
Covington, A. D., 303
Covington, A. K., 303
Cowley, A. H., 71, 73, 160, 163
Cox, R. H., 303, 313
Coxon, B., 56, 65
Cozzone, P. J., 232
Crabb, T. A., 75
Craig, A. C., 312
Craig, L. C., 68
Craig, R. E. R., 312
Cramer, F., 220
Cramer, R. E., 300, 309
Crane-Robinson, C., 217
Crapo, L. M., 10
Creel, R. B., 141, 249
Cremer, D., 40, 63, 96
Cremer, S. E., 69, 114
Crespi, H. L., 216
Creswell, D. J., 268, 269
Crisan, M., 259
Croft, T. S., 75
Crompton, M. W., 215
Cronin, D. L., 53
Cross, R. J., 72
Crothers, D. M., 220
Csopak, H., 232, 234, 277
Cullen, W. R., 85, 155, 156
Cullis, P. R., 287
Culty, J.-L., 138, 262
Culvenor, C. C. J., 81, 156
Cunliffe, A. V., 154
Curci, R., 173, 179
Curry, E. R., 161
Curtis, J. L. S., 76
Cutnell, J. D., 102
Cyr, N., 67, 155
Czeisler, J. L., 282
Czieslik, W., 41

Dadok, J., 138, 191, 198
Dähne, S., 62
Dahlberg, E., 168
Dahlqvist, F. W., 191
Dahlstrom, P. L., 135, 300
Dailey, B. P., 49, 50, 51, 66, 72, 79, 117, 314, 315, 323, 324
Dale, B. J., 220
Dalling, D. K., 41, 42, 52
Damadian, R., 265, 277, 282
Damle, S. D., 258
Danby, S. E., 217
Danchura, W., 62
Daneshrad, A., 53
Danneels, D., 160
d'Annibale, A., 315, 317, 319
Danyluk, S. S., 296
Darke, A., 225, 274
Darville, J., 256
Das, A. K., 254
Dauchna, W., 157
Daunt, J. G., 268
Davidson, D. W., 246
Davidson, J. R., 40

Davies, A. M., 6, 48
Davies, D. W., 36
Davies, J. S., 298
Davies, T. K., 150
Davis, D. G., 229
Davis, M., 26
Dawson, D. A., 30
Day, R. O., 182
Day, V. W., 182
Deady, L. W., 18, 26, 95
Dean, A. M., 61
Deber, C. M., 52, 306
De Bie, M. J. A., 53, 66, 75
De Boer, E., 187
Decker, J. A., 200
Decora, A. W., 267
Deeming, A. J., 181
de Fontaine, D. L., 137, 202
Dega-Szafran, Z., 299, 301
Degelaen, J., 316, 317
de Graaf, A. M., 4
de Haan, J. W., 212, 294
Deininger, D., 16, 271
De Jonge, W. J. M., 259, 260
de Kievit, W., 315
De Kowalewski, D. G., 317
de Lange, C. A., 315, 318
Delayre, J., 150
Delfini, M., 213
Del Gaudio, J., 55, 73
Delmau, J., 67, 105, 138
Delpuech, J. J., 57, 183
Dembech, P., 84
Demco, D., 282
Demco, D. E., 135
Demerseman, P., 77, 312
Dempster, A. B., 167
Dence, J. B., 41
Deneke, M., 212
Denis, J.-M., 22
Dennis, N., 67
Den Otter, G. J., 317, 320, 323
Deranleau, D. A., 301
Derbyshire, W., 226, 273, 278
Dereppe, J. M., 316, 317
Deslauriers, R., 222
Des Marteau, D. D., 90, 155
Desreaux, J. F., 307
Desseyn, H. O., 295
Deutch, J. M., 177, 236
Deverell, C., 49
de Villardi, G. C., 311
Deville, G., 248
Devillers, J., 155, 156
Devreux, F., 258
de Waele, R., 88
Dewar, M. J. S., 16, 18, 54
DeW. Horrocks, W., jun, 306
De Wit, J., 42
De Zwaan, J., 108
Dhingra, M. M., 306
Di Blasi, R., 86, 215, 222
Diegnan, G. A., 44
Diehl, P., 314, 315, 316, 317, 319, 321
Diemart, K., 157
Diez, E., 75, 160
Di Furia, F., 173
Dijkema, C., 285
Dillon, K. B., 55
Dimitrov, V. S., 170
Dimroth, K., 25, 96

Dingle, T. W., 157
Di Nola, A., 95, 317
Dirlikov, S. K., 208
Di Stefano, S., 295
Ditchfield, R., 8, 12, 13
Dixon, M., 9
Dixon, K. R., 88, 157
Djerassi, C., 41
Dmitriew, J. I., 276
Dobosh, P. A., 16
Dobson, C. M., 151, 194, 219, 308
Doddrell, D., 18, 20, 44, 52, 68, 78, 86
Dolphin, D., 230
Dolzine, T. W., 157
Dong, R. Y., 124
Donvito, T. N., 52, 295
Dopper, J. H., 155
Dorn, H. C., 52, 63, 67, 149, 193
Doroshev, V. D., 261
Doskocilova, D., 271, 272
Doty, L. F., 73
Doucet, J. P., 25
Douek, Z., 180
Douglass, D. C., 245
Douglass, I. B., 295
Douris, J., 18, 93
Doyle, M. J., 182
Drago, R. S., 316
Drake, J. E., 56
Drakenberg, T., 88, 164, 174, 230, 276, 284
Drapeau, G. R., 219
Dreeskamp, H., 70, 289, 317
Dresden, M. H., 225, 274
Druck, S. J., 107
Drysdale, J. J., 22
Dubault, A., 269
Dubbers, D., 251, 252
Dubois, R., 309
Duch, M. W., 42
Duchamp, D. J., 176
Dudzik, Z., 209
Duff, I. D., 226, 273, 278
Duflot, C., 215
Dufourcq, J., 284
Dugas, H., 23
Dugdale, D. E., 237
Dumler, V. A., 167
Duncan, R. M., 190
Dunmur, R. E., 72, 159
Dunn, M. B., 316
Duplan, J. C., 67, 105, 138, 262
Durcansky, G., 130
Duret, C., 146
Duret, J., 319
Durham, D. L., 279, 282
Durham, N. N., 282
Dutasta, J. P., 75, 157, 158
Du Vernet, R. B., 37, 38, 191
Dwek, R. A., 215, 228, 275, 276
Dybowski, C. R., 211, 245
Dye, J. L., 303
Dziergwa, U., 140

Eaborn, C., 53
Eastmond, G. C., 210
Easton, N. R., jun., 175
Ebdon, J. R., 208
Ebisawa, H., 248

Author Index

Ebraheem, K. A. K., 32
Economou, J. S., 282
Edmonds, D. T., 117
Edwards, J. M., 53
Edzes, H. T., 136
Egan, R. S., 158
Ege, G., 37
Eggert, H., 41
Eggleston, J. C., 282
Egorov, Y. P., 43
Eisenberg, R., 182
Eisenstadt, M., 121, 177, 178, 276
Elbaum, C., 259
Elbe, H.-L., 306
El Borai, M., 154
Eley, D. D., 275
Elgad, U., 73
Elgert, K.-F., 212
Elguero, J., 35, 81, 154, 170
El-Hanany, U., 145
Eliel, E. L., 42
Elke, D., 249
Elleman, D. D., 133, 139
Ellenburger, M., 160
Ellett, J. D., 200
Ellis, G. E., 32
Ellis, P. D., 1, 12, 57, 60, 67, 70, 119, 148
Elmqvist, C., 210
Els, H., 18
El Seoud, O. A., 291, 305
Elvidge, J. A., 48, 65, 150, 189
Emanuel, R. V., 8
Emelin, V. P., 111
Emid, S., 236, 247, 262
Emori, S., 254
Emsley, J. W., 64, 70, 87, 190, 292, 295, 301, 314, 315, 317, 318, 320, 324
Endom, L., 117
Enga, E., 286
Engel, R., 187
Engleberg, M., 142
Englehardt, G., 31, 56, 175
Engler, E. M., 295
Englert, G., 18, 49
Englesberg, M., 261, 262
Englich, J., 258
Englman, R., 8
Englosh, A. D., 181
Epperlein, B. W., 7
Epstein, S. T., 2
Ernst, L., 42, 78, 94, 97, 152, 167
Ernst, R. R., 30, 134, 137, 188, 191, 193, 196, 197, 198, 200
Ershov, B. A., 27
Esteban, A. L., 75, 160
Etienne, G., 230
Ettinger, M. J., 234
Eujen, R., 66
Evans, D. F., 307, 311
Evans, D. H., 131
Evans, E. A., 48, 150
Evans, F. E., 89
Evans, J., 31, 181
Evans, S. A., 44
Ewers, U., 77

Fabry, M. E., 276
Fairhurst, M. T., 185

Faix, O., 220
Fajt, B., 211
Falconer, W. E., 253
Falius, H., 155
Faller, J. W., 107, 306
Fallick, C. J., 61
Fan, S., 226
Farnell, L. F., 189
Farnier, M., 88, 173
Farquhar, B. Y., 302
Farrar, T. C., 107
Fasig, K. M., 73
Faure, J. P., 209
Fay, R. C., 180
Featherman, S. I., 52, 55, 70
Fedarko, M.-C., 97
Fedders, P. A., 256
Fedin, E. I., 25, 81, 226, 243, 267, 301
Feeney, J., 77, 83, 155, 194, 195, 201, 219, 292
Fehér, F., 67
Feigenson, G. W., 285, 286
Fel'dman, E. B., 166
Feldman, I., 187
Feldmann, R. J., 232
Fellmeth, B., 287
Fendler, E. J., 290, 291, 304
Fendler, J. H., 290, 291, 305
Ferber, G. J., 173
Ferguson, R. C., 162, 207
Fermandjian, S., 68, 70
Ferretti, J. A., 138, 156, 199
Fesenko, D. A., 155
Fiat, D., 185, 186
Fild, M., 70, 73
Filippova, T. M., 81, 156, 159
Filleux-Blanchard, M. L., 174
Finch, E. D., 279
Finer, E. G. 225, 274, 285, 286
Finkenstadt, W. R., 219
Firkins, E., 118
Firl, J. R., 19
Fischer, E. O., 70
Fischer, G., 299
Fischer, J. J., 277
Fischer, P., 67
Fischer, R. D., 309
Fish, R. H., 182
Fleischer, D., 213
Floyd, R. A., 282
Flygare, W. H., 40, 110
Fochi, R., 156
Fölsch, G., 232
Folland, R., 244
Fomin, V. N., 210
Forchioni, A., 210, 215
Foreman, M. I., 300
Forrest, J. W., 244
Forrest, T. P., 80, 311
Forsén, S., 189, 228, 233, 276, 288
Forsyth, D. A., 21
Foster, H. R., 258
Foster, R., 293, 301
Fournari, P., 154
Fox, L. E., 307
Fox, W. B., 155
Fradin, F. Y., 258
Fraenkel, G., 165, 183, 184, 185

Frahm, J., 101
Franconi, C., 227, 276
Franks, F., 111
Freed, J. H., 163, 236
Freedman, C. D., 171
Freedman, M. H., 20
Freeman, R., 103, 104, 137, 191, 192, 196, 304
Freeman, W. J., 69
Freitag, W., 129
Frenz, B. A., 180
Fretz, R. E., 52
Freude, D., 270
Freund, R., 67
Fridovich, I., 227
Friedman, H. L., 121
Frijters, G. A. M., 256
Fringuelli, F., 67, 315
Fripiat, J. J. 266
Frisch, K. H., 63
Fritz, G., 31, 66, 67
Fritz, H., 75, 91, 152, 157
Fromageot, P., 70
Frommer, M. A., 233
Fronza, G., 74, 75, 157, 158, 164, 317
Frosini, V., 207
Fruchier, A., 35, 81, 154, 170
Fry, D. J. I., 204
Fu Chen, 170
Fueno, T., 24, 25
Fuerst, A., 18
Fung, B. M., 190, 275, 279, 281, 282
Fujieda, K., 23
Fujishige, S., 206
Fujiwara, F., 284, 289, 316
Fujiwara, F. Y., 11, 53, 66, 297
Fujiwara, H., 258
Fujiwara, S., 57, 106, 155
Fukui, H., 302
Fukui, K., 155
Fukunaga, J., 30
Furie, B., 232
Furukawa, J., 213
Furuyama, H., 155
Fyfe, C. A., 244

Gabbay, E. J., 233, 277
Gabriel, H., 253
Gabuda, S. P., 54
Gadian, D. G., 130, 233, 279
Gagarin, S. G., 24
Gagnaire, D., 71
Galasso, V., 62
Gale, B. N., 134
Galts, K., 155
Gamba, A., 157
Ganellin, C. R., 158
Ganguli, S., 256
Garbay-Jaureguiberry, C., 158
Garber, A. K., 55
Garg, S. K., 246, 256
Garnett, M. W., 122
Garreau, M., 32, 67
Garrou, P. E., 76
Garroway, A. N., 133, 144, 214, 239, 262, 283
Gasic, M. J., 26
Gasparro, F. P., 300

Gasser, R. P. H., 304
Gatto, K., 93
Gazzard, V. J., 54, 79
Geanangel, R. A., 25
Geckle, M., 20, 78
Geisel, M., 56, 225, 282
Gelan, J., 80
Gelbaum, L., 187
Gent, M. P. N., 287
George, R., 67
Georgieva, Z., 208
Gerhart, G. R., 251
Gerig, J. T., 93, 139, 155, 159
Gero, S. D., 52
Gerritsen, J., 317
Gerritsma, C. J., 136
Geschke, D., 16, 271
Gey, E., 62, 63
Ghatikar, M. N., 258
Giacometti, G., 323
Giannini, D. D., 70
Giao, D. H., 155
Gibb, M., 261
Gibbons, W. A., 68, 219
Gibby, M. G., 49, 140, 285
Gibson, J. A., 161
Gielen, M., 75
Gierer, A., 119
Giessner-Prettre, C., 62
Gilbert, A., 295
Gilje, J. W., 73, 163
Gilles, J.-M., 179
Gillespie, R. J., 54, 73
Gilon, C., 215
Gilson, D. F. R., 238, 319
Gindin, V. A., 44
Ginzburg, A. G., 77
Giral, L., 95
Girard, H., 207
Girard, J. P., 310
Girardet, J. L., 286
Gladkii, Yu. G., 56
Glättli, H., 131
Glasel, J. A., 151, 189, 194
Glickson, J. D., 172, 191
Glonek, T., 79
Glotsmann, C., 40
Glowinkiwski, S., 209
Godici, D. E., 286
Godwin, A. D., 307
Görlitz, M., 75, 80
Goethals, E. J., 207
Goggin, P. L., 73
Gogte, V. N., 17
Golden, R., 18, 54
Gol'dfarb, E. I., 55
Goldman, M., 236, 260, 261
Goldstein, J. H., 53, 70, 315
Goldwhite, H., 55, 295
Goll, G. C., 283
Gonord, P., 146, 319
Goodfellow, R. J., 73, 79, 153
Goodisman, J., 110
Goodman, M., 170, 215
Goodwin, B. W., 195
Gordon, M. C., 55
Gordon, M. D., 44
Gordon, R., 132
Gore, J., 75
Goren, S., 241
Goren, S. D., 250, 322

Gorenstein, D. G., 54, 229, 230
Gorin, P. A. J., 223
Goscinski, O., 293
Gossard, A. C., 256
Goto, Y., 269
Gough, S. R., 246
Gould, D. E., 155
Gould, R. O., 57
Gounelle, Y., 317, 319
Gourdji, M., 253
Gourévitch, M., 230
Govil, G., 221
Grabowska, A., 56
Grach, E., 301
Gradstajn, S., 226, 276
Grätzel, M., 305
Graham, J. D., 244
Gramstad, T., 156
Grand, A., 158
Granger, P., 158
Granger, R., 310
Grannell, P. K., 133, 144, 262, 283
Granot, J., 185, 186
Grant, D. M., 13, 18, 21, 41, 42, 52, 100, 114, 202
Grant, M. W., 155
Grasdalen, H., 226
Grasselli, J. G., 308
Graveron-Demilly, D., 67
Gray, A. I., 310
Gray, G. A., 53, 69, 114, 231
Green, C. H., 155
Green, R. G., 54, 159
Greene, R. L., 52
Greenberg, A., 166
Greenberg, E. S., 109
Greenblatt, M., 254
Gregson, M. J., 59
Greifenstein, L. G., 45, 47
Griffin, J. H., 229, 232
Griffin, R. G., 50, 240
Griffith, R. C., 45, 174, 298
Grigor'ev, V. P., 209
Grim, S. O., 55, 73
Grimaldi, J. J., 97, 130, 190, 226, 230
Grimme, W., 159
Grimmer, D. P., 268
Grinberg, V. Ya., 226
Grindley, T. B., 173
Grisham, C. M., 234
Grobet, P., 254
Groeneveld, W. L., 56
Grollman, A. P., 219, 276
Gronowitz, S., 67
Gronski, W., 208
Grosescu, R., 239
Grosse, J., 71
Grover, S. H., 41
Grüner, M., 114
Grupp, M., 252
Grutzner, J. B., 41
Gruznov, A. G., 313
Gryff-Keller, A., 165
Gudlin, D., 57
Günther, H., 40, 52, 63, 75, 77, 80, 96, 157, 179, 317
Guggenheim, H. J., 258
Guibe, L., 253
Guidi, M., 317
Guilard, R., 154
Guillot, J., 208

Guimaraes, A. G., 85, 158
Guiochon, G., 95
Gupta, B. D., 20, 44, 78
Gupta, C. M., 154, 157
Gupta, L. C., 245, 258
Gupta, R. K., 138, 148, 192, 193, 199, 218, 225, 234, 276
Gurd, F. R. N., 221, 223
Gusarova, N. K., 25, 27
Gustavsson, H., 288
Gustyakova, M. P., 53, 79
Guthrie, D. R., 52
Guthrie, J. P., 41
Gutowsky, H. S., 5, 45, 101, 117
Guy, J., 6
Gylling, R. G., 129
Gysegen, P., 55

Haas, H., 256
Haasnoot, J. C., 56
Hackeloer, H. J., 251
Haddock, S. R., 73
Hadži, D., 110
Haeberlen, U., 50, 133, 239
Hägele, G., 54, 72, 78, 81, 91, 97, 157, 160, 161, 162
Häkli, H., 310
Hänssgen, D., 77
Hagaman, E. W., 26
Hagen, K., 298
Hahn, E. L., 196
Haigh, C. W., 37, 160, 297
Haikala, M. T., 248
Halder, N. C., 255
Halfon, M., 130
Halford, S. E., 230
Hall, D. W., 160
Hall, G. G., 59
Hall, J. R., 84, 184
Hall, L. D., 85, 149, 155, 156, 191, 193, 311, 316
Hall, L. W., 60, 70
Hall, M. J., 75
Hall, P. W., 57, 70
Hall, R. H., 88
Haloui, E., 67, 316
Halstead, J. A., 106
Halstead, T. K., 122, 256
Halton, M., 193
Ham, N. S., 81, 156, 184
Hamer, G. K., 62
Hamilton, J. A., 231
Hammes, G. G., 226
Hampson, M., 29
Hanabusa, M., 255
Hanay, V. A., 208
Handy, P. R., 57, 303
Hanebeck, H., 67, 155
Hanes, G. R., 74
Hansen, E. A., 95
Hansen, M., 157, 201
Hansen, P. E., 19, 70, 77, 83, 155
Hansen, R. S., 157
Hansen, W. W., 1
Hara, K., 299
Hara, K.-I., 254
Hara, T., 27
Hardy, W. N., 124, 126
Harmon, J. F., 123
Harper, W. C., 250
Harrell, J. W., jun., 102

Author Index

Harris, D. C., 53, 231
Harris, D. H., 31, 77
Harris, J. M., 18
Harris, R. K., 54, 56, 57, 67, 70, 72, 73, 78, 79, 109, 138, 154, 156, 158, 159, 160, 161, 164, 195, 203, 214
Harris, R. L., 135
Harrison, J. M., 85
Harrison, L. W., 226
Harrison, P. M., 18, 95
Hartman, J. S., 68
Hartmann, W., 19, 155
Hartwell, G. E., 76
Harwood, H. J., 207, 208
Hasebe, T., 247
Haselbach, E., 16
Haseltine, R., 179
Hashimoto, T., 208
Hatada, K., 206, 212
Hatch, G. F., 140
Hatton, M., 149
Haubold, W., 56
Hausen, R. S., 201
Hausser, R., 123
Hawk, R. M., 130
Hawkes, G. E., 19, 56, 67, 77, 184, 306
Hawkins, E., 221
Hayanizu, K., 182
Hayes, M. B., 216
Hayter, J. B., 132, 288
Hazlewood, C. F., 265, 277, 278, 280, 281, 282
Healey, M. A., 294
Heberhold, M., 180
Hecht, A. M., 288
Hechtfischer, D., 140, 255
Hedberg, K., 298
Hedge, S. G., 268
Heffernan, M. L., 155
Hegeman, G. D., 228
Heggelund, S., 157
Hehre, W. J., 12
Heidberg, J., 98, 166
Heijser, W., 317
Heine, H. G., 155
Heitjans, P., 252
Heller, R., 157
Hellier, D. G., 155
Hellmuth, E., 155
Hendrick, K., 182
Hendrickson, J. R., 252
Hendriks, B. M. P., 187
Henkens, R. W., 231
Henneke, H., 271
Hennig, H. J., 62, 226, 271, 273
Henis, J. M. S., 270
Herman, G. T., 132
Herman, R. M., 47
Hermanek, S., 55
Hermann, G., 209
Hernandez, M. C., 53, 155
Hernandez, O., 35
Herrig, W., 317
Hersh, K. E., 179
Hertz, H. G., 101, 103, 106, 114, 117, 118, 120, 122
Hesse, J., 256
Hester, R. K., 257
Hewitt, R. C., 320
Hewson, M. J. C., 72, 159
Hey, M. J., 275

Heya, H., 300
Hickernell, D. C., 268
Hietan, J., 160
Higashimura, T., 214
Highet, R. J., 53, 231
Hikita, T., 245
Hilbers, C. W., 220, 293
Hildenbrand, K., 70, 317
Hill, H. D. W., 103, 191, 192, 196
Hill, J. R., 247
Hillenbrand, D. F., 65
Hills, B. P., 164
Hilt, R. L., 247
Himes, R. H., 228, 276
Hindman, J. C., 108
Hines, W. A., 229
Hinshaw, W. S., 133, 242, 244, 255, 283
Hinton, J. F., 54, 67, 160, 156, 158
Hioka, T., 258
Hirai, A., 260
Hirao, K., 59, 64
Hirayama, M., 308
Hirohashi, T., 40
Hiroike, E., 59
Hirota, M., 171, 299
Hirschfelder, J. O., 2
Hirsekorn, F. J., 182
Ho, C., 216
Hoarau, J., 58
Hobson, T., 247
Höfler, M., 73
Höhener, A., 137, 197
Hörnfeldt, A.-B., 67
Hoffmann, R. W., 52
Hoffmann, W., 212, 213
Hoffmann, W. D., 271
Hoffman, R. A., 165
Hogan, R. J., 35
Hohner, A., 197
Hohorst, F. A., 90, 155
Hokin, L. E., 233, 276
Hollenstein, R., 52
Hollis, D. P., 227, 276, 282
Holloway, C. E., 53
Holloway, J. H., 73
Hollstein, U., 52, 222
Holz, M., 118
Holzer, H., 26
Homer, L-D., 279
Homfeld, E., 26
Honerjaeger, R., 6
Honeybourne, C. L., 39
Hoogasian, S., 166
Hooper, D. L., 171, 311
Hooper, H. O., 251
Hopf, H., 92, 317
Horiguchi, T., 259
Horlbeck, G., 67, 155
Horn, H.-G., 214
Horne, J. E., 277
Horvitz, E. P., 125
Hosemann, R., 210
Hoshino, R., 265
Hotta, K., 302
Hoult, D. I., 130, 142, 194, 279, 287
Howard, J. W., 67, 119
Howard, T. J., 297
Howarth, O. W., 115
Howes, P. D., 39
Huang, E., 179

Hubbard, P. S., 100, 102, 247
Huber, H., 115, 133, 136, 160
Hudson, H. W., 55, 77
Huestis, W. H., 224
Hüther, H., 67, 155
Hughes, D. G., 98, 137, 165, 237
Hughes, L. M., 301
Hughes, R. P., 32
Huisman, P., 282
Hull, W. E., 223
Hunt, E. R., 259
Hunter, D., 313
Hunter, D. L., 52, 180, 181, 187
Hunter, G., 182
Hunstman, W. D., 41
Huntress, W. T., 99, 112
Hutchings, M. G., 242, 244
Hutchison, J. M. S., 283
Hsi, E., 275
Hsu, C.-J., 217
Hsu, M.-F., 313
Husa, D. L., 268
Husar, J., 306

Ibusuki, T., 81
Ido, M., 265
Ihrig, A. M., 70
Iida, S., 261
Iimuro, H., 207
Iizuka, T., 185, 216
Ikegami, Y., 171
Il'in, E. G., 72, 79
Imaizumi, S., 52
Imanari, M., 27, 56
Imanishi, Y., 214
Immer, H., 35
Inaba, S., 40
Inamoto, N., 55, 65
Inch, T. D., 85
Inch, W. R., 225, 280, 282
Indovina, P. L., 156
Infante, G. A., 290, 304
Ingman, L. P., 247
Inoue, M., 254
Irie, Y., 207
Isawa, K., 24
Ishigure, K., 214
Ishiguro, E., 6
Ishii, C., 74, 310
Istrate-Robilă, G., 206
Itkina, L. S., 120
Ito, H., 209
Ito, S., 186
Itoh, J., 258
Ittel, S. D., 181
Ivanov, B. E., 55
Ivin, K. J., 74, 157
Izuhara, M., 208
Izumiya, N., 219
Izydore, R. A., 52

Jaccarino, U., 259
Jackman, L. M., 103, 109
Jackson, R. R., 278
Jacobsen, E. S., 55
Jacobsen, J. P., 93, 155, 157, 158, 316
Jacobson, S. E., 73
Jacquinot, J. F., 236, 260
Jaenicke, L., 77
Jagodic, F., 211

Jakobsen, H. J., 25, 48, 94, 96, 98, 151, 157, 158, 201
James, P. K., 258
James, T. L., 192, 228, 233, 276
Jameson, A. K., 5, 150
Jameson, C. J., 5, 150, 189
Jameson, K. A., 189
Jammer, M., 236
Jancke, H., 56, 175
Janese, W. W., 282
Janousek, Z., 55
Januszewski, H., 14, 56
Janzen, A. F., 168
Janzen, W. R., 140
Jaques, L. W., 54, 67, 156, 158, 160
Jardetzky, O., 221, 232
Jarvis, A. C., 54, 79
Jautelat, M., 26, 41
Jeffrey, K. R., 144, 244, 251, 262
Jeminet, G., 194
Jen, J., 101
Jenkin, G. T., 132
Jenning, W. B., 24
Jennings, H. J., 223
Jentoft, J. E., 233
Jerina, D. M., 24
Jerome, D., 258
Jesson, J. P., 55, 73, 79, 159, 166, 173, 181, 195, 201
Jikeli, G., 157
Jimbo, T., 259
Jochims, J., 169
Johannesen, R. B., 54, 73
John, K. P., 54, 73, 159
Johnsen, A., 208
Johnson, B. F. G., 180, 181, 308
Johnson, C. S., jun., 39, 135, 246
Johnson, J., 67
Johnson, M. E., 216
Johnson, P. D., 262
Johnson, T. W., 232
Johnson, R. N., 159
Johnston, M. D., jun., 22, 159, 300, 307
Johnston, R. B., 230
Jokisaari, J., 62, 95, 155, 156
Jolicoeur, J., 118
Jolley, K. W., 18, 53, 79, 301
Jonas, J., 103, 108, 117, 134, 202
Jonas, V., 306
Jones, A. A., 211
Jones, A. J., 179
Jones, C. R., 220
Jones, D. W., 155, 220
Jones, J. R., 48, 150
Jones, R., 228, 276
Jones, R. G., 32
Jordan, D. 44
Jordan, F., 302
Jordan, R. B., 186
Joseph-Nathan, P., 53, 155
Jost, R., 173
Jouany, C., 65
Jugie, G., 165
Julia, M., 81
Juliano, P. C., 109, 203

Jung, G., 222
Jung, H., 73
Jurga, K., 262
Jurga, S., 262

Kabanov, V. A., 207
Kablinska, K., 74
Kadaba, P. K., 242
Kahl, W., 156, 159
Kainosho, M., 286, 308
Kaiser, R., 150, 196, 200, 204
Kaczyński, J. A., 155
Kalabin, G. A., 25, 27, 155
Kalder, H. J., 70
Kalechstein, W., 126
Kalinowski, H. O., 52, 174
Kalman, J. R., 85
Kalyanasundaram, K., 305
Kamen, M. D., 216
Kamenzawa, N., 156
Kan, L. S., 220, 302
Kan, S., 146
Kan, S. K., 319
Kanai, T., 212
Kanashiro, T., 252
Kandil, S. A., 53
Kanert, O., 251, 252
Kantschovska-Dimicoli, I., 170
Kaplan, J., 165
Kaplan, J. I., 198
Kaplan, N. O., 219
Kaplan, R., 131
Kaplan, S., 50
Kappauf, K. A., 81
Kaptein, R., 192
Karcher, R., 140, 255
Karger, J., 270, 271
Karmilov, A. Yu., 187
Karnezos, N., 258
Karpel, R., 186
Karplus, M., 10, 14
Karra, J. S., 125
Kasturi, S. R., 282
Katayama, S., 154
Kato, G., 229, 277
Kato, H., 59, 64
Kato, S., 44
Katritzky, A. R., 173
Katz, J. J., 53
Kaufman, R. J., 177
Kawakami, M., 258
Kawamori, A., 254
Kawamura, T., 212
Kearns, D. R., 220, 221
Keat, R., 78, 161
Kehiaian, H., 301
Keim, P., 221, 223
Keiter, R. L., 55, 73
Keller, F., 210, 213, 245
Keller, H. U., 258
Kelley, P. L., 236
Kelsey, D., 307
Kemmitt, R. D. W., 54, 79
Kempf, J., 50
Kennedy, J. D., 50, 51, 68, 73, 321
Kenny, C. P., 223
Kenny, M. E., 297
Kenyon, G. L., 228
Kepert, D. L., 310
Keskineu, R., 80
Kessenikh, A. V., 56
Kessler, H., 52, 174

Kessler, Yu. M., 111
Khaddar, M. R., 57
Khalife el Saleh, M., 95
Khan, S. A., 35, 81
Khetrapal, C. L., 315, 316, 319, 320, 323
Khong, P. W., 52
Khor, T. C., 44
Khripun, M. K., 117
Kidd, R. G., 57
Kiewiet, A., 42
Kim, K. S., 57, 238, 253
Kimber, B. K., 56, 57, 67, 109, 138, 151, 195, 203, 214
Kimmer, W., 212
Kimtys, L., 301
Kimura, A., 39
Kind, R., 251
King, A. R., 259
King, J., 313
Kintzinger, J. P., 105, 106, 190, 303
Kiricuta, I. C., 282
Kirk, W. P., 248
Kirsh, Yu. E., 207
Kisel'nik, V. V., 103
Kisin, A. V., 171
Kisliuk, R. L., 219
Kisman, K. E., 126, 144
Kitching, W., 20, 44, 68, 78, 86
Klaassen, T. O., 258
Klein, E., 258
Klein, H., 159
Klemm, L. H., 157
Klemm, P., 305
Klemperer, W., 49
Kleschick, W. A., 52
Klesper, E., 208
Klochan, V. A., 261
Kloosterziel, H., 17
Klose, G., 291
Klotz, I. M., 232
Kneissel, G., 201
Knispel, R. R., 242, 267, 279, 280
Knott, J. C. A., 215
Knutsson, L., 170
Kobayashi, E., 213
Kobayashi, H., 208
Kobayashi, S., 255
Kobrina, L. S., 94
Kocienski, P. J., 179
Köbrich, G., 306
Koehler, F. H., 55, 67, 70
Koenig, S. H., 224, 231, 282
Koenigsberger, R. N., 297
Köpf, H., 156, 159
Kohara, T., 259
Kohler, F., 301
Koi, Y., 258
Koide, S., 6
Koike, Y., 57
Kok, H. B., 256
Kokunov, Yu. V., 53, 79
Kolbah, D., 310
Kolker, H. J., 10
Kollman, P. A., 70
Kolokol'tsev, V. B., 117
Kolos, W., 6
Kol'tsov, A. I., 27
Komoroski, R. A., 106, 188
Kondo, M., 16

Author Index

Konovalov, E. V., 43, 54
Kontani, M., 258
Kopanev, V. D., 53, 79
Kopp, L. D., 42
Kopple, K. D., 86
Kornprobst, J. M., 25
Kortzeborn, R. N., 11
Kosfeld, R., 229
Kosman, D. J., 234
Kostelnik, R. J., 303
Kostina, N. G., 157
Kostka, A. G., 216
Kostowycz, G., 186
Kovton, N. M., 261
Kowalewski, V. J., 317
Krane, J., 86
Krasnaya, Zh. A., 174
Krasnoshchek, A. P., 43
Krchnak, V., 54
Kreilick, R. W., 140, 189
Kreissl, F. R., 114
Kreiter, G. G., 180
Kresge, A. J., 183
Kricheldorf, H. R., 213
Kricka, L. J., 81, 176
Krische, B., 273
Krishnarmurthy, B., 262
Kristoff, J. S., 182
Kroon, P. A., 71, 286
Kroth, H. J., 57, 66, 88
Krow, G. R., 19
Kruck, T., 73
Kruczynski, L., 180, 181
Krueger, H., 7
Krugh, T. R., 194
Kruk, G., 311
Krustus, M., 248
Krynicki, K., 107
Kubo, R., 236, 237, 254
Kuchen, W., 72, 157
Kucherov, V. F., 174
Kudryavtseva, L. A., 55
Kühnel, W., 62
Kugimiya, K., 214
Kuivila, H. G., 57, 88
Kulkarni, A. B., 312
Kulkarni, B. A., 298
Kulnevich, V. G., 156
Kulpe, S., 62
Kumar, A., 134, 135
Kumar, N. G., 208, 219
Kummer, R. B., 248
Kundla, E., 196
Kuntz, G. P. P., 186
Kuntz, I., 213
Kuntz, I. D., 273, 282
Kuntz, I. D., jun., 300
Kunwar, A. C., 315, 316, 319, 320, 323
Kuramshin, I. Ya., 76
Kurashova, E. K., 44
Kurkijian, C. R., 249
Kurland, R. J., 136, 234, 282
Kursanov, D. N., 77
Kurz, H., 52
Kushida, K., 26, 65, 67, 98
Kushida, T., 255
Kushnarev, D. F., 25, 27
Kustin, K., 186
Kuszmann, J., 155, 160
Kuzay, P., 212
Kwart, L. D., 26
Kydon, D. W., 242

Labaziewicz, H., 81
Labes, D., 62
Labisch, L., 271
Laczynski, S. F., 131
Lafond, C., 146
Lagodzinskaya, G. V., 166
Lahajnar, G., 110, 224, 245, 276
Laht, A., 44
Lahuerta, P., 180, 181
Lala, A. K., 312
Lalita, K., 127, 135
Lallemand, J. Y., 181
Laloe, F., 248, 269
La Londe, R. T., 52, 295
Lam, Y-F., 186
La Mar, G. N., 107, 187, 217 306
Lambert, J. B., 35, 45, 47, 81
Landesman, A., 248
Lando, J. B., 208
Landsberger, F. R., 286
Lang, C., 312
Lange, Y., 287
Langer, E., 40
Langford, C. H., 303
Lanir, A., 225, 226, 276
Lapin, I. I., 167
Lapinski, R. L., 212
Lapper, R. D., 22
Lappert, M. F., 31, 77, 182
Laskorin, B. N., 76
Laszlo, P., 166, 181, 295
Latimer, M., 210
Lau, A. L. Y., 289
Lauer, D., 44
Laurent, J.-P., 65
Laurenz, R., 258
Lauterbur, P. C., 132, 283
Lavrukhin, B. D., 81, 156, 159
Lawler, R. G., 130
Lawson, J. A., 37
Lazaro, R., 170
Lazzeretti, P., 29, 36
Lechert, H., 226, 271, 273
Le Dang Khoi, 130, 258, 261
Leduc, M., 248
Leduc, P., 219
Ledwith, A., 81, 176
Lee, A. G., 99, 286
Lee, C. H., 89, 224
Lee, F. H. W., 157
Lee, L., 138, 191
Lee, S., 254
Lee, S. O., 44, 52, 70
Lee, Y., 103, 317
Lee, Y. K., 117
LeGeyt, M. R., 53
Leggett, A. J., 248
Le Goff, E., 39
Legrand, A. P., 269
Legrand, L., 166
Lehn, J. M., 106, 176, 182, 190, 303
Lehner, H., 40
Leigh, J. S., jun., 282
Leipert, T. K., 56, 183
Leitch, L. C., 190, 284
Lemieux, R. U., 44
Lengeler, B., 130
Lenk, R., 189
Lenz, R. W., 213
Leonard, J., 206

Lepoivre, J. A., 295
Leppert, E., 213
Lequan, R.-M., 67
Lerner, E., 268
Letendre, L. J., 166, 175
Levesque, D., 108
Levine, B. A., 308
Levine, Y. K., 99, 237, 238, 261, 286
Levstik, A., 245
Levy, D., 224
Levy, G. C., 22, 106, 109, 114, 138, 145, 188, 202, 203
Levy, J. B., 157
Levy, L. C., 203
Lewa, C. J., 123
Lewellyn, M., 72, 159
Lewis, E. S., 22
Lewis, J., 180, 181, 308
Lewis, K. G., 22
Leyte, J. C., 233, 272
Li, H.-J., 217
Li, N. C., 302
Liang, G., 21, 52
Librando, L., 170
Lichtenberg, D., 286
Lichter, R. L., 27, 28, 30, 63
Lidén, A., 167
Liebermann, E., 301
Lightonlers, D., 238
Lilja, H., 232
Liljefors, T., 167, 172
Lilich, L. S., 117
Lillie, E. D., 74, 157
Limbach, H.-H., 182
Limouzin, Y., 57, 65
Lin, H. C., 21
Lincoln, D. N., 95, 97, 98, 152, 160, 190
Lincoln, S. F., 186
Lind, A. C., 239
Lindberg, B. J., 7, 54
Lindberg, J. G., 44
Lindblom, C., 137, 284, 285, 288
Linde, S. A., 48, 98, 151, 201
Lindon, J. C., 70, 87, 190, 314, 315, 317, 318, 320, 324
Lindman, B., 232, 234, 277, 288
Lindmark, A. F., 180
Lindoy, L. F., 312
Lindsay, G. A., 208
Lindskog, S., 218
Lindstrom, T. R., 224, 282
Linschied, P., 176
Lipnick, R. L., 66
Lippmaa, E., 27, 41, 56
Lipwick, R. L., 155
Li Shing Mar, L. K. K., 180
Litchman, W. M., 13, 52
Litt, M., 208
Liu, K-T., 313
Live, D. H., 115, 136
Lizina, V. P., 52
Loboda-Čačković, J., 210
Locher, P. R., 249
Lochmüller, C. H., 231
Lode, E. T., 222
Lodge, N. W., 255
Löwdin, P.-O., 8
Loewenstein, A., 318

Loewenthal, E., 116
Loftus, P., 161
Loginova, E. I., 71
Lohman, A. B., 189
Long, E. R., 36
Long, K. R., 315
Long, R. C., jun., 53, 70, 315, 317
Long, T. C., 95
Longenecker, D. U., 125
Longeri, M., 317, 319
Longmuir, K. J., 191
Loo, J. L., 135
Looyestijn, W. J., 258
Loucheux-Lefebvre, M.-H., 215
Lourandos, M. Z., 84
Lovas, A., 256
Low, G. D., 273
Lowe, B. M., 57
Lowe, I. J., 142, 261, 262
Lowenstein, A., 82
Lowman, D. W., 60
Lozach, R., 159, 166
Lucchini, V., 93, 154, 173, 179
Lucken, E. A. C., 189
Luckhurst, G. R., 190, 314
Luders, K., 140
Lück, R., 62
Luedemann, H. D., 25, 105, 106, 223
Lueders, K., 255
Luedtke, E., 157
Lueg, V., 91
Lukacs, G., 52
Lukmanov, V. G., 43
Lukovkin, G. M., 207
Lumbroso-Bader, N., 26, 302
Lunazzi, L., 173, 315, 317, 319
Lunde, B. K., 141
Lussau, C., 284
Lustig, E., 95, 190
Luszcynski, K., 268
Luther, G. W., 27, 95, 155
Lutz, O., 7, 49, 57
Luz, Z., 101, 121, 163, 275, 288, 289, 319, 322
Lyakhovich, I. S., 209
Lyerla, J. R., jun., 20, 52, 100, 115, 222
Lynch, L. J., 274
Lynch, R. J., 55
Lynden-Bell, R. M., 100
Lynn, J. L., 183, 290

Mabon, F., 174
McAlister, A. J., 256
McAmis, L. L., 275
McArdle, P., 308
McBrierty, V. J., 206, 245
McCammon, J. A., 116
Macciantelli, D., 173, 319
McClanahan, J. L., 159
McClung, R. E. D., 107, 110
McConnell, H. M., 33, 62, 286, 306
McCoy, L. L., 155
McCredie, J. A., 225, 280, 282
McDonald, F. R., 267
McDonald, G. G., 192

McDonald, I. R., 206
McDowell, C. A., 245
Macdowell, F. D. H., 283
McFarlane, W., 31, 37, 50, 51, 68, 70, 73, 77, 161, 321
McGaughy, T. W., 279
MacGilp, N. A., 57
McGuire, R. R., 93
Maciel, G. E., 44, 52, 57, 67, 149, 193
McIntyre, H. M., 115
McIvor, M. C., 317
McKinney, J. D., 291
McLaughlan, K. A., 49
McLaughlin, A. C., 215, 287
MacLean, C., 189, 293, 317, 318, 320, 322, 323
McLeod, R. S., 159
McMahon, V. A., 295
McMillin, D. R., 316
McNeff, L. C., 230
McNeil, J. A., 142
McVicker, E. M., 70
Mägi, M., 56
Maestro, M., 14, 59
Magagnini, P. L., 207
Maggio, E. T., 228
Magnuson, J. A., 232
Magyar, E. S., 81
Mahanti, S. D., 255
Mahendroo, P. P., 278
Mahnke, H., 50
Maier, L., 65
Maire, J. C., 57, 65
Makela, J., 243
Maki, K., 248
Maklakov, A. I., 209
Malcolm, R. B., 316
Malhotra, S. L., 206
Mali, M., 261
Malik, S. K., 256
Malinowsky, E. R., 68
Mallard, J. R., 283
Mallion, R. B., 2, 37, 297
Mallory, C. W., 97, 98, 159
Mallory, F. B., 97, 98, 159
Malm, S. M., 49
Malnar, M., 310
Malyuk, N. G., 103
Mamaev, V. M., 40
Mametyuk, V. I., 157
Man, M., 186
Manatt, S. L., 71
Mandel, F. S., 313
Mandema, W., 242
Manelis, G. B., 166
Mangeot, B., 6
Mangras, M., 158
Manke, H., 70
Mann, B. E., 53
Mann, G., 175
Manning, B. C., 150
Mannschreck, A., 306
Manojlovic-Muir, L., 72
Mansfield, P., 133, 144, 214, 239, 262, 283
Mantch, H. H., 109
Manuel, A. J., 210
Manzer, L. E., 32, 70
Maravigna, P., 167
Marcelja, S., 284
Marchal, J.-P., 98
Marcus, D., 276
Mareev, Yu. M., 156

Mares, J., 53, 155
Maričić, S., 224, 225, 276
Markley, J. L., 205, 219
Marks, S. B., 322
Marks, T. J., 307
Marraud, M., 62
Marsden, C. J., 54
Marsden, K. H., 227
Marsh, D., 215
Marshall, A. G., 100, 149, 189, 193, 311
Marshall, G. R., 191
Marshall, J. L., 67, 70, 75, 77, 155, 159
Marshall, T. W., 45
Marsmann, H. C., 66
Martin, A., 223
Martin, C. J., 280
Martin, G. J., 16, 32, 67, 174
Martin, J. S., 11, 36, 53, 66, 84, 85, 158, 297
Martin, M. L., 16, 32, 67
Martin, R. B., 78, 308
Martin, R. L., 60
Martinelli, L. C., 158
Martinez-Carrion, M., 233
Martinsson, P., 167, 168
Marwedel, B. J., 234
Marzilli, L. G., 302
Masahiro, O., 27
Maskasky, J. E., 297
Mason, J., 29
Mason, P., 219
Massara, C. I., 245
Massey, C., 182
Massol, M., 156
Masson, A., 305
Masuda, S., 65
Masuda, Y., 258
Matheson, T. W., 181
Mathews, R. S., 155
Mathias, A., 29
Mathieu, A., 18, 93
Matson, G. B., 139
Matsumoto, S., 212
Matsumura, M., 256
Matsuzaki, K., 208, 209, 212
Matthews, R. W., 57
Matula, S., 130
Matveev, V. V., 56
Matyjaszewski, K., 209
Maudsley, A. A., 283
Mautsch, H. H., 190
Mauzer, L. E., 70
Maxwell, G. P., 280
Mazitov, R. K., 120
Mazzola, E., 295
Meakin, P., 79, 159, 166, 173, 181, 195, 201
Meakins, G. D., 173
Medina, D., 282
Mehlhorn, A., 55
Mehring, M., 138
Meiboom, S., 101, 275, 319, 320, 322
Meier, A., 18
Meisenheimer, H., 75
Meissner, P., 252
Meister, W., 18
Melendres, C. A., 120
Mellgren, R. L., 44
Melnikov, N. N., 119
Melson, G. A., 57, 304
Memory, J. D., 36

Author Index

Menafra, L., 111
Menger, F. M., 183, 290
Merbach, A., 186
Meriel, P., 131
Merle, Y., 207
Merle-Aubry, L., 207
Mermet-Bouvier, R., 70
Merrick, T., 97, 142, 262
Merrill, R. E., 157
Méry, J., 230
Messe, M. T., 26
Metcalle, J. C., 99, 286
Metzger, J., 167
Meyer, G. H., 112
Michel, D., 271
Mihailov, M. H., 208
Mihailovski, A., 91
Miiller, D. E., 67, 70
Mikhailov, B. M., 56, 67
Mikhaleva, A. I., 155
Miknis, F. P., 267
Miko, M., 282
Milano, J. C., 18, 93
Mildvan, A. S., 225, 227, 228, 234, 276
Miller, D. L., 220
Miller, D. P., 13
Miller, J. M., 68
Miller, J. S., 181
Miller, T. A., 130
Millet, F. S., 117
Millett, F., 232, 276
Milne, G. W. A., 28
Milone, L., 181
Minott, C., 250
Mins, W. B., 189
Mirtsch, S., 271
Mishov, S. M., 154
Mishra, P. K., 100
Mishustin, A. I., 111
Mitani, K., 206, 208
Mitchell, H. L., 177
Mitchell, J., jun., 205
Mitchell, L. W., 53, 222
Mitchell, T. D., 109, 203
Mitchell, T. N., 67
Miyabayashi, T., 52
Miyahara, Y., 261
Miyajima, G., 23, 67
Miyoshi, M., 219
Mlochowski, J., 17
Moccia, R., 14, 59
Modelli, R., 164
Modena, G., 179
Mogharab, I., 223
Mognaschi, E. R., 245
Mohanty, S., 323
Mokarram, M., 250
Molenda, R. P., 155, 73
Molin, Yu. N., 56
Mollier, Y., 54
Monasterios, J. R., 44, 156
Mondelli, R., 74, 75, 157, 158, 317
Mondovi, B., 227, 276
Moniz, W. B., 204
Monjol, P., 207
Monoi, H., 277
Montaudo, G., 170, 311
Montginoul, C., 95
Montgomery, J. A., 65
Mooberry, E. S., 194
Moody, D. C., 190
Moody, W. E., 312

Moon, R. B., 53, 231, 276
Moore, C. J., 68
Moore, W. S., 137
Mooser, A., 52
Moran, G. F., 24
Morel, J., 32, 67
Moreland, C. G., 106
Morgan, R. J., 308
Morgan, S. D., 227
Morgan, W. E., 79
Mori, H., 154
Mori, Y., 208
Morimoto, K., 245
Morishima, I., 185, 216
Morishima, Y., 207
Moritani, I., 23
Moritani, T., 213
Morris, A., 294
Morris, D. G., 88
Morris, H. P., 282
Morrow, J. S., 221, 223
Morrow, J. K., 301
Mortimer, M., 237, 261
Morton, T. H., 167
Moss, R. E., 4
Motell, E. L., 44
Moss, K. C., 88
Motono, H., 27
Mottley, C., 246
Mouse, A. M., 272
Müller, A., 105
Müller, K., 179
Müller, V., 144
Muenter, J. S., 49
Muertterties, E. L., 182
Muir, K. W., 72
Mukhomorov, V. K., 30
Mukoyama, Y., 207
Mulay, L. N., 214
Mulcay, M., 148, 193
Muller, B. H., 123
Muller, N., 290, 299
Muller, R., 239
Muller, W. H. G., 258
Muntz, R. L., 305
Muratova, A. A., 76
Murday, J. S., 270
Murray, A. M., 88
Murray, A. S., 24
Murray, M., 155
Muaev, I. A., 44
Musher, J. I., 36
Musker, W. K., 57
Muzette, C., 181
Myers, S. M., 254
Mynott, R. J., 73

Nadkarni, J. S., 282
Nagata, S., 155, 213
Nagel, M., 270
Nageswara Rao, B. D., 164
Naider, F., 215
Nair, P. M., 17
Nair, V., 67
Nakabayashi, M., 208
Nakagawa, T., 289
Nakajima, R., 27
Nakamura, M., 166, 258
Nakanishi, H., 74, 157, 166, 170, 179
Nakanishi, K., 52
Nakashima, Y., 36, 224
Nakatsuji, H., 59, 64
Nanassy, A. J., 283

Nanda, R. K., 221
Narath, A., 145, 254, 256
Nardin, R., 85, 158
Narten, A. H., 238
Narumi, H., 6
Nascret-Barciszewska, M. Z., 299, 301
Natterstad, J., 149, 193
Navech, J., 81, 155, 156, 158
Navon, G., 165, 226, 228, 276
Nechtschein, M., 258
Neda, A., 259
Neel, J., 62
Neely, J. W., 140
Negita, H., 243
Negrebetskii, V. V., 56, 67
Nelson, D. J., 221, 232
Nelson, G. L., 22
Nesmeyanov, A. N., 25
Neszmelyi, A., 158
Netzer, A., 54, 73
Neuman, C. H., 280
Neville, M. C., 278
Newman, B. A., 207
Newmark, R. A., 75
Newsoroff, G. P., 88
Newton, M. D., 60
Ng, S., 297, 298, 299, 300, 301
Ng, T. C., 141, 262
Nichols, B. L., 278
Nicolau, C. I., 289
Nicole, D., 183
Niebuhr, H. H., 254
Niederberger, V. W., 284, 314, 316, 319, 321
Niedoba, H., 258
Nielsen, J. T., 155, 158
Niemela, L. K. E., 243
Nigen, A. M., 223
Nikkilä, A., 80
Nikonorov, Yu. I., 54
Nilsson, B., 167, 168
Nimz, H., 25, 223
Nishimoto, K., 67
Nishioka, A., 16, 206, 209
Niwa, J., 27
Nixon, J. F., 54, 73
Noack, F., 123, 211, 243
Noble, P. N., 11
Noe, E. A., 172, 175
Noeth, H., 55
Noggle, J. H., 56, 112, 183, 233, 276
Nolle, A., 7
Nome, F., 290
Nomura, M., 258
Norbury, A. H., 313
Nordio, P. L., 323
Nores, K. O., 129
Norne, J.-E., 234
Northey, H. L., 278
Norton, J. R., 31, 308
Norton, R. S., 216
Norton, R. V., 295
Nozakura, S., 207
Nummelin, A. J., 275
Nuretdinov, I. A., 71
Nutt, W. R., 52

Oakes, J., 229
O'Brien, C., 48

Odberg, G., 262
Odberg, L., 262
Odiot, S., 16
Odom, J. D., 57, 60, 67, 70, 119
Oehme, W., 270
Ogawa, M., 212
Ohguri, N., 206
Ohnishi, T., 53, 215, 222
Ohno, T., 252
Ohta, H., 65
Ojima, J., 39
Okada, K., 259, 261
Okada, M., 207
Okada, T., 208
Oki, M., 40, 166, 168, 310
Oksenenko, B. G., 54, 95
Okuda, H., 206
Okulevich, P. O., 77
Okuyama, T., 24, 25
Olah, G. A., 21, 22, 52
Olari, P., 156
Oldfield, E., 193
Oliver, J. P., 35, 75, 76, 157
Olsen, R. W., 149
Olsson, K., 167, 168
Olszanski, D. J., 57, 304
O'Neil, J. W., 177
Onishi, T., 73
Opella, S. J., 221
Oppelt, A., 258
Oppenheim, I., 125, 126
Orahavato, A. S., 154
Orchard, M. J., 133, 144
O'Reilly, D. E., 103, 121, 122, 238
Osborn, J. A., 180
Osheroff, D. D., 248
Oshima, K., 214
Osredkar, R., 245, 261
Ostlund, N. S., 61
Ostroy, F., 233, 276
Otsubo, T., 37
O'Sullivan, W. J., 227
Oth, J. F. M., 179
Ottinger, R., 43
Oudman, D., 155
Ovenall, D. W., 79, 159, 173, 212
Ozawa, H., 106
Ozier, I., 10

Paasivirta, J., 310
Pachler, K. G. R., 96, 98, 150, 201, 308
Packard, M. E., 1
Packer, E. L., 222
Packer, K. J., 264, 278
Paddock, N. L., 53
Padhye, S. B., 298
Pagani, G. A., 74, 75, 157, 158
Paiva, A. C. M., 222
Pajak, Z., 111, 209, 262
Palau, J., 217
Palazzotto, M. C., 183
Palke, W. E., 9
Panayotova, M., 208
Pang, T. S., 300, 301
Panosyan, G. A., 77
Pant, P. M., 254
Pantankar, A. V., 316
Panunzi, A., 175
Paolillo, L., 175

Papadopoulos, M. G., 32
Papay, J. J., 81
Papke, G., 66, 67
Paquette, D., 259
Parello, J., 230
Parfondry, A., 67
Park, M.-G., 41
Parker, D. S., 123
Parks, S., 145
Parlebas, J. C., 256
Parr, R. G., 8
Parrish, R. G., 136, 282
Partington, P., 99, 201
Pascual, C., 35, 160
Paselk, R. A., 224
Pastor, G., 95
Pastore, E. J., 219
Pastukhova, E. D., 54
Patankar, A. V., 315
Patel, D. J., 40, 217, 218, 221
Paterson, C. A., 278
Patt, S. L., 230
Patterson, L. K., 290, 304
Patterson, R. L., 270
Paukstelis, J. V., 90
Paul, E. G., 41
Paulmier, C., 32, 67
Pauling, L., 34
Paulson, D. R., 24
Pausak, S., 50, 124, 240
Paust, J., 25
Paviot, J., 58
Pawliczek, J. B., 317
Peach, M., 54, 81, 157, 161
Peake, S. C., 54, 73
Pearce, H. L., 307
Pearson, G. A., 140
Pearson, H., 115, 136
Pearson, R. M., 266
Pearson, R. T., 278
Peat, I. R., 20, 25, 62, 138, 145, 202, 203
Pecora, R., 113
Pedersen, B., 92, 317
Pedersen, C. Th., 67
Pedersen, E. J., 155
Peeling, H., 167, 168
Peeva, N., 208
Peguy, A., 57
Pehk, T., 27, 41, 44
Pekary, A. E., 217
Pelka, B. P., 24
Pellisson, J., 255
Penczek, S., 209
Pendlebury, M. H., 46, 76
Peneau, A., 253
Penko, M., 211
Pepper, E. S., 158
Perlin, A. S., 67, 155
Permin, A. B., 76
Perrin, C. L., 183
Persson, N. O., 288
Pesando, J. M., 218, 219
Pescia, A., 214
Petch, H. E., 242, 267
Peters, C. S., 1, 148
Peters, W., 97, 160
Petersen, I. H., 74, 157
Peterson, E. M., 121, 238
Peterson, G. E., 249
Peterson, M. R., jun., 52
Peterson, N. O., 286
Petriceli, V., 132
Petrocine, D. V., 307

Petrosyan, V. S., 76, 79
Petrosyants, S. P., 57, 73, 119
Petrov, Yu. A., 71
Petrovskii, P. V., 67
Peverelli, K. J., 318
Pfeifer, H., 264, 269, 270, 271
Pfleidere, W., 169
Pham, Q.-T., 208
Phaovibul, O., 210
Phil, D., 142
Phillips, L., 46, 48, 76, 160
Phillips, M. C., 285
Piesvaux, J., 131
Pifat, G., 224, 225, 276
Pignolet, L. H., 183
Pihlaja, K., 80, 154, 156
Pinell, R., 54
Pines, A., 49, 50, 97, 140, 142, 200, 240, 262, 266, 285
Pinhey, J. T., 85
Pinot, M., 131
Pintar, M. M., 241, 279, 280, 282
Pinto, T. P., 191
Piotrowska, H., 56
Piott, J. E., 268
Pirkle, W. H., 305
Pirnat, J., 242
Pirs, J., 245
Pitner, T. P., 172
Platt, S.-L., 194
Platzer, N., 25, 77, 312
Plaute, L. T., 219
Pleiter, F., 256
Plesek, J., 55
Podo, F., 156
Pogliami, L., 160
Pohl, H. H., 96
Polak, M., 165, 237
Poland, J. S., 31, 77
Polnazek, C. F., 163, 284
Pomerantz, M., 65
Pommier, C., 95
Pop, I., 259
Pope, J. M., 138, 287
Popkova, N. V., 94
Pople, J. A., 12, 13, 14, 16, 33, 36, 46, 60, 164
Popov, A. I., 57, 303
Port, C. N. J., 158
Porter, R., 307
Porter, R. F., 67, 158
Potashnik, R., 322
Pouet, M.-J., 67, 89
Poulin, D. D., 79
Poulis, N. J., 256, 258
Poulsen, O. K., 19, 70
Pound, R. V., 1
Pourquie, J. F. J. M., 247
Powell, J., 32
Powles, J. G., 107
Pozdnyakovich, Yu. V., 53, 94
Praeger, D., 68, 86
Prange, T., 158
Pratt, J. C., 144
Pregosin, P. S., 28, 73
Preissing, G., 211
Press, E. M., 215
Prestegard, J. H., 287, 288
Pretsch, E., 35

Author Index

Price, E., 171
Price, J. T., 155
Price, K., 167
Price, N. C., 215
Price, R., 317
Prins, D. Y. H., 233
Prokof'ev, E. P., 174
Pronanski, C. F., 204
Proshutinskii, V. I., 167
Prostakov, N. S., 155
Proulx, T. W., 307
Pruckmayr, G., 213
Prud'homme, J., 206
Pruniski, M. J., 61
Puddephatt, R. J., 57, 70, 160
Pudovik, A. N., 76
Pugmire, R. J., 52
Puigdoménech, P., 230
Pumpernik, D., 110, 211, 319, 321, 324
Punkkinen, M., 247, 261
Purcell, E. M., 1
Purcell, G. V., 273
Purcell, J. M., 273
Purcell, K. F., 60
Purdum, W. R., 54
Pyper, N. C., 72, 164

Quack, G., 212
Quattrone, A. J., 222
Quentrec, B., 108
Quilley, P. J., 205
Quin, L. D., 44, 55, 70
Quinn, R. K., 256

Raban, M., 172, 176
Rabenstein, D. L., 185
Raber, H., 138
Rabinowitz, J. C., 222
Radda, G. K., 130, 233, 279, 287
Radeglia, R., 31, 52, 55, 56, 62, 63, 189, 299
Radics, L., 156, 158
Radley, K., 289
Rae, I. D., 35
Rae, J. L., 278
Rädle, C., 103, 117
Raftery, M. A., 223, 224, 232, 276
Raghunathan, P., 245
Ragle, J. L., 250
Rahkamaa, E. J., 107, 156, 158, 317
Rahnanskaya, A. A., 207
Rajan, S., 127, 135
Ralph, E. K., 287
Ramadan, B., 141, 226
Ramage, R., 311
Ramirez, J. E., 273
Ramsey, N. F., 2, 6, 10
Ranade, S. S., 282
Randall, E. W., 56, 158, 164, 181, 189, 317
Randall, J. C., 212
Rao, B. D. N., 100
Rao, G. V., 296
Rao, K. C., 156
Rao, U. R. K., 258
Rapp, B., 56
Rathke, J. W., 77
Rattle, H. W. E., 217
Ray, J. G., 183

Ray, S., 311
Raynes, W. T., 2, 6, 10, 48
Reardon, E. J., jun., 161
Redfield, A. G., 138, 191, 193, 287
Reed, G. H., 227, 228, 276
Rees, D., 59
Rees, R. G., 77
Reeve, R. N., 55
Reeves, L. W., 284, 289, 315, 316, 317, 318, 319
Regel, W., 213
Rehfeld, S. J., 282
Reich, H. J., 26
Reichert, B. E., 57, 70
Reichmais, E., 179
Reid, R. V., jun., 5
Reilley, C. N., 307
Reilly, J. L., 19
Reinheimer, J. D., 93, 155
Reisse, J., 43
Reiter, R. C., 299
Reitz, R. R., 190
Remeika, J. P., 256
Remijnse, J. D., 75
Renade, S. S., 282
Renner, E., 270
Reny, J., 177
Repmann, H., 26
Resing, H. A., 244, 264, 266, 267, 270
Rest, A. J., 317, 318
Reuben, J., 205, 233, 277
Reuton, O. A., 76, 79
Revel, M., 158
Reyes, J. M., 257
Reyes-Zamora, C., 24
Reynhardt, E. E., 252
Reynolds, W. F., 20, 30, 62
Rhee, W. Z. M., 182
Rhim, W.-K., 133, 139
Riand, J., 26
Rice, M., 160
Richard, H., 284
Richards, A. J., 131
Richards, E. E., 173
Richards, G. N., 223
Richards, J. H., 53, 231, 276
Richards, K. H. B., 133
Richards, M. G., 125, 269
Richards, P. M., 260
Richards, R. E., 56, 130, 142, 194, 233, 275, 279, 287, 304
Richards, W. G., 158
Richer, J.-C., 75
Richter, J., 54, 81, 157, 161
Richtering, H., 251
Rico, M., 75
Riddell, F. G., 80, 81
Rigatti, G., 323
Rigny, P., 53, 74, 104, 108, 253
Rigo, A., 227, 276
Rijacek, J., 258
Ripmeester, J. A., 246
Ritchie, R. G. S., 67, 155
Ritschl, F., 63
Riveros, J. M., 318
Rizvi, R. K., 154, 157
Roach, E. T., 57, 303, 304
Robb, I. D., 211, 272, 290
Robert, J. B., 71, 75, 84, 85, 157, 158

Roberts, A. W., 306
Roberts, B. W., 176
Roberts, D. W., 306
Roberts, G. C. K., 77, 83, 99, 155, 194, 195, 218, 219
Roberts, H. G. Ff., 36, 37
Roberts, J. C., 55
Roberts, J. D., 19, 22, 26, 27, 29, 41, 45, 56, 67, 69, 94, 114, 115, 136, 298, 306
Roberts, N. K., 278
Roberts, T. B., 158
Robertson, R. E., 306
Robiger, V., 271
Robillard, G., 45, 231
Robins, R. K., 158
Rockelmann, H., 233, 272
Rodinov, E. S., 171
Rodriguez, M., 295
Roe, A. M., 63
Röder, O., 106
Roeder, S. B. W., 102, 105
Roelle, W., 77
Röschenthaler, G. V., 161
Rogers, H. R., 44, 70
Rogers, M. T., 138
Rogers, N. H., 313
Roggero, A., 213
Rollefson, R. J., 267
Rolt, J. S., 268, 269
Romanov, L. M., 313
Roos, J., 251
Roques, B. P., 95, 158, 173
Rorschach, H. E., 282
Rosanske, R., 70, 145
Roseberry, T., 190
Rosen, W., 311
Rosenberg, E., 181
Rosenberger, H., 239
Ross, D. K., 137, 202
Rossetti, Z. L., 295
Rossi, J.-C., 310
Rotilio, G., 227, 276
Rotter, M., 258
Roussel, C., 167
Roussel, J., 158
Rowan, R., 96, 116
Rowbotham, J. B., 45, 62, 64, 91, 156, 157, 159, 167, 168, 298
Roy, N., 206
Ruben, H., 175
Rubin, P., 57
Rudyk, V. I., 54
Rueger, C., 55
Rummens, F. H. A., 189
Runge, W., 19
Ruotselainen, H., 156
Rupp, H. H., 53, 57, 73
Rushworth, F. A., 235
Rusnak, L. L., 186
Ruthenberg, M., 62
Rycroft, D. S., 70, 161

Sabine, R. M., 318
Sadee, W., 176
Sadlej, A. J., 10
Sadler, P. J., 218
Safiullin, R. K., 60
Sagnowski, S., 51
Sahm, W., 7, 120
Saika, A., 6
Saillant, R. B., 50
St-Jacques, M., 48, 175

Saitô, H., 109, 190, 223
Saito, K., 212
Saito, Y., 81
Sakaguchi, U., 36
Sakamoto, K., 40, 168, 310
Sala, E., 70
Sallos, J., 149, 193
Salset, J., 146
Salvatori, T., 213
Samitov, Yu. Yu., 60, 76, 156
Samulski, E. T., 214, 229
Sanches de Cara, J., 317
Sanctuary, B. C., 124, 125
Sandercock, A. C., 186
Sandström, J., 166, 167, 170
Sandu, H. S., 142
Sanin, P. I., 44
Santee, E. R., jun., 207, 208
Santry, D. P., 46, 59, 60
Sardella, D. J., 161
Sarma, R. H., 89, 224
Sarrazin, M., 176
Sarrif, A., 233, 276
Sarteaux, J.-P., 98
Saryan, L. A., 282
Sataty, I., 295
Sato, H., 206, 212
Sato, M., 308
Satoh, M., 252
Satoh, S., 26, 65, 67, 98
Saul, D., 290
Saupe, A., 49, 319
Sauter, H., 75, 152
Sautière, P., 217
Savory, C. G., 55, 65, 190
Sawyer, D. W., 134
Sayer, M., 257
Schacher, G. E., 122
Schacht, E. H., 207
Schaefer, H., 31, 66, 67
Schaefer, J., 141, 194, 232, 270
Schaefer, T., 45, 52, 62, 64, 67, 91, 93, 156, 157, 159, 161, 167, 168, 292, 298
Schaeffer, C. D., 57, 70, 88
Schaeffer, R., 190
Schäublin, S., 197
Schaffner, K., 35
Schalman, E. M., 202
Schano, R., 301
.Schäublin, S., 137
Schaug, J., 92, 317
Schaumburg, K., 49, 67, 87, 93, 155, 157, 158, 159, 222, 316
Schechter, A. N., 232
Scheinbeim, J., 249
Schejter, A., 225, 276
Scheler, G., 239
Schell, F. M., 42
Schellenberger, A., 299
Schemp, E., 249
Schenetti, M. L., 84, 157
Scherr, P. A., 35
Scheve, H., 98, 166
Schiebel, H. M., 73
Schiemenz, G. P., 305
Schilling, G., 213
Schindler, H., 284
Schirber, J. E., 256
Schirmer, R. E., 112
Schlak, O., 72, 73, 156

Schlegel, J., 229
Schmickler, H., 40, 96, 157
Schmidbaur, H., 67
Schmidbaur, M., 55
Schmidt, A., 55
Schmidt, F., 130, 246
Schmidt, H. V., 249
Schmiedel, H., 50, 241, 270
Schmutzler, R., 54, 65, 71, 72, 73, 156, 159, 161
Schneider, B., 208, 209, 211, 271
Schneider, H., 57
Schneider, H. J., 129
Schneider, W. G., 292
Schnorr, G. K., 109
Schoenberger, R. J., 141
Schoep, G. K., 256
Scholl, R. L., 57
Schollner, R., 271
Schommer, M., 129
Schouteeten, A., 81
Schow, S., 55
Schraml, J., 56, 154
Schreiber, E., 231, 286, 289
Schreiber, L. B., 51, 139, 239
Schriever, J., 272
Schrobilgen, G. J., 54, 73
Schrock, R. R., 181
Schröder, G., 179
Schroer, W., 16
Schug, K., 122
Schulman, E. M., 18
Schulman, J. M., 60
Schulte-Frohlinde, D., 289
Schulz, R. C., 213
Schumann, A., 317
Schumann, C., 70, 90, 159, 317
Schumann, H., 66, 88, 57
Schuster, I. M., 69
Schwartz, D., 250
Schwartz, I. L., 219
Schweers, W., 220
Schweiger, J. R., 71, 160
Schweitzer, D., 51, 110
Schweizer, E. E., 69
Schweizer, M. P., 158
Schwenk, A., 7, 57, 73, 120
Schwering, H. U., 67
Schwetlick, K., 55
Schwind, H., 16
Schwyzer, R., 305
Scott, A. I., 98, 205
Scrutton, M. C., 227
Sears, B., 286
Sears, R. E. J., 53, 241
Seeley, P. J., 130, 279
Seelig, A., 284, 285
Seelig, J., 284, 285
Seeman, J. I., 114
Seiffert, W., 182
Seiter, C. H. A., 285, 286
Seiwell, R., 77, 159
Seff, K., 309
Segal, J. A., 180
Segard, C., 95
Segel, S. L., 257
Segre, A. L., 95, 213, 317
Segre, U., 323
Sehgal, R. K., 297
Selegny, E., 207
Selezneva, V. A., 54
Selig, H., 73

Seliger, J., 261
Senda, Y., 52
Seppelt, K., 53, 57, 73
Sepulchre, A. M., 52, 207
Séquin, U., 98, 205
Serforzo, G., 256
Sergeyev, N. M., 26, 61, 70, 106
Serpone, N., 179
Serra, A. M., 317
Servis, K. L., 44, 70, 74, 175, 176, 310
Setkina, V. N., 77
Shafer, K., 93, 155
Shafer, M. W., 258
Shaham, M., 255
Shalaby, S. W., 212
Shapely, J. R., 180
Shapiro, B. L., 159, 307
Shapiro, M. J., 307
Sharp, R. R., 130
Sharpless, N. E., 156
Shashkov, A. S., 210
Shattuck, T. W., 240
Shaw, D., 188, 190
Shaw, E., 231
Shaw, P., 159
Shcherbakov, V. A., 117
Shcherbakova, M. N., 72
Sheahen, T. P., 98, 199
Sheinblatt, M., 117, 237
Sheline, R. K., 70
Sheludyakov, V. D., 171
Sheppard, N., 167
Sheppard, W. A., 79, 159, 173
Sher, A., 257
Sherry, A. D., 227
Sheline, R. K., 50
Shigeru, Y. N., 52
Shih, P. S., 290, 304
Shimazaki, K., 209
Shimko, T. M., 208
Shimokawa, S., 302
Shimomura, K., 243
Shin, J. H., 26
Shizgal, B., 123
Shkrebets, A. I., 156
Shmueli, U., 237
Shmyrev, I. K., 156, 159
Shoffner, J. P., 313
Shohet, S. B., 282
Sholl, C. A., 255
Shoolery, J. N., 53, 155
Short, S., 47
Shoup, R. R., 107
Shporer, M., 121, 233, 277, 279, 288, 289
Shriver, D. F., 307
Shtark, A. A., 157, 160
Shteingarts, V. D., 53, 54, 94, 95, 157, 160
Shull, H., 8
Shulman, R. G., 45, 220, 231
Shuvalov, L. A., 245
Shyoukh, A., 63
Siedle, A. R., 55
Sieg, H.-G., 62
Siegel, B. M., 40
Sillescu, H., 113, 233, 272
Silvidi, A., 245
Simeral, L., 149, 193
Simms, J. A., 148, 149, 193
Simon, H. E., 136, 137

Author Index

Simon, W., 35
Simonnin, M.-P., 67, 89
Simons, W. W., 27
Sîmplăceanu, V., 135, 282
Sinha, K. P., 262
Sinher, K. O. P., 156
Singh, R. D., 67
Singh, S. N., 67
Siverns, T. M., 77, 295
Skjaeveland, S. M., 243
Skrzelewski, A., 169
Skubnevskaya, G. I., 56
Skvortsov, N. K., 52
Slak, J., 250
Slichter, C. P., 241
Slivnik, J., 242
Sliwa, W., 17
Sloan, D. L., 227
Slomp, G., 176
Slonim, I. Y., 313
Smidsrod, O., 226
Smidt, J., 111, 282
Smiriga, S., 97, 142, 262
Smirnov, V. N., 56
Smirnova, F. M., 40
Smith, D., 242
Smith, E., 227
Smith, E. G., 211, 272
Smith, G. D., 312
Smith, G. M., 216
Smith, I. C. P., 109, 190, 222, 223, 284
Smith, J. A. S., 144
Smith, J. B., 210
Smith, K. M., 22, 77, 184
Smith, M. A. R., 88
Smith, M. L., 290
Smith, M. R., 98, 137, 165, 237
Smith, P. J., 31, 90
Smith, R., 290
Smith, R. A., 19, 41, 67
Smith, R. S., 102
Smith, U. H., 216
Smith, W. B., 88
Smyth, D. G., 219
Snerling, O., 155
Snider, R. F., 123, 124, 126
Snowden, B. S., jun., 102, 112, 267
Snyder, L. C., 8, 320
Sobottka, J., 210, 245
Soda, G., 247, 254
Soenko, E. P., 43
Sørensen, S., 94, 98, 201
Soest, J. F., 257
Sogn, J. A., 68
Sohar, P., 155, 160
Sojka, S. A., 204
Sokal'skaya, L. I., 76
Sokoloski, E. A., 221, 231, 232
Sokolovsky, M., 228
Solkan, V. N., 26, 61
Söll, D., 220
Solomon, I., 275
Solomon, P. H., 52
Soma, J., 302
Somin, I. N., 44
Sommer, J. M., 173
Sondheimer, F., 39
Sone, T., 23
Sonoda, N., 73
Soon, Ng., 156

Sorensen, T. S., 179
Sorenson, S., 158
Sosnkowska-Kehiaian, K., 301
Sowerby, D. B. 158
Spassky, N., 207
Spassov, S. L., 154, 170
Spear, R. J., 22, 61, 89
Speight, P. A., 117, 144, 262
Spence, R. D., 260
Spendijan, G., 155
Spěváček, J., 208, 209, 211
Spialter, L., 30
Spielvogel, B. F., 52
Spiess, H. W., 50, 51, 110, 239
Spjut, H. J., 282
Spotswood, T. M., 155
Spragg, R. A., 318
Sprecher, R. F., 27, 35, 138, 198
Sprinzl, M., 220
Squires, R. T., 237, 238
Srinivasan, P. R., 305
Srinivasan, R., 245
Srivastava, B. B. P., 154, 157
Stahkiewicz, T., 73
Stalker, D. C., 133, 144, 214, 239
Stamberg, J., 272
Stanescu, L., 257
Stange, J. H., 244
Stanislowski, A. G., 181
Stanney, G., 10
Staral, J. S., 52
Staudte, B., 266
Stec, W. J., 85, 158
Steele, W. A., 100
Stefaniak, L., 14, 56
Steiger, T., 62, 63
Steinberger, H., 72
Steiner, P. R., 208
Steininger, H., 16
Steinkilberg, W., 57
Stejskal, E. O., 102, 141, 194, 232, 270
Stelzer, O., 65
Stelzner, F., 291
Stengle, T. R., 121, 202, 303
Stenhouse, I. A., 297
Stephen, M. J., 50
Stephens, K. R., 43
Stephens, R. S., 233
Stephenson, D. S., 317, 318
Stepisnik, J., 250
Sterk, H., 16, 26
Sternhell, S., 61, 67, 85, 89
Sternglanz, H., 172
Sternlicht, H., 222, 276
Sternson, L. A., 158
Stevens, R. M., 10
Stevenson, P. E., 166, 177
Stibr, B., 55
Stierl, M., 62
Stiles, P. J., 33, 297
Stock, A. D., 139
Stockis, A., 181
Stockmayer, W. H., 211
Stockton, G. W., 284
Stoeckmann, H. J., 251, 252
Stoffel, W., 231, 286, 289
Stoll, M. E., 50
Stone, N. J., 258, 262
Storek, W., 299

Stothers, J. B., 32, 41, 43, 70, 93, 154
Strack, H., 180
Stranks, D. R., 186
Streefkerk, D. G., 75
Streever, R. L., 258
Strehlow, H., 101
Streicher, H. J., 26
Streitwieser, A., 23
Strelenko, Yu. A., 171
Stricker, G., 7, 49
Strom, E. T., 102
Stromar, M., 310
Stubbs, M. E., 182
Stützel, B., 212
Sturm, W., 40, 96
Styles, G. A., 255
Subbotin, O. A., 70
Subramanian, M. S., 306
Subramanian, N. V., 296
Suchanski, W., 111
Suganok, H., 219
Sugimoto, R., 243
Sugiura, M., 305
Sugiyama, G. Y., 44
Sullivan, N., 248
Sullivan, R. H., 171
Sumi, M., 208
Sumitomo, H., 207
Sundaralingam, M., 89
Suprunchuk, T., 207
Sutcliffe, L. H., 53, 54, 79, 159, 160, 292
Sutter, D. H., 41
Suzuki, F., 166
Suzuki, K. T., 70
Suzuki, M., 317
Suzuki, T., 56, 207, 208
Svanson, S. E., 210
Svare, I., 226, 243
Svirmickas, A., 108
Svoboda, J., 254
Sweeney, M. A. J., 94, 316
Swift, T. J., 185, 275
Swile, G. A., 84, 184
Swinton, P. F., 77, 295, 317
Switzer, D. A. M., 98, 165, 237
Swuste, C. H. W., 259, 260
Swyke, C., 55
Sykes, B. D., 96, 97, 116, 130, 190, 194, 223, 226, 230
Sýkora, S., 314, 315
Sysoeva, N. A., 187
Szafran, M., 299, 301
Szafraniec, L. L., 72
Szarek, W. A., 27
Szeverenyi, N. M., 109
Szmant, H. H., 111
Szmuszkovics, J., 176
Szuba, Z., 209
Szyecs, Z., 256
Szymanski, S., 165

Tabata, Y., 214
Tabony, J. M., 64, 295, 318, 324
Taddei, F., 29, 36, 76, 84, 157
Taft, R. W., 29, 30, 299
Tagami, S., 186
Taieb, M., 158
Takahashi, K., 23, 67

Takagi, S., 150
Takats, J., 180
Takeda, Y., 167
Takegami, Y., 208
Takeuchi, Y., 67
Takoshima, T., 245
Talkowski, C., 231
Tamura, Y., 65
Tan, C. T., 41, 43
Tan, H.-W., 75, 157, 159
Tanabe, M., 70
Tang, F. Y. N., 24
Tanaka, H., 207
Tanaka, S., 52
Tanaka, T., 73, 167, 207
Tanaka, Y., 206, 212
Tanida, H., 26, 67
Tanzawa, H., 208
Tao, D. D., 272
Tapia, O., 293
Tarasov, V. P., 57, 64, 73, 119
Taticchi, A., 67, 315
Tatsuzaki, I., 245
Taylor, B., 54, 159
Taylor, B. F., 79, 153
Taylor, D. G., 145, 246, 247
Taylor, P., 73, 232, 276
Taylor, P. C., 238
Taylor, R. C., 55, 73, 313
Taylor, R. P., 300
Tazeeva, N. K., 156
Tchao, Y. H., 254
Tegenfeldt, J., 50, 240
Teichmann, H., 31, 55
Templeton, D. H., 175
Temussi, P. A., 175
Terawaki, Y., 206, 212
Terenzi, M., 227, 243, 276
Ternai, B., 137, 202
Ternay, A. L., 44
Terry, H. W., jun., 303
Tewari, R., 221
Texier, A., 95
Thoai, N., 310
Thoennes, D. J., 136, 148, 202
Thomas, J. K., 305
Thomas, N. A., 56
Thomas, R. J., 298
Thomas, W. A., 159
Thompson, B. C., 225, 227, 282
Thompson, J. K., 270
Thompson, R. T., 242, 267, 279, 280
Thomson, A. L., 268, 269
Thorpe, M. C., 65
Thül, B., 117
Tiddy, G. J. T., 288, 290
Tiffen, R. S., 255
Tignor, S. L., 102
Tipper, C. F. H., 57, 70, 160
Tischer, R., 6
Todd, L. J., 32, 53
Todokoro, K., 171
Todorova-Mancheva, R., 170
Tokoyama, T., 39
Tokuhiro, T., 111
Tokumaru, K., 65
Tolan, J. W., 130
Tolman, C. A., 55, 73
Tolstoguzov, V. B., 226

Tomita, K., 237
Tomiyasu, H., 186
Tomkins, I. B., 84, 159
Tomlinson, B. L., 191
Tomlinson, D. J., 105
Tompa, K., 256
Tonelli, A. E., 40, 221
Tong, J. P. K., 303
Tonge, A. P., 158
Topart, J., 75
Topsom, R. D., 18, 95, 173
Torchia, D. A., 52, 115, 222
Torgeson, D. R., 141, 250
Tori, K., 26, 65, 67, 308
Torocheshnikov, V. N., 57
Torreilles, E., 95
Torrey, H. C., 1, 123
Torri, J., 26
Toryanik, A. I., 103
Tracey, A. S., 284, 289, 315, 316, 317, 318, 319
Tracey, M. M., 289, 315
Traficante, D. D., 148, 149, 177, 193
Trahanovsky, W. S., 41
Tran Dinh Son, 68, 70
Tranfield, G., 255
Trappeniers, N. J., 136, 242
Trekoval, J., 271
Trexler, F. D., 214
Trinh-Minh-Chan, 310
Tripodi, M. K., 270
Trofimov, B. A., 25, 27, 155
Troitskaya, V. L., 54
Trontelj, Z., 242
Troup, G. J. F., 287
Trska, P., 54
Tsai, A. I.-M., 295
Tsau, J., 238
Tseng, C. K., 91
Tsitsishvili, V. G., 226
Ts'O, P. O. P., 220
Tsuchima, T., 67
Tsuda, T., 260, 261
Tsuge, M., 52
Tsuneto, T., 248
Tsuchima, T., 26
Tuck, D. G., 184
Tucker, E. E., 114
Tucker, J. N., 311
Tucker, J. W., 254
Tulloch, A. P., 284
Tumanov, V. S., 97, 98, 154
Tunggal, B. D., 231, 286, 289
Tunstall, D. P., 235
Tuohi, J. E., 243
Tupciauskas, A., 57
Turbini, L. J., 67, 158
Turchin, K. F., 40
Turner, N. H., 270
Tward, E., 141, 262
Tyrell, H. M., 295
Tzalmona, A., 116

Ueyama, N., 215
Uhl, A., 7
Ulmen, J., 179
Umezawa, Y., 155
Ulmius, J., 265
Ulrich, S. E., 316
Underhill, M., 181
Unsworth, J. F., 22
Urbina, J., 285

Urman, Ya. G., 313
Urry, D. W., 53, 215, 219, 222
Ursu, I., 135
Usui, T. T., 52
Utikal, H. P., 19
Utochka, T., 27
Utton, D. B., 253

Valentine, K. M., 227, 282
Valentini, J., 95, 155
Valet, J., 160
Valic, M. I., 286
Vaida, M. L., 5
Van Acker, L., 160
van Bekkum, H., 75
van Broekhoven, J. A. M., 212
Vande Griend, L. J., 66
van den Enden, J., 130, 263
van der Goot, S., 293
van der Haar, F., 220
van der Hart, D. L., 134
van der Kelen, G. P., 31
van der Klink, J. J., 233, 272
Vanderkooi, J. M., 192
van der Toorn, J. M., 168
van der Touw, F., 233
van der Valk, H. J., 256
van Derveer, D. G., 182
van der Wen, L. J. M., 294
van Dijkman, H. W. D., 53, 66
van Dongen, J. P. C. M., 53, 66
Van Gauwenberghe, R., 80
Van Gerven, L., 254
van Hecke, P., 135
Vanin, J. A., 317, 318
van Meerssche, M., 317
van Ooteghem, D., 207
Van Putte, K., 262
van Putte, K. P. A. M., 130
Van Vleck, J. H., 237
van Wazer, J. R., 79
Van Woert, H. C., 290
Vasquez, E. M., 24
Vaughan, R. W., 50, 51, 133, 139, 211, 239, 245
Vdovin, G. P., 43
Veillet, P., 130, 258, 261
Venanzi, L. M., 73
Venteicher, R. F., 306
Veracini, C. A., 317, 319
Verdini, A. S., 215, 222
Verhoeven, J. W., 305, 311
Verhoeven-Schoff, A., 305
Verkade, J. G., 66, 77
Verlinden, R., 254
Verma, S. M., 156
Versmold, H., 99, 103
Verstakov, E. S., 111
Vibet, C., 146
Vidal, J.-P., 310
Vigna, R. A., 223
Vijayaraghavan, R., 245, 256, 258, 282
Vikane, O., 158
Villafranca, J. J., 227, 228, 276
Vincent, E. J., 176
Vinogradov, L. I., 76
Viola, R. E., 228, 276

Author Index

Virtanen, I., 156
Viscio, D. B., 187
Visscher, R. B., 221
Viti, V., 156
Vivarelli, P., 84, 157
Vladimiroff, T., 68
Vlasova, L. V., 94
Vliegenthart, J. F. G., 75
Voelkel, R., 113
Voelter, W., 26
Vogel, E., 40, 96
Vogl, O., 207
Vogler, H., 37
Vogt, J., 317
Vold, R. L., 57, 73, 101, 104, 119, 136, 137
Vold, R. R., 57, 77, 104, 119, 136, 137
Vollmann, W., 253
Vollmer, J. J., 176
Vollmers, K. W., 261
Volz, H., 26
von Ammon, R., 309
von Dreele, P. H., 297
von Goldammer, E., 105, 106
Von Philipsborn, W., 189
von Voithenberg, H., 169
Voronovich, A. N., 117
Vorontsova, L. G., 81
Vu Hoang Chan, 236
Vuk-Pavlović, S., 225
Vyas, D. M., 27

Wada, S., 256
Waddington, T. C., 55
Waegell, B., 81
Waelder, S., 138, 191
Wagner, G., 185
Wagner, T. E., 217
Wahl, G. H., jun., 52
Waigh, R. D., 310
Walker, I. O., 228
Walker, S. M., 54, 159
Wallace, R., 195
Walling, C., 18, 202
Walsh, H. C., 1, 148
Walter, A., 174, 271
Walter, R., 219, 222
Walter, R. I., 140
Walter, S. R., 75, 155
Walton, D. R. M., 53
Wampler, W. R., 130
Wang, C. H., 100
Wang, C. Y., 177
Wang, S. M., 302
Wangermann, K., 210
Ward, A. J. I., 275, 294
Ward, A. R., 219
Ward, I. M., 210
Ward, J. E. H., 70, 85, 156
Wardell, G. E., 245
Warren, B., 215
Warren, J. P., 29
Wassil, D. A., 279, 282
Wasylishen, R. E., 30, 52, 63, 67, 93, 161
Waterman, M. R., 225, 282
Waterman, P. G., 310
Watson, D., 185
Watson, G. C., 24
Watson, I. D., 301
Watson, R. E., 75

Waugh, J. S., 49, 50, 124, 133, 135, 140, 240, 243, 267, 285
Wazeer, M. I. M., 72, 73, 78, 156, 158, 159, 161
Weaver, H. T., 256
Webb, G. A., 14, 32, 205
Webster, D. S., 274
Weder, H. G., 77, 295
Wedler, F. C., 227, 276
Wee, V., 187
Weekes, J. E., 77
Wehner, R., 52
Wehrli, F. W., 173, 208
Wei, C., 245
Weibel, A. T., 75, 76
Weichelman, K. J., 216
Weidenbruch, M., 97, 160, 161
Weidlein, J., 56, 67
Weigert, F. J., 26, 94
Weill, G., 209
Weisenthal, L., 4
Weiser, C. J., 283
Weisman, I. D., 256
Weiss, V. W., 102
Weissberger, E., 36
Weisz, G., 257
Wells, M. A., 291
Wells, P. R., 18, 93, 154
Welsh, L. B., 258
Welti, D., 134
Wen, W. Y., 117
Wendisch, D., 155
Wendt, J., 243
Wenkert, E., 26, 42
Wennerbeck, I., 171
Wennerstrom, H., 285, 288
Wepster, B. M., 75, 168
Werbelow, L. G., 100, 311
Weringa, W. D., 42
Werner, H., 175
Wessels, P. L., 96, 98, 150, 201
West, B. O., 57
West, R. J., 186
Westerman, P. W., 22, 61, 67
Westfelt, L., 213
Westhead, R., 29
Westover, C. J., 225, 274
Wettermark, G., 171
Wetzel, J. C., 179
Wheeler, B. A., 290
Wheeler, D. H., 304
White, A. I., 215, 218
White, A. J., 181, 187
White, H. F., 155
White, J. W., 132, 288
White, R. F. M., 31, 90
Whitesides, G. M., 177, 178
Whitesides, T. H., 180
Whittle, D., 210
Wickberg, B., 52
Widen, K. G., 310
Wieder, M. J., 28
Wikowski, J. T., 158
Wilbur, D. J., 202
Wild, S. B., 182
Wiles, D. M., 207
Wilhite, D. L., 30
Wilkins, B., 54, 73
Wilkins, C. L., 136, 148, 202
Wilkinson, A., 288

Wilkinson, J. R., 53
Willer, R. L., 42
Williams, D. H., 48, 308
Williams, E., 231
Williams, E. A., 97, 150
Williams, L. F., 319
Williams, M. K., 298
Williams, R. J. P., 151, 194, 219, 305, 308
Williamson, K. L., 53, 79
Williamson, J. P., 258
Willis, E., 290
Wilson, A. R. N., 294
Wilson, D. M., 149
Wilson, G. E., jun., 155
Wilson, L. A., 284
Wilson, N. K., 41, 93, 154
Wilson, S. T., 180
Wind, R. A., 247, 262
Wing, R. M., 297
Winkler, H., 270
Winkler, T., 91, 157
Winstead, J. A., 93
Winstein, S., 179
Winther, C. R., 142
Wirtz, K., 119
Witanowski, M., 14, 56
Witschel, J., jun., 275
Wittekoek, S., 104
Witten, L., 115
Wittern, K. P., 271
Wittich, E. K. H., 225
Woessner, D. E., 102, 105, 111, 112, 116, 267, 276, 278
Wofsy, S. C., 49
Wojcicki, A., 154
Wolf, D., 241
Wolff, M. A., 186
Wolff, M. E., 70
Wolfsberger, W., 54, 70
Wolkowski, Z. W., 310
Wollam, D. S., 254
Wolniewicz, L., 6
Wong, C., 295
Wong, C.-P., 306
Wong, K. F., 299, 300
Wong, K. L., 220
Wong, Y. P., 155, 220, 221
Wood, D. J., 62, 157
Wood, M., 108
Woodhouse, D. R., 278
Woodward, C., 217
Woolfenden, W. R., 42
Woolley, R. G., 5
Woplin, J. R., 159
Workman, T. D., 245
Worvill, K. M., 195
Wrackmeyer, B., 55
Wray, V., 48, 76, 160
Wright, D. A., 138
Wright, J. M., 219
Wright, P. E., 305
Wristers, H. J., 41
Wróbel, F., 74
Wu, T. K., 212, 213
Wüthrich, K., 185, 205
Wullschleger, E., 160, 315
Wunderlich, K., 210, 245
Wuyts, L. F., 31
Wyatt, M., 307
Wynberg, H., 155
Wyrwicz, A., 229, 230
Wyssbrod, H. R., 219

Xavier, A. V., 219, 308
Yagupol'skii, L. M., 43, 54
Yajima, F., 57
Yakshin, V. V., 76
Yalcin, T., 137
Yamada, A., 185
Yamada, H., 97, 134
Yamada, Y., 52
Yamaguchi, T., 56
Yamake, T., 155
Yamamoto, H., 40
Yamamoto, O., 74, 157, 166, 170, 179, 182
Yamamoto, Y., 23
Yamaoka, K., 282
Yamasaki, A., 57
Yamashita, A., 299
Yamazaki, M., 27
Yanagisawa, M., 182
Yang, P. P., 165
Yang, S. K., 220
Yashina, N. S., 76
Yastremskii, P. S., 111
Yasufuku, K., 70
Yasuoka, H., 258, 259
Yathindra, N., 89
Yavari, I., 176
Yeagle, P. L., 231
Yeda, A., 208

Yee, K. C., 75
Yesinowski, J. P., 317, 318
Ylinen, E. E., 243
Yokoyama, Y., 207
Yoneda, H., 86
Yoneyama, T., 55
Yonezawa, T., 59
Yoon, B., 236
Yoshida, M., 65
Yoshimura, Y., 308
Yost, F. J., jun., 227
Young, A. C., 280
Young, G. B., 72
Young, J. A., 308
Young, J. M., 227
Younger, D., 55
Yu, K. C., 157
Yu, S. H., 184
Yurchenko, A. G., 27

Zachmann, H. G., 210, 245
Zahra, J. P., 81
Zakharov, V. F., 155
Zalkin, A., 175
Zamir, D., 255
Zandomeneghi, M., 59
Zanger, H., 27
Zaucer, M., 324
Zeb, M. A., 54
Zegarski, B. R., 130

Zeidler, M. D., 100, 105, 117
Zeisberg, R., 27, 91
Zeiss, R., 239
Zelikman, Z. I., 156
Zeltmann, A. H., 66
Zemel, H., 288
Zemskov, S. V., 54
Zeya, M., 179
Zheglova, D. Kh, 27
Zhidomirov, G. M., 24
Zielen, A. J., 108
Zierenberg, O., 231, 286, 289
Ziessow, D., 150, 200
Zimmermann, D., 43
Zimmermann, H., 50
Zimmerman, J. P., 276
Žinić, M., 310
Zipp, A., 282
Zisapel, N., 228
Zuckerman, D., 189
Zuckerman, J. J., 57, 70, 88, 182
Zukin, R. S., 227, 276
Zumbulyadis, N., 66, 79, 314, 315, 324
Zumer, S., 245, 261
Zune, A. E., 52
Zupančič, I., 224, 250, 276
Zvolinskii, V. P., 155
Zwolle, S., 233

QC
762
N88
v.5

JAN 28 1977